Colour Vision Deficiencies XII

AF785586

Documenta Ophthalmologica Proceedings Series

VOLUME 57

The titles published in this series are listed at the end of this volume.

Colour Vision Deficiences XII

Proceedings of the twelfth Symposium
of the International Research Group on
Colour Vision Deficiencies, held in
Tübingen, Germany
July 18-22, 1993

Edited by
B. Drum

with associate editors:
A.J. Adams, C.R. Cavonius, S.J. Dain, G. Haegerstrom-Portnoy,
K. Kitahara, K. Knoblauch, A. Kurtenbach, B.B. Lee, J. Mollon,
J.D. Moreland, J. Pokorny, L.T. Sharpe, H.A. Sperling, W.H. Swanson,
E. Zrenner

Springer Science+Business Media, B.V.

Library of Congress Cataloging-in-Publication Data

International Research Group on Colour Vision Dificiencies. Symposium (12th : 1993 : Tübingen, Germany)
Colour vision deficiencies XII : proceedings of the Twelfth Symposium of the International Research Group on Colour Vision Deficiencies, held in Tübingen, Germany, 18-22 July 1993 / edited by B. Drum ; with associate editors, A.J. Adams ... [et al.].
p. cm. -- (Documenta ophthalmologica. Proceedings series ; v. 57)
Includes bibliographical references and index.
ISBN 978-94-010-4226-0 ISBN 978-94-011-0507-1 (eBook)
DOI 10.1007/978-94-011-0507-1
1. Color blindness--Congresses. I. Drum, B. II. Adams, A. J. III. Title. IV. Series.
[DNLM: 1. Color Vision Defects--congresses. W3 DO637 v. 57 1994 / WW 150 I602 1994c]
RE921.I57 1993
617.7'59--dc20
DNLM/DLC
for Library of Congress 94-13946

ISBN 978-94-010-4226-0

Printed on acid-free paper

All Rights Reserved
© 1995 Springer Science+Business Media Dordrecht
Originally published by Kluwer Academic Publishers in 1995
Softcover reprint of the hardcover 1st edition 1995
No part of the material protected by this copyright notice may be reproduced or utilized in any form or by any means, electronic or mechanical, including photocopying, recording or by any information storage and retrieval system, without written permission from the copyright owner.

Contents

PART TWO: GENETICS AND CONGENITAL RED/GREEN COLOUR DEFICIENCIES

PART THREE: BASICS OF COLOUR VISION

PART SIX: ELECTROPHYSIOLOGY OF COLOUR VISION

PART SEVEN: COLOUR VISION TESTING

Preface

The twelfth Symposium of the International Research Group on Colour Vision Deficiencies (IRGCVD) was held July 18–22, 1993 in Tübingen, Germany. The scientific program included 61 talks and 37 posters for a total of 98 presentations. Papers based on 64 of these presentations are included here, in *Colour Vision Deficiencies XII.*

Much of the scientific contents of the meeting reflected the research interests of the local organizer, Eberhart Zrenner, with many presentations on basic questions of the anatomical and electrophysiological organization of the neural pathways underlying color vision (especially those involving short-wavelength-sensitive cones) and on ways in which disturbances of these pathways can produce acquired color vision deficiencies. Added to sessions on these topics were sessions on genetics and congenital red-green color deficiencies and on color vision testing. The resulting proceedings therefore contain much of interest to basic vision scientists as well as to specialists in color vision deficiencies.

With three exceptions, the titles of the major sections of this volume follow those of the talk sessions at the Symposium. First, multiple sessions on the same topic are combined here in the same section. Second, the session on cortical color deficiencies is included here in the section on acquired color deficiencies Third, the special workshop session on measuring color surfaces and monitors is included here in the section on color vision testing. Also, for the reader's convenience, similar papers and similar topics are grouped together more rigorously here than was possible in the Symposium; thus, papers from the poster session are intermingled with those from the appropriate talk sessions, the order of sections and papers within sectons are rearranged, and several papers are switched to different sections.

In addition to the usual author index following the text, *Colour Vision Deficiencies XII* contains a cumulative IRGCVD Proceedings Author Index, compiled by associate editor Joel Pokorny. This is the first comprehensive index published since the Group was formed in 1968, and it covers all twelve of the biennial IRGCVD Symposium Proceedings volumes, including the present one. The index will be an invaluable reference resource for both the IRGCVD

B. Drum (ed.), Colour Vision Deficiencies XII, pp. xi–xii.
© 1995 *Kluwer Academic Publishers, Dordrecht.*

membership and the entire vision community, as the Proceedings are not included in computerized bibliographic databases.

All manuscripts submitted for publication in *Colour Vision Deficiencies XII* were reviewed by one or two expert referees who were instructed to employ the high acceptance standards of peer-reviewed journals. This review process was greatly facilitated by a board of 15 associate editors, each of whom was given editorial responsibility for several manuscripts in his or her areas of expertise. The efforts of the associate editors and anonymous reviewers were instrumental in speeding the publication process, reducing my own workload to managable proportions and improving the clarity and scientific quality of the contributors in this volume. On behalf of both the authors and myself, I am grateful to these volunteers for their essential contribution to *Colour Vision Deficiencies XII.*

The Editor

PART ONE

Acquired colour vision deficiencies

1. Peripheral colour contrast sensitivity in patients with inherited retinal degenerations

G.B. ARDEN,[1] J. WROBLESKI,[1] S. BHATTACHARYA,[1] F. FITZKE,[2] C.J. HOGG,[1] A. ECKSTEIN,[2] C.H. HOGG[1] and A.C. BIRD[1]

([1] London, U.K. and [2] Tübingen, Germany)

Abstract

A method is described which permits measurement of colour contrast sensitivity in the retinal periphery. It has been applied to patients with autosomal dominant retinitis pigmentosa, in which the defect is caused by a defect in the gene coding for rhodopsin. In early and in mild cases, in which retinal function as judged by quantitative perimetry, threshold measurement and electroretinography is normal or nearly normal, there is a profound loss of colour contrast sensitivity outside the rod-free area. This may occur although cone thresholds are normal. An explanation may be that affected rods can inject 'noise' signals into the cone pathways.

Introduction

Inherited retinal degenerations of the type called retinitis pigmentosa (RP) are often thought of as rod diseases, due to the frequency of night blindness as an early symptom. In the last few years, more than 70 different point mutations have been described in the gene coding for rhodopsin, which are associated with autosomal dominant RP (McWilliam *et al.*, 1989; Inglehearn *et al.*, 1992; Wells *et al.*, 1993). However, RP causes not only night blindness, but also constriction of the visual field, and ultimately loss of visual acuity, which must be due to the associated loss of cones or cone pathways (Berson, 1992). The progress of RP has been followed by perimetry, normally at a low photopic or mesopic state. Massof and Finklestein (1981) and Lyness *et al.* (1985) first determined dark-adapted perimetric thresholds using narrow-band red and blue-green stimuli to determine the sensitivity and boundaries of the visual fields to rods and to cones separately. These studies demonstrated that RP could be separated into two classes. In one, *Diffuse* rods were selectively affected, while cone thresholds in night-blind regions might be relatively normal. In such persons, there is a regional distribution of loss of rod sensitivity (and also cone sensitivity), so that areas of relatively normal function may surround absolute scotomata. In

B. Drum (ed.), Colour Vision Deficiencies XII, pp. 3–12.
© 1995 *Kluwer Academic Publishers, Dordrecht.*

Regional disease, all rods and cones are initially affected to much the same degree.

These investigations do not refer to suprathreshold cone activity, and it is known that much of the disability suffered by RP patients is due to 'whiteout' – an inability to maintain contrast sensitivity over as large a brightness range as normals. So analysis of cone function might be of interest in assessing individuals disability, the type of disease, and also, investigating the mechanism(s) of cone loss. We have therefore measured colour contrast sensitivity in the retinal periphery of such persons.

Methods

We have used the computer-generated colour displays previously described (Arden *et al.*, 1988; Gündüz *et al.*, 1990; Tak Yu *et al.*, 1991). A special driver card, the TSL206, working at 100 MHz, provides separate signals for the R,G, and B guns of a high-quality 'multisynch' monitor. Each gun can provide 255 intensity levels. The display runs with a refresh rate of 94 Hz. The definition is 980 x 760 pixels. A data collection board and special photodiode are used to construct look-up tables relating the bits set in the digital-to-analogue converters of the TSL206 to the light output from each gun. The contrast and luminance controls of the monitor are set so that at maximum output, the colorimetric properties of the light emitted corresponds to the specification of the phosphors: and under these circumstances, the computer-calculations of the setting required to produce any particular colour (and luminance) correspond to the actual screen output, within the limits of measurement with a Minolta CS1000. The program assumes that the spectral response properties of the human eye correspond to that of the CIE standard observer, but when the system is used to test an individual's colour performance, deviations from the standard observer are assessed in a preliminary test, in which first the red and green guns, and then the green and blue, alternately illuminate (at 22 Hz) a 7° central patch of the screen and the brightness of the green is reduced until, at equiluminance, the sensation of flicker disappears. These individual settings – deviations from the standard observer – are then incorporated into all the subsequent calculation of luminosity and colour, so that for all individuals, tests of colour vision can be carried out with equiluminous images and backgrounds. The images used for testing foveal function consist of alphabetic letters, although gratings and Gaussians can also be employed. The program modulates the colour of an image along a colour confusion line for protan, deuteran or tritan defects, against a background which consists of the average (in XY space) of the extreme equiluminant colours which can be generated. The image can appear briefly, or can reverse in colour contrast – for example change from red to green and back against a desaturated yellow background. Threshold is expressed as the fraction of the maximum modulation at which the image becomes visible. It is measured with a modified binary search technique.

In the present work the programs were modified (using the built-in window facilities for changing experimental parameters) to test the colour-contrast thresholds of progressively more eccentric annuli of retina. Foveal threshold was determined with letters subtending 1°, and then the region 2-3° was sampled, with a coloured annulus surrounding a central fixation spot. The annulus inner margin subtended 2° and the outer, 3°. The annulus was then enlarged to 4–5, 6–7, 8–9, 10–11, and 12–13° from fixation, and new thresholds determined. The patient was asked to fixate (and fixation checked with a TV camera) while 45° of the annulus was removed at random from one of the quadrants: the patient was required to correctly identify the quadrant, so a 4-way forced choice test was administered. In this way the average colour contrast sensitivity of the retinal annulus was determined.

Electroretinograms were obtained using gold foil electrodes. In all patients, a Grass stimulator was used, with various filters. Blue and neutral filters were employed to obtain ERGs driven by rods, and the intensity of light required to generate a half-maximal response was determined. Dim red flashes generated both cone and rod responses. Bright white flashes evoked maximal ERGs, with oscillatory potentials, and 30 Hz flashes evoked cone-based flicker responses, in which timing delays are often an early sign of abnormality in the photopic system. The techniques are standard in this clinic and have been described in detail (Arden *et al.*, 1981). In addition ERGs were obtained with a stimulator which consisted of an 11 cm hemisphere lined with red and green light-emitting diodes. These could be energised separately, and provided stimulus intensity variation over 6.5 log units, up to a maximum green intensity of log 4 quanta absorbed/rod/10 msec flash, and an estimated red intensity log 5.4 quanta/3 msec flash for long wavelength cones. Variation in duration from 50 μsec to infinity is possible. Flash repetition rate varies from 0.01 Hz to 100 Hz (if stimulus duration permits). For full descriptions of the stimulator and results obtained see Spileers *et al.* (1993a, b).

Results

Normals: Figure 1 shows the mean and standard deviation of the colour contrast thresholds in 9 normals. The results are similar to those previously found for larger targets (Tak Yu *et al.*, 1992). In that paper it was found that pupillary constriction or reduction in retinal illumination in young observers (but not in an older person) scarcely affected the colour thresholds. In addition, at the normal higher luminance, there is almost no change in colour contrast with age (Ruddock, 1965). The results are obtained with stimuli flashed for 200 msec once a second, and there is no peripheral fading (Troxler effect).

In cases in which there is a loss of cone sensitivity of more than 0.6 log unit, one would expect apparent loss of colour vision. The investigations have therefore focussed on patients with very early disease in which dark-adapted perimetry

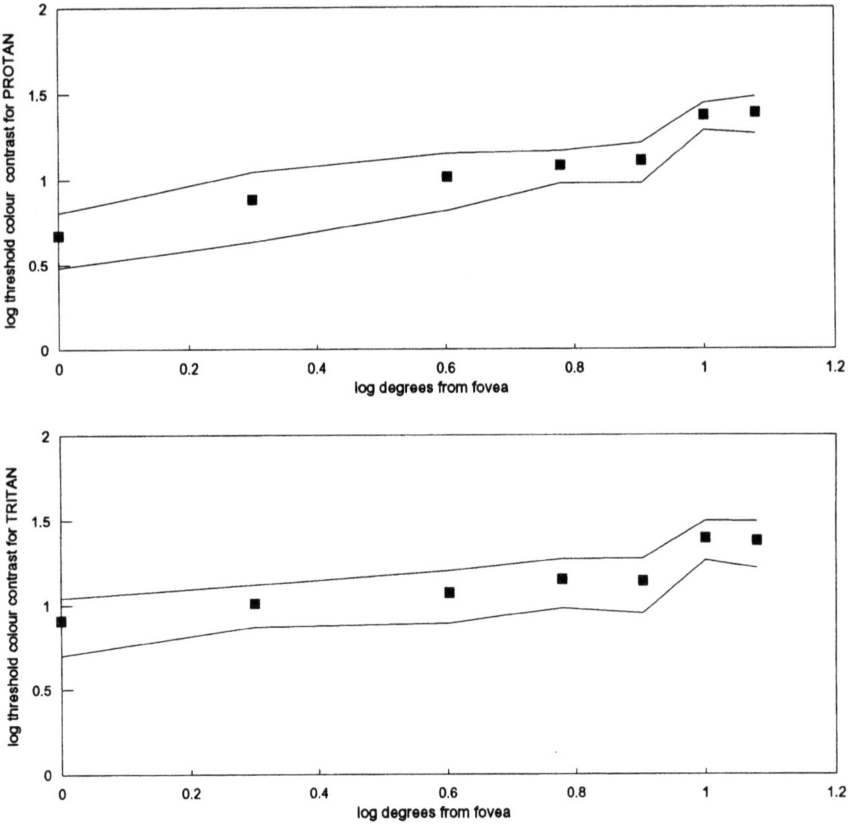

Fig. 1. Relationship between eccentricity and colour contrast thresholds, for colours on protan and tritan colour-confusion axes. The lines show ± 2SDs from the mean. The logarithmic abscissa is used to facilitate comparison as described in Tak Yu, *et al.* (1991).

→

Fig. 2. Patient PB, a 20-yr old member of an ADRP family with a mutation at codon 296 of the gene coding for Rhodopsin. The mutation has been confirmed in this patient, who denies all visual loss. A 'regional' type of RP is illustrated. (a) Visual field, light adapted, obtained with a Humphrey Automated Static Perimeter: The visual field shows minimal change in cone thresholds from 0 dB loss (centrally) to 6 dB (at 10° off axis) to 15 dB (darker shading at edge of fields). (b) Similar perimetry, with blue target and dark adapted eye, to determine rod thresholds. Thresholds elevated by 20 dB. (cl) ERG obtained in dark adapted eye with green LED stimulus. The a-wave is absent, the b-wave delayed and the amplitude is that expected from a normal eye with a stimulus 2 log units less intense. (2c) ERG obtained with a dim red light and a minimal, rod-suppressing green background. This light produces cone b-waves in the normal observer, and here there is a very small

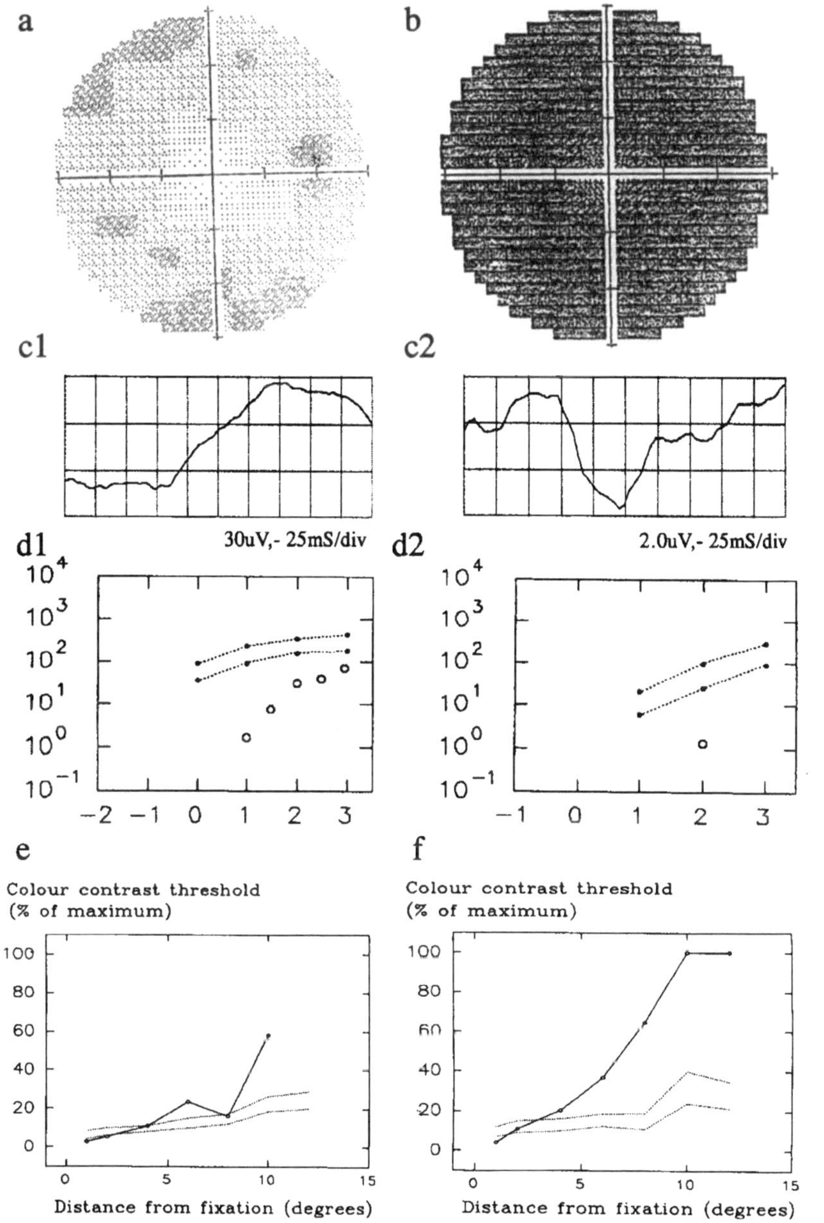

b-wave and following it photopic threshold response. In normal subjects, the b-wave would be larger, and obscure the PTR, so some desensitisation of the cone mechanism may be present. (d1, d2) Normal b-wave amplutude v. light intensity graphs for green and red light responses. Solid symbols and dashes: normal values, open symbols, patient's results. (e) Colour contrast threshold for colours lying on a protan colour confusion line, as a function of retinal eccentricity. The dotted lines indicate the limits of normal. The patient has defects at 5–10 degrees. (f) Peripheral retinal colour contrast thresholds, for colours on a tritan confusion line. The loss of colour vision is gross.

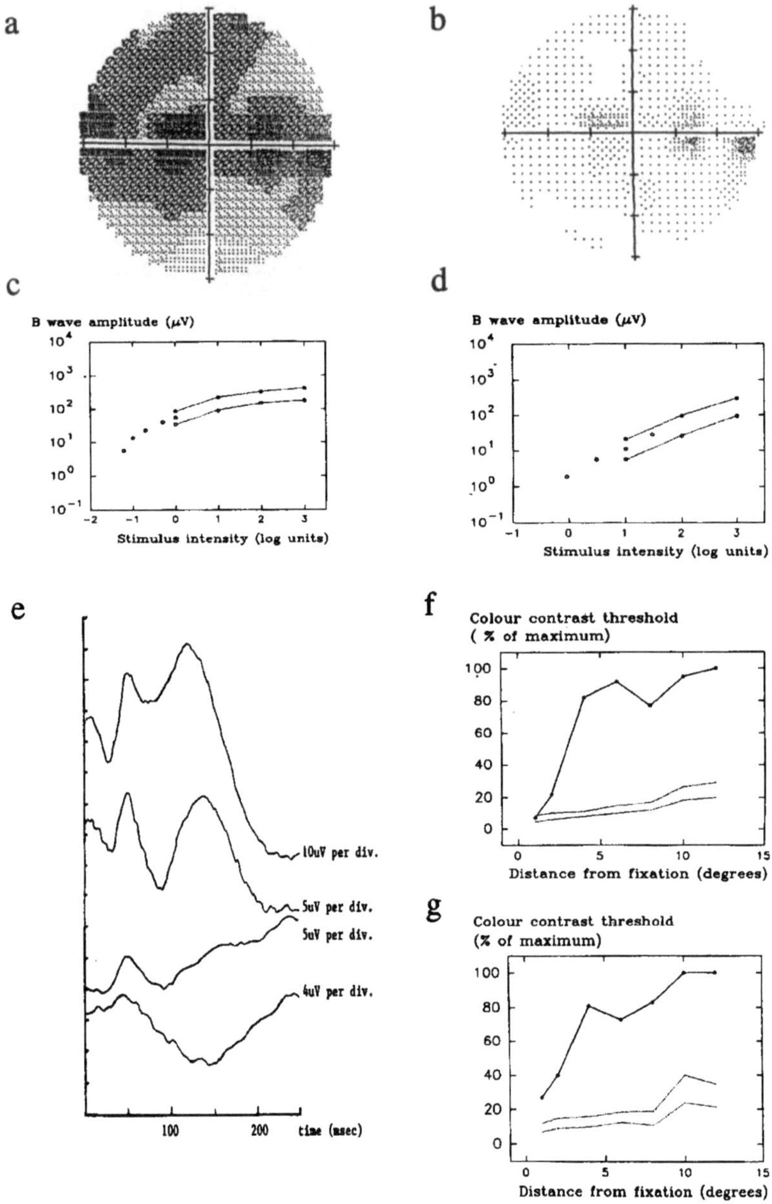

Fig. 3. Patient RV male, age 46 Autosomal Dominant RP. There are small relative scotomata near the fovea in the light adapted state, but rod loss appears minimal. (a) Automated perimetric field, red light, dark adapted showing small areas close to fovea with relative scotoma. (b) Automated perimetric field, dark adapted, blue light showing minimal loss of rod sensitivity. (c) Graph of rod ERG amplitude, elicited by green light of various intensities: dashed lines represent upper and lower limits of normal (see comment below for intensities used). (d) Graph of cone ERGs elicited by red light of differing intensities in dark adapted eyes: dashed lines represent limits of normal; note that

with red and blue targets has shown that cone thresholds are nearly normal, at least within the 12–13° region investigated. Therefore all the patients have autosomal dominant disease. Illustrative results are shown in Fig. 2, 3 and 4.

PB is asymptomatic, from a family with a characterised defect on the gene coding for rhodopsin. All the family are said to be night-blind from birth, but the patient has some residual rod vision, and his fields by kinetic perimetry are full and visual acuity normal. The fundi have only slight, non-characteristic pigmentation, and vessels are normal. The EOG is flat, but there is a slow rod b-wave. This can only be seen with relatively intense flashes: the scotopic system appears to be desensitised by about 2 log units. The cone system, on the other hand, when stimulated with dim red light, maximally attenuated, produces small but distinct 'photopic threshold responses' (PTRs) (Spileers *et al.*, 1993) which are indistinguishable from normal responses although the sensitivity may be slightly reduced (see Fig. 2 legend). Colour vision is abnormal, with both protan, deuteran (not shown) and especially tritan losses.

Figure 3 shows another, older patient, (RV) with more advanced disease. There are small regions of loss demonstrable with a Goldman perimeter, using a small target. Automated dark-adapted static perimetry shows a widepread mild uniform loss of rod sensitivity to blue light, with variation correlated to loss of sensitivity to red. The relative losses of rods and cones are similar at similar loci. The visual acuity is 6/6. The EOG is flat, but the ERG although small (the maximal rod b-wave is 220 μV with a peak time of 66 msec) is otherwise is nearly normal. Stroboscopic flashes evoke rod b-waves at the usual intensities, and the semisaturating flash intensity is normal. The cone responses to dim red light, are small, but not delayed (70 μV, peak time 53 msec). The oscillatory potentials are small, but flicker responses are within normal limits (60 μV peak time 31 msec). The figure shows LED ERGs, (Spileers *et al.*, 1993) to dim red flashes. The normal relationship between rod and cone responses is preserved, and the intensity of red light required to evoke 'PTR' is normal. Nevertheless, colour vision is grossly abnormal. In the central 1 degree (30′ off axis) the protan threshold is normal, but the tritan is grossly elevated. Outside this area, the thresholds can scarcely be determined, they are so high.

Figure 4 shows results from a patient with obvious regional disease. JB, aged 37 has a large scotoma in the upper field. The ERG obtained with a stroboscope shows reduction in rod sensitivity, and cone responses are very small. Oscillatory responses are absent. LED ERGs (not illustrated) show the scotopic threshold response to be developed only when the flash duration is increased from the normal 50 μsec duration to 0.3 msec, while when red light is used, the PTR is developed by light which is ten times as bright as is needed in a normal eye.

---

in this patient (seen early in the series) only weak flashes were used, and only for the more intense were normal values (subsequently) obtained. (e) Examples of red light ERGs demonstrating (from below upwards) the scotopic threshold response, the minimum cone ERG and the cone and rod b-waves elicited by more intense flashes. (f,g) Colour contrast thresholds in peripheral retina, along protan (f) and tritan (g) axes.

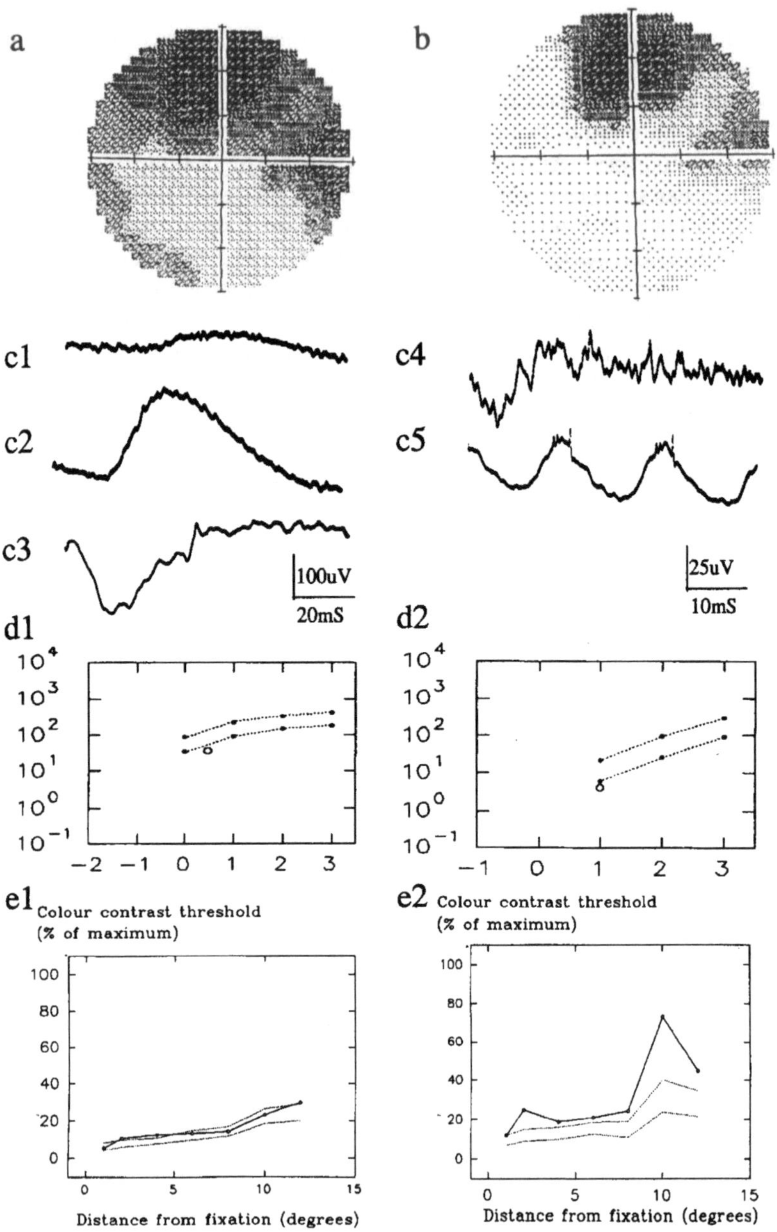

Fig. 4. Patient JB, age 37 has a typical regional ADRP, with lower retinal damage. (a, b) Dark adapted fields determined by automated static perimetry with red and blue targets respectively. (c) Standard ERGs obtained with stroboscopic flashes and filters: (c1) response to blue stimulus which gives half maximal responses in normal observer; (c2) response to blue stimulus which gives maximum amplitude b-waves in normal observer and a large rod a-wave, which is absent in this patient; (c3) response to intense white stimuli showing 'negative ERG'; (c4), oscillatory potential recording: the responses are absent; (c5) 30 Hz flicker responses are normal. (d) ERGs obtained with

There is loss of the oscillatory potentials, which are generated in the inner retina, and the ERG to bright white flashes is abnormal. Colour vision is within normal limits for red – green discriminations, and only slightly elevated for tritan discrimination except at one point (10°). It is not certain whether this represents a real loss.
In other cases results similar to the above have been found. If cone thresholds are increased, or there are scotomata, the method of testing for colour vision, using annular stimuli, might yield abnormal results as a result of local scarring. However, it seems to be the rule that in most diffuse cases colour vision is abnormal at all retinal locations except the fovea, and even slightly outside this region, there are gross losses of colour vision discrimination of a predominantly tritan type of defect. In the case of regional disease, it is possible to find patients with quite profound circumscribed defects, but with normal colour contrast thresholds in the central 12 x 12°.

Other types of RP

In X-linked disorders, colour vision approximates to the regional pattern. In simplex and recessive cases, there is also gross loss of peripheral colour vision, more than would be expected from the perimetric defects. In patients with point mutations in the gene coding for peripherin, who show RP like fundus changes, there is gross loss of colour vision. In one family, with a point mutation at codon 172 (glu-arg) in which the condition is very mild, in asymptomatic children with apparently normal thresholds fields and fundi, and nearly normal electroretinograms, there were clear defects of colour vision.

Discussion

It is known that a tritan type of defect can be found in many patients with RP, but the severity of abnormalities in the retinal periphery surprised us. Because the peripheral colour vision losses do not occur in some patients (e.g. Fig. 4) they are not an inevitable consequence of disease, but they are seen in genetically determined conditions where the mutation should only affect a pigment found in rods (Fig. 2). The method has been use in diagnosis and investigations of inherited disease – for example it has confirmed that an anomaly in the gene is a point mutation causing disease and not just a polymorphism – but in most patients would not act as an aid to diagnosis. It remains to determine whether the continuing elevation of colour thresholds in the periphery, and the encroachment of achromatopsia on the foveal region, helps chart the progress of the condition in the individual. Colour vision tests are simpler and quicker

---

LED stimulator, showing (d1) rod responses to green light and (d2) cone responses to red light. The intensities should be compared to those used for patients in Fig. 2 and 3. (e) Peripheral colour contrast sensitivity along (e1) protan and (e2) tritan colour confusion axes.

and more acceptable than very detailed perimetry. The mechanism of loss of colour vision is also of interest. Perhaps the death of rods causes morphological changes which damage cones. This could be mechanical, or due to processes of repair and debris removal secondarily damaging cones. However, in some of our cases, colour vision is affected, although cone threshold is normal. This suggests that cones continue to transduce effectively, and if so, colour vision may be disturbed by abnormal signals from rods, mimicking the effect of light, which are injected into the cone pathway, and sum with and dilute the chromatic signal. These abnormal signals would join the pathway subsequent to any point where sensitivity is regulated.

References

Arden, G.B., Carter, R.M., Hogg, C.R., Powell, D.J., Ernst, E.J.K., Clover, G.M., Lyness, A.L. and Quinlan M.P. (1981). A modified ERG technique and the results obtained in X-linked Retinitis Pigmentosa. Brit. J. Ophthalmol 67: 419–430.

Arden, G.B., Gündüz, K. and Perry, S. (1988). Colour vision testing with a computer graphics system. Clin. Vis. Sci. 2: 303–320.

Berson, E.L. (1992). Principles and Practice of Clinical Electrophysiology of Vision. Heckenlively, J. and Arden, G.B. (eds.), chapter 6, 505–509, Mosby, St. Louis.

Gündüz, K., Arden, G.B. and Perry, S. (1990). Seeing, Contour and Colour. In: Kulikowski, J.J., Dickinson, C.M. and Murray, I.J. (eds.), IInd Internat. Symposium Neurosci: 373–380, Pergamon, Oxford.

Inglehearn, C.F., Keen, T.J., Bashir, R., Jay, M., Fitzke, F., Bird, A.C., Crombie, A. and Bhattacharya, S. (1992). A completed screen for mutations of the rhodopsin gene in a panel of patients with autosomal dominant retinitis pigmentosa. Hum. Mol. Genet. 1: 41–45.

Lyness, A.L., Earnest, W., Quinlan, M.P., Clover, G., Arden, G.B., Carter, R.M., Bird, A.C. and Parker, J.A. (1985). A clinical, psychophysical and electrophysiological study of patients with familial dominant retinitis pigmentosa. Brit. J. Ophthalmol. 69: 326–3396.

Massof, R.W. and Finklestein, D. (1981) Two forms of autosomal dominant Retinitis Pigmentosa. Doc Ophthalmol. 51: 289–346.

McWilliam, P., Farrar, G.J., Kenna, P., Bradley, D.G., Humphries, M.M., Sharp, E.M., McConnell, D.J., Lawler, M., Sheils, D. and Ryan, C. (1989). Autosomal dominant retinitis pigmentosa (ADRP): localization of an ADRP gene to the long arm of chromosome 3. Genomics 5: 619–622.

Ruddock, K.H. (1965). The effect of age upon colour vision. I. Response in the receptoral system of the human eye. Vision Res. 5: 37-45.

Spileers, W., Falcao-Reis, F., Hogg, C., Smith, R. and Arden G.B. (1993). The human ERG evoked by a Ganzfeld Stimulator powered by Red and Green Light emitting diodes. Clin. Vis. Sci. 8: 21–39.

Spileers, W., Falcao-Reis, F., Hogg, C., and Arden G.B. (1993). Evidence for Human Electroretinogram A- and Off Responses that Color processing occurs in the Cones. Invest. Ophthalmol. Vis. Sci. 34: 2079–2091.

Tak Yu, Falcao-Reis, F., Spileers, W. and Arden, G.B. (1991). Peripheral colour contrast. A new screening test for preglaucomatous visual loss. Invest. Ophthalmol. Vis. Sci. 32: 2779–2789.

Wells, J., Wrobleski, J., Keen, J., Inglehearn, C., Jubb, C., Eckstein, A., Jay, M., Arden, G.B., Bhattacharya, S., Fitzke, F. and Bird, A. (1993). Mutations in the human retinal degeneration slow (rds) gene can cause either retinitis pigmentosa or macular dystrophy. Nature Genetics 3: 213–218.

Arden, G.B.
Moorfields Eye Hospital
City Rd.
London, U.K.

2. Color vision and retinitis pigmentosa

MAIJA MÄNTYJÄRVI and KAIJA TUPPURAINEN
(Kuopio, Finland)

Abstract

Color vision of 31 patients (13 women, 18 men) with retinitis pigmentosa was studied with the Standard Pseudoisochromatic Plates part 2 (SPP2), Farnsworth-Munsell 100 hue test (FM 100), and Nagel anomaloscope. Three of the patients could not read the SPP2 plates, and 5 patients could not complete the FM 100 test. The Lanthony Tritan Album and Farnsworth Panel D 15 were used for these patients. The age of the patients varied from 8 to 69 years (mean 40 ± 19, SD), the duration of the disease from 0.5 to 60 years (mean 12 ± 15), and the visual acuity from 0.1 to 1.2 (mean 0.8 ± 0.3). Thirteen of the patients (42%) had defective color vision, a tritan defect, and 4 of them had also an additional red defect. The color vision test results of the right eyes correlated significantly with the visual acuity and with the dark-adapted cone threshold but not with the age of the patients, with disease duration, with the dark-adapted rod threshold or with the extent of the visual fields.

Introduction

Retinitis pigmentosa is classified with those retinal dystrophies that predominantly involve the rods (Goodman *et al.*, 1963). The first symptoms of retinitis pigmentosa are deterioration of dark adaptation and changes in peripheral visual fields. Later in the progress of the disease, also visual acuity and color vision are affected (Newsome, 1988). Color vision in retinitis pigmentosa has been studied earlier with different kinds of test (François and Verriest, 1962; Birch *et al.*, 1979a; Fishman *et al.*, 1981). However, not all of these patients can perform color vision tests because of the often very constricted visual fields. The purpose of the present study was to test retinitis pigmentosa patients with several color vision tests to find out which would be the most suitable for the examination of the patients. Moreover, correlation of the color vision test results to the age and visual acuity of the patients, duration of the disease, dark adaptation thresholds, and extent of visual fields was studied.

B. Drum (ed.), Colour Vision Deficiencies XII, pp. 13–19.
© 1995 *Kluwer Academic Publishers, Dordrecht.*

Subjects and methods

In the registers of the University Hospital of Kuopio, 70 patients with the diagnosis of Retinitis pigmentosa were found. According to the registers, of these, one had died, six had moved out of the area, and 28 had visual acuity less than 0.1 or visual field diameter constricted to less than 10°. This left 35 patients to be invited for eye examination and color vision testing. The eye examination revealed in one of these patients congenital protanomaly and in another congenital deuteranomaly; they were excluded from the study. One patient (a child) refused to cooperate in the color vision examination, and one patient refused to come to the examination. Thus, there were 31 patients, 13 women and 18 men (58 eyes) in the final study group. Patient age varied from 8 to 69 yr (mean 40 yr ± 19, SD) and disease duration from 0.5 to 60 yr (mean 12 yr ± 15). Nine of the patients had an autosomal dominantly inherited disease, 5 patients most probably had an autosomal recessively inherited disease, and 17 patients were simplex (sporadic) cases; of these, 4 had Usher's syndrome.

The eye examination consisted of testing visual acuity, refraction, phorias and tropias, as well as biomicroscopy, ophthalmoscopy and retinoscopy with the aid of tropicamid 0,5 % + phenylephrine 5% or cyclopentolate HCl 1% drops. The tension was measured with the Goldmann applanation tonometer in the persons aged 40 years or older.

The visual fields were tested with the full field 120 point three zone strategy in the Humphrey Visual Field Analyzer (Humphrey Instruments, San Leandro, California, U.S.A.). In analyzing the results, the evaluation of visual fields was done in three different ways: using four groups from I to IV according to François and Verriest (1962), calculating the percentage of the remaining field (only the absolute scotoma points were considered as visual field loss), and calculating in degrees the diameter of the central field in those patients who had an absolute ring scotoma or a tubular field. If the Humphrey visual fields could not be tested, the Friedman Central Field Analyzer (Clement Clark, Ltd, London, U.K.) was used.

The dark adaptation examination was done with the Goldmann-Weekers Adaptometer (Haag-Streit, Bern, Switzerland) according to the instructions in the method I: The absolute threshold. The normal thresholds and recovery times with this particular machine are: the cone threshold 4.2 ± 0.2 log units and the cone time 7 ± 1 minutes; the rod threshold 2.2 ± 0.1 log units and the rod time 28 ± 3 minutes. In the logarithmic scale, the luminance of 10^7 (7 log units) corresponds to 1 cd/m^2, 10^6 to 0.1 cd/m^2 etc.

Color vision was tested with the Standard Pseudoisochromatic Plates part 2 (SPP2) (1983), Farnsworth-Munsell 100 hue test (FM 100) (1957), and Nagel anomaloscope (Schmidt & Haensch, Berlin, Germany). If SPP2 plates or FM 100 test could not be accomplished, the Lanthony Tritan Album (1985), the Farnsworth Panel D15 test (1947), and the Lanthony Desaturated Panel (1974) were shown. All of these tests were administered and interpreted according to the instructions of the tests. Error plots for the FM 100 test were drawn according to the suggestion of Kinnear (1970). The normal values of the error scores were defined according to Verriest *et al.* (1982), and the axis direction according to Smith *et al.* (1985). Nagel anomaloscope testing was done according to the method of Linksz, as referred to by Birch *et al.* (1979b). To compare the FM 100 test results at different ages, the number of SDs differing from the mean normal age-matched error score (Verriest *et al.*, 1982) was calculated. For instance, in the age group of 40–49 years, the normal score is 9.28 ± 2.33 (in square roots). If a patient in this age group had a score of 12.20, his age-matched result would be:

$$\frac{12.20 - 9.28}{2.33} = 1.25.$$

His result would thus be 1.25 SDs above the normal mean.

As a statistical test, linear regression was used.

Results

Altogether 13 patients (42%) and 23 eyes (40%) had a color vision defect. All had a tritan defect, and 4 had an additional red-green defect (Table 1).

Table 1. The number of color vision defects in the retinitis pigmentosa patients.

Type of color vision defect	Patients no. 31		Eyes no. 58	
	no.	%	no.	%
Tritan (FM 100, Panel D15)	9	29	16	28
Tritan and pseudo-protanomalous (AQ 0.5–0.6)	4	13	7	12
All color vision defects	13	42	23	40

The SPP2 plates could be read by 28 patients (52 eyes). Three patients (6 eyes) with a central visual field of 10° could read only the Lanthony Tritan Album plates out of all the plates tests. Two did not pass the Album test.

The FM 100 test could be performed by 26 of the patients (48 eyes). The three above mentioned patients and 2 other patients (10 eyes) could not arrange the FM 100 caps, but they could work with the Panel D 15 test. Three of them passed the Panel D 15 test and also the Desaturated Panel test and were considered to have normal color vision. Two patients made several tritan confusions in the Panel D 15 test.

The Nagel anomaloscope did not give trouble to any of the retinitis pigmentosa patients, they could reliably interpret the color fields. Four patients (7 eyes, Table 2) had a red-shift (pseudo-protanomaly) with anomaly quotients (AQ) of 0.5–0.6, but with normal matching ranges (MR) of 2–7.

Table 2. The performance of the 31 patients (58 eyes) in the different color vision test

	Plate tests		Cap tests		Anomaloscope
	SPP2	Lanthony Tritan Album	FM100	Panel D15	Nagel
No. of patients (eyes) tested	28 (52)	3[a] (6)	26 (48)	5[b] (10)	31 (58)
Fail no.	11 (19)	2 (4)	11 (19)	2 (4)	4 (7)
Fail total	13 (23)		13 (23)		4 (7)

[a] These 3 patients could not perform the SPP2 test. [b] These 5 patients (including the 3 patients mentioned above) could not perform the FM 100 test.

The results of the eye examination were as follows:

Visual acuity of the patients varied from 0.1 to 1.2 (mean 0.8 ± 0.3).

In the visual fields, 3 patients had only partial relative ring scotomas, 10 had partial absolute ring scotomas, 5 had full absolute ring scotomas, and 11 had tubular central fields. Two of the patients (children) could be examined only with the Friedman Central Field Analyzer, their results have not been included in the calculations. The diameter of the central field in the patients with full absolute ring scotomas or tubular fields was from 10° to 50° (mean 22° ± 14). The extent of the remanining visual field varied from 9% to 100% (mean 48% ± 28). According to the classification of François and Verriest (1962), 3 patients had type I, 4 patients type II, 11 patients type III, and 11 patients type IV visual fields.

Dark adaptation could be examined in 26 patients (49 eyes). The cone threshold varied from 4.4 to 5.6 log units (mean 4.9 ± 0.4) and the rod threshold from 2.0 to 5.8 log units (mean 3.7 ± 1.1.).

The lens was clear in 37 of the eyes, 20 had a small star or spot shaped posterior subcapsular opacity, and one had nuclear sclerosis.

In the fundus, typical retinitis pigmentosa changes were seen. Two of the patients had cystic maculopathy and 4 patients had atrophic macular changes. The other 25 patients had no remarkable changes in the macular area.

In linear regression (right eyes), the number of correctly interpreted blue figures in the SPP2 test correlated significantly with visual acuity and with cone threshold of the dark adaptation ($y = 0.031x + 0.547$, $r = 0.40$, $p = 0.04$; $y = -0.051x + 5.241$, $r = 0.62$, $p = 0.004$, respectively). Age corrected FM 100 scores also correlated significantly with visual acuity and with cone threshold ($y = -0.069x + 0.804$, $r = 0.57$, $p = 0.004$; $y = 0.081x + 4.811$, $r = 0.61$, $p = 0.008$, respectively). The corresponding correlations with age, disease duration, rod threshold or visual field size were not significant.

Discussion

It was difficult for some of the retinitis pigmentosa patients to perform the usual color vision tests. The figures in pseudoisochromatic plates were too large for viewing with a central tubular field of 10°. The Lanthony Tritan Album plate test with the colored dots in one corner of a grey dotted square was the only pseudoisochromatic test reliably seen by these patients.

Of the cap tests, the FM 100 boxes with 21 caps were too long to be reliably arranged by these patients, too. The shorter boxes of the standard and desaturated Panel D 15 were easier and more reliable.

The Nagel anomaloscope with its field of 2° was the best tool for color vision testing the small intact central visual field in retinitis pigmentosa patients. All of the patients in the present study could reliably compare the colored fields in this instrument.

The most frequent color vision defect in retinitis pigmentosa patients was a tritan defect, 42%. This has been observed also earlier (Birch *et al.*, 1979a; Fishman *et al.*, 1981). An additional pseudo-protanomaly was detected only in 12%. Earlier studies have also observed pseudo-protanomaly in retinitis pigmentosa patients (Young and Fishman, 1980; Fishman *et al.*, 1981). Thus, color vision testing might be valuable in differential diagnosis early in the disease when diagnostic symptoms cannot be observed, or the eye fundus does not yet show typical changes. An early red defect would most probably be a sign of a cone dystrophy or Stargardt's disease (Birch *et al.*, 1979a; Pinckers, 1971; Pinckers, 1982; Mäntyjärvi and Tuppurainen, 1992) while normal color vision or an early blue defect could indicate retinitis pigmentosa.

Lens opacities and macular changes, often found in retinitis pigmentosa patients, can further deteriorate color vision affecting especially blue color vision. However, some of our patients with small subcapsular lens opacities had normal color vision. All of those with macular affection had a tritan defect.

Comparing the status of color vision to the progress of the disease, the color vision test results correlated significantly with visual acuity and with cone threshold of dark adaptation. This is quite understandable, color vision as well as visual acuity and cone threshold represent central cone functions. It was, however, surprising that the color vision test results did not correlate with disease duration or with age. Earlier, correlation with age, visual acuity and visual fields has been found (François and Verriest, 1962; Massof *et al.*, 1979).

Conclusion

Color vision defects in retinitis pigmentosa patients could be best followed with an anomaloscope. If the visual field later deteriorates to a small central tubular field, the anomaloscope testing can still be done. In addition to a red-green anomaloscope, a blue anomaloscope is then also needed.

In the early stages of the disease, color vision testing could help in the differential diagnosis between retinitis pigmentosa and cone dystrophies or Stargardt's disease.

Acknowledgements

The authors would like to thank the Finnish Eye Foundation for supporting this study.

References

Birch, J., Chisholm, I.A., Kinnear, P., Marré, M., Pinckers, A.J.L.G., Pokorny, J., Smith, V.C. and Verriest, G. (1979a). Acquired color vision defects. In: Pokorny, J., Smith, V.C., Verriest, G. and Pinckers, A.J.L.G. (eds.), Congenital and Acquired Color Vision Defects: 243–348, Grune and Stratton, New York.

Birch, J., Chisholm, I.A., Kinnear, P., Pinckers, A.J.L.G., Pokorny, J., Smith, V.C. and Verriest, G. (1979b). Clinical testing methods. In: Pokorny, J., Smith, V.C., Verriest, G. and Pinckers, A.J.L.G. (eds.), Congenital and Acquired Color Vision Defects: 83–135, Grune and Stratton, New York.

Farnsworth, D. (1947). The Farnsworth Dichotomous Test for Color Blindness, Panel D 15. The Psychological Corporation, New York.

Farnsworth, D. (1957). The Farnsworth-Munsell 100-Hue Test. Munsell Color Company, Baltimore.

Fishman, G.A., Young, R.S., Vasquez, V. and Lourenco, P. (1981). Color vision defects in retinitis pigmentosa. Ann. Ophthalmol. 13: 609–618.

François, J. and Verriest, G. (1962). Étude biométrique de la rétinopathie pigmentaire. Ann. Oculist. (Paris). 195: 937–951.

Goodman, G., Ripps, H. and Siegel, I.M. (1963). Cone dysfunction syndromes. Arch. Ophthalmol. 70: 214–231.

Ichikawa, H., Hukami, K. and Tanabe, S. (1983). Standard Pseudoisochromatic Plates part 2. Igaku-Shoin Ltd., Tokyo.

Kinnear, P.R. (1970). Proposals for scoring and assessing the 100-hue test. Vision Res. 10 : 423–433.

Lanthony, P. (1974). Lanthony's Desaturated Panel 15 Hue Test According to Farnsworth-Munsell. Luneau Ophthalmologie, Paris.

Lanthony, P. (1985). The Tritan Album. Luneau Ophthalmologie, Paris.

Mäntyjärvi, M. and Tuppurainen, K. (1992). Color vision in Stargardt's disease. Int. Ophthalmol. 16: 423–428.

Massof, R.W., Finkelstein, D., Starr, S.J., Kenyon, K.R., Fleischman, J.A. and Maumenee, I.H. (1979). Bilateral symmetry of vision disorders in typical retinitis pigmentosa. Br. J. Ophthalmol. 63: 90–96.

Newsome, D.A. (1988). Retinitis pigmentosa, Usher's syndrome, and other pigmentary retinopathies. In: Newsome, D.A. (ed.), Retinal Dystrophies and Degenerations: 161–194, Raven Press, New York.

Pinckers, A. (1971). La maladie de Stargardt. Ann. Oculist. (Paris) 204: 1331–1346.

Pinckers, A. (1982). X-linked progressive cone dystrophy. In: Verriest, G. (ed.), Colour Vision Deficiencies. VI. Doc. Ophthalmol. Proc. Ser. 33: 399–403, Dr. W. Junk Publishers, The Hague.

Smith, V.C., Pokorny, J. and Pass, A.S. (1985). Color-axis determination on the Farnsworth-Munsell 100-hue test. Am. J. Ophthalmol. 100: 176–182.

Verriest, G., Van Laethem, J. and Uvijls, A. (1982). A new assessment of the normal ranges of the Farnsworth-Munsell 100-hue test scores. Am. J. Ophthalmol. 93: 635–642.

Young, R.S.L. and Fishman, G.A. (1980). Color matches of patients with retinitis pigmentosa. Invest. Ophthalmol. Vis. Sci. 19: 967–972.

Maija Mäntyjärvi, MD.
Department of Ophthalmology
University Hospital of Kuopio
P.O. Box 1777
70211 Kuopio
Finland

3. Colour vision after surgery for retinal detachment: Scleral buckling versus vitrectomy

A. SERRA, I. ZUCCA, C.M. DESSY, M. FOSSARELLO, A. TANDA and V. PIRAS

(Cagliari, Italy)

Abstract

Various parameters of visual function, i.e. visual acuity, colour vision and visual field, and fundus appearance were studied in seventeen selected patients who had undergone anatomically successful surgery for unilateral primary rhegmatogenous retinal detachment by scleral buckling or by vitrectomy. All patients had a normal contralateral eye. Colour vision and visual sensitivity scores were significantly worse in the operated eye as compared to the contralateral eye in both groups. Moreover, these parameters were significantly worse in operated eyes from the vitrectomy group as compared to the operated eyes from the scleral buckling group. No remarkable fundus changes were observed. The potential hazard of subclinical phototoxic lesions to rods and cones and nuclear sclerotic changes to the crystalline lens during vitrectomy should be kept in mind by the surgeon. For evaluating visual recovery after this type of surgery functional testing should include colour vision and visual sensitivity testing.

Introduction

Anatomically successful surgery for primary rhegmatogenous retinal detachment may be obtained either by scleral buckling or by pars plana vitrectomy, and the choice of the type of surgery to be performed depends on the clinical judgement, the availability of instrumentation for vitrectomy, and the surgeon's skill. Although visual recovery after scleral buckling or pars plana vitrectomy for retinal detachment may be comparable at short term when visual acuity alone is used as a criterion, it may be argued that pars plana vitrectomy may be more prone to cause subtle anatomical, biochemical and functional changes to the crystalline lens and to the retina. These changes may be due to the mechanical stress caused by the infusion mechanism, but also to the light toxicity hazard related to the operating microscope and the fiber optic illumination (Fuller *et al.*, 1978; Schachat *et al.*, 1983; Novak *et al.*, 1984; de

B. Drum (ed.), Colour Vision Deficiencies XII, pp. 21–27.
© 1995 *Kluwer Academic Publishers, Dordrecht.*

Bustros *et al*, 1988; McDonald and Harris, 1988; Michels and Sternberg, 1990; Kuhn *et al*., 1991; Michels *et al*., 1992).

In the present study we report the data on visual acuity, colour vision and visual sensitivity following successful scleral buckling and pars plana vitrectomy procedures to correct rhegmatogenous retinal detachment.

Patients and methods

For this study we selected 17 patients (9 males and 8 females; age varying between 15 and 71 years) affected by unilateral phakic rhegmatogenous retinal detachment, which were operated for the first time either by scleral buckling (12 eyes) or by a three-port pars plana vitrectomy (5 eyes). The clinical characteristics of the retinal detachments were by and large very similar in both groups, involving no more than 2 retinal quadrants and the macula. Criteria for selection of patients included: a clinically normal contralateral eye, lack of any systemic disease, no intraoperative or postoperative complications and anatomically successful retinal surgery. For scleral buckling, localization of retinal breaks was made with the indirect ophthalmoscope, cryotherapy was applied to the retinal breaks, solid silicone was used for buckling, and subretinal fluid was drained. Pars plana vitrectomy was performed with a Vitreovisc unit (Optikon, Italy) using light source at maximum illumination for about 90 minutes,and for infusion and vitreous gel replacement a balanced salt solution was used: retinal breaks were photocoagulated by Argon endolaser. The cornea was covered when no intraocular operation was being performed to minimize retinal exposure to the operating microscope.

The patients were tested at different time intervals following surgery (1–30 months: mean = 9 months; S.D. ± 7.34) for visual acuity (Snellen chart), colour vision (FM 100 hue at 200 lux C-illuminant, New Colour Test, Lanthony D-15), automatic visual field (Octopus 500 E automatic perimeter, Interzeag). In all the patients we performed contact-lens fundus biomicroscopy, indirect ophthalmoscopy, and fluorescein angiography. The clinical parameters evaluated statistically were the following: visual acuity, FM 100 hue total score (TS), Octopus automatic perimeter mean defect (MD) and loss variance (LV). For these parameters we calculated the mean value (and SD) for operated and contralateral eyes in all the patients; then for each patient we calculated the difference of value between operated and contralateral eye, and we applied to such differences the paired Student t-test in order to compare the scleral buckling with the vitrectomy groups.

Results

The values of visual acuity, colour vision, and visual field recorded in the operated and in the contralateral eye from all the patients are shown in Tables 1 and 2. It is noteworthy that all the contralateral eyes have remarkably homogeneous values.

In Table 3 we report the mean values of the different parameters of visual function found in the two different groups of surgery, and in Table 4 the Student's t-test estimates of the means of the differences of value between operated eyes and contralateral eyes for the scleral buckling and the vitrectomy group.

Visual acuity is impaired in the majority of the operated eyes. As it is shown in Table 3 and Table 4, the difference of value between the scleral buckling group and the vitrectomy group is not significant ($p = 0.081$).

Colour vision. The FM 100 hue TS shows an higher impairment in the vitrectomy group than in the scleral buckling group ($p = 0.162$) as it is shown in Table 4. The profile shows a tritan axis in the vast majority of cases, in both groups of surgery (Table 1 and Figure 1), as it was expected on the basis of previous works (Pokorny et al., 1979).

Automatic perimetry. The MD and LV are the parameters whose impairment shows the highest difference in significance between the two groups of surgery ($p = 0.0007$ and $p = 0.035$, respectively), as it is shown in Table 4.

Fundus examination. No obvious fundus abnormalities were detected by contact-lens biomicroscopy or by fluorescein angiography in the operated eyes.

Table 1. Age, visual acuity, and colour vision data for patients operated of rhegmatogenous retinal detachment (scleral buckling or vitrectomy[a]).

Patient and age	operated eye				contralateral eye			
	visual acuity	FM axis and TS	L-15 D	NCT	visual acuity	FM axis and TS	L-15 D	NCT
1) 57	1	Tri-140	Tri	Tri	1	Tri-132	Tri	Tri
2)[a] 15	1	Tri-89	Tri	Tri	1	Tet-80	Tri	Tri
3) 71	0.4	Tri-239	Tri	Tri	0.9	Tri-161	Tri	Tri
4)[a] 48	0.4	Tri-448	Tri	Tri	1	Tet-97	nor	nor
5) 63	0.5	Tri-212	Tri	Tri	0.9	Tri-119	Tet	Tri
6) 63	1	Ana-148	Tet	Ana	1	Tri-158	nor	Tri
7) 70	0.6	Tri-291	Tri	Tri	1	Tri-162	Tri	Tri
8) 63	1	Tri-186	Tri	Tet	1	Tri-148	Tet	Tri
9) 68	0.5	Ana-222	Tri	Tri	0.9	Tri-176	Tet	nor
10) 68	0.7	Tri-263	Tri	Tri	1	Tri-120	nor	Tri
11) 44	0.5	Tri-221	Tri	Tri	1	Tet-89	Tet	Tri
12) 66	1	Tri-283	Tri	Tri	1	Tri-93	nor	nor
13) 45	1	Tri-64	nor	nor	1	Tri-44	nor	nor
14)[a] 62	0.4	Tri-224	Ana	Tri	1	Tri-72	nor	nor
15) 64	0.8	Tri-186	Tri	Tri	1	Tri-107	nor	nor
16)[a] 64	0.3	Tri-191	Tri	nor	1	nor-77	nor	nor
17)[a] 63	0.3	Ana-399	Ana	Ana	1	nor-156	nor	nor

Tri = tritan; Tet = tetartan, Ana = anarchic; nor = normal.

Table 2. Automatic perimetry (mean defect and loss variance) of patients operated for rhegmatogenous retinal detachment (scleral buckling or vitrectomy[a]).

Patient	operated eye		contralateral eye	
	MD	LV	MD	LV
1)	3.9	4.4	2.5	5.8
2)[a]	3.4	29.6	0.0	2.5
3)	9.8	6.6	7.8	15.5
4)[a]	15.2	5.5	3.1	4.6
5)	7.4	13.9	2.6	6.6
6)	5.9	7.1	2.7	3.9
7)	9.4	15.7	5.6	8.4
8)	6.1	5.4	4.2	5.4
9)	9.5	16.4	9.3	15.9
10)	9.1	24.2	6.4	13.5
11)	7.4	4.3	3.7	4.4
12)	15.6	53.2	6.1	17.9
13)	5.7	3.2	3.8	4.6
14)[a]	14.9	26.8	7.6	8.2
15)	4.9	17.4	0.8	2.4
16)[a]	15.5	16.6	5.4	13.1
17)[a]	16.3	64.3	5.0	10.7

MD = mean defect; LV = loss variance.

Table 3. Mean values of the different functional parameters found in the two different groups of surgery.

Parameter	Scleral buckling		Vitrectomy	
	Operated eye	Fellow eye	Operated eye	Fellow eye
Visual acuity	6.75	9.75	4.8	10
FM total score	204.6	125.8	270.2	96.4
Visual field: mean defect (dB)	7.89	4.62	13.06	4.22
Loss variance	14.32	8.69	28.56	7.82

Table 4. Student's t-test estimates of the means of the differences of value between operated eyes and contralateral eyes from scleral buckling and vitrectomy group.

Parameter	Scleral buckling (mean)	Vitrectomy (mean)	t-Student	P value
Visual acuity	-3	-5.20	-1.469	0.081
FM total score[a]	14.10	15.87	1.016	0.162
Visual field: mean defect (dB)	3.267	8.840	3.846	0.0007
Loss variance	5.625	20.74	1.938	0.035

[a] Square root of the FM 100-hue scores.

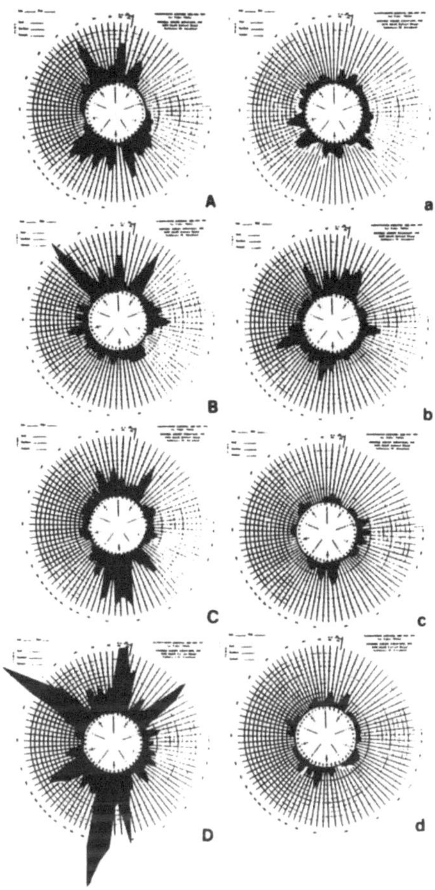

Fig. 1. Example of FM 100 hue profiles from patients with scleral buckling or vitrectomy. A and B eyes with scleral buckling; a and b: contralateral, normal eyes; C and D eyes with vitrectomy; c and d: contralateral, normal eyes.

Discussion

The results of this study indicate that after anatomically successful surgery for retinal detachment certain parameters of visual function are affected by the surgical procedure employed. In particular, visual sensitivity appear to be significantly worse following vitrectomy. The difference in visual acuity outcome is of higher significance than the difference in colour vision between the two groups. It is noteworthy that the difference in automatic perimetry MD is highly significant.

The poorer colour vision and visual sensitivity performance observed following vitrectomy may be related to mechanical and/or phototoxic effects to the lens and the retina induced by surgical instruments. Scleral buckling is basically an external procedure not requiring the operating microscope. The procedure utilizes the indirect ophthalmoscope to localize retinal breaks for only a short time. Pars plana vitrectomy includes a three-port intraocular access, intraocular infusion, vitreous removal, fiberoptic light endoillumination for about 90 minutes, and the use of the operating microscope for sclerotomy opening and closure.

Although a relatively small number of documented cases of endoilluminator photoxicity are reported in the literature, phototoxic retinal damage from both the operating microscope and the intraocular fiberoptic probe are well recognized operative occurrences (Fuller et al., 1978; McDonald and Harris, 1988; Michels and Sternberg, 1990; Kuhn *et al.*, 1991; Michels *et al.*, 1992). In our patients no fundus changes suggestive of retinal phototoxicity were observed. However, many of the characteristic angiographic findings of excessive light exposure visible at the level of the retinal pigment epithelium may not be present when phototoxicity occurs in detached retina (Zilis and Machemer, 1991).

In our study the type and incidence of postoperative cataract was not specifically evaluated. However, in one case we could document a severe nuclear sclerosis of the crystalline lens developing three months after vitrectomy. The cause of postoperative nuclear sclerosis is unknown: it may be caused by altered lens metabolism after removal of the vitreous gel, or it may be linked to osmotic stress to the lens capsule induced by intraocular irrigating solution. (Schachat *et al.*, 1983; Novak *et al.*, 1984; de Bustros *et al.*, 1988; Cherfan *et al.*, 1991). Direct damage to the lens may also be caused by the endoilluminator through a mechanism of light toxicity, or by increasing the intraocular temperature locally (Cherfan, 1991).

Colour vision and vision sensitivity testing alone are not sufficient to discriminate with certainty between a loss of perception related to opacities of the ocular media or to impairment of photoreceptors. Our study suggests that discrete impairment to psychophysical testing probably can be recorded in a high number of patients after vitrectomy, as a consequence of the combined effect of diffuse subclinical phototoxic lesions to photoreceptors and/or nuclear sclerotic changes to the crystalline lens.

Acknowledgments

This work was supported in part by Research Grant from the Italian Ministero della Pubblica Istruzione (40%), and in part by a grant from the Assessorato Sanità Regione Sardegna.

References

Cherfan, G.M., Michels, R.G., deBustros, S., Enger, C., and Glaser, B.M. (1991). Nuclear sclerotic cataract after vitrectomy for idiopathic epiretinal membranes causing macular pucker. Am. J. Ophthalmol. 111: 434–438.

de Bustros, S., Thompson, J.T., Michels, R.G., Enger, C., Rice, T.A., and Glaser, B.M. (1988). Nuclear sclerosis after vitrectomy for idiopathic epiretinal membranes. Am. J. Ophthalmol. 105: 160–167.

Fuller, D., Machemer, R., and Knighton, R.W. Retinal damage produced by intraocular fiber optic light (1978). Am. J. Ophthalmol. 85: 519–537.

Kuhn, F., Morris, R., and Massey, M. (1991). Photic retinal injury from endoillumination during vitrectomy. Am. J. Ophthalmol. 111: 42–46.

McDonald, H.R., and Harris, M.J. (1988). Operating microscope induced retinal phototoxicity during pars plana vitrectomy. Arch. Ophthalmol. 106: 521–524.

Michels, M., and Sternberg, P. (1990). Operating microscope-induced retinal phototoxicity. Pathophysiology, clinical manifestations and prevention. Surv. Ophthalmol. 34: 237-251.

Michels, M., Lewis, H., Abrams, G.W., Han, D.P., Mieler, W.F., and Neitz, J. (1992). Macular phototoxicity caused by fiberoptic endoillumination during pars plana vitrectomy. Am. J. Ophthalmol. 114: 287–296.

Novak, M.A., Rice, T.A., Michels, R.G., and Auer, C. (1984). The crystalline lens after vitrectomy for diabetic retinopathy. Ophthalmology 91: 1480-1489.

Pokorny, J., Smith, V.C., Verriest, G., and Pinckers, A.J.L.G. (1979). Congenital and Acquired Color Vision Defects. Grune and Stratton, New York.

Schachat, A.P., Oyakawa, R.T., Michels, R.G., and Rice, T.A. (1983). Complications of vitreous surgery for diabetic retinopathy. II. Postoperative complications. Ophthalmology 90: 522–530.

Zilis, J.D., and Machemer, R. (1991). Light damage in detached retina. Am. J. Ophthalmol. 111: 47-55.

Antonina Serra
Via Belgrado 5
I-09047 Selargius (Cagliari)
Italy

4. Pharmacology of a colour coding synapse in the retina

KONRAD KOHLER, YVONNE SCHMITZ and EBERHART ZRENNER
(Tübingen, Germany)

Abstract

In the fish retina dendritic terminals of second order neurons invaginate into the synaptic endings of the photoreceptors. Within this very specific synaptic arrangement alterations in the ambient light conditions are accompanied by remarkable morphological alterations of the synaptic connections between cones and horizontal cells. During light-adaptation fingerlike new neurites, called spinules, grow out from the dendritic terminals of the horizontal cells. During dark-adaptation these spinules are retracted. This synaptic plasticity is correlated with the expression of colour opponency in horizontal and ganglion cells. Spinules are therefore considered to be the site of the colour coding feedback synapse between cones and horizontal cells. The formation of spinules can be examined in the electron microscope and their expression can be influenced by drugs. In this report we give an overview of the known neurotransmitters and intracellular mechanisms involved in spinule formation. In addition, we present data to show that the spinules of the fish retina seem to be a valuable model to study the cellular mechanisms of colour vision defects caused by certain pharmaca in humans.

Introduction

In the last decades we have learned much about the organisation of colour pathways in the human retina from psychophysical and electrophysiological studies. Compared to these studies, ultrastructural and cell biological studies of the colour coding pathways have not kept up. This is mostly due to the limited availability of appropriate human retina samples required to demonstrate physiological changes on the cellular level which occur subsequent to intoxication or drug application. Therefore we have to look for animal models which generate one or more of the same psychophysical and electrophysiological phenomena found also in man. These animal models will then allow one to design experiments which give closer insights in the colour coding circuitries

B. Drum (ed.), Colour Vision Deficiencies XII, pp. 29–36.
© 1995 *Kluwer Academic Publishers, Dordrecht.*

and to reveal, at a cellular level, pathological processes induced by a certain pharmacon.

Colour coding in horizontal cells

In the vertebrate retina, horizontal cells are second order neurons that mediate the lateral spread of inhibition in the outer plexiform layer. They contribute to spatial and temporal acuity, and to the control of absolute sensitivity. In the teleost fish retina, axon bearing horizontal cells are exclusively connected with cones and play an important role in the colour coding process. They are subdivided into two types by the characteristics of their intracellularly recorded responses to light stimuli of different wavelengths: the luminosity (L)-type hyperpolarizes to all wavelengths whereas the chromaticity (C)-type generate a very specific pattern of hyper- and de-polarization depending on chromatic stimulation. The colour opponent behaviour of the C-type horizontal cells is thought to be generated by a feedback mechanism acting between horizontal cells and cones: horizontal cells receive synaptic input from the photoreceptors but they also make inhibitory synapses back onto cones. The circuit of these colour coding interaction was first described by Stell and coworkers (1975); these authors established a morphological model which was recently paralleled by a new picture of cone-horizontal cell wiring based on electrophysiological data (Kamermans *et al.*, 1991).

Where does the feedback between horizontal cells and cones take place? The feedback synapses are most likely located somewhere along the dendritic terminals of horizontal cells that invaginate into the synaptic pedicle of the cones. Figure 1 gives an overview of the morphological situation in the fish retina and especially of its cone pedicles. In the fish retina, cone pedicles are bell-shaped and the central cavity is filled with the dendrites of the second order neurons, i.e. horizontal and bipolar cells. The cone cytoplasm contains conspicuous ribbon synapses which can be easily identified in the electron microscope. The arciform ridge of a synaptic ribbon is orientated towards the pedicle's cavity and along the ridge dendrites of the second order neurons are arranged in triads. In the fish retina, the central element of the triad could either be a bipolar or a horizontal cell but the synaptic terminals laterally to the ribbon always belong to horizontal cells.

The laterally located horizontal cell terminals show a remarkable light-dependent synaptic plasticity. During light-adaptation, numerous finger-like processes grow out from the horizontal cell terminals and protrude further into the cone pedicle. These newly formed neurites are about 0.1 μm in diameter and up to 0.5 μm long and bear characteristic membrane densities (Fig. 2). As those neurites look like small spines, they were called spinules (Wagner, 1980). Their development is initiated by light and they are completely withdrawn in darkness; both, spinule formation and retraction is performed within 45 minutes (Wagner, 1980). Spinules are exclusively found in cones and are never

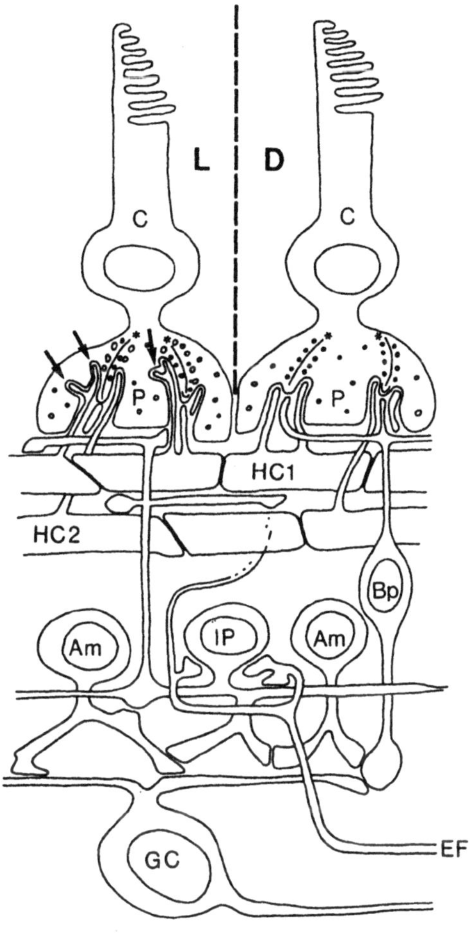

Fig. 1. Diagram of a radial section through the fish retina emphasizing the synaptic alterations in light (L)- and dark (D)-adapted cone pedicles. In the L-adapted cone, spinules (arrows) protrude from horizontal cell dendrites located laterally to a synaptic ribbon (asterisk); in the D-adapted cone, spinules are retracted and the terminals of the HC dendrites rounded. Abbreviations: L, light; D, dark; C, cone; P, cone pedicel; HC, horizontal cells type 1 and 2; Bp, bipolar cell; Am, amacrine cell; IP, inter plexiform cell; GC, ganglion cell; EF efferent fibre. (Modified from Dowling, 1987)

observed in the synaptic endings of rods.

The appearance of spinules is correlated with the appearance of colour opponency in horizontal and ganglion cells (Raynauld *et al.*, 1979; Weiler and Wagner, 1984; Djamgoz *et al.*, 1988). They were, therefore, considered to be the site of the colour coding feedback synapse.

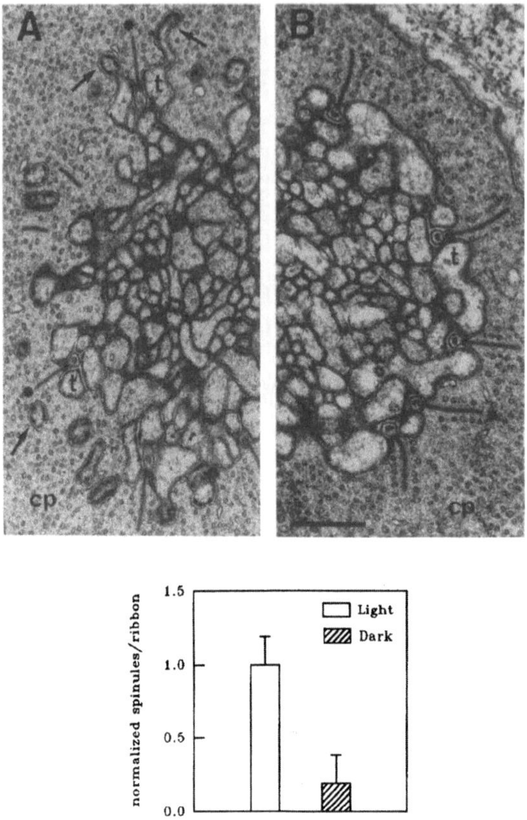

Fig. 2. Spinule formation in light- (A) and dark-adapted (B) control retinas. Electron micrographs from tangential sections at the level of the cone pedicles. The cytoplasm of the cone pedicle (cp) is densely filled with synaptic vesicles. Arrows mark spinules of horizontal cell terminals (t) lateral to the synaptic ribbon (asterisk); spinules bear characteristic membrane densities. Some of the spinules are cross cut and appear as more or less spherical structures wihtin the pedicle. The histogram summarizes the effects of light and dark on spinule expression. The number of spinules/ribbon is normalized to the light-adapted values and the standard deviation is included. Calibration bar: 0.5 μm

Spinule observation and pharmacological treatment

Spinule formation can be documented in the electron microscope only. After decapitaton, the animals' eyes are enucleated, cornea and lens are removed and the retina is dissected from the pigment epithelium; the retina is then prepared for electron microscopy by conventional protocols. Ultrathin sections of 80–90 nm are cut tangentially to the retinal surface and spinule formation is observed in cross sections of the cone pedicles (Fig. 2). Typically, about 20 spinules can be found in a cone pedicle. Quantification of spinule formation is done by calculating the spinule/ribbon ratio which is about 2.5 in a light-adapted retina and 0.5 or less in a dark-adapted retina. For easy comparison of the rates of spinule formation in light and dark, the calculated values can be normalized to the values of a fully light-adapted retina as shown in the histogram of Fig. 2. To influence spinule formation pharmacologically, drugs are either injected in the vitreous of an anaesthetized animal or the dissected retina is incubated in a appropriate solution containing the drug. From a single animal, one retina is always treated with the drug whereas the second retina serves as a control. Two principally different series of experiments are normally carried out: when the tested substance is applied to dark- or light-adapted retinas then (1) the effect of the drug on an already established morphological situation can be examined or (2) with a subsequent change in the adaptational state the drug's effect on development of spinules can be observed.

Inter- and intra-cellular mechanisms of spinule formation

Since spinules had been described for the first time by Wagner (1980), considerable progress has been made with regard to the inter- and intra-cellular mechanisms involved in spinule formation. Glutamate, the putative neurotransmitter of cones, is released during darkness and it is a glutamate receptor that is responsible for the retraction of the spinules (Weiler *et al.*, 1988). Recently this receptor has been characterized as an ionotropic non-NMDA type (Weiler and Schultz, 1993). For the outgrowth of spinules, dopamine plays some role (Weiler *et al.*, 1988; Kohler *et al.*, 1990). However, unlike glutamate, dopamine acts as a sort of neuromodulator rather than as a neutransmitter, inducing the light adaptive state of spinules (Kohler and Weiler, 1990). This idea is supported by the observation that the dopamine-linked second messenger cAMP is not able to induce spinule formation *in vivo* (Kohler and Weiler, 1990) but has the same effects in an *in vitro* preparation (Kisch *et al.*, 1991). The most vigorous signal for spinule formation is the activation of protein kinase C (PKC) which is sufficient for spinule outgrowth even in a retina depleted of dopamine (Weiler *et al.*, 1991). Until now, we do not know the receptor and signal cascade by which PKC is activated to trigger spinule formation. An actin mediated 'gel-sol' transformation in the dentritic terminal is responsible for the movement and protrusion of spinules (Schmitz and Kohler, 1993) whereas the retraction of spinules does not require actin. Spinule retraction seems to be a passive process that mainly depends on a change in membrane rigidity of the pedicle (Kohler, unpublished).

Acquired colour vision defects and spinule formation

Ethambutol is a drug commonly used in tuberculosis therapy. About 3% of Tb-patients treated with Ethambutol develop optic neuropathy, central scotoma and changes in red-green colour vision (Zrenner and Krüger, 1981). In some cases, Ethambutol even leads to absolute colour blindness. Usually, all these failures recover after the drug is suspended. ERG and VEPs are not altered by the drug. Therefore, it seems that Ethambutol induced colour blindness is due to a abnormal function at the level of the colour opponent neurons in the retina which are the horizontal and ganglion cells.

When fishes were fed with Ethambutol, they developed the same red-green colour blindness observed in man (Spekreijse *et al.*, 1991). Moreover, when a fish retina is isolated and superfused with an Ethambutol containing solution, intracellular recordings from ganglion cells show changes in colour coding. In goldfish, Ethambutol suppresses the colour opponent signals in light adapted horizontal cells (Van Dijk *et al.*, 1983). It seems that some of the pathological effects of Ethambutol are located very distally in the retina and may, therefore, correlate with alterations in spinule expression.

In already light adapted retinas a high dose of Ethambutol (10 mM) reduces the

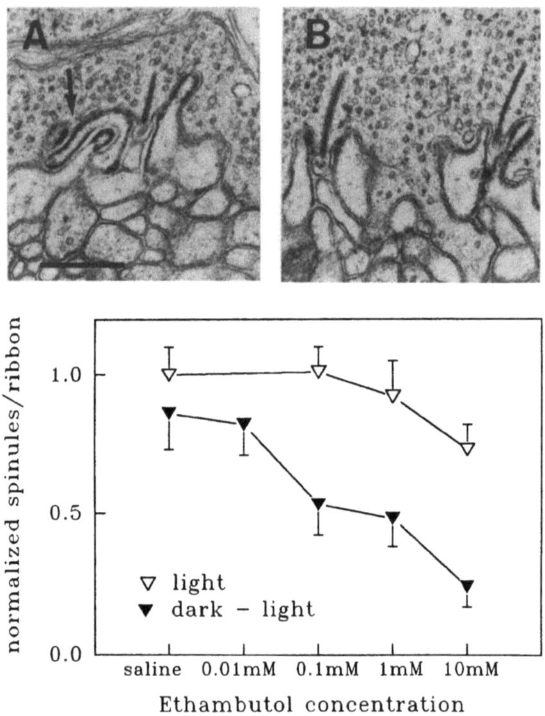

Fig. 3. (A) In a light-adapted retina a high concentration of Ethambutol (10 mM) reduces the number of already formed spinules. Spinules are also occasionally bent away from the synaptic ribbon and even sometimes overlap one another (arrow). (B) Application of Ethambutol in the dark, 10 minutes prior to light-adaptation, results in a severe inhibition of light-induced spinule formation. The histogram summarizes the dose-response data for Ethambutol injected into the vitreous of either light-adapted (open triangles) or dark-adapted animals with subsequent light-adaptation (filled triangles). Calibration bar: 0.5 μm

number of spinules by 30% and already formed fingers are bent away from the synaptic ribbon (Fig. 3a). Ethambutol application in the dark with subsequent light-adaptation results in severe dose-related inhibition of light-induced spinule formation (Fig. 3b). In these experiments, low doses of Ethambutol (0.1 mM) cause a 40% and high doses (10 mM) a 70% inhibition. These dose-related effects are summarized in the histogram of Fig. 3. Aside from affecting spinules, Ethambutol occasionally induced a complete degeneration of cone pedicles. This neurotoxicity only occurred in cones exposed to light.

The results of the Ethambutol experiments show that, in the fish retina, the drug is able to alter synaptic connections between horizontal cells and cones in a dose-related fashion. Furthermore, Ethambutol treatment can be toxic for cone pedicles and can cause their degeneration. There is a very specific toxicity because rods are never affected by the drug. The results indicate that Ethambutol influences the colour coding process already at the level of the cone-horizontal cell synapse.

Concluding remarks

There is an ongoing discussion whether spinules are pre- or postsynaptic structures and there are serious morphological, pharmacological, and electrophysiological evidences which favour their postsynaptic character (Kohler, 1990; Kamermans *et al.*, 1991). However, the correlation between spinule formation and colour coding is striking. Some new ideas about the wiring in the outer retina suggest that horizontal cells in fish and primates may not be so functionally different (Kamermans *et al.*, 1991). Even though spinules are unique to the fish retina, they may represent an ancient type of synapse which is no longer expressed as a distinct morphological structure in higher vertebrates but still operates in the horizontal cell's dendritic terminal. This hypothesis is supported by the close similarity of Ethambutol induced vision defects in fish and man (Spekreijse *et al.*, 1991). Our own experiments with Ethambutol show that the spinules of the fish retina may be a valuable model for studying the cellular mechanisms of colour vision defects caused by certain pharmaca in humans.

References

Djamgoz, M.B.A., Downing, J.E.G., Kirsch, M., Prince, D.J. and Wagner, H.-J. (1988). Plasticity of cone horizontal cell functioning in cyprinid fish retina: effects of background illumination of moderate intensity. J. Neurocytol. 17: 701–710.

Dowling, J.E. (1987). The Retina: An Approachable Part of the Brain. Harvard University Press, Cambridge.

Kamermans, M. Van Dijk, B.W. and Spekreijse, H. (1991). Color opponency in cone-driven horizontal cells in carp retina. J. Gen. Physiol. 97: 819–843.

Kirsch, M., Wagner, H.-J. and Djamgoz M.B.A. (1991). Dopamine and plasticity of horizontal cell function in the teleost retina: Regulation of spectral mechanism through D1-receptors. Vision Res. 31: 401–412.

Kohler, K. (1990). Neuronal and intracellular regulation of synaptic plasticity in horizontal cells of the carp retina. PhD. Thesis, University of Oldenburg, Germany.

Kohler, K., Kolbinger, W., Kurz-Isler, G., and Weiler, R. (1990). Endogenous dopamine and cyclic events in the fish retina, II: Correlation of retinomotor movement, spinule formation, and connexon density of gap junctions with dopamine activity during light/dark cycles. Visual Neurosci. 5: 417–428.

Kohler, K. and Weiler, R. (1990). Dopaminergic modulation of transient neurite outgrowth from horizontal cells of the fish retina is not mediated by cAMP. Eur. J. Neurosci. 2: 788–794.

Raynauld, J.P., Laviolette, J.R. and Wagner, H.-J. (1979). Goldfish retina: a correlate between cone activity and morphology of the horizontal cell in cone pedicles. Science 204: 1438–1438.

Schmitz, Y. and Kohler, K. (1993). Spinule formation in the fish retina: Is there an involvement of actin and tubulin? An electronmicroscopic immunogold study. J. Neurocytol. 22: 205–214.

Spekreijse, H., Wietsma, J.J. and Neumeyer C. (1991). Induced color blindness in goldfish: A behavioral and electrophysiological study. Vision Res. 31: 551–562.

Stell, W.K., Lightfoot, D.O., Wheeler, T.G. and Leeper, H.F. (1975). Goldfish retina: functional polarization of cone horizontal cell dendrites and synapses. Science 190: 989–990.

Van Dijke, B.W. and Spekreijse, H. (1983). Ethambutol changes the color coding of carp retinal ganglion cells reversibly. Invest. Ophthalmol. Vis. Sci. 24: 128–133.

Wagner, H.-J. (1980). Light-dependent plasticity of the morphology of horizontal cell terminals in cone pedicles of fish retinas. J. Neurocytol. 9: 573–590.

Weiler, R., Kohler, K. and Janssen U. (1991). Protein kinase C mediates transient spinule-type neurite outgrowth in the retina during light adaptation. Proc. Natl. Acad. Sci. USA 88: 3603–3607.

Weiler, R., Kohler, K., Kirsch, M. and Wagner H.-J. (1988). Glutamate and dopamine modulate synaptic plasticity in horizontal cell dendrites of fish retina. Neurosci. Letters 87: 205–209.

Weiler, R. and Schultz, K. (1993). Ionotropic non-N-methyl-D-aspartate agonists induce retraction of dendritic spinules from retinal horizontal cells. Proc. Natl. Acad. Sci. USA 90: 6533–6537.

Weiler, R. and Wagner, H.-J. (1984). Light-dependent change of cone-horizontal cell interactions in carp retina. Brain Res. 298 : 1–9.

Zrenner, E. and Krüger, C.J. (1981). Ethambutol mainly affects the function of red-green opponent neurons. Doc. Ophthalmol. Proc. Ser. 27: 13–25.

Konrad Kohler
Department of Pathophysiology of Vision and Neuroophthalmology
Division of Experimental Ophthalmology
University Eye Hospital, Röntgenweg 11, 72075 Tübingen, Germany

5. Dopamine antagonists impair 'red-green' discrimination in goldfish after intravitreal injection

CARLOS MORA-FERRER and CHRISTA NEUMEYER
(Mainz, Germany)

Abstract

Color vision in goldfish is tetrachromatic. However, under mesopic illumination conditions the L-cones do not contribute to wavelength discrimination, and color vision becomes trichromatic. To investigate the role of dopamine in light adaptation, we tested the effect of the unspecific dopamine antagonist haloperidol and of the specific dopamine D1 receptor antagonists SCH 23390 and SKF 83566-biotinyl in behavioral wavelength discrimination experiments. After injection of dopamine antagonists into the vitreous, a reduction in wavelength discrimination was observed in the range between 560 nm and 630 nm but not in the range between 404 nm and 501 nm.

As SCH 23390 and SKF 83566-Biotinyl are D1 dopamine receptor antagonists, we suppose that the D1 dopamine receptors on horizontal cells (see Wagner and Behrens, 1993) are blocked, and, thus, the modulatory activity of dopamine on horizontal cells is inhibited. We assume that H2 horizontal cells are affected as they have been demonstrated to lose their biphasic responses to chromatic stimuli after application of the D1 dopamine receptor antagonist SCH 23390 (Kirsch, *et al.*, 1991). As a result of the changed response behavior of horizontal cells, the color coding in the 'red-green' could be impaired

Introduction

Goldfish possess tetrachromatic color vision which is based on four cone types maximally sensitive at 350 nm, 450 nm, 530 nm and 620 nm (UV-, S-, M-, and L-cones, Neumeyer, 1992). Behavioral experiments indicated a lack of L-cone contribution to wavelength discrimination under white room illumination of about 1 lx (Neumeyer and Arnold, 1989). Under this condition no wavelength discrimination was observed in the range between 550 nm and 700 nm, and spectral sensitivity was reduced by 1.5 log units in the long wavelength range. Similar results have been obtained after intravitreal injection of bicuculline, a

B. Drum (ed.), Colour Vision Deficiencies XII, pp. 37–42.
© 1995 *Kluwer Academic Publishers, Dordrecht.*

GABAa antagonist (Wietsma and Spekreijse, 1991), and after feeding goldfish with ethambutol (Spekreijse *et al.*, 1991).

Retinomotor movement as well as horizontal cell coupling and synaptic plasticity seem to be controlled by an endogenous circadian dopamine release (Kohler *et al.*, 1990). As shown recently, dopamine has a prominent influence on horizontal cells: after application of the D1 dopamine receptor anatagonist SCH 23390, H2 horizontal cells respond to chromatic stimuli as if dark adapted, i.e. the response is monophasic instead of biphasic (Kirsch, *et al.*, 1991).

If the contribution of horizontal cells to color coding is essential and restricted to the light adapted state, an inhibition of dopaminergic activity on horizontal cells should reduce wavelength discrimination. To test this hypothesis, we performed behavioral experiments with the unspecific dopamine antagonist haloperidol, and the specific D1 dopamine receptor antagonists SCH 23390 and SKF 83566-Biotinyl-Cl.

Material and method

Goldfish were kept separately at a 12-hour light-dark rhythm. The illumination intensity during the measurements was 20 lx (Osram L 36W/19, daylight 5000 de luxe).

The goldfish were trained to discriminate between a training wavelength and a second comparison wavelength which were presented simultaneously on two test fields. The test fields were illuminated by quasi monochromatic light slide projectors and interference filters (Schott and Gen., DIL). The intensities were adjusted to equal brightness according to the spectral sensitivity function of the goldfish (Neumeyer, 1984) by neutral density filters (Schott and Gen., type NG). Relative choice frequency was obtained by counting the number of bites of the goldfish on both testfields. A relative choice frequency of 70% was chosen as threshold criterion (see for details: Neumeyer, 1986).

Haloperidol (Sigma), SCH 23390 and SKF 83566-biotinyl (both RBI) were dissolved in DMSO (haloperidol) or ethanol (SCH 23390, SKF 83566) and diluted with ringer or phosphate buffer. The calculated concentrations at the retinal level were 500 μM of haloperidol, 100 nM of SCH 23390 and 45 nM of SKF 83566-biotinyl-Cl.

Prior to injection, fish were dark adapted for at least one hour. The diluted substance (5μl) was injected into the vitreous of both eyes under dim red light. Thirty minutes after injection the room illumination (20 lx) was switched on and the experimental session started.

Results

As shown in Figure 1 for fish G 114, as one example of six tested animals, haloperidol reduced choice frequency to values below threshold in the range

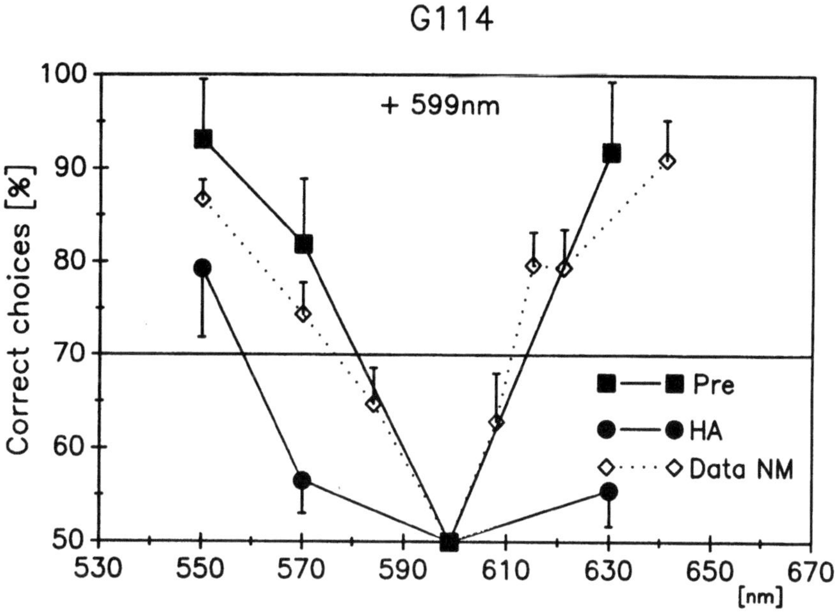

Fig. 1. Choice behavior of fish G114, trained on 599 nm, before (■) and after (●) injection of haloperidol. (◇) reference data from Neumeyer (1986). Ordinate: relative choice frequency on training wavelength 599 nm; abscissa: wavelength shown for comparison.

between 560 and 630 nm. Injections of ringer or phosphate buffer had no effect on wavelength discrimination ability.

However, if trained on wavelength 404 nm instead of 599 nm, and tested against longer wavelengths, haloperidol did not change the discrimination ability significantly (Fig. 2; one experiment). Similarly, in the range between 450 nm and 500 nm, the choice behavior did not change after injection of haloperidol (not shown; one experiment).

As shown in Fig. 3 for one of the three fish tested, the specific D1 dopamine receptor antagonist SCH 23390 reduced choice frequency in the wavelength range 560 nm – 630 nm in a way similar to haloperidol. In two experiments no change in choice behavior was observed after injection of SCH 23390 when fish were trained on 404 nm (not shown).

One experiment was performed with a second D1 dopamine receptor antagonist, SKF 83466-biotinyl-Cl, which led to similar results as for SCH 23390 (not shown).

Discussion

After injection of both unspecific and specific D1 dopamine receptor antagonists, the ability to discriminate wavelengths in the range between 560 nm

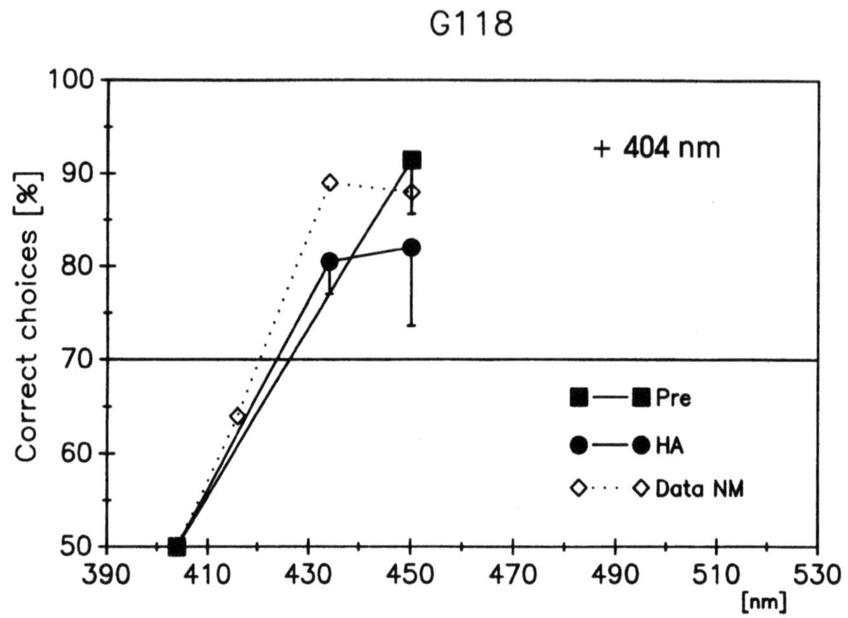

Fig. 2. Choice behavior of fish G118, trained on 404 nm, before and after injection of haloperidol.

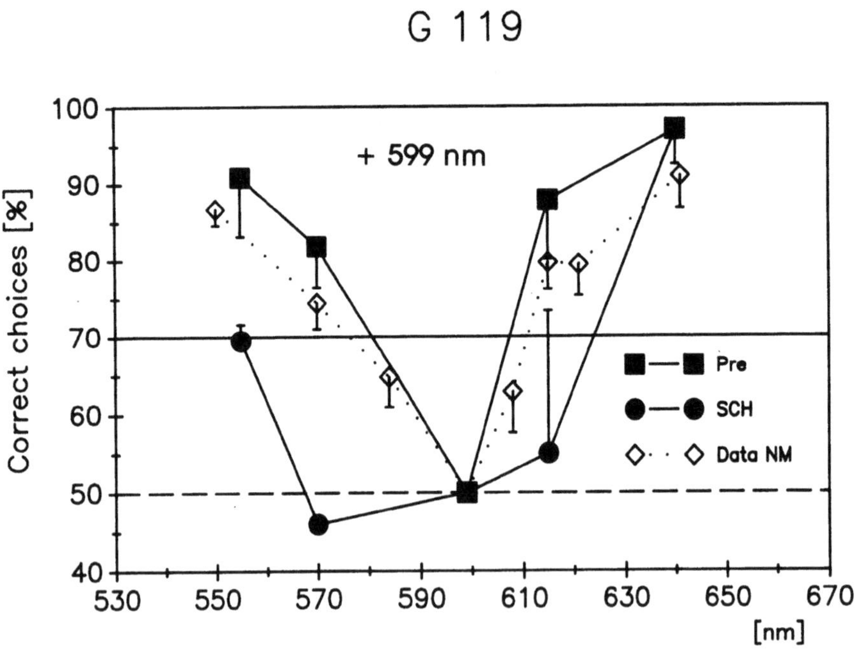

Fig. 3. Choice behavior of fish G119, trained on 599 nm, before and after injection of SCH 23390.

and 630 nm was reduced. No significant change in wavelength discrimination was observed in the range between 404 nm and 500 nm.

It is widely accepted that horizontal cells possess dopamine receptors of the D1 type exclusively (Wagner and Behrens, 1993). Therefore, we conclude that the dopaminergic modulation of horizontal cells is mediated by D1 dopamine receptors, and injections of SCH 23390 mainly affect horizontal cells. An inhibition of the effect of dopamine on horizontal cells by dopamine antagonists seems to change the biphasic response of H2 cells, which is based on feedback from horizontal cells to cones mediated by GABA, into a monophasic response (Kirsch, *et al.*, 1991). This is in line with the finding that injection of the GABAa antagonist bicuculline into the vitreous yielded similar results to the dopamine antagonists (Wietsma and Spekreijse, 1991) on wavelength discrimination.

Dopamine seems to control the transition of the retina from the dark- into the light-adapted state (Kohler, *et al.*, 1990). Especially the horizontal cells show a remarkable plasticity in response behavior and morphology (electrical coupling, spinule formation) during light-adaptation. The dopamine antagonists seem to prevent the retina from becoming entirely light-adapted. This is inferred from the comparison with the behavioral data obtained under reduced with illumination of 1 lx (Neumeyer and Arnold, 1989).

Experiments with ethambutol, a tuberculostatic drug which induces 'red-green' color blindness in man, revealed a selective reduction of wavelength discrimination in the long wavelength range in goldfish, which is due to a reduced contribution of L-cones to color vision. It was also shown that ethambutol only affects the contribution of L-cones to color vision and visual acuity, but not to brightness and movement perception (Neumeyer *et al.*, 1991; Neumeyer and Schaerer, 1992; Spekreijse *et al.*, 1991; Schaerer and Neumeyer, 1993). From these findings we concluded that the L-cone contribution to different visual functions is processed separately.

The finding that 'red-green' discrimination but not 'violet-blue' and 'blue-green' discrimination is impaired by ethambutol, GABAa and dopamine receptor antagonists shows that essential steps of 'red-green' color coding are located in the retina. The result with specific D1 dopamine receptor antagonists indicate that horizontal cells (probably the H2 type) play a prominent role in 'red-green' discrimination.

Acknowledgements

We are very grateful to Neil Beckhaus for improving the English. The study was financially supported by the Deutsche Forschungsgemeinschaft (Ne 215/9-1) and the Human Frontier Science Programme.

References

Kirsch, W., Wagner, H.-J. and Djamgoz, M.B.A. (1991). Dopamine and plasticity of horizontal cell function in the teleost retina: Regulation of a spectral mechanism through D1-receptors. Vision Res. 31: 401–412.

Kohler, K., Kolbinger, W., Kurz-Isler, G. and Weiler, R (1990). Endogenous dopamine and cyclic events in the fish retina II: Correlation of retinomotiric movement, spinule formation and connexon density of gap junctions with dopamine activity during light/dark cyclus. Visual Neurosci. 5: 415–428.

Neumeyer, C. (1984). On spectral sensitivity in the goldfish. Evidence for neural interactions between different 'cone mechanisms'. Vision Res. 24: 1223–1231.

Neumeyer, C. (1986). Wavelength discrimination in the goldfish. J. Comp. Physiol. 158: 203–218.

Neumeyer, C. (1992). Tetrachromatic color vision in goldfish: Evidence from color mixture experiments. J. Comp. Physiol. 171: 639–649.

Neumeyer, C. and Arnold, K. (1989). Tetrachromatic color vision in the goldfish becomes trichromatic under white adaptation light of moderate intensity. Vision Res. 829: 1719–1727.

Neumeyer, C. and Schaere, S. (1992). Two separate pathways in the processing of L-cone type information in goldfish: Color and acuity vs. brightness and flicker. Invest. Ophthalmol. Vis. Sci. Sup. 33(4): 703.

Neumeyer, C., Wietsma, J.-J. and Spekreijse, H. (1991). Separate processing of 'color' and 'brightness' in goldfish. Vision Res. 31(3): 537–549.

Schaerer, S. and Neumeyer, C. (1993). Parallel processing of motion and color in goldfish. Invest. Ophthalmol. Vis. Sci. Sup. 34(4): 1033.

Spekreijse, H., Wietsma, J.-J. and Neumeyer, C. (1991). Induced color blindness in goldfish: A behavioral and electrophysiological study. Vision Res. 31:551–562.

Wagner, H.-J. and Behrens, U.D. (1993). Microanatomy of the dopaminergic system in the rainbow trout retina. Vision Res. 33: 1345–1358.

Wietsma, J.-J. and Spekreijse, H. (1991). Bicuculline produces reversible red-green color blindness in goldfish, as revealed by monocular behavioral experiments. Vision Res. 31: 2101–2107.

Christa Neumeyer
Institut für Zoologie III
Universität Mainz
55099 Mainz, Germany

6. Visual defects in subjects with Down's syndrome

J. PÉREZ-CARPINELL, M.D. DE FEZ and V. CLIMENT
(Burjassot, Spain)

Abstract

Because the incidence of other visual defects is much higher in cases of Down's syndrome than in individuals with other forms of retardation (Gardiner, 1967; Aitchison *et al.*, 1990), we tested the color vision of a population of Down's syndrome patients, using Ishihara plates and the Davico anomaloscope. The Down's syndrome group had a significantly higher proportion of color deficiencies than either normal controls or a matched mentally retarded population. Refractive errors were corrected with static retinoscopy, and spatial contrast sensitivity was measured with the Vistech test.

Method

Seventy-two subjects, 38 males (M) and 34 females (F), of the Koynos School of Special Education (Godella, Valencia, Spain) were studied. Table 1 gives the descriptive statistics of the sample, which included 34 moderately retarded and 38 severely retarded subjects.

Table 1. Description of the total sample.

Group	Sex	N	IQ		CA	
			Mean	SD	Mean	SD
Moderate	Male	19	40.1	3.0	11.6	3.5
	Female	15	41.4	4.6	14.3	3.7
Severe	Male	19	27.4	5.7	16.6	4.5
	Female	12	29.6	3.9	14.8	3.4

The subjects were tested individually by one of the authors in the presence of a member of the school staff. Each test was performed on a different day using methods described elsewhere (Pérez-Carpinell *et al.*, 1993). Each session began

B. Drum (ed.), Colour Vision Deficiencies XII, pp. 43–46.
© 1995 *Kluwer Academic Publishers, Dordrecht.*

by establishing rapport with the subject. For the Ishihara test the subject was shown a chart with numbers in random order, and was asked to name them. If the subject could not do this, he or she traced the numbers and symbols of the Ishihara plates with a fine brush. The end points of the subjects' matching ranges on the Davico anomaloscope were determined three times, taking care to make sure that the subject understood the concept of matching, and the difference between brightness and color. Contrast sensitivity was measured with the Vistech test (Ginsburg, 1987) by showing the subjects a chart with gratings that could be oriented in three directions, and blank patches; the subject reported whether there were bars in the patch, and if there were the subject held his or her hand in the orientation of the grating. No time limits were imposed. Refractive errors were corrected by static retinoscopy before testing.

Results and Conclusions

Table 2 gives the results of the ophthalmic examination.

Table 2. Eye symptoms found among 65 Down's syndrome patients.

	N eyes	% eyes
Lens opacities	45	35
Strabismus	36	28
Nystagmus	16	12
Phoria	5	4
Amblyopia	4	3
Emmetropia	8	6
Hypermetropia	68	52
Myopia	35	27
Astigmatism	99	76

Our examination corfirms earlier reports that a large number of Down's syndrome subjects have lens opacities (35%), strabismus (28%), nystagmus (12%), hypermetropia (52%), high myopia (21 of the eyes examined had myopia of > 6D, and of these, 11 were > 8.5D), and a high incidence of astigmatism (76%).

Only 57 subjects (79%) were able to perform the Ishihara test, and of these only 48 (67%) could be tested with the anomaloscope. The validity of both these tests had been established in an earlier study of miscellaneous mentally retarded subjects (Pérez, Carpinell *et al.*, 1993), in which testing procedures were adapted to the capabilities of retarded subjects, so as to minimize problems of cooperation and concentration. With those Down's syndrome subjects whose color vision could be tested, 13 (23%; 7F, 6M) had defective color vision. Of these, 4 (14%; 1F, 3M) were from the 28 moderately retarded subjects and 9 (31%; 6F, 3M) were from the 29 severely retarded subjects. Thus, our Down's

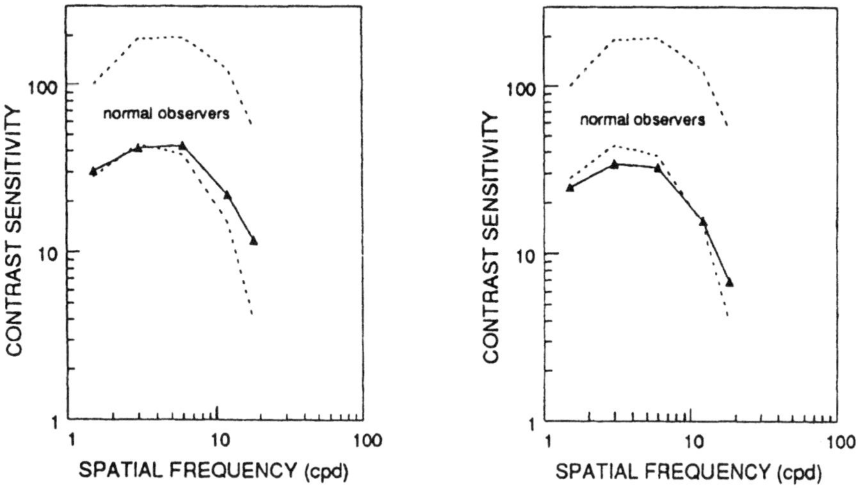

Fig. 1. Triangles: contrast sensitivity for moderately retarded (left) and severely retarded (right) Down's syndrome subjects. The dotted lines show the limits for 90% of the normal population.

syndrome subjects show a substantially higher incidence of color vision deficiency than other subjects with similar mental retardation, for whom we have found no appreciable deviation from the normal population (Pérez-Carpinell *et al.*, 1993). This differs other studies of Down's syndrome subjects, which did not find an abnormally high incidence of color defects (Lowe, 1949; Stratford and Mills, 1984). We are aware of only two other reports of a high incidence of color defects (between 20 and 30%) among retarded subjects, but the type or cause of retardation in these groups was not given (Archer, 1964; Krause, 1967). Using Pickford's classification, among our 13 subjects 10 eyes were protan (5 simple, 3 extreme and 2 deviant), one eye was simple deuteranomalous, and the remaining eyes were normal; tested binocularly, 4 were classified as protan (2 simple and 2 deviant), 2 were deutan (1 simple and 1 deviant), and the rest were normal.

Of the 72 subjects, 55 were capable of performing the contrast-sensitivity test. Figure 1 shows the contrast-sensitivity function for the moderately retarded (left) and severely retarded (right) groups. The range for 90% of the normal population falls between the dotted lines. It can be seen that as a group the Down's syndrome subjects have a considerable loss of contrast sensitivity, particularly at lower spatial frequencies, and that this loss is greater in the severely retarded group.

References

Aitchison, C., Easty, D.L. and Jancar, J. (1990). Eye abnormalities in the mentally handicapped. J. Mental Deficiency Res. 34: 41–48.

Archer, R.E. (1964). Color discrimination and association of educable mentally retarded children. Doctoral Dissertation, Colorado State College.

Gardiner, P.A. (1967). Visual defects in cases of Down's syndrome and in other mentally handicapped children. Brit. J. Ophthal. 51: 469–474.

Ginsburg, A.P. (1987). Clinical findings from a new Contrast Sensitivity Test Chart. In: Fiorentini, A., Guyton, D.L. and Siegel, I.M. (eds.), Advances in Diagnostic Visual Optics: 132–140, Springer-Verlag, Berlin.

Krause, I.B., (1967). A study of the relationship of certain aspects of color vision to educable mentally retarded children. Techn. Rpt., U.S. Office of Education Grant No. OEG 2-6-068528-1593.

Lowe, R.F. (1949). The eyes in mongolism. Brit. J. Ophthal. 33: 131–174.

Pérez-Carpinell, J., de Fez, M.D. and Climent, V. (1993). Color blindness and mental retardation. Atti Fond. G. Ronchi 48: 231–238.

Stratford, B and Mills, K. (1984). Colour discrimination in mentally handicapped children with particular reference to Down's Syndrome. Austral. N.Z.J. Developm. Disab. 10: 151–155.

J. Pérez-Carpinell, M.D. de Fez
Departament d'Optica
Universitat de Valéncia
C/Dr. Moliner 50
SP-46100 Burjassot

and

V. Climent
Departament de Ciències Experimentals
Universitat Jaume I
SP-12080 Castelló
Spain

7. Heterochromatic brightness matching and wavelength discrimination in juvenile diabetics: A three-year study

ANNE KURTENBACH, LUKAS RÜTTIGER, ULRICH SCHIEFER, EBERHART ZRENNER and ANDREAS NEU

(Tübingen, Germany)

Abstract

We tested the brightness sensitivity and colour vision of 10 juvenile diabetics with no (8) or very mild (2) retinopathy once a year over a three-year period. We used the Farnsworth-Munsell 100 hue Test (FM-100), heterochromatic brightness matching (HBM), and wavelength discrimination. Their results were compared to an age-matched control group.

The FM-100 results from the diabetic group showed a general increase in error scores only in the first year of testing, compared to that of the control group. In the HBM task, a decrease in sensitivity between 480 and 600 nm was found in the diabetic group, which became most conspicuous in the third year of testing. In the wavelength discrimination task, an increasing deficiency was found at 460 nm during the second and third years, compared to the control group.

The results show that in diabetics, over the three-year period, an increasing deficit in sensitivity and colour vision occurs, before the onset of a clinically visible retinopathy.

Introduction

Diabetes mellitus is associated with an acquired colour vision deficit which can occur before the onset of a clinically detectable retinopathy. Testing colour vision in diabetics is therefore a method which can, in some cases, offer an early detection of impaired retinal function. The aim of this study was to investigate the brightness sensitivity and colour vision in diabetics in a long-term study using the Farnsworth-Munsell 100 hue test (FM-100), heterochromatic brightness matching (HBM) and wavelength discrimination. Ten juvenile diabetic patients were studied, 8 with no retinopathy and 2 with a mild retinopathy. Their results were compared to an age-matched control group.

B. Drum (ed.), Colour Vision Deficiencies XII, pp. 47–52.
© 1995 *Kluwer Academic Publishers, Dordrecht.*

Method

Subjects

Ten insulin-dependent diabetics, aged between 11 and 18 years (mean 13.7 ± 2.21, SD) were tested once a year over a three-year period. They exhibited no (8 patients) or mild (2 patients) retinopathy (up to six microaneurysms with no cotton-wool spots) as evidenced from colour fundus photographs, and had suffered from diabetes for between 2 and 13 years (mean 7.54 ± 3.74, SD). Visual acuity and intraocular pressure were normal. In addition, blood and urine were analysed: Urine was glucose-free in all patients and all showed normal liver (cholesterin) and kidney function (beta2-microglobulin; serum creatin). Glycosylated hemoglobin (gHb) values ranged between 7.2% and 18.4%. The results were compared to those from an age-matched control group of 20 subjects (aged between 11 and 22 years, mean 17.8 ± 4.23, SD), with normal colour vision, as tested by the FM-100 hue test.

Apparatus

A three-channel Maxwellian view system (see Reitner *et al.*, 1991) was used to generate a bipartite field of 2° diameter. The two halves could be adjusted independently in luminance and wavelength (4 nm half-width) and were separated by a black vertical line of 5 min arc. Calibration, experimental procedure and data processing were computer controlled. The subject's head was held steady using a chin and head rest.

Procedure

Ophthalmological examinations were performed directly before the experimental procedure outlined below; fundus examination (indirect and direct ophthalmoscope) and intraocular pressure measurements were carried out immediately after. Blood and urine were analysed on average within 1 month of experiment. Subjects were first tested with the FM-100, presented under standard conditions (colour temperature 6,000 K; 500 cd/m^2). For the HBM, subjects viewed the circular bipartite field and brightness matches were made against a comparison hemifield of 540 nm and 9.88 log quanta s^{-1} cm^{-2}. An ascending and descending method of limits was used and four measurements were made for each of 27 wavelengths between 420 and 680 nm, presented in steps of 10 nm. A homogeneous adaptation field (colour temperature 6,000 K) presented for 20 sec between each wavelength presentation maintained a standard adaption level throughout the experiment.

These equal-brightness luminances were then used for the wavelengths presented in the colour discrimination task. The adaptation field was presented for 5 sec alternately with a 1 sec presentation of the bipartite field, after which the subject had to indicate whether a difference in hue could be perceived between the two hemi-fields. An ascending method of limits was used. Wavelength was increased in 1, 2 or 5 nm steps towards the longer wavelength. Four measurements were made for 11 wavelengths between 440 and 640 nm, presented in 20 nm steps.

Results

In none of the tests did the results of the 2 patients with mild retinopathy differ significantly from the 8 who had no retinopathy.

Table 1 shows the mean results of the FM-100 for the control group in comparison to those of the diabetic group over the three year period. The ratio of (B + Y) to (R + G) shows a tendency towards the tritan axis. Only in the first year was there a significant difference between the diabetic and control scores ($^{t}p < 0.01$, Mann-Whitney U-test). The differences in the BY/RG ratio are not significant: Both groups make slightly more mistakes in the tritan axis.

Table 1. Mean error scores ± 1SD obtained for the control group and the diabetic group over 3 years in the FM-100 Hue test (*$p < 0.01$).

		FM-100 score	Ratio BY/RG
Control		51.9 ± 17.48	1.14
Diabetic	1990–91	84.3 ± 37.1*	1.18
Diabetic	1991–92	48.0 ± 38.1	1.08
Diabetic	1992–93	49.8 ± 33.1	1.22

We did not find any significant correlation between the FM-100 scores, blood glucose level or duration of disease. We also found no correlation between glucose level during the first 5 years of disease (measured at three-month intervals) and FM-100 error scores in the 7 patients for whom these data were available.

The mean results of the HBM can be seen in Figure 1. In order to detect relative changes in sensitivity, all curves for both diabetics and control subjects have been normalised at 680 nm. At 680 nm, there is little interaction between different cone types, which, if it occurred, could contaminate the results; thus vision is predominantly based on long-wavelength cone sensitivity. The control group (continuous line) shows an inverted U-shaped function. For the juvenile diabetic group (broken lines) the curves are similar in shape to that of the control, but there is a relative decrease in sensitivity between 480 nm and 600 nm, which is most marked in the third year of testing.

Results of wavelength discrimination can be seen in Figure 2. The control group (continuous line) shows two minima in the delta-lambda curve, at between 480–500 nm and 560–580 nm. Maxima are found at 460 and 520 nm. There is also an increase in delta-lambda at longer wavelengths. The curves for

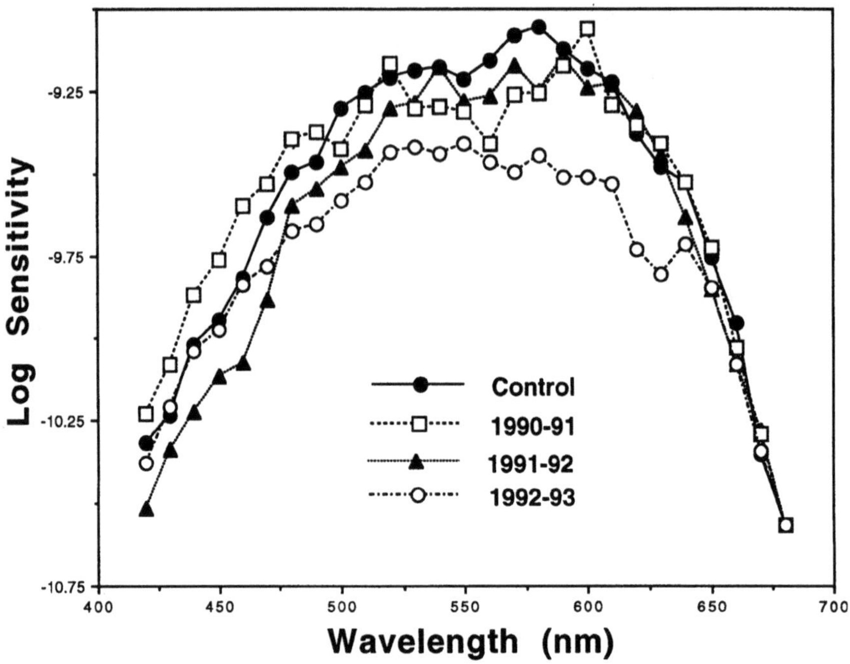

Fig. 1. Results obtained for the HBM task for the control group (continuous line) and for the diabetic group (broken lines) over a 3 year period. The relative sensitivity (inverse of the luminance at the equal brightness match) is plotted as a function of wavelength. Curves are normalised at 680 nm.

the diabetic group are similar in shape to that of the control, except for a larger increase found at the longer wavelengths. In addition, during the second and third years there is a marked increase in delta lambda at 460 nm.

Discussion

In this study, the FM-100 test scores of juvenile diabetics are significantly different from the control group only for the first year, although the means of both groups lie within the normal range (Table 1). In the second year, all patients gave an improved score. The reduction in the mean error score in the second and third years indicates that there may be an effect of learning in this task. Both controls and diabetics made marginally more errors along the tritan axis, which may be due to a lack of specificity for the FM-100 hue test.

We find no correlation between duration of disease and FM-100 score confirming the results of Roy *et al.* (1986), nor do we find significantly increased scores in the two patients with retinopathy.

Compared to the control group, the HBM curves of the diabetics (Fig. 1) show a relative decrease in sensitivity between 480 and 600 nm, although this is

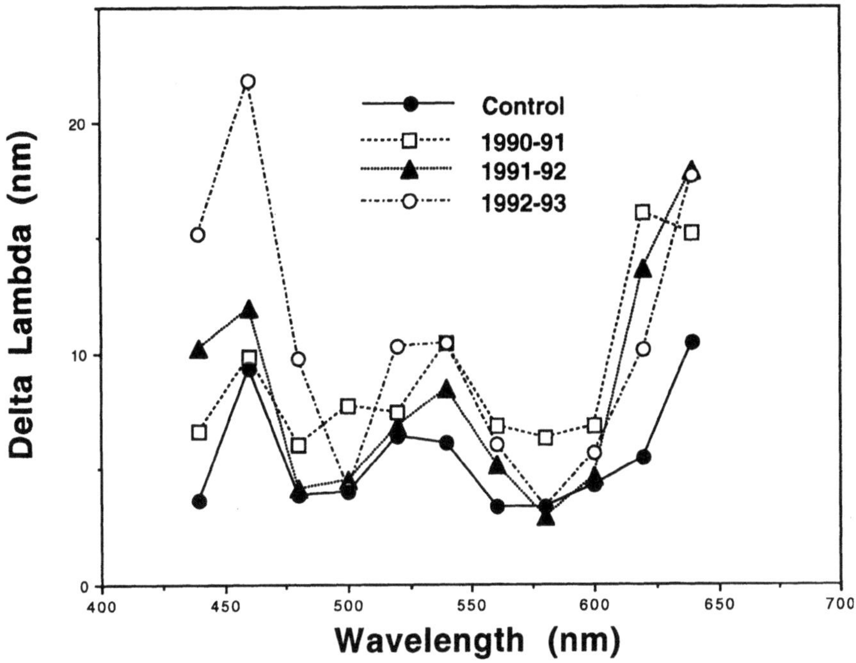

Fig. 2. Results for the colour discrimination task, obtained for the control group (continuous line) in comparison to the juvenile diabetic group (broken lines) during 3 years of testing. The just noticeable difference in colour (delta-lambda) is plotted as a function of wavelength.

not directly related to blood glucose levels or FM-100 scores. This result may relate to those of Elsner *et al.* (1987) who found that more light was required to bleach the photopigment of some diabetics. They explain their results as due to either an increased rate of photopigment regeneration or to a decreased ability to absorb light.

The results of the wavelength discrimination experiment show that the diabetics exhibit a marked increase in delta-lambda at 460 nm, during the second and third years of testing. At 460 nm, the short wavelength cone pathway is predominantly responsible for wavelength discrimination (Mollon and Estévez, 1988). These results could therefore be explained by a decrease in the sensitivity of the short wavelength mechanism, occuring before the onset of a clinical retinopathy, as found in spectral sensitivity measurements (Marré and Marré, 1972) and in some FM-100 hue test results (Kinnear *et al.*, 1972: Lakowski *et al.*, 1972/3).

Conclusion

This experiment shows that for the tasks of FM-100, HBM, and wavelength discrimination, differences are found between a juvenile diabetic group with no or mild retinopathy and an age-matched control group. We find a progressive sensitivity and colour deficiency in diabetics before the onset of visible retinopathy.

Acknowledgements

We thank Hoechst AG for supporting part of this study.

References

Elsner, A.E., Burns, A., Lobes, L.A. and Doft, B.H. (1987). Cone photopigment bleaching abnormalities in diabetics. Invest. Ophthalmol. Vis. Sci. 28: 718–724.
Kinnear, P.R., Aspinall, P.A. and Lakowski, R. (1972). The diabetic eye and colour vision. Trans. Ophthalmol. Soc. U.K. 92: 69–78.
Lakowski, R. Aspinall, P.A. and Kinnear, P.R. (1972/3). Association between colour vision losses and diabetes mellitus. Ophthalmic Res. 4: 145–159.
Marré, E. and Marré, M. (1972). The influence of the three colour vision mechanisms on the spectral sensitivity of the fovea. Mod. Probl. Ophthalmol. 11: 219–223.
Mollon, J.D. and Estéves, O. (1988). Tyndall's paradox of hue discrimination. J. Opt. Soc. Am. A. 5: 151–159.
Reitner, A., Sharpe, L.T. and Zrenner, E. (1991). Is colour vision possible with only rods and blue sensitive cones? Nature 352: 798–100.
Roy, M.S., Gunkel, R.D., Podger, M.J. (1986). Color vision defects in early diabetic retinopathy. Arch. Ophthalmol., 104: 225–228.

Dr. Anne Kurtenbach
Department of Pathophysiology of Vision
and Neuro-Ophthalmology,
University Eye Hospital 72076 Tübingen
Germany

8. Opponent-color detection threshold asymmetries in subjects with optic nerve abnormalities

VINCENT A. BILLOCK, ALGIS J. VINGRYS, SCOTT S. GRIGSBY, SUSAN C. BENES, and P. EWEN KING-SMITH
(Columbus, OH, U.S.A.)

Abstract

Thresholds for equiluminous opponent-colors are almost always similar. King-Smith *et al.* (1989a) summarized four cases where a gross asymmetry was manifested. In one case the subject had Leber's optic atrophy; in another, the subject had a microinfarct of the optic nerve fiber layer. In the other cases no pathology was known. However, we subsequently found that the third subject has optic nerve hypoplasia, and the fourth case has megalopapilla of the optic nerve. Additional subjects with Leber's optic atrophy and optic nerve hypoplasia have been examined and opponent-color asymmetries seem to be relatively common in these populations. Also, many normal observers and many observers with diverse visual pathologies have been studied. Although there are differences between the asymmetries found in these groups (e.g. Leber's patients show only blue/yellow asymmetries, while hypoplasics show red/green asymmetries as well), it is interesting that in our experiments, strong asymmetries are always associated with an optic nerve abnormality. Moreover, the asymmetries are often manifested by patients with no other obvious symptoms of their conditions (e.g. hypoplastic subjects with good acuity and subtle double rings can show asymmetries in excess of .300 log units). Some of these asymmetries may be explained by selective losses of some color-sensitive ganglion cell subpopulations, others may be due to nonlinearities.

Introduction

Detection thresholds for equiluminous opponent-colors are almost always similar. Recently we (King-Smith *et al.*, 1989a) reported four observers who had asymmetries in detection thresholds for equiluminous red and green or equiluminous blue and yellow. In two of these cases the subjects were known to have an optic nerve pathology (retinal ischemia and Leber's optic atrophy). In the other two cases the subjects appeared normal. Those cases were extremely interesting because in addition to the asymmetries, they both showed unusual

B. Drum (ed.), Colour Vision Deficiencies XII, pp. 53–62.
© 1995 *Kluwer Academic Publishers, Dordrecht.*

temporal integration properties. Both subjects were examined by a neuro-ophthalmologist, and fundus photography revealed mild bilateral congenital optic nerve defects. One of these subjects has optic nerve hypoplasia, the other has megalopapilla. We subsequently discovered several other subjects with significant opponent-color detection threshold asymmetries. All of these subjects have either optic nerve hypoplasia or Leber's optic atrophy. Here, we present a taxonomy of optic nerve defects that accompany detection threshold asymmetries, along with some speculations on how some of the asymmetries may arise.

Methods

The methods we used are described in detail elsewhere (King-Smith *et al.*, 1983, 1987; Sellers *et al.*, 1986). Circular test spots were created on a computer controlled color television display with a 50 cd/m^2 white screen. Stimuli were adjustable for diameter, duration, eccentricity and relative contrast of the three phosphors (phosphor contrast is defined as $\delta L/L$ were L is the background luminance of the display and δL is the difference in luminance for that phosphor for the test and background locations). The phosphors' voltage-intensity nonlinearities were corrected during stimulus generation. Thresholds were measured using a modified QUEST yes/no staircase procedure (King-Smith, 1984). Blank trials were randomly interlaced with stimulus trials to monitor the subject's false positive rate. Viewing was monocular, with a natural pupil. All stimuli were presented with a square-wave temporal profile. Subjects were either students encountered in the course of instructional experiments or patients referred to us by the Ohio State University Department of Ophthalmology and the OSU College of Optometry.

To establish a baseline for asymmetry, we measured detection thresholds for equiluminous red, green, blue and yellow in 19 normal observers. We found that for 1 deg, 0.2 sec duration spots (eight trial QUEST staircase), that the average green/red asymmetry was $0.028 \pm .076$ (SD) log units and the average blue/yellow asymmetry was $0.056 \pm .068$ log units.

Previously recognized acquired defects that accompany asymmetries

King-Smith *et al.* (1989a,b) and Grigsby *et al.* (1991) described a subject with optic retinal ischemia caused by closure of an inner arteriole. The area of the retina affected had a general increase of thresholds. More interestingly, some regions of the defect showed an increase in yellow thresholds relative to the thresholds for equiluminous blue (as much as .30 log units asymmetry), other regions showed much smaller asymmetries, suggesting a random variation in the numbers of blue and yellow sensitive ganglion cells surviving the infarct.

King-Smith *et al.* (1989a, 1992) have studied the changes in blue/yellow color asymmetries in a 23 year old white male suffering from Leber's optic atrophy.

Following the initial acute phase, which reduced the subject's acuity to counting fingers, the subject was treated with hydroxycobalamin and returned to 20/40 acuity after 7 months, yet he displayed an enormous blue/yellow asymmetry of .84 log units (thresholds for blue higher than for yellow). At 14 months the subject's acuity was 20/25 and his asymmetry was reduced to 0.37 log units and at 29 months, his acuity was normal and his asymmetry was reduced to a clinically insignificant 0.14 log units. More recently, several other Leber's patients with comparable asymmetries have been observed. King-Smith *et al.* (1989a, 1992) suggested that the asymmetry in Leber's could be produced by a disease induced nonlinearity in the blue/yellow pathway, perhaps due to changes in membrane properties.

Congenital defects that accompany asymmetries

King-Smith *et al.* (1989a) described a subject (a 24-year-old white male) who had a blue/yellow asymmetry for brief flashes in peripheral vision. The subject had 20/10 acuity and near perfect performance on the FM-100 hue test, so no optic nerve defect was suspected. Ophthalmoscopic examination has since been made and the subject was found to have very large optic disks, consistent with prior descriptions of a little understood condition – optic nerve megalopapilla. The most remarkable aspect of this subject's data is his temporal integration for blue and yellow spots (Fig. 1b). He shows a unique violation of Bloch's law for equiluminous blue spots (initial slope of -1.329 ± 0.063; $p < .001$) compared to a normal subject (-0.973 ± 0.047). This unusual violation of Bloch's law (which is greatest in peripheral retina; see Fig. 1a) could be explained using the asymmetric nonlinear compression model described in Fig. 2. This nonlinearity could arise from unusual membrane properties (the optic disks of subjects with megalopapilla are pale – perhaps due to excessive myelination).

The other subject whose optic nerve defect went unrecognized was subsequently found to have optic nerve hypoplasia. This subject showed large asymmetries for both opponent systems. Figure 3a shows the foveal nature of the defect. Optic nerve hypoplasia is believed to result from a reduction in the number of developing retinal ganglion cells. All electrophysiological and anatomical evidence suggests that those cells which remain function normally (Lambert *et al.*, 1987). In this light, Fig. 3a is interesting. In normals, there is a foveal minimum for thresholds that corresponds to the increased density of ganglion cells (Schein, 1988). The data in Fig. 3 are consistent with a relatively low and constant density of ganglion cells that mediate detection of blue, yellow and green. It is as if the subject did not develop a fovea for anything but red detecting mechanisms. The most interesting result to come from this subject is a violation of Bloch's law for detection of green spots (Fig. 3b). This result is readily interpretable if one assumes that the color detection mechanism consists of lowpass spatial filtering (useful for getting rid of P-cell luminance signals), followed by a rectifier and a temporal integrator (Fig. 4). The number of cells pooled by the spatial integrator determines the magnitude of the signal seen by

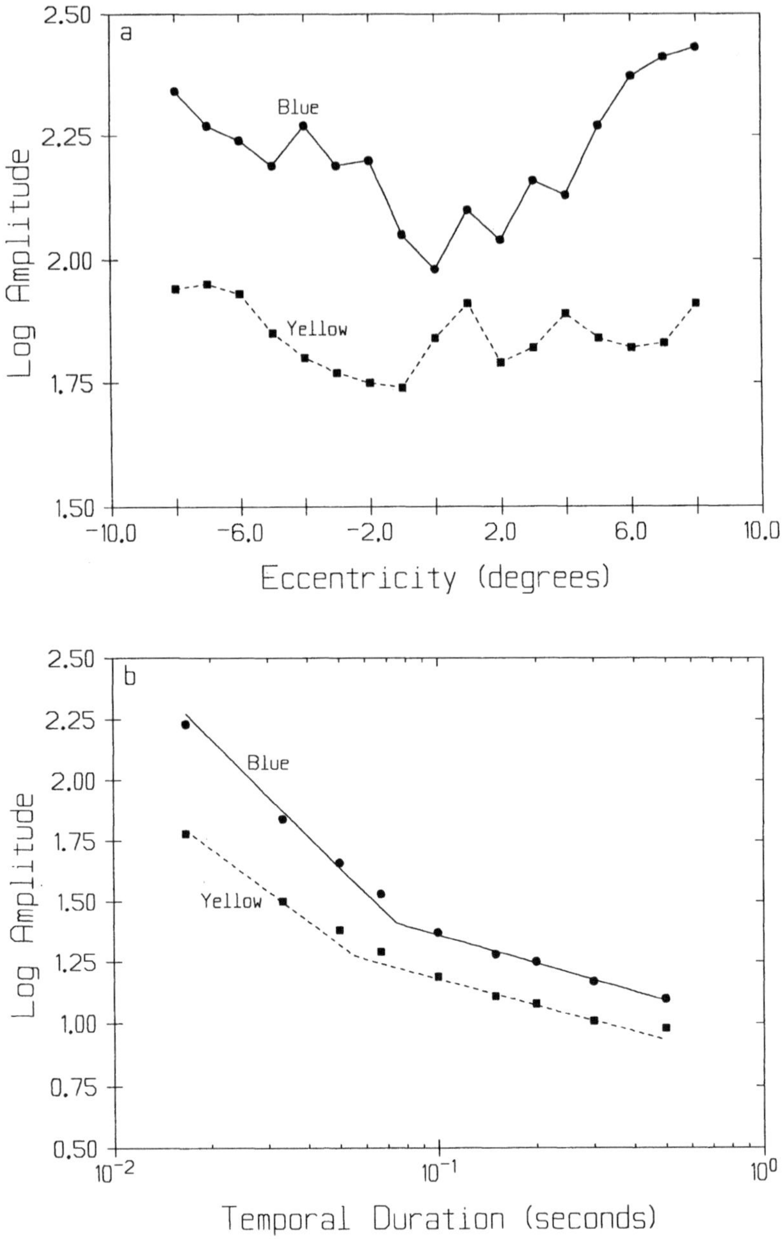

Fig. 1. Blue/yellow asymmetries for a subject with optic nerve megalopapilla. (a) Detection threshold as a function of horizontal distance from the fixation point. Threshold amplitudes are plotted as phosphor contrast ($\delta P/P$) needed to reach threshold. Yellow consists of equal contributions from the red and green phosphors, so yellow amplitude is $\sqrt{2}$ times red or green phosphor contrast. Blue thresholds are much higher than yellow for some positions in peripheral retina. For normal data see Fig. 3a. (b) Detection threshold as a function of temporal duration for

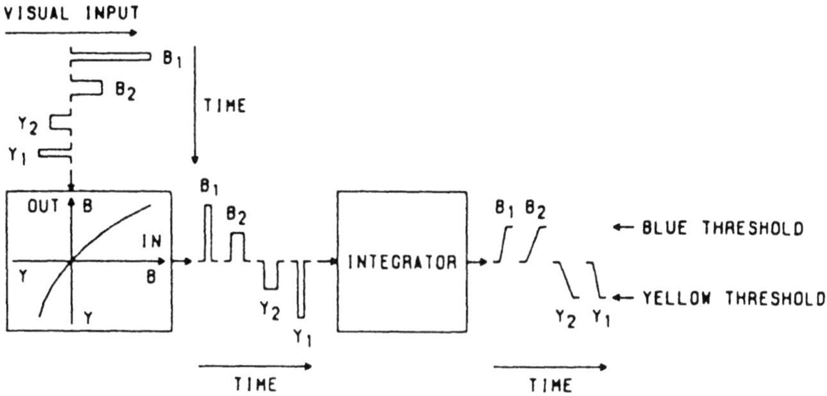

Fig. 2. A model of the temporal integration asymmetry in Fig. 1a. A compressive nonlinearity precedes temporal integration. The nonlinearity compresses blue signals more than yellow, giving rise to a general increase in thresholds. Moreover, a short duration, high amplitude blue signal (B_1) is compressed more than an equal 'energy' long duration, low amplitude (B_2 signal, leading to a violation in Bloch's law for blue.

the rectifier. In the hypoplastic, too few green sensitive units result in a small signal relative to the rectifier's threshold. This means that more of a small amplitude long duration stimulus will be attenuated relative to an equal integrated magnitude short duration stimulus.

Asymmetries have been noted in the data of other hypoplastics. Out of eleven hypoplastics measured in our laboratory, five had significant red/green asymmetries ($p < .01$, 2-tailed t-test), four of which were significant at the $p < .001$ level. Three of the eleven hypoplastics showed significant blue/yellow asymmetries ($p < .01$), but only the subject previously discussed was significant at the $p < .001$ level.

---

1 deg. spots falling on nasal retina 5 degrees from the fovea. The blue data are well fit ($r = .997$) with a low duration slope of -1.329, a breakpoint of 74.6 msec, and a long duration slope of -0.382. The yellow data are well fit ($r = .995$) by a low duration slope of -1, a breakpoint of 55.9 msec, and a long duration slope of -0.355. Comparison normals (not shown) have slopes of -1 and 0 (Bloch's law).

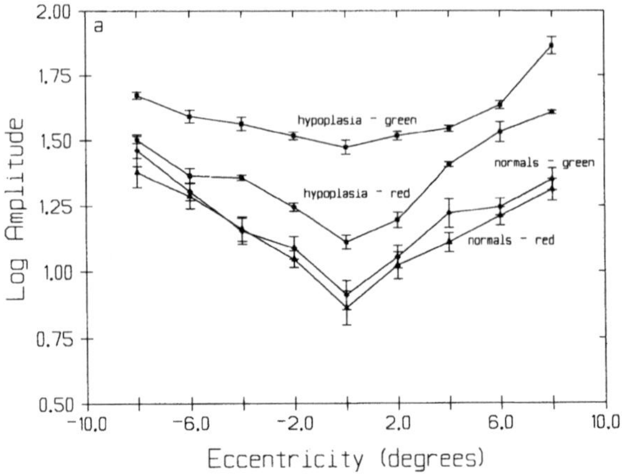
a
Log Amplitude
Eccentricity (degrees)
hypoplasia - green
normals - green
hypoplasia - red
normals - red

(a)

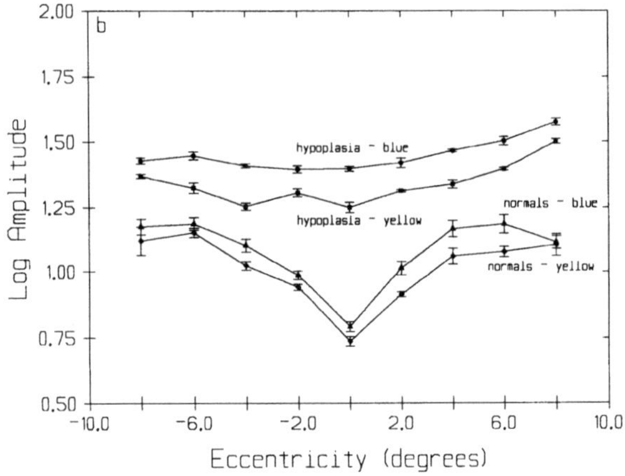
b
Log Amplitude
Eccentricity (degrees)
hypoplasia - blue
normals - blue
hypoplasia - yellow
normals - yellow

(b)

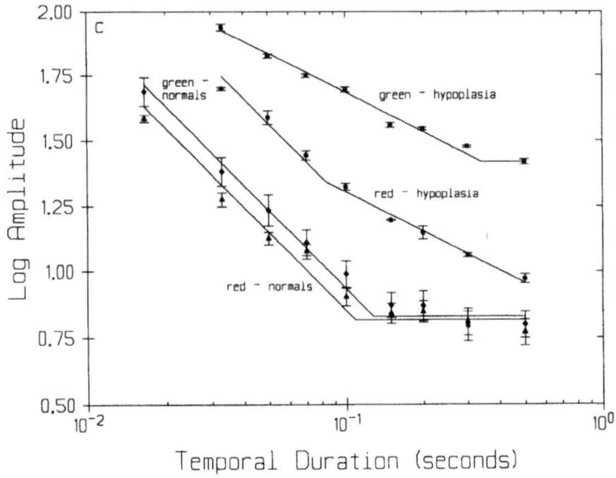
c
Log Amplitude
Temporal Duration (seconds)
green - hypoplasia
red - hypoplasia
green - normals
red - normals

(c)

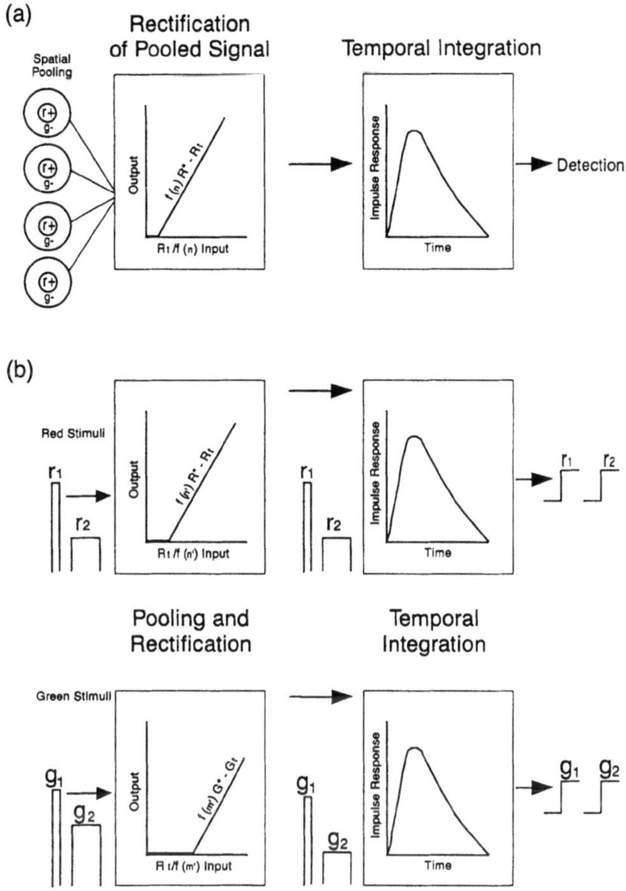

Fig. 4. A model of the temporal integration asymmetry in Fig. 3b. The detection system consists of lowpass spatial pooling, a rectifying nonlinearity, and temporal integration (Billock *et al.*, 1994). The hypoplastic has fewer cells to contribute to pooling, especially cells sensitive to green increments. With a smaller pooled signal, more of the signal is subject to rectification, which affects long duration, low amplitude signals more than equal 'energy' short duration high amplitude signals.

←

Fig. 3. Equiluminous red/green and blue/yellow asymmetries in optic nerve hypoplasia, compared to three normals. (a) Detection threshold as a function of horizontal distance from the fixation point. Normals have V-shaped data with a threshold minima at the fovea. The hypoplastic observer follows this trend only for red; for blue, green and yellow, thresholds are flat. Hence his largest red/green asymmetry is in the fovea. (b) Temporal integration data for red and green stimuli. Normal subjects obey Bloch's law with a short duration slope of -1 and a long duration slope of 0 ($r > .98$ for both red and green). The hypoplastic observer obeys Bloch's law for short duration red stimuli with a transition to DeVries-Rose (slope of -0.5) behavior at longer durations, but follows the DeVries-Rose law for all but the longest duration green stimuli ($r > .99$ for both red and green). Additional data on this subject's color vision can be found in Billock *et al.*, 1994).

Conclusions

It is interesting to note that a similar asymmetry has been found for saturation discrimination on Lanthony's New Color Test. Subjects with central serous retinopathy, retinitis pigmentosa and optic neuritis sometimes show 'monopolar' (one color direction only) saturation discrimination defects (Lanthony, 1975). To date, the only subjects to show significantly asymmetrical detection thresholds for equilumious opponent-colors have been patients with congenital or acquired defect of the optic nerve. Out of the many possible defects that could lead to such asymmetries, only four have shown this symptom: retinal ischemia, Leber's optic atrophy, megalopapilla, and optic nerve hypoplasia. Out of these, Leber's responds to treatment while the others seem to remain stable. Only in optic nerve hypoplasia does the defect lead to both red-green and blue-yellow asymmetries, sometimes (as in the case detailed here) in the same observer. There are several mechanisms by which defects could arise. For Leber's optic atrophy and megalopapilla the asymmetries may stem from nonlinearities. For retinal ischemia and optic nerve hypoplasia, the most likely mechanisms are asymmetries in the numbers of retinal ganglion cell subpopulations.

Acknowledgements

This research was supported by NIH grant EY 04948 and by grants from the Ohio Lions Eye Research Foundation.

References

Billock, V.A., Vingrys, A.J. and King-Smith, P.E. (1994). Opponent-color detection threshold asymmetries may result from reduction of ganglion cell subpopulations. Visual Neurosci. (11: 99–109).

Grigsby, S.S., King-Smith, P.E. and Benes, S.C. (1991). Evidence for selective loss of Type I ganglion cells in optic nerve ischemia: Selective spatial and red/green losses. Invest. Ophthalmol. Vis. Sci. 32 supp.: 1115.

King-Smith, P.E. (1984). Efficient threshold estimates from yes-no procedures using few (about 10) trials. Am. J. Opt. Physiol. Opt. 61: 119P.

King-Smith, P.E., Chioran, G.M., Sellers, K.L. and Alvarez, S.L. (1983). Normal and deficient colour discrimination analyzed by colour television. In: Mollon, J.D. and Sharpe, L.T. (eds), Colour Vision: Physiology and Psychophysics: 167–172, Academic Press, London.

King-Smith, P.E., Vingrys, A.J. and Benes, S.C. (1987). Visual thresholds measured with color video monitors. Color Res. Appl. 12: 73–80.

King-Smith, P.E., Vingrys, A.J. and Benes, S.C. (1992). Chromatic and achromatic thresholds during recovery from probable Leber's optic atrophy. Invest. Ophthalmol. Vis. Sci. 33, supp.: 1220.

King-Smith, P.E., Vingrys, A.J., Benes, S.C., Grigsby, S.S. and Billock, V.A. (1989a). Detection of light and dark, red and green, blue and yellow. In: Kulikowski, J.J., Dickinson, C.M. and Murray, I.J. (eds.), Seeing Contour and Colour: 379–389, Pergamon Press, Oxford.

King-Smith, P.E., Vingrys, A.J., Benes, S.C. and Havener, W.H. (1989b). Differences between perimetric thresholds for white and equiluminous red, blue and yellow in a nerve fiber bundle defect. In: Drum, B. and Verriest, G. (eds.), Colour Vision Deficiencies IX, Doc. Ophthalmol. Proc. Ser. 52: 301–308, Kluwer Acad. Publ., Dordrecht

Lambert, S.R., Hoyt, C.S., and Narahara, M.M. (1987). Optic nerve hypoplasia. Surv. Ophthalmol. 32: 1–9.

Lanthony, P. (1975). New Color Test de Lanthony Selon Munsell. Luneau, Paris.

Schein, S.J. (1988). Anatomy of macaque fovea and spatial densities of neurons in foveal representation. J. Comp. Neurol. 269: 479–505.

Sellers, K.L., Chioran, G.M., Dain, S.J. Benes, S.C., Lubow, M., Rammohan, K. and King-Smith, P.E. (1986). Red-green mixture thresholds in congenital and acquired color defects. Vision Res. 26: 1083–1097.

Vincent A. Billock
404 Melbourne Dr.
Enterprise, AL 36330
U.S.A.

9. S-cone contrast sensitivity in glaucoma as a function of mean luminance

WILLIAM H. SWANSON, RONALD L. FELLMAN, JOHN R. LYNN and RICHARD J. STARITA

(Dallas, TX, U.S.A.)

Abstract

Evaluation of tritan defects is potentially useful for early diagnosis and evaluation of treatment of glaucoma, but the clinical utility of tests such as the Farnsworth-Munsell 100 hue has been limited due in part to effects of individual variations in pupil size and extent of lens yellowing. S-cone contrast sensitivity has been proposed as a more useful tritan test, but it is not known precisely how S-cone contrast sensitivity measured with a computer monitor is influenced by variations in mean S-cone quantal catch due to individual differences in pupil size and lens density. We measured S-cone contrast sensitivity of patients with glaucoma, using 1.0 cycle/degree blue grating superimposed on a bright yellow background. Sensitivity was measured at a range of mean luminances, and pupil size was monitored with a closed-circuit video system. Inter-subject variability was reduced by plotting S-cone contrast sensitivity vs. retinal illuminance rather than stimulus luminance, indicating that much of the variability was due to differences in pupil size. Each patient's data were fit with a threshold versus retinal illuminance (TVR) function. The TVR analysis suggested that additional variability was due to prereceptoral filters. These results indicate that reductions in S-cone contrast sensitivity at luminances available from computer monitors may reflect, in part, effects of pupillary miosis or lens yellowing, rather than optic nerve damage.

Introduction

It is well-known that many patients with glaucoma have tritan color vision defects on the Farnsworth-Munsell (FM) 100 hue test (Pokorny *et al.*, 1979), and it has been suggested that a tritan color vision defect may be the earliest sign of glaucomatous damage (Lakowski and Drance, 1979). However, tritan defects can also be due to non-glaucomatous factors such as small pupil diameter or high lenticular density (Sample *et al.*, 1988).

Reduction of mean retinal illuminance can produce tritan defects in data

B. Drum (ed.), Colour Vision Deficiencies XII, pp. 63–71.
© 1995 *Kluwer Academic Publishers, Dordrecht.*

obtained from normal observers with the FM 100 hue and other tests (Smith *et al.*, 1985; Knoblauch *et al.*, 1987; van Everdingen *et al.*, 1991; Yeh *et al.*, 1993a), and decrease in retinal illuminance can increase the severity of tritan defects in patients with glaucoma (Ourgaud *et al.*, 1972). Since yellowing of the lens and pupillary miosis decrease mean retinal illuminance for blue light, they could lead to tritan errors (Moreland, 1993). For example, there is a strong effect of age on FM 100 hue results: the mean normal score for the 60–65 year age group falls outside the 99.9% confidence limit for the 30–35 year age group (Verriest *et al.*, 1982). Much of this age dependence may be due to lens yellowing and pupillary miosis.

A promising test for detecting glaucomatous loss is measurement of spatial contrast sensitivity under conditions for which the stimuli can only be detected by the short-wavelength-sensitive (S-) cones (Gündüz *et al.*, 1988; Rosenshein and Cyrlin, 1991). Previous studies of S-cone contrast sensitivity in glaucoma have controlled for variations in pupil size and have used flicker photometry to compensate for variations in lens density. However, in a typical clinical setting it may be necessary to test patients who have small pupils, and it may not be possible to estimate lens density. The purpose of the current study was to determine how the variability in pupillary miosis and lens yellowing present in a clinical population could affect variability of S-cone contrast sensitivity. This was accomplished by measuring S-cone contrast sensitivity for 1 cycle/degree gratings as a function of mean luminance in a group of patients with glaucoma.

Materials and methods

S-cone contrast sensitivity was measured using a system described in detail elsewhere (Swanson, 1989; Swanson *et al.*, 1993). Square-wave gratings were presented on a white monochrome computer monitor (we found that this Apple™ monochrome monitor provided more blue light than a standard Sony Trinitron™ color monitor), using a video attenuator (Pelli and Zhang, 1991) which allows lower contrasts than available with a standard video card. Light from the monitor was passed through a blue cutoff filter and then reflected off a dichroic beamsplitter, which superimposed the gratings on a yellow background. The background was produced by passing light from a diffused tungsten source through a yellow cutoff filter. The test field was rectangular, subtending 9° by 12°, and the background was circular, with a diameter of 15°. The mean luminance of the stimulus was controlled with neutral density filters placed after the dichroic beamsplitter.

The test and background were calibrated with a Pritchard spectroradiometer. The CIE coordinates were $x = 0.56$, $y = 0.43$ for the yellow background and $x = 0.15$, $y = 0.03$ for the blue test field. The mean luminances measured with no neutral density filter after the dichroic beamsplitter were 1439 cd/m^2 for the yellow adapting field and 1.25 cd/m^2 for the blue test field. Using the Smith-Pokorny cone fundamentals (Smith and Pokorny, 1975) a grating at

100% contrast on the monitor would yield a cone contrast of 0.1% for the L-cones, 0.4% for the M-cones, and 64% for the S-cones. Control experiments showed that the L- and M-cone contrasts were below contrast threshold for normals at all spatial frequencies.

Subjects were asked to state whether the orientation of the grating was horizontal or vertical, and contrast was adjusted with a 2-down-1-up staircase which used a step size of 1 octave from 100% contrast until the first reversal, then 1/2 octave for the remainder of the staircase. The staircase continued until a total of 10 reversals were obtained, then data were analyzed with maximum likelihood estimation (Swanson and Birch, 1992).

A video camera, mounted next to the apparatus, was focused on the patient's pupil. The image of the pupil was displayed on a monitor, where it was enlarged by a factor of 8.8. This allowed precise measurement of pupil diameter for each mean luminance. Pupil diameter was used to compute mean retinal illuminance in trolands (Td), using LeGrand's method for incorporating the Stiles-Crawford effect (LeGrand, 1968).

Ten patients with primary open-angle glaucoma, ages 41 – 79 years (mean ± 1 SD = 66 ± 12), were recruited from the glaucoma practice of three of the authors (JRL, RLF, RJS). Patients were selected who had not significant cataract, had no other visual disorder besides glaucoma, and were stable under current treatment. Patients who were being treated with miotic agents were asked to refrain from using their medications for 24 hours before the test. For the brightest adapting field used, mean pupil diameter (± 1 SD) was 2.1 (± 0.6) mm.

Results

S-cone contrast sensitivity data for the patients with glaucoma are shown in Figures 1 and 2. The inter-subject variability was much greater when data were plotted in terms of mean luminance (Fig. 1) than when they were plotted in terms of mean retinal illuminance (Fig. 2). This indicates that variations in pupil size can have measurable effects on S-cone contrast sensitivity.

These data demonstrate that S-cone contrast sensitivity can be highly dependent on mean S-cone quantal catch, which in turn is affected by the amount of blue light absorbed by the macular pigment and crystalline lens. To estimate the effect of these prereceptoral filters, the relation between S-cone contrast sensitivity and mean retinal illuminance was modelled with threshold vs. retinal illuminance (TVR) functions. TVR functions are typically used to analyze increment threshold data, which are plotted as log threshold vs. log adapting luminance. We fit our data with an equation of the form

$$\log(T) = \log(T_0) + \log((A + A_0)/A_0), \qquad (1)$$

where T is the increment threshold, A is the mean retinal illuminance for the blue test field, T_0 is dark-adapted threshold and A_0 is the mean retinal

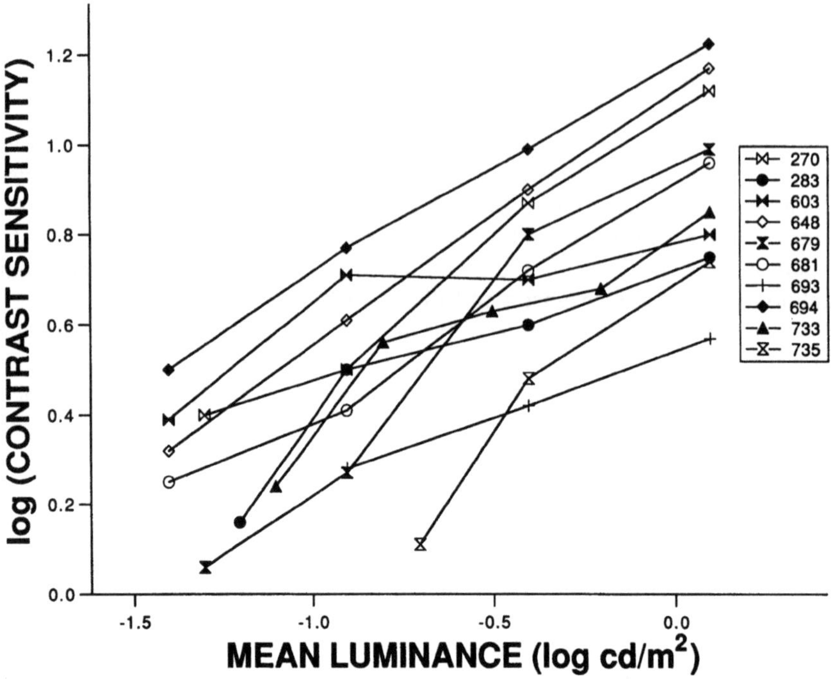

Fig. 1. S-cone contrast sensitivity for a 1.0 c/deg. grating is shown as a function of mean luminance for the blue test field, for 10 patients with glaucoma. The legend lists patient identification codes. Note that for most patients S-cone contrast sensitivity continues to increase with mean luminance up to the highest mean luminance used.

illuminance at which threshold begins to rise. Typically, A_0 is considered to reflect the combined effects of prereceptoral filters, photoreceptor optical density, and photoreceptor sensitivity (Hood and Greenstein, 1990; Pokorny *et al.*, 1990), although it can also reflect effects of second-site adaptation (Yeh *et al.*, 1993b). Since the site of disease action in glaucoma is primarily at the retinal ganglion cell level, we considered variation in A_0 across patients as an index of the variation in the density of prereceptoral filters.

In order to fit our data with equation (1), we considered the increment threshold to be the incremental difference at contrast threshold between the light and dark bars of the grating, so that

$$\log(T) = \log(2A/S), \qquad (2)$$

where A is the mean retinal illuminance for the blue test field and S is the contrast sensitivity of the patient. Figures 3 and 4 show the data from Fig. 2 replotted as increment threshold vs. mean retinal illuminance, and fit with TVR functions. For the 10 patients, the mean values ($\pm$ 1 SD) were -0.58 ($\pm$ 0.16) for T_0 and 0.44 ($\pm$ 0.32) for A_0.

At high retinal illuminances TVR functions asymptote to a straight line with

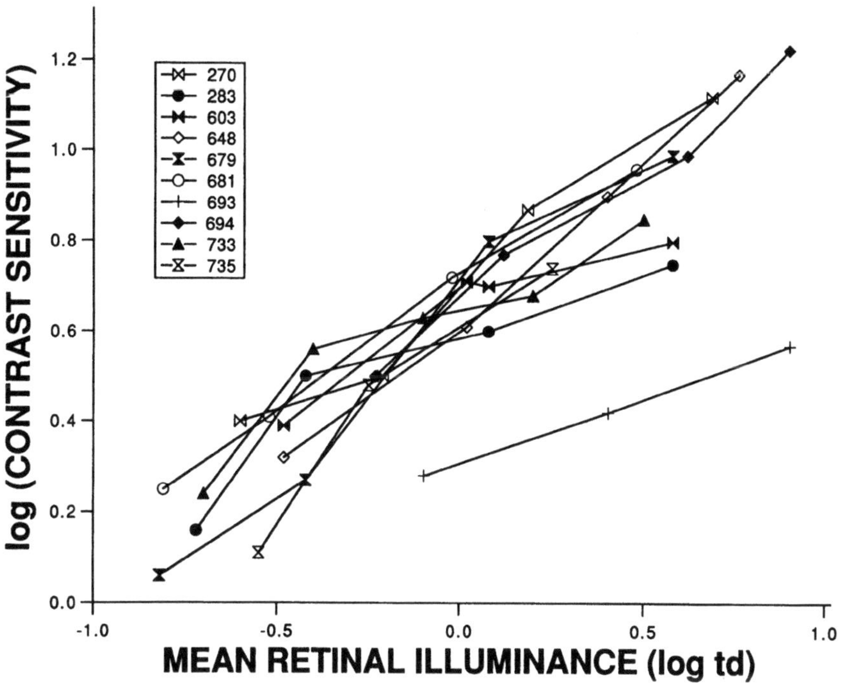

Fig. 2. Data from Fig. 1 plotted as a function of mean retinal illuminance for the blue test field. Note that the variability across patients is in general much smaller than in Fig. 1, where effect of pupil size were not accounted for.

a slope of 1. This is the Weber region, in which increment threshold is directly proportional to retinal illuminance and contrast threshold is independent of retinal illuminance. For most patients, even at the highest retinal illuminance the data were not in the Weber region of the TVR functions, indicating that asymptotic contrast sensitivity had not been reached. For each patient, we computed the contrast sensitivity for the Weber region of the best-fit TVR function as $\log(2A_0/T_0)$, and compared this computed asymptotic sensitivity with the contrast sensitivity measured at the highest mean luminance used. For all patients the computed contrast sensitivity was greater than the measured contrast sensitivity. The mean difference (± 1 SD) was 0.15 (± 0.16) log unit.

Discussion

These data indicate that non-glaucomatous factors such as pupillary miosis and lens yellowing can affect S-cone contrast sensitivity by decreasing the mean S-cone quantal catch. Analysis with TVR functions indicated that for all 10 patients the measured S-cone contrast sensitivity was lower than the value expected for the Weber region.

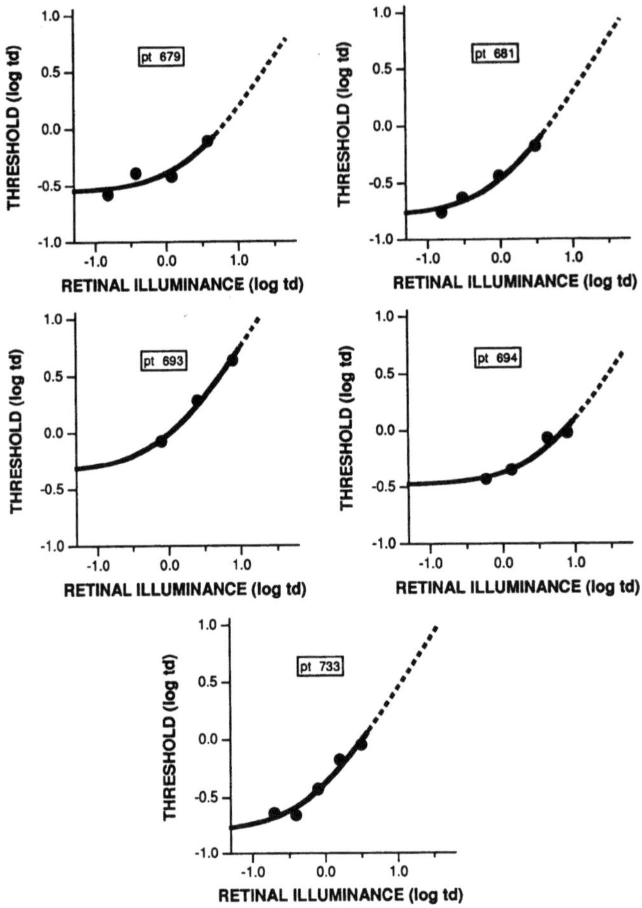

Fig. 3. Data for 5 patients replotted from Fig. 2 as increment thresholds, and fit with TVR functions. The solid curve is the portion of the TVR curve fit to the data. The dashed curve shows the extrapolation of the TVR curve to higher luminances. For most patients the maximum retinal illuminance is not in the Weber region, requiring extrapolation to determine asymptotic sensitivity.

We considered the parameter A_0 to be an index of the optical density of prereceptoral filters. We obtained a standard deviation for A_0 of 0.3 log unit, with individual values varying by as much as 0.8 log unit. This range is consistent with the 1 log unit range in optical densities of human ocular media at short wavelengths reported for individual age groups (Savage *et al.*, 1993).

In order for S-cone contrast sensitivity to be relatively independent of variations in density of the prereceptoral filters, the mean retinal illuminance should be high enough for increment thresholds to be in the Weber region. To ensure that contrast sensitivity is within 0.05 log unit of the asymptotic value of the TVR function, mean retinal illuminance must be at least 0.9 log unit above A_0. Our data indicate that values of log (A_0) in a clinical populatiion can range

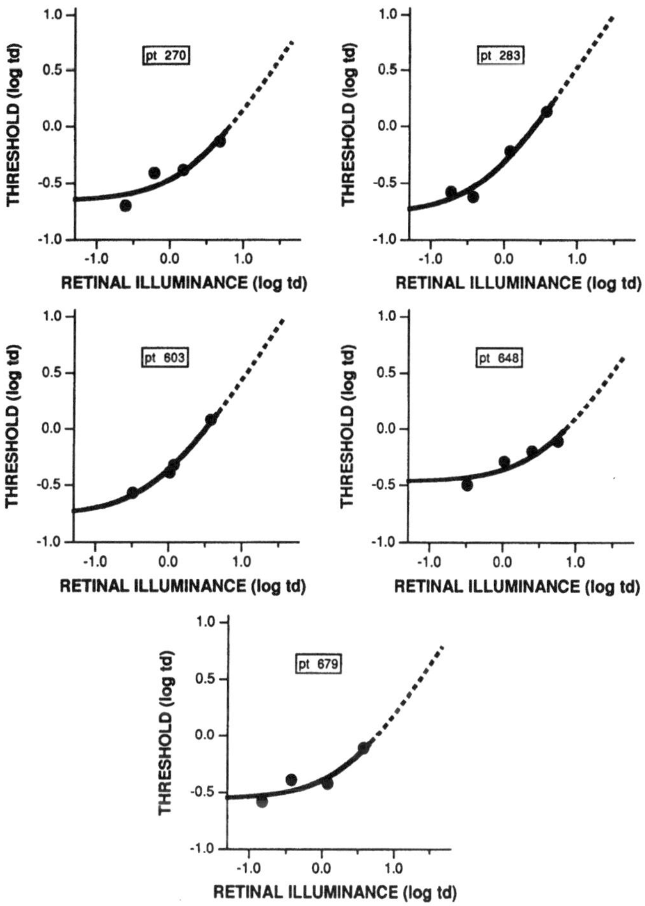

Fig. 4. Data for 5 remaining patients from Fig. 2 replotted as increment threshold and fit with TVR functions. Details as in Fig. 3.

at least as high as 0.6 log td. This suggests that mean retinal illuminance should be at least 1.5 log td for clinical use. With pupil diameter near 1 mm (as measured in some of our patients), this would require a mean luminance 25 times greater than we were able to obtain with a standard computer monitor.

We emphasize that caution must be exercised in interpreting S-cone contrast sensitivity data gathered with standard computer monitors. However, optical systems which provide greater retinal illuminances should be able to produce S-cone contrast sensitivity tests that are relatively immune to prereceptoral factors.

Acknowledgements

This work was supported by Public Health Service grant EY07716 (WHS) from the National Eye Institute of the United States and by a grant from Alcon Laboratories, Ft. Worth, Texas.

References

Gündüz, K., Arden, G.B., Perry, S., Weinstein, G.W. and Hitchings, R.A. (1988). Color vision defects in ocular hypertension and glaucoma. Arch. Ophthalmol. 106: 929–925.

Hood, D.C. and Greenstein, V. (1990). Models of the normal and abnormal rod system. Vision Res. 30: 51–68.

Knoblauch, K., Saunders, F., Kusuda, M., Hynes, R., Podgor, M., Higgins, K.E. and deMonasterio, M. (1987). Age and illuminance effects in the Farnsworth-Munsell 100-hue test. Appl. Opt. 26: 1441–1448.

Lakowski, R. and Drance, S.M. (1979). Acquired dyschromatopsias: The earliest functional losses in glaucoma. Doc. Ophthalmol. Proc. Ser. 19: 159–165.

LeGrand, Y. (1968). Light, Color and Vision. Chapman and Hall, London.

Moreland, J.D. (1993). Matching range and age in a blue-green equation. In: Drum, B. (ed.), Colour Vision Deficiencies XI, Doc. Ophthalmol. Proc. Ser. 56: 129–134, Kluwer Acad. Publ., Dordrecht.

Ourgaud, A.G., Vola, J.L., Jayle, G.E. and Baud, C.E. (1972). A study on the influence of the illumination level and pupillary diameter on chromatic discrimination in glaucomatous patients. Mod. Probl. Ophthalmol. 11: 141–144.

Pelli, D.G and Zhang, L. (1991). Accurate control of contrast on microcomputer displays. Vision Res. 31: 1337–1350.

Pokorny, J., Smith, V.C., Verriest, G. and Pinckers, A.J.L.G. (1979). Congenital and Acquired Color Vision Defects. Grune and Stratton, New York.

Pokorny, J., Smith, V.C. and Yeh, T. (1990). Analysis of tritan detection thresholds and discrimination errors. In: Ohta, Y., (ed.), Color Vision Deficiencies, Proceedings of the IRGCVD, Tokyo, Japan, 1990, 191–197, Kugler and Ghedini, Amsterdam.

Rosenshein, J.S. and Cyrlin, M.N. (1991). Glaucoma multi-testing. Invest. Ophthalmol. Visual Sci. 32: 811.

Sample, P.A., Boynton, R.M. and Weinreb, R.B. (1988). Isolating the color vision loss in primary open-angle glaucoma. Am. J. Ophthalmol. 105: 686–691.

Savage, G.L., Haegerstrom-Portnoy, G., Adams, A.J. and Hewlett, S.E. (1993). Age changes in the optical density of human ocular media. Clin. Vis. Sci. 8: 97–108.

Smith, V.C. and Pokorny, J. (1975). Spectral sensitivity of the foveal cone photopigments between 400 and 500 nm. Vision Res. 15: 161–171.

Smith, V.C., Pokorny, J. and Pass, A.S. (1985). Color-axis determination on the Farnsworth-Munsell 100-hue Test. Am. J. Ophthalmol. 100: 176–182.

Swanson, W.H. (1989) Short-wavelength-sensitive cone acuity: Individual differences and clinical application. Appl. Opt. 28: 1151–1157.

Swanson, W.H., Birch, D.G. and Anderson, J.L. (1993). S-cone function in patients with retinitis pigmentosa. Invest. Ophthalmol. Visual Sci. 34: 3045–3055.

Swanson, W.H. and Birch, E.E. (1992). Extracting thresholds from noisy psychophysical data. Percept. Psychophys. 51: 409–422.

van Everdingen, J.A.M., Smith, V.C. and Pokorny, J. (1991). Sensitivity of tritan screening tests as evaluated in normals at reduced levels of illumination. In: Drum, B., Moreland, J., and Serra, A. (eds.), Colour Vision Deficiencies X, Doc. Ophthalmol. Proc. Ser. 54: 167–175, Kluwer Acad. Publ., Dordrecht.

Verriest, G., Laethem, J. v. and Uvijls, A. (1982). A new assessment of the normal ranges of the Farnsworth-Munsell 100-hue test scores. Am. J. Ophthalmol. 93: 635–642.
Yeh, T., Pokorny, J. and Smith, V.C. (1993a). S-cone discrimination sensitivity and performance on arrangement tests. In: Drum, B. (ed.), Colour Vision Deficiencies XI, Doc. Ophthalmol. Proc. Ser. 56: 293–302), Kluwer Acad. Publ., Dordrecht.
Yeh, T., Pokorny, J. and Smith, V.C. (1993b). Chromatic discrimination with variation in chromaticity and luminance: data and theory. Vision Res. 33: 1835–1845.

William H. Swanson
Retina Foundation of the Southwest
9900 N. Central Expressway, Suite 400
Dallas, Texas 75231–3303
U.S.A.P

10. Motion thresholds of coloured stimuli of different luminance contrasts are increased in ocular hypertension and early primary-open-angle-glaucoma (POAG)

HENDRIK P.N. SCHOLL and EBERHART ZRENNER
(Tübingen, Germany)

Abstract

The purpose of this study was to assess the functional integrity of the pathway mediating motion perception (i.e. the magnocellular pathway) in ocular hypertension and early POAG in relation to the pathway mediating colour vision (i.e. the parvocellular pathway).

We have developed a psychophysical technique to measure motion thresholds at various luminance contrasts (max., 95%, 78% and isoluminant) in patients with ocular hypertension and POAG and age matched controls. We employed a dynamic random dot display that contained varying degrees of a coherent motion signal embedded within a background of random noise and varying degrees of luminance contrast of the red array of dots and the green background. The motion stimulus was produced under control of an IBM-PC compatible computer, using software developed by the authors.

We found statistically significant differences between patients with ocular hypertension/POAG and normal subjects in motion sensitivity at the full range of luminance contrast. Differences were particularly high near isoluminance. We also found a shift of the point of isoluminance by flicker-photometry.

Our findings indicate that motion threshold testing in combination with testing of colour vision may reveal preclinical optic nerve disease in early POAG. The specific deficits in motion sensitivity near isoluminance in an early stage of the disease have to be proved by using larger groups of patients.

Introduction

The loss of the visual field associated with early optic nerve damage often goes undetected in primary-open-angle-glaucoma (POAG), because a large proportion of the optic nerve fibres must be lost before a visual deficit can be found with traditional perimetric techniques (Quigley *et al.*, 1989). Therefore there is great interest in detecting early signs of this insidious disease: structural and functional changes (Drance and Airaksinen, 1986). Psychophysical

B. Drum (ed.), Colour Vision Deficiencies XII, pp. 73–85.
© 1995 *Kluwer Academic Publishers, Dordrecht.*

disturbances which have been identified as early signs of glaucoma include anomalies of the colour sense (Drance *et al.*, 1981; Adams *et al.*, 1982; Gündüz *et al.*, 1988), temporal and spatial contrast sensitivity (Atkin *et al.*, 1979 and 1980; Falcao-Reis *et al.*, 1990: Lachenmayr and Drance, 1992), the differential light sense (Flammer and Drance, 1984), receptive field-like functions (Enoch *et al.*, 1981) and motion sensitivity (Silverman *et al.*, 1990).

Several lines of evidence suggest that the large optic nerve fibers, which form the magnocellular retinocortical pathway, are preferentially susceptible to early glaucomatous damage. Quigley *et al.* (1988) showed that human eyes with glaucoma appear to lose large optic nerve fibers more rapidly than smaller fibers. This is also true in monkeys with experimental glaucoma (Quigley *et al.*, 1987; Glovinsky *et al.*, 1993).

Drum *et al.* (1986) and Glovinsky *et al.* (1992), in a different approach to the same issues, found scotopic sensitivity loss by applying psychophysical tests which they had worked out for this purpose. They suggested that Y-like M ganglion cells, having large cell bodies and axons and being particularly sensitive to luminance contrast under mesopic and scotopic conditions, are optimally sensitive to the scotopic stimulus.

Livingstone and Hubel's (1987 and 1988) experiments of the functional architecture of the visual system demonstrate that the magnocellular pathway underlies the global perception of motion and the parvocellular pathway underlies the perception of colour. Schiller *et al.* (1990) examined the visual capacities of rhesus monkeys before and after interrupting the colour-opponent or the broad-band channel. They reported that the colour-opponent channel is essential for the processing of colour, texture, fine pattern and fine stereopsis, whereas the broad-band channel is crucial for the perception of fast flicker and motion. No deficits were found in brightness and coarse-shape discrimination, low spatial frequency stereopsis and contrast sensitivity after the disruption of either of the channels. Gorea and Papathomas (1989) proposed a complicated model of three movement pathways: a pure chromatic, a chromatic-plus-luminance and a luminance pathway.

Recalling these research works and following the lines of Silverman *et al.* (1990), we thus developed a psychophysical technique to assess the interdependence of these pathways and the changes caused by glaucomatous damage. For this purpose we employed a dynamic random dot display that contained varying degrees of a coherent motion signal embedded within a background of random motion noise and two colours (green for the random dots and red for the background) of varying degrees of luminance contrast. Recently it has become clear that the yellow/blue and blue/green parts of the colour sense show the early subtle disturbances of POAG, whereas the red/green part was relatively spared (Drance *et al.*, 1981). Therefore, we have chosen the red/green part assuming its relative intactness.

We used this technique to measure motion thresholds of different luminance contrasts in patients with ocular hypertension and early stages of POAG and age-matched controls.

The aim of this study was to assess the functional integrity of the pathway mediating motion perception (i.e. the magnocellular pathway) in early POAG in relation to the pathway mediating colour vision (i.e. the parvocellular pathway).

Material and methods

Subjects

The control group consisted of 21 eyes (20 subjects) aged between 41 and 74 years (Table 1). Visual acuities ranged from 0.4 to 1.0. All subjects had normal ocular histories. None of them was receiving miotic eyedrops, and none had undergone previous cataract surgery. Cataract due to age was admitted. The subjects were volunteers and not associated with ophthalmology; they were intelligent and well motivated.

Table 1. Characteristics of the four clinical groups. Number of individuals, average age, average visual acuity and average cup/disk ratio (CDR) for normal subjects and for patients with elevated intraocular pressure, with physiologic papilla and normal visual field (low-risk), for patients with elevated intraocular pressure, with pathologic papilla and normal visual field (high-risk) and for patients with primary-open-angle-glaucoma (elevated intraocular pressure, pathologic papilla and early stages of visual field loss).

Characteristics	Controls	OHT (low-risk)	OHT (high-risk)	POAG
N	21	6	10	9
Mean (± SD) age (yrs)	54.6 ± 8.3	54.8 ± 12.8	58.1 ± 11.3	63.4 ± 13.9
Mean (± SD) visual acuity	0.82 ± 0.24	0.99 ± 0.34	0.78 ± 0.35	0.93 ± 0.19
Mean (± SD) CDR	–	0.34 ± 0.07	0.77 ± 0.10	0.58 ± 0.23

Twenty-five eyes of 22 patients were tested. The 22 patients (age between 36 and 81 years, Table 1) with ocular hypertension and glaucoma were all under clinical care. All eyes had a complete ophthalmologic examination, including best-corrected visual acuity, slit-lamp biomicroscopy, applanation tonometry (diurnal pressure profile), dilated funduscopy and perimetry on the Tübingen Automated Perimeter (suprathreshold threshold-related examination strategy). Visual acuities ranged from 0.2 to 1.25. All eyes had open anterior chamber angles. No restrictions were placed on patient medications or surgical history for glaucoma.

Cup/disc ratio and perimetric results were the criterions for classification of the patients into groups (Table 1): The first group (OHT low-risk) had elevated intraocular pressure (i.e. history of intraocular pressure > 21 mm Hg), normal visual field and physiologic papilla, the second group (OHT high-risk) had elevated intraocular pressure, normal visual field and pathologic papilla (cup/disc ratio > 0.6) and the third group had elevated intraocular pressure, pathologic papilla and early stages of visual field loss (stage I or II by Aulhorn and Karmeyer, 1976).

Random dot motion display and procedure

The motion stimulus was produced under the control of a 14 Mhz IBM-PC compatible computer, using software developed by the authors (Turbo-Pascal 5.0). The monitor was a Princeton Ultra 14 (70 Hz). The display measured 25.0 × 17.2 cm and subtended 60° × 43° of the visual field. Each dot measured 11.25 min in diameter. The array of dots was green (peak of relative radiation density at 539 nm) and the background was red (peak at 629 nm). In the random dot display, an array of 150 randomly located dots is plotted in rapid succession on the CRT. This array of dots is replaced by a new array of randomly located dots, with each successive replot occuring at a frequency of 32 Hz. When viewing this display, a perception of random noise is produced since there is no net motion in this display. If, however, a subset of the dots is replotted at a fixed spatial offset in a common direction (up, down, left, or right), then a coherent motion signal is produced, embedded wihtin the background of random noise (Silverman *et al.*, 1990). The percentage of dots in coherent motion and the luminance of dots and background could be varied.

The subjects were seated in a darkened room and viewed the display monocularly at 21.7 cm. One eye was tested while the other was covered with an opaque patch. In the beginning the point of isoluminance was determined for every subject with flicker-photometry. For each luminance contrast (measured with Minolta Luminance Meter LS 100): maximum (≤ 100% (dots, 98.0 cd/m^2, background 0.0 cd/m^2), 95% (dots 38.6 cd/m^2, background 0.9 cd/m^2), 78% (dots 32.9 cd/m^2, background 4.2 cd/m^2) and isoluminance (dots variable, background 11.1 cd/m^2) 10 trials were presented at each of 7 motion coherence levels whose ranges were determined by a short test before each trial. Subjects were required to indicate the direction of perceived motion with a joystick, using a four-alternative, forced choice technique with a stimulus presenting time of 8 sec.

Results

The motion thresholds of each luminance contrast were estimated from the curve (correct responses at motion coherence level) by the Maximum-Likelihood-Method (Le Cam, 1990). The average motion thresholds of each luminance contrast and of all luminance contrasts for each group were then calculated. In Fig. 1 all measured motion thresholds averaged for each group at each luminance contrast are shown.

Table 2 shows the sum of average motion thresholds at all luminance contrasts for the control, OHT (low-risk), OHT (high-risk) and the POAG group. This indicates motion sensitivity over the full range of luminance

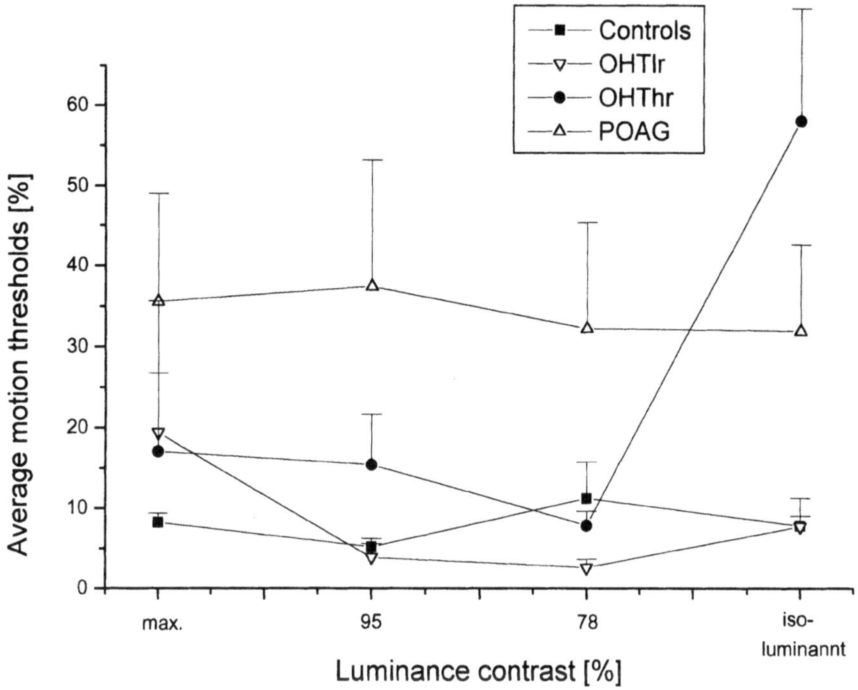

Fig. 1. Average motion thresholds at each luminance contrast in: Normals, OHT (low-risk), OHT (high-risk) and POAG patients. Luminance contrast is defined as: $\frac{L_{max} - L_{min}}{L_{max} + L_{min}} \times 100$, where L stands for luminance (of the random dots or the background respectively) in cd/m^2. Vertical bars indicate 1 SEM.

Table 2. Sum of average motion thresholds for the control, OHT (low-risk), OHT (high-risk) and the POAG group (Mean ± SD).

Group	N	Sum of average motion thresholds (± SD)
Controls	21	32.5 ± 25.5
OHT (low-risk)	6	33.5 ± 51.0[a]
OHT (high-risk)	10	98.6 ± 76.5[b]
POAG	9	137.0 ± 134.3[b]

[a] Not significantly different from control by t-test.
[b] Significantly different from control ($p < 0.005$) by t-test.

contrast (Fig. 2). We used Student's t-test on two groups (independent) and found statistically significant differences to the controls for the OHT (high-risk) and for the POAG group ($p < 0.005$ by t-test). To rule out the functional criterion of perimetry results for drawing a comparison based only on the morphologic criterion we found a statistically significant difference for the

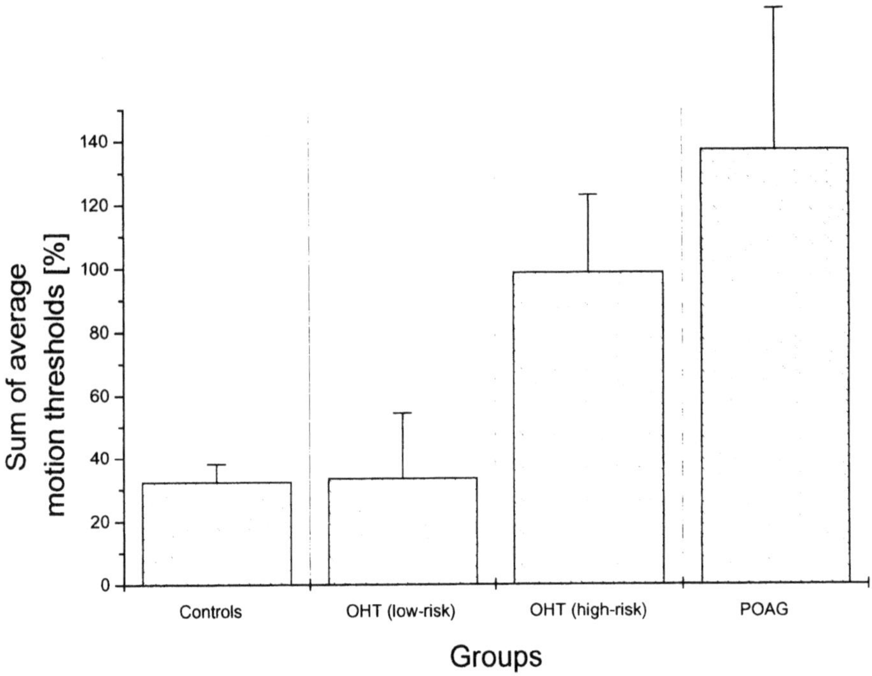

Fig. 2. Sum of average motion thresholds at isoluminance in: Normals, OHT (low-risk), OHT (high-risk) and POAG patients. Vertical bars indicate 1 SEM.

patients with pathologic papilla (OHT high-risk and POAG group) and the controls ($p < 0.001$ by t-test).

Table 3 shows the average motion thresholds at isoluminance for the control, OHT (low-risk), OHT (high-risk) and the POAG group. This indicates motion sensitivity at isoluminance (Fig. 3). We found statistically significant differences to the controls for the OHT (high-risk) and for the POAG group ($p < 0.00005$ and $p < 0.005$ by t-test, respectively), the difference for the OHT (high-risk)

Table 3. Average motion thresholds at isoluminance for the control, OHT (low-risk), OHT (high-risk) and the POAG group (Mean ± SD).

Group	N	Average motion thresholds at isoluminance (± SD)
Controls	21	7.8 ± 5.7
OHT (low-risk)	6	7.7 ± 8.6[a]
OHT (high-risk)	10	57.9 ± 44.7[b]
POAG	9	31.8 ± 32.1[c]

[a] Not significantly different from control by t-test.
[b] Significantly different from control ($p < 0.00001$) by t-test.
[c] Significantly different from control ($p < 0.005$) by t-test.

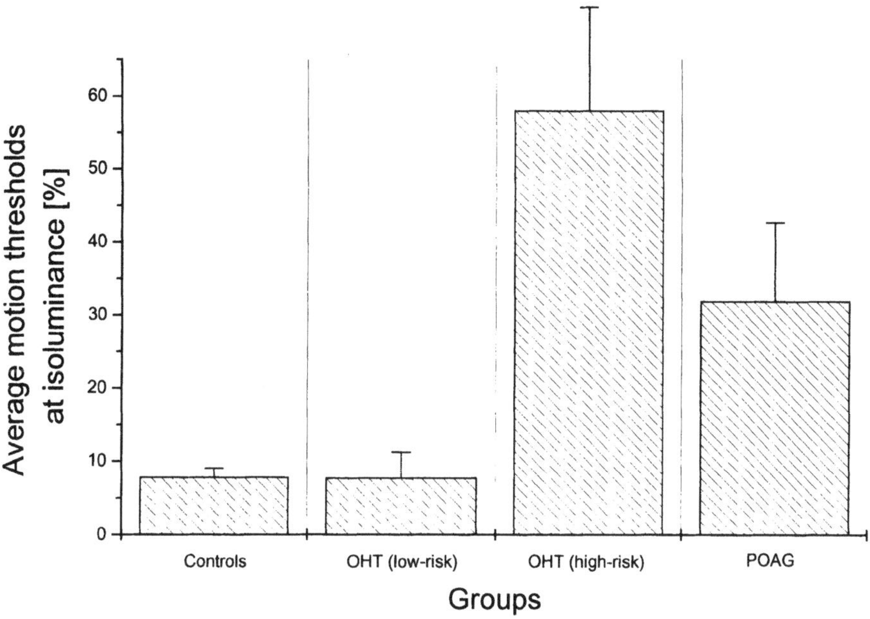

Fig. 3. Average motion thresholds at isoluminance in: Normals, OHT (low-risk), OHT (high-risk) and POAG patients. Vertical bars indicate 1 SEM.

group being larger than for the POAG group. Disregarding perimetry results we found a statistically significant difference for the patients with pathologic papilla (OHT high-risk and POAG group) and the controls ($p < 0.0005$ by t-test). Calculated from the results of the OHT (high-risk) and the POAG group we found a sensitivity of 80% and specificity of 90.5%.

Average luminance contrast (physically defined) at the individual point of isoluminance was calculated for each group (Fig. 4). We observed increasing values for OHT (low-risk), OHT (high-risk) and POAG (i.e., increased luminance of the green stimulus for isoluminance in flicker-photometry). Table 4 shows that there was a statistically significant difference between the controls and the POAG group ($p < 0.05$ by t-test).

Table 4. Luminance contrast at isoluminance for the control, OHT (low-risk), OHT (high-risk) and the POAG group (Mean ± SD).

Group	N	Luminance contrast at isoluminance (± SD)
Controls	21	4.3 ± 4.3
OHT (low-risk)	6	6.2 ± 4.7[a]
OHT (high-risk)	10	7.0 ± 6.9[a]
POAG	9	9.1 ± 5.5[b]

[a] Not significantly different from control by t-test.
[b] Significantly different from control ($p < 0.05$) by t-test.

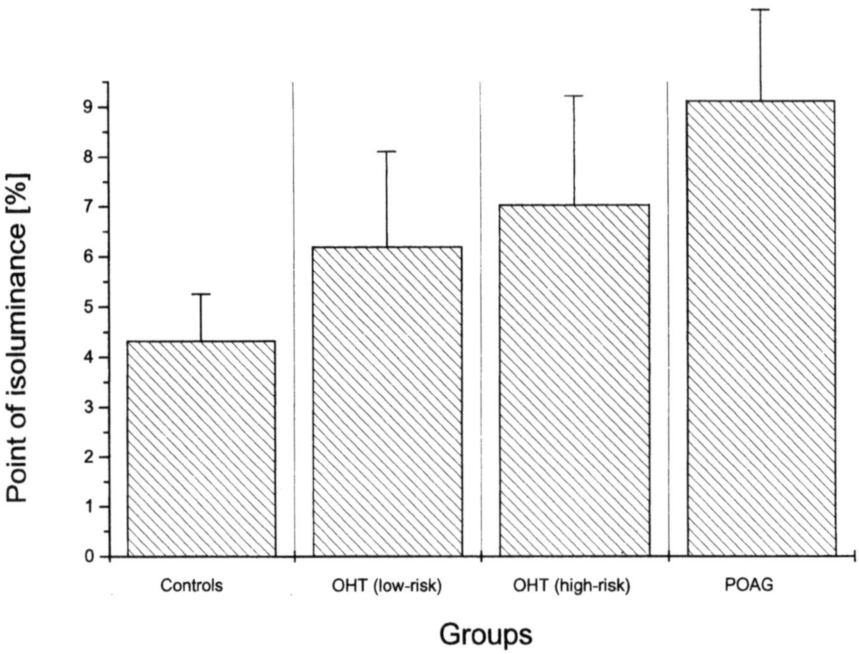

Fig. 4. Shift of the point of isoluminance. Columns show the point of isoluminance (% luminance contrast) in each group: Normals, OHT (low-risk), OHT (high-risk) and POAG patients. Vertical bars indicate 1 SEM. The shift means that the luminance of the green stimulus increases for isoluminance in flicker-photometry.

The association between motion thresholds and intraocular pressure was assessed using linear regression analysis. No significant relationship was evident.

A similar regression analysis was used to investigate the association between subject age and motion thresholds. For the control group, age did not show a statistically significant association ($r = 0.12$, $p = 0.61$) with the sum of average motion thresholds.

Nevertheless we used covariance analysis for all groups to consider influence of subject age. We found a statistically significant difference of motion thresholds at isoluminance ($p = 0.0002$).

Discussion

The results of this test show the dependence of motion sensitivity on luminance contrast of two colours in normal vision and the specific changes in early POAG*.

* The small sample of patients led to even smaller subdivision groups. Therefore random variance could bias the results. The study has to be taken as preparatory work to an explicit study of motion sensitivity at isoluminance in early POAG.

Random dot displays have been employed extensively in the investigation of motion perception and have many advantages for the study of sensitivity of motion. When random dot displays are used, measurements of motion perception are not confounded by the influence of form and position dependent mechanisms (De Bruyn and Orban, 1988). This type of display also allows for the intensity of the motion signal to be precisely varied while keeping other stimulus parameters constant (e.g. mean luminance, and contrast, as well as spatial and temporal frequency). Drum *et al.* (1989) developed a pattern discrimination perimetry test by employing random-dot displays. Pattern discrimination perimetry proved to be substantially better than convential visual fields at distinguishing between the glaucoma suspect and normal distributions.

Flicker-photometry is suitable to determine the point of isoluminance. We tested the entire field (foveal and extrafoveal) and the condition for the motion signal is therefore only near isoluminance (Troscianko and Fahle, 1988). Beyond that, pseudo-convergence errors of the random dot display restricted the condition.

Morphologic changes usually occur before functional deficits (in particular, visual field loss) are observable (Cloux-Fey *et al.*, 1986; Iwata, 1989). For quantification of the pathologic morphology of the optic disc we used the cup/disc ratio. There is a great variation in the cup/disc ratio of the physiologic papilla; and the cup/disc ratio of the pathologic papilla is not necessarily correlated with the actual morphologic damage of the optic disc (Gloor, 1990). Nevertheless, the prognostic significance of the cup/disc ratio was demonstrated in retrospective and prospective studies on ocular hypertensive patients (Armaly, 1969; Wilensky *et al.*, 1974; Yablonski *et al.*, 1980).

We found a significant dissociation of the results between ocular hypertensive patients with physiologic and pathologic cup/disc ratios, but we could not find an association between motion thresholds and cup/disc ratios. Silverman *et al.* (1990) found a remarkable dissocation of their results in motion sensitivity in ocular hypertensives, but they made no subdivision concerning morphologic criteria.

The results show that motion thresholds at all luminance contrasts are increased in early POAG, especially at (or near) isoluminance in comparison with normal subjects (Fig. 3). The fact that motion perception in glaucoma is depressed agrees with Silverman *et al.* (1990).

Surprisingly, isoluminant stimuli are best suitable for detecting early POAG, even though the parvocellular pathway should be less damaged. But the functional reserve has to be considered. The M cells have large receptive fields and a great overlap (Perry and Cowey, 1985). Because of the considerably lower functional reserve of the P cells there may be a great potential of diagnostic possibilities (Glovinsky *et al.*, 1991). Korth and Horn (1990) showed that the greatest changes in the red/green coloured pattern ERG; these changes were stronger than with blue/yellow patterns, because the blue/yellow sensitive ganglion cells represent only 6% and the red/green sensitive ganglion cells represent 46% of the ganglion cell population in monkeys (Zrenner, 1983; Zrenner and Gouras, 1981). Moreover, the size-dependent glaucomatous damage cannot be explained by early death of M cells alone, since they comprise only a small minority of the ganglion cells in the retina. The results of Glovinsky *et al.* (1991) provide some evidence for the greater vulnerability to glaucomatous damage among the larger P cells. The large P cells may be an appropriate cell type for psychophysical assessment in early glaucoma along with the M cells.

The fact that the method appears to be more sensitive to the OHT (high-risk) group than to the POAG group is significant and surprising. An explanation, however, might be quite difficult. Disregarding random variance due to the small number of samples, perimetry could be faulty and, as a criterion for group division, inappropriate. A plausible explanation is that testing of motion sensitivity (at isoluminance) detects some abnormalities missed by standard perimetry, but there were also cases where motion sensitivity showed no conspicious results whereas perimetry detected beginning field defects. In addition it is conceivable that the shift of the point of isoluminance is restricted to flicker stimuli, and therefore the measurement with non-flicker stimuli does not take place at the required point for isoluminance.

A great difference (in peripheral colour contrast sensitivity) between patients with ocular hypertension with a cup/disc ratio greater and smaller than 0.6 was also found by Yu *et al.* (1991) who suggested that measurement of colour contrast sensitivity could thoroughly be a suitable method for detecting early POAG.

It is not easy to explain the influence of luminance contrast on motion sensitivity (see Fig. 1) in detail. The results agree with the work of Drum and Bissett (1991), showing that low contrast is as good as or better than high contrast for motion sensitivity. Motion sensitivity saturates at low contrast levels in the periphery, and high contrast may even impair motion sensitivity under certain conditions. It may be that the lower limits of the magno system contrast sensitivity were tested because of the restricted precision of the determination of isoluminance. In contrast to the P system, the M system is predominantly sensitive for stimuli of low spatial and high temporal frequency.

We found no significant association between motion thresholds and actual intraocular pressure. However, this result does not rule out the possibility of a relationship between the average long-term intraocular pressure and the motion thresholds (Silverman *et al.*, 1990).

Although there is a significant effect of age on the nerve populations of the human optic nerve (Balazsi *et al.*, 1984) we did not find a statistically significant age effect on motion sensitivity and agree with Silverman *et al.* (1990). Nevertheless covariance analysis was used to consider the influence of age.

The increased thresholds show statistically significant differences from the controls, and are even strongly marked in OHT (high risk). However, there was considerable overlap in motion thresholds with individual control and OHT/POAG patients. Further testing (i.e. improving the biomathematical part of the software in combination with enlargement of the groups) will be necessary to adequately evaluate this issue.

Moreover a study of only peripheral vision without foveal functions could follow this experiment of global vision. Global vision testing is an adequate method, because there is diffuse and localized nerve fibres loss in glaucoma (Airaksinen *et al.*, 1984). But recent experimental work showed that foveal functions are spared a long time by glaucomatous damage and therefore testing of peripheral vision could improve the results (Falcao-Reis, 1990). In addition, Drum and Bissett (1991) showed that the fovea is highly specialized for P system activity and the periphery is specialized for M system activity.

Besides, testing of motion perception of early POAG with yellow and blue stimuli would be of much interest. Airaksinen *et al.* (1986) suggested a significant correlation between yellow/blue and green/blue anomaloscopic matching ranges and diffuse retinal nerve fiber loss. Unfortunately the yellow/blue part is more susceptible to cataract then the red/green. Also age and small pupils can disturb psychophysical tests of colour vision (Grützner and Schleicher, 1972; Marmion, 1978). These factors occur frequently in primary-open-angle-glaucoma. Sample and co-workers were able to verify that colour vision losses associated with glaucoma cannot be attributed to these variables (Sample *et al.*, 1988a, b). They showed that short-wavelength colour visual fields detect progressive loss sooner than automated visual fields in eyes with glaucoma (Sample and Weinreb, 1992) and colour visual fields show deficits in suspect eyes when automated visual fields are normal (Sample and Weinreb, 1990). Finally there is evidence for the ability of short-wavelength colour visual fields to predict future glaucoma in suspect eyes (Sample *et al.*, 1993; Johnson *et al.*, 1993). However, Lakowski *et al.* (1972) found disturbances in red/green colour vision in ocular hypertensives.

Our findings indicate that motion threshold testing in combination with testing of colour vision may reveal preclinical optic nerve disease in early POAG.

References

Adams, A.J., Rodic, R., Husted, R. and Stamper, R. (1982). Spectral sensitivity and color discrimination changes in glaucoma and glaucoma-suspect patients. Invest. Ophthalmol. Vis. Sci. 23: 516–524.

Armaly, M.F. (1969). Cup/disc ratio in early open-angle glaucoma. Doc. Ophthalmol. 26: 526–533.

Aulhorn, E. and Karmeyer, H. (1976). Frequency distribution in early glaucomatous visual field defects. Doc. Ophthalmol. Proc. Ser. 14: 75–83.

Airaksinen, P.J., Drance, S.M., Douglas, G.R. and Mawson D.K. (1984). Diffuse and localized nerve fiber loss in glaucoma. Am. J. Ophthalmol. 98: 566–571.

Airaksinen, P.J., Lakowski, R., Drance, S.M. and Price, M. (1986). Color vision and retinal nerve fiber layer in early glaucoma. Am. J. Ophthalmol. 101: 208–213.

Atkin, A., Bodis-Wollner, I., Wolkstein, M., Moss, A. and Podos, S.M. (1979). Abnormalities of central contrast sensitivity in glaucoma. Am. J. Ophthalmol. 88: 205–211.

Atkin, A., Wolkstein, M., Bodis-Wollner, I., Anders, M., Kels, B. and Podos, S.M. (1980). Interocular comparison of contrast sensitivities in glaucoma and suspects. Br. J. Ophthalmol. 64: 858–862.

Balazsi, A.G., Rootmen, J., Drance, S.M., Schulzer, M. and Douglas, G.R. (1984). The effect of age on the nerve fiber population of the human optic nerve. Am. J. Ophthalmol. 97: 760–766.

Cloux-Fey U., Gloor, B., Jaeggi, P. and Hendrickson, P. (1986). Papille und Gesichtsfeld beim Glaukom. Klin. Monatsbl. Augenheilkd. 189: 92–103.

De Bruyn, B. and Orban, G.A. (1988). Human velocity and direction discrimination measured with random dot patterns. Vision Res. 28: 1323–1335.

Drance, S.M. and Airaksinen, P.J. (1986). Signs of early damage in open angle glaucoma. In: Weinstein, G.W. (ed.), Contemporary Issues in Ophthalmology: Open Angle Glaucoma: 17–30 Churchill Livingstone, New York.

Drance, S.M., Airaksinen, P.J., Price, M., Schulzer, M., Douglas, G.R. and Tansley, B. (1987). The use of psychophysical structural and electro-diagnostic parameters to identify glaucomatous damage. Graefes Arch. Clin. Exp. Opthalmol. 225: 365–368.

Drance, S.M., Lakowski, R., Schulzer, M. and Douglas, G.R. (1981). Acquired color vision changes in glaucoma: Use of the 100-hue test and Pickford anomaloscope as predictors of glaucomatous field change. Arch. Opthalmol. 99: 829–831.

Drum., B., Armaly, M.F. and Huppert, W. (1986). Scotopic sensitivity loss in glaucoma. Arch. Ophthalmol. 104: 712–717.

Drum, B. and Bissett, R. (1991). Optimizing dot size and contrast in pattern discrimination perimetry. In: Mills, R.P. and Heijl, A. (eds.), Perimetry Update 1990/91. Proceedings of the IXth International Perimetric Society Meeting, Malmö, Sweden, 1990: 373–380, Amsterdam, New York.

Drum, B., Severns, M., O'Leary, D., Massof, R., Quigley, H., Breton, M. and Krupin, T. (1989). Pattern discrimination and light detection test different types of glaucomatous damage. In: Heijl, A. (ed.), Perimetry update 1988/89. Proceedings of the VIIIth International Perimetric Society Meeting, 341–346, Amsterdam, Berkeley, Milano.

Enoch, J.M., Fitzgerald, C.R. and Campos, E.C. (1981). Quantitative layer-by-layer perimetry. Current ophthalmology monographs. Grune and Stratton, New York.

Falcao-Reis, F., O'Donoghue, E., Buceti, R., Hitchings, R.A. and Arden, G.B. (1990). Peripheral contrast sensitivity in glaucoma and ocular hypertension. Br. J. Ophthalmol. 74: 712–716.

Falcao-Reis, F.M., O'Sullivan, F., Spileers, W., Hogg, C. and Arden, G.B. (1991). Macular colour contrast sensitivity in ocular hypertension and glaucoma: evidence for two types of defect. Br. J. Ophthalmol. 75: 598–602.

Flammer, J., Drance, S.M. (1984). Correlation between color vision scores and quantitative perimetry in suspected glaucoma. Arch. Ophthalmol. 102: 38–39.

Gloor, B. (1990). Verlaufskontrolle beim Glaukom. Fortschr. Ophthalmol. 87: 163–171.

Glovinsky, Y., Quigley, H.A. and Dunkelberger, G.R. (1991). Retinal ganglion cell loss is size dependent in experimental glaucoma. Invest. Ophthalmol. Vis Sci. 32: 484–491.

Glovinsky, Y., Quigley, H.A., Drum, B., Bissett, R.A. and Jampel, H.D. (1992). A whole-field scotopic retinal sensitivity test for the detection of early glaucoma damage. Arch. Ophthalmol. 110: 486–490.

Glovinsky, Y., Quigley, H.A. and Pease, M.E. (1993). Foveal ganglion cell loss is size dependent in experimental glaucoma. Invest. Ophthalmol. Vis. Sci. 34, no. 2:395–400.

Gorea, A. and Papathomas, T.V. (1989). Motion processing by chromatic and achromatic visual pathways. J. Opt. Soc. Am. A. 6: 590–602.

Grützner, P. and Schleicher, S. (1972). Acquired color vision defects in glaucoma patients. Mod. Probl. Ophthalmol. 11: 136–140.

Gündüz, K., Arden, G.B., Perry, S., Weinstein, G.W. and Hitchings, R.A. (1988). Color vision defects in ocular hypertension and glaucoma. Arch. Ophthalmol. 106: 929–935.

Iwata, K. (1989). Ophthalmoscopy in the detection of optic disc and retinal nerve fiber layer changes in early glaucoma (Summary). Surv. Ophthalmol. (Suppl.) 33: 447–448.

Johnson, C.A., Adams, A.J., Casson, E.J. and Brandt, J.D. (1993). Bluw-on-yellow perimetry can predict the development of glaucomatous visual field loss. Arch. Ophthalmol. 111: 645–650.

Korth, M. and Horn, F. (1990). Helligkeitskontrast- und Farbkontrastmuster-evozierte Elektroretinogramme von normalen und Glaukomaugen. Fortschr. Ophthalmol. 87: 403–408.

Lachenmayr, B.J. and Drance, S.M. (1992). Diffuse field loss and central visual function in glaucoma. German J. Ophthalmol. 1: 67–73.
LeCam, L. (1990). Maximum Likelihood: An introduction. International Statistical Review 58: 153–171.
Livingstone, M.S. and Hubel, D.H. (1987). Psychophysical evidence for separate channels for the perception of form, color, movement and depth, J. Neurosci. 7: 3416–3468.
Livingstone, M.S. and Hubel, D.H. (1988). Segregation of form, color, movement and depth: anatomy, physiology and perception. Science 240: 740–749.
Marmion, V.J. (1987). The colour vision deficiency in open angle glaucoma. Mod. Probl. Ophthalmol. 19: 305–307.
Perry, H.G. and Cowey, A. (1985). The ganglion cell and cone distributions in the monkey's retina: Implications for central magnification factors. Vision Res. 25: 1795–1810.
Quigley, H.A., Dunkelberger, G.R. and Green, W.R. (1988). Chronic human glaucoma causing selectively greater loss of large optic nerve fibers. Ophthalmology 95: 357–363.
Quigley, H.A., Dunkelberger, G.R. and Green, W.R. (1989). Retinal ganglion cell atrophy correlated with automated perimetry in human eyes with glaucoma. Am. J. Ophthalmol. 107: 453–458.
Quigley, H.A., Sanchez, R.M., Dunkelberger, G.R., L'Hernault, N.L. and Baginski, T.A. (1987). Chronic glaucoma selectively damages large optic nerve fibers. Invest. Ophthalmol. Vis. Sci. 28: 913–920.
Sample, P.A., Boynton, R.M. and Weinreb, R.N. (1988a). Isolating the color vision loss in primary open-angle glaucoma. Am. J. Ophthalmol. 106: 686–691.
Sample, P.A., Esterson, F.D., Weinreb, R.N. and Boynton, R.M. (1988b). The aging lens: *In vivo* assessment of light absorption in 84 human eyes. Ophthalmol. Vis. Sci. 29: 1306–1311.
Sample, P. A. and Weinreb, R.N. (1990). Color perimetry for assessment of primary open angle glaucoma. Invest. Ophthalmol. Vis. Sci. 31: 1869–1875.
Sample, P.A. and Weinreb, R.N. (1992). Progressive color visual field loss in glaucoma. Invest. Ophthalmol. Vis. Sci. 33: 2068–2071.
Sample, P.A., Taylor, J.D.N., Martinez, G.A. Lusky, M. and Weinreb, R.N. (1993). Short-wavelength color visual fields in glaucoma suspects at risk. Am. J. Ophthalmol. 115: 225–233.
Schiller, P.H., Logothetis, N.K. and Charles, E.R. (1990). Functions of colour-opponent and broad-band channels of the visual system. Nature 343: 68–70.
Silverman, S.E., Trick, G.L. and Hart, W.M., Jr (1990). Motion perception is abnormal in primary open-angle glaucoma and ocular hypertension. Invest. Ophthalmol. Vis. Sci. 31: 722–729.
Trick, G.L., Nesher, R., Cooper, D.G., Kolker, A.E. and Bickler-Bluth, M. (1988). Dissociation of visual deficits in ocular hypertension. Invest. Ophthalmol. Vis. Sci. 29: 1486–1491.
Troscianko, T. and Fahle, M. (1988) Why do isoluminant stimuli appear slower? J. Opt. Soc. Am. A. 5: 871–880.
Wilensky, J.T., Podos, S.M. and Becker, B. (1974) Prognostic indicators in ocular hypertension. Arch. Ophthalmol. 91: 200–202.
Yablonsky, M., Zimmerman, T., Kass, M. and Becker, B. (1980). Progressive significance of optic disk cupping in ocular hypertensive patients. Am. J. Ophthalmol. 89: 585–592.
Yu, T.C., Falcao-Reis, F., Spileers, W. and Arden, G.B. (1991). Peripheral color contrast. Invest. Ophthalmol. Vis. Sci. 32: 2779–2789.
Zrenner, E. (1983). Neurophysiological aspects of color vision in primates. Springer, New York, Berlin, Heidelberg.
Zrenner, E. and Gouras, P. (1981). Characteristics of the blue sensitive cone mechanism in primate retinal ganglion cells. Vision Res. 21: 1605–1609.

Eberhart Zrenner
Department of Pathophysiology of Vision and Neuro-Ophthalmology
University Eye Hospital
72076 Tübingen, Germany

11. Visual functions in a patient with acquired achromatopsia

A.B. MORLAND, C. KENNARD, M. LAWDEN and K.H. RUDDOCK
(London, U.K.)

Abstract

We describe an investigation of visual responses in a patient, whose MRI brain scans reveal bilateral lesions affecting the fusiform and lingual gyri. Subjectively, he reports that colours appear abnormal to him and his descriptions of surface colours are similar to those of other patients who suffer cerebral achromatopsia. His colour matching and two colour increment threshold data establish that his cone photoreceptors have normal spectral responses, and his spectral sensitivity measured against a white background exhibits normal colour-opponent characteristics. His colour discriminations are, however, impaired but other visual response characteristics, including lightness discrimination, velocity discrimination, stereoscopic fusion and binocular rivalry are normal. He does, however, suffer from two neurological conditions, prosopagnosia and topographical agnosia, commonly found in association with cerebral achromatopsia.

Introduction

Impairment of colour vision associated with cortical lesions has been described in a number of reports published over the last hundred or so years, and these have been recently reviewed by Meadows (1974), Zeki (1990) and Plant (1991). Patients with this condition, which is known as achromatopsia, frequently misidentify colours and describe them as 'dull', 'grey' or 'washed out' in appearance. More formal examination reveals numerous errors made in performing the Farnsworth-Munsell 100 hue discrimination test, and various other abnormal chromatic responses have also been described. One important question which has received particular attention concerns the extent to which the impairment of colour vision in achromatopsia is dissociated from disturbances of other visual functions. Prosopagnosia, topographical agnosia and altitudinal visual field losses occur frequently, but not invariably, in patients with achromatopsia, and it has been established that other visual

B. Drum (ed.), Colour Vision Deficiencies XII, pp. 87–94.
© 1995 *Kluwer Academic Publishers, Dordrecht.*

capacities, such as lightness discrimination, motion sensitivity and spatial resolving power, are essentially unimpaired. Brain scans and autopsies both indicate that lesions which gives rise to achromatopsia involve damage to the fusiform and lingual gyri of the ventro-medial occipital lobe.

In this paper, we present the results of an experimental study on chromatic and other visual responses of a patient, BL, who has well established lesions involving the fusiform and lingual gyri. We report that BL responds normally in several tests of non-chromatic visual function, but his colour discriminations are impaired and his identification of colours is both inaccurate and unstable under changes of illumination.

Methods

Colour matching and wavelength discrimination were investigated with the tristimulus colorimeter designed by Wright (1946), fitted where required with a large field attachment (Clarke, 1963). Spectral sensitivies were recorded with a Maxwellian view optical system (Barbur and Ruddock, 1980), the collimated beams of which were passed through interference filters (Balzer B40) to obtain quasi-spectral stimuli. Threshold sensitivities were determined by a staircase method, to an accuracy of 0.03. Saturation discrimination relative to a white reference (chromaticity co-ordinates $x = 0.305$; $y = 0.323$) were measured by the method described by Barbur *et al.* (1992). Lightness discrimination was measured with a set of 46 ceramic tiles, the luminance factors of which ranged from 1% to 82% (Robertson and Wright, 1965). Surface colours were calibrated *in situ* with a telespectroradiometer (Gamma Scientific).

Subjects

We present data for BL, a 54-year-old male, who reported complete loss of colour vision after a period of unconsciousness caused by encephalitic illness of viral origin. Over the following twenty months he achieved partial recovery of colour vision, but still complains that colours 'no longer appear as I remember them'. We have examined BL regularly over the last 14 months, during which period his visual responses have remained consistent. Comparison data are given for a normal control, JW, who is a 25-year-old female, and for CT, a 35-year-old male with congenital red-green dichromacy (deuteranopia).

Results

Ophthalmic examination

The examination revealed that the patient's retinae were entirely normal in appearance.

MRI scans

The scans with a GEC Signa 1.5 Tessla scanner showed bilateral infarctions involving the infracalcarine lobes. The infarcts predominantly involve the fusiform gyri, but extend into the lingual gyri and the posterior part of the parahippocampal gyri. The calcarine fissures are, however, spared.

Visual fields

These, as assessed with a Goldmann perimeter, are normal except for a scotoma in the upper left quadrant, homonymous for the two eyes, observed only under static presentation of the smallest ($1/4mm^2$) field.

Spatial vision

BL's pin hole visual acuity is 6/9 in either eye and his descriptions of a number of illusory patterns, including those due to Ehrenstein, Kanisza, Müller-Lyer, Poggendorff and Zöllner (Gregory, 1970), are normal. He also perceives the apparent changes in orientation and periodicity of a grating induced by appropriate pre-adaptation (Gibson and Radner, 1937; Blakemore and Sutton, 1969). He does, however, suffer mild prosopagnosia and topographical agnosia.

Binocular vision

BL has normal stereoscopy as measured with Julesz (1971) random dot anaglyphs; he detects binocular rivalry between crossed red and green gratings viewed through colour separation filters and he perceives the reversal of apparent depth in the Necker cube (Gregory, 1970).

Velocity discrimination

Velocity discrimination was measured for a pair of moving spots (3.5 deg diameter) presented sequentially, the first travelling at constant velocity, 48 deg s^{-1}, and the second at different velocities v, in random order. He achieved above 95% correct discrimination of the velocity difference when v differed from v_0 by ± 10 deg s^{-1}, which is normal performance (Barbur *et al.,* 1980).

Lightness discrimination

Lightness discrimination was assessed by asking BL to arrange 46 grey tiles in order of apparent brightness (see Methods). The close spacing of their luminance factors produces, on average, three or four reversals of the correct order in sequences made by normal subjects, and BL's performance was very similar.

BL responded slowly to the Ishihara psudoisochromatic plates, but made only between 0 and 3 errors on different occasions. His error scores on different occasions with the FM 100 hue test was between 400 and 600, and he commented that he judged the sequences mainly on *perceived* differences in brightness. His just noticeable wavelength discrimination steps for a 1° 20′ bipartite square field were about five times the normal size and his saturation discrimination steps were also greater than the corresponding normal values, but with the same axis of elongation in the chromaticity chart (Fig. 1). His Rayleigh colour match is normal, albeit with an abnormally large range, and he accepts the normal match. He was unable to establish an accurate match with a blue-green test, but we were able to isolate a blue-sensitive Π_3 spectral response (Stiles, 1978). BL's spectral sensitivity measured against a white background (Sperling and Harwerth, 1971; King-Smith and Carden, 1976) was normal, and showed clearly the maxima and minima attributed to detection by post-receptoral, colour-opponent mechanisms. BL's colour naming for surface

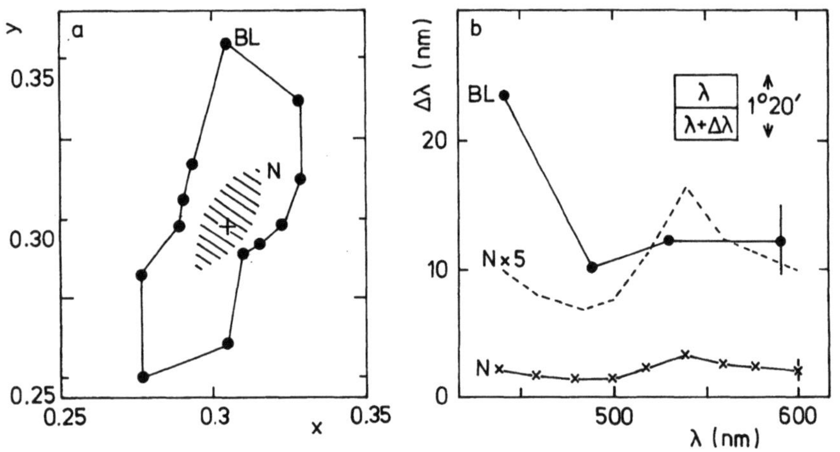

(a) (b)

Fig. 1. Colour discrimination data for normal vision (N) and for the patient BL. (a) Saturation discrimination values plotted on the C.I.E. 2° x-y chromaticity diagram. The full circles denote the chromaticities of colours which BL can just discriminate from a reference white, the chromaticity co-ordinates of which are denoted by the central cross. The hatched area denotes the normal data. The target consisted of vertical bars (width 15′) which occupied the central 4° × 4° of the field (20° vertical × 16° horizontal) and was presented for 0.5 s. The screen luminance was 34 cd m^{-2} and the target grating was masked by luminance noise. Thresholds were determined by a staircase method (Barbur *et al.*, 1992). (b) Wavelength discrimination steps,.Δλ, measured by a forced choice 'same' or 'different' discrimination between the two halves of a 1° 20′ square field which were matched for apparent brightness at all wavelength settings. The error bars show the range of settings for the discrimination steps, Δλ, made at different settings, which were approximately equal for a given observer at all wavelengths λ.

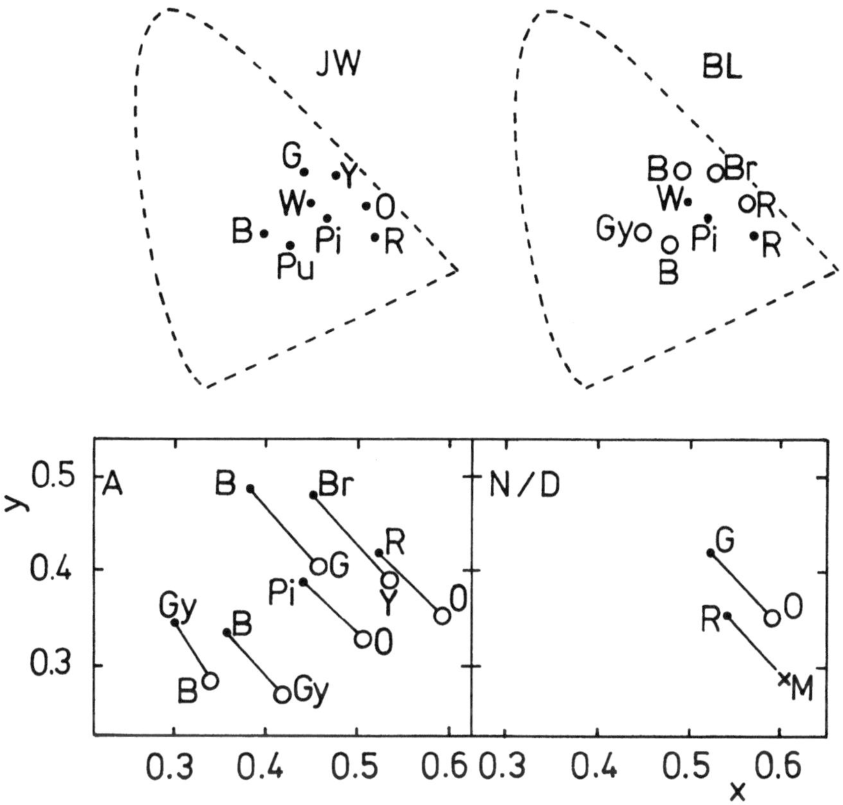

Fig. 2. Colour naming data for a Mondrian pattern (16° vertically by 20° horizontally) which was composed of 20 patches, with each of eight individual colours represented by two or three patches. The colour names selected by each of the three subjects were consistent for patches of a given colour viewed under a given illuminant. The full C.I.E. diagrams show the chromaticity co-ordinates under white light (3100 K) for each of the eight colours which made up the Mondrian. The names chosen by the normal, JW, and the patient BL are denoted by the letters (see below for code). For the patient BL, open circles denote colours given names different from, and full circles given names the same as those chosen by the normal. The deuteranope chose the same names as the normal, except that the patches called 'orange' (O) by the normal were called 'green' by the deuteranope.
The lower panels represent portions of the C.I.E. 2° x-y chromaticity diagram. Points plotted as full circles denote chromaticity co-ordinates of the Mondrian patches under white light, and other symbols those under a 'red' illuminant. Pairs of symbols connected by lines refer to the same patch, and the letters denote the names chosen to describe the patches. Only those colours for which the chosen colour name changed with change in illuminant are shown. The left hand panel, A, refers to the achromatopsic patient BL, and the right hand panel, N/D, to the normal (cross) and the deuteranope (open circle). *Code to colour names* B, blue; Br, brown; G, green; Gy, grey; M, maroon, O, orange; Pi, pink; Pu, purple; R, red; W, white; Y, yellow.

colours was inaccurate, particularly for yellows and greens, but was always correct for reds and whites. The names given by BL to the component patches for a Mondrian figure, viewed under two different illuminants, are shown in Fig. 2, together with comparison data for a normal control and for a deuteranope.

Discussion

The principal features of BL's chromatic responses are characteristic of those associated with achromatopsia and the lesions revealed by MRI scanning are consistent with the findings for other achromatopsic patients. BL's Rayleigh colour match is normal, which implies that both the red and green sensitive photopigments have normal spectral sensitivities, and isolation of Π_3 shows that his blue-sensitive cones also have normal spectral sensitivity. Normal trichromatic colour vision was demonstrated in one achromatopsic patient by Mollon *et al* (1980) but it was reported that the blue sensitive mechanism was absent in two other patients (Pearlman *et al.,* 1979; Young and Fishman, 1980). BL's spectral sensitivity measured with a white background is normal, which establishes that the outputs from the photoreceptors combine to give colour-opponent response channels.

BL responds abnormally in colour discrimination tasks, including the FM 100-hue test, wavelength discrimination and saturation discrimination (Fig. 1), and his colour naming is inaccurate for certain groups of colours (Fig.2). Impaired colour discrimination establishes that the visual pathways are directly implicated in the functional abnormalities, and that they cannot be attributed to a linguistic dysfunction of the kind that gives rise to colour anomia (Meadows, 1974).

BL's colour naming is accurate for reds, pinks and white (Fig. 2) and he is confident of his ability to identify these colours. The instability in BL's colour naming which accompanies a change in illumination of the Mondrian figure contrasts with the highly stable responses given by a normal trichromat and a deuteranope (Fig. 2). The data illustrated refer to a change in illuminant which produces shifts in chromaticity in a direction almost parallel to the red-green arm of the spectrum locus, and close to that corresponding to the deuteranopic confusion lines (Wright, 1946, Chapter 26). BL's discrimination in this part of the spectrum is sub-normal, but much superior to that of the deuteranope, whereas his colour naming is much less stable under change of illumination than that of the normal or of the deuteranopic subject. We conclude that BL's colour naming performance cannot be attributed to his impaired colour discrimination. The pre-striate area V4 of the macaque monkey contains cells which are responsive to colour appearance, regardless of the spectral composition of the physical light stimulus, and thus they express colour constancy in response to changes in illumination (Zeki, 1978, 1980, 1990). If the colour sensitive complex in the fusiform and lingual gyri in man (Lueck *et al.,* 1989) is, indeed, analogous to area V4 in monkey, then lesions of the kind suffered by BL should disrupt colour constancy, leading to response instability of the kind seen in Fig. 2. We note that BL's responses for other changes in illumination which produce displacements in other directions across the chromaticity chart, are similar to those illustrated in this paper.

BL displays homonymous altitudinal losses in the left hemifield, mild prosopagnosia and topographical agnosia, all of which are characteristic symptoms accompanying achromatopsia. Nonetheless, he responds normally in a variety of visual tasks which test spatial vision and binocularity. He also has normal velocity discrimination, and Rizzo *et al.* (1992) also found that responses to moving targets were normal in achromatopsia. Our test of lightness discrimination was performed with a series of neutral samples which were separated by lightness steps of magnitude similar to, and in some cases smaller than, that required for visual discrimination. Consequently, normal subjects make some errors in arranging the samples according to lightness and BL's performance indicates that he has normal lightness discrimination, in marked contrast to his responses on colour discrimination tasks. Heywood *et al.* (1987) and Rizzo *et al.* (1993) both report that lightness discrimination is preserved in achromatopsia.

We conclude that BL suffers achromatopsia caused by cortical lesions, which is manifested in impaired colour discriminations and colour naming which is inaccurate and unstable under change of illumination. He exhibits certain other neurological losses, specifically prosopagnosia and topographical agnosia, which are frequently observed in association with achromatopsia. Nonetheless, his normal responses in a variety of tests involving non-colour specific aspects of vision demonstrate that his lesion is highly selective in its effects on visual function.

Acknowledgements

We are grateful to the patient, BL, who performed the observations which form the basis of this report. We are also grateful to the Wellcome Trust for the award of research grants to CK and KHR which support this study.

References

Barbur, J.L. and Ruddock, K.H. (1980). Spatial characteristics of movement detection mechanisms in human vision I Achromatic vision. Biological Cybernetics 37: 77–92.

Barbur, J.L., Ruddock, K.H. and Waterfield, V.A. (1980). Human visual responses in the absence of the geniculo-calcarine projection. Brain 103: 906–928.

Barbur, J.L., Birch, J. and Harlow, A.J. (1992). Colour vision testing using spatiotemporal luminance masking: Psychophysical and pupillometric methods. In: Drum, N. (ed.), Colour Vision Deficiencies XI, Doc. Ophthalmol. Proc. Ser. 56: 417–426, Kluwer Acad. Publ., Dordrecht.

Blakemore, C. and Sutton, P. (1969). Size adaptation: a new after effect. Science N.Y. 166: 245–247.

Clarke, F.J.J. (1963). Further studies of extra-foveal colour metrics. Optica Acta 10: 257–284,

Gibson, J.J. and Radner, M. (1937). Adaptation, after effects and contrast in the perception of tilted lines. J. exp. Psychol. 20: 453–467.

Gregory, R.L. (1970). The Intelligent Eye. Weidenfeld and Nicolson, London.

Heywood, C.A., Wilson, B. and Cowey, A. (1987). A case study of cortical colour 'blindness' with relatively intact achromatic discrimination. J. Neurol., Neurosurg. and Psychiatry 50: 22–29.

Julesz, B. (1971). Foundations of Cyclopean Perception. University Press, Chicago.

King-Smith, P.E. and Carden, D. (1976). Luminance and opponent-color contributions to visual detection and adaptation and to temporal and spatial integration. J. Opt. Soc. Am. 66: 709–717.

Lueck, C.J., Zeki, S., Friston, K.J., Dieber, M.P., Cope, P., Cunningham, V.J., Lammerstsina, A.A., Kennard, C. and Frakowiak, R.S.J. (1989). The colour centre in the cerebral cortex of man. Nature (London) 340: 386–389.

Meadows, J.C. (1974). Disturbed perception of colours associated with localized cerebral lesions. Brain 97: 615–632.

Mollon, J.D., Newcombe, F., Polden, P.G. and Ratcliff, G. (1980). On the presence of three cone mechanisms in a case of total achromatopsia. In: Verriest, G. (ed.), Colour Vision Deficiencies V: 130–135, Hilger, Bristol.

Pearlman, A.L., Birch, J. and Meadows, J.C. (1979). Cerebral color blindness; an acquired defect in hue discrimination, Annals of Neurology 5: 253–261.

Plant, G.T. (1991). Disorders of colour vision in diseases of the nervous system. In: Foster, D.H. (ed.), Vision and Visual Dysfunction 7. Inherited and Acquired Colour Vision Deficiencies: 173–198, Macmillan, Basingstoke.

Rizzo, M., Nawrot, M., Blake, R. and Damasio, A. (1992). A human visual disorder resembling area V4 dysfunction in the monkey. Neurology 42: 1175–1180.

Rizzo, M., Smith, V., Pokorny, J. and Damasio, A.R. (1993). Color perception profiles in central achromatopsia. Neurology 43: 995–1001.

Robertson, A.R., and Wright, W.D. (1965). International comparison of working standards for colorimetry. J. Opt. Soc. Am. 55: 694–706.

Sperling, H.E. and Harwerth, R.S. (1971). Red-green cone interactions in the increment-threshold spectral sensitivities of primates. Science N.Y. 180–184.

Stiles, W.S. (1978). Mechanisms of colour vision. Academic, London.

Wright, W.D. (1946). Researches on Normal and Defective Colour Vision. Kimpton, London.

Wright, W.D. (1952). The characteristics of tritanopia. J. Opt. Soc. Am. 42: 509–521.

Young, R.S. and Fishman, G.A. (1980). Loss of color vision and Stiles π_1 mechanism in a patient with cerebral infarction. J. Opt. Soc. Am. 70: 1301–1305.

Zeki, S.M. (1978). Uniformity and diversity of structure and function in rhesus monkey prestriate cortex. J. Physiol. (London). 227: 275–290.

Zeki, S. (1980). The representation of colours in the cerebral cortex. Nature (London). 284: 412–418.

Zeki, S. (1990). A century of cerebral achromatopsia. Brain 113: 1721–1777.

K.H. Ruddock
Biophysics Section
Physics Dept.
Imperial College
London SW7 2BZ, U.K.

12. The effects of sub-cortical and cortical damage on colour vision

VINCENT WALSH and JANUS J. KULIKOWSKI
(Oxford and Manchester, U.K.)

Abstract

Fundamentally different colour vision deficiencies result from the damage to retinal and cortical stages of colour processing. Damage to colour-opponent retinal units, at the ganglion cells or optic nerve fibres, abolishes all aspects of colour vision. Conversely, cortical lesions of primate visual area V4 affect only one aspect of colour vision, namely colour constancy leaving colour discrimination and simple categorization intact. The chromatic visual evoked potentials in macaques with bilateral V4 lesions are normal.

It is argued on the basis of these differences that wavelength information can bypass area V4, that constancy functions should be different according to categorical location of the stimuli, and that V4 contains a colour constancy centre. This suggests both hierarchical and parallel nature of colour analysis: damage to a high-level centre destroys only its function, and does not obstruct the flow of information about lower-level colour vision.

We discuss the anatomical basis of these findings, as well as similarities and differences between the results of experiments on monkeys and on brain damaged human subjects.

Introduction

In this paper we outline the anatomy and physiology of colour vision and discuss the dysfunctions of colour vision caused by lesions to various stages along the visual pathway. The main body of the paper is devoted to a review of the most important lesion studies of colour vision. On the basis of some recent findings it is argued that categorization of hues is achieved by post-geniculate mechanisms operating in areas V1 (striate cortex) and V2, and that V4 is important for normal colour constancy. These effects of cortical lesions are contrasted with the effects of pre-chiasmal (eye-specific) damage which have less specific effects and in the extreme abolish all aspects of colour vision.

B. Drum (ed.), Colour Vision Deficiencies XII, pp. 95–108.
© 1995 *Kluwer Academic Publishers, Dordrecht.*

Anatomy and physiology of colour processing

There is good agreement that post-receptoral chromatic information processing is subserved by P-neurons in the dorsal, parvocellular layers of the LGN, whose responses show spectral opponency. The inputs to these cells are the optic nerve fibres (P) which are thin and originate mostly from the P-beta retinal ganglion cells which, relative to P-alpha cells, have small and compact dendritic fields, or from 'bistratified' cells (cf. Leventhal *et al.,* 1981 vs. Rodieck, 1991; and Calkins *et al.,* Chapter 32, this volume). The majority of chromatic-opponent cells typically have a cone-opponent receptive field organization: most of these have a concentric centre-surround (Type 1) receptive field organization; others (Type 2) are centre-only with segregated cone inputs (Reid and Shapley, 1992) and may have 'bistratified' dendritic structure. Relative to P-alpha cells (which project mainly to the two ventral laminae of LGN and sparsely to the midbrain (Leventhal *et al.,* 1981), all chromatic-opponent cells and fibres have low luminance contrast gain (low achromatic contrast sensitivity), moderate temporal resolution, sustained (tonic) responses to long temporal presentations, slow conduction velocities, and linear spatial summation (Kolb and Lipetz, 1991). On the other hand, the responses of cells in the ventral, magnocellular laminae lack wavelength opponency, but are very sensitive to changes in luminance contrast, show transient response profiles, moderate spatial resolution, high temporal resolution and fast conductance velocities (Lee, 1991).

To avoid caricaturing the segregation of pathways some caveats must be given. For example, most reviews stress that the projections from the four dorsal, parvocellular, P, and from two magnocellular, M, laminae of LGN to V1 are segregated. The main P-projection targets the lamina 4Cbeta (but also laminae 1, 4A and 6). The M projections to V1 are to lamina 4Calpha (and also but less so to lamina 6). However, the recipient neurons in these laminae have dendritic arbors not confined to beta and alpha sub-laminae, thereby suggesting some overlap (Lund, 1988, 1991), rather than strictly segregated P- and M-streams.

The distributions of temporal response properties of P- and M-LGN cells (and even more so those of their striate recipients) do overlap (see Kaplan *et al.,* 1989 for review) especially when luminance modulation is used, as though pattern analysis was adapted to a broad range of temporal frequencies. However, optimal temporal frequencies for both P- and M-cells (Hicks *et al.,* 1983) tend to be higher than psychophysical limits, e.g. P-LGN cells respond to chromatic flicker beyond the psychophysical resolution limit of the chromatic system (Lee, 1991).

The substrate colour processing in V1 and V2 is shown by staining with the metabolic enzyme cytochrome oxidase (CO). In layers 2 and 3, and to a lesser extent in laminae 5 and 6, there are pronounced blobs of CO-staining both in monkey (Horton and Hubel, 1981) and in human post-mortem brains (Burkhalter and Bernardo, 1989; Hockfield *et al.,* 1990).

The receptive fields of cells in the CO blobs are monocular, wavelength selective, and 'orientation blind' (Dow, 1974; Gouras and Kruger, 1979; Livingstone and Hubel, 1984a; Michael, 1989; Ts'o, 1989), whereas cells in interblob regions are orientation selective but do not have conspicuous wavelength opponency. Cells aligned with the blobs, in lamina 4C, have blob-like properties (Ts'o and Gilbert, 1988), suggesting a columnar organization of colour in V1 (Dow, 1974; Michael, 1981).

The columns of wavelength selective cells appear to analyze the outputs of only one of the two opponent systems. Thus there are red/green (R/G) columns and blue/yellow (B/Y) columns, but there are no mixed R/G and B/Y columns (Vautin and Dow, 1985; Dow and Vautin, 1987; Ts'o and Gilbert, 1988; Ts'o, 1989; Dow, 1991). The ratio of R/G to B/Y columns is about 3:1 (Ts'o and Gilbert, 1988), a finding presaged by Michael (1978b). There is also a horizontal organization of colour and orientation processing in V1, and the dominant principle appears to be 'like connects with like'. Thus blobs tend to connect with blobs and interblobs with interblobs (Livingstone and Hubel, 1984b; McGuire *et al.*, 1985). More specifically, R/G CO blobs talk to other R/G blobs and B/Y blobs talk to other B/Y blobs, via what appear to be 'bridges' of CO (Ts'o, 1989). Unfortunately, this monocular organization fails to explain some perceptual phenomena in which same colours from two eyes are combined and suppress the opponent colour (Kulikowski, 1991b).

Some chromatic responsive cells in V1 can be classified according to the system developed for LGN cells (Wiesel and Hubel, 1966). Michael (1978a) and Livingstone and Hubel (1984a) reported that the majority of chromatically responsive cells in V1 blobs were double opponent. Ts'o and Gilbert (1988; see also Creutzfeldt *et al.*, 1987), however, found only 1% of their sample to be double opponent cells, and 26% to be modified type 2 cells which have centre colour opponency and broadband surround spatial opponency. Kulikowski *et al.*, (1989a) have noted that some colour-opponent cells in V1 are difficult to classify unambiguously due to their low levels of spontaneous activity and poor OFF responses. It may be, then, that one man's Modified Type 2 is another man's double-opponent cell. Nevertheless, all V1 cells with double chromatic-opponency in the receptive field centre show uniform (slow) temporal properties, consistent with psychophysics (Kulikowski and Walsh, 1993).

The chromatic-blob/achromatic-interblob dichotomy in V1 is not absolute but some colour/orientation integration occurs in V1: A fraction of cells (usually borders of the CO blobs) have been found which have combined chromatic and orientation selectivity (Ts'o and Gilbert, 1988; Michael, 1989), and the anatomical substrates of P and M interaction are beginning to be identified in detail (Yoshioka and Lund, 1991; Kritzer *et al.*, 1992).

Area V1 sends direct projections to several extra-striate areas. In V1, CO-dense areas called Thick Stripes receive magno-inputs, whereas the blob regions of V1 project to Thin Stripes, and the interblob regions to the inter-stripes (Livingstone and Hubel, 1983; Zeki and Shipp, 1988). Thus some segregation of colour, form and movement seen in V1 appears to be maintained in the CO architecture of area V2 (Hubel and Livingstone, 1987; Zeki and Shipp, 1989). Chromatic response properties of V2 cells (Yoshioka *et al.*, 1988) are particularly interesting: some cells appear to respond to combinations of wavelength and luminance, thus one finds cells responding to brown (a contrast colour), but not to yellow or orange, and cells responding to maroon or olive, but not to red or green. The important point about these results is that they demonstrate the existence of a level of analysis in between simple wavelength processing and colour constancy. This level, as we shall argue later, is categorization.

Cortical visual area V4 receives its major input from the CO thin and inter-stripes of V2 (Zeki and Shipp, 1988). In contrast to the cells in V3 and V5, visual area V4 shows a marked absence of motion-specific cells, but a great number of cells concerned with analyzing information about form and colour. The most interesting property of cells in area V4 is that of colour constancy – the ability to identify a coloured surface despite changes in the spectral distribution of the illuminant (Zeki, 1983) a property which has not been demonstrated in any other area.

Effects of sub-cortical damage on colour vision

The sub-cortical substrates of most interest are the tonic retinal ganglion cells, their optic nerve fibres, and the lateral geniculate nucleus.

Lesions of primate retinal ganglion cells

Lesions of a population of retinal ganglion cells have been carried out by dosing monkeys with acrylamide monomer which, unlike in the rat (Wild and Kulikowski, 1984) is said to cause complete and selective degeneration of P-cells in the monkey LGN (see Merigan, 1991). Acrylamide poisoning does not discriminate between colour and pattern/texture processing units; it leads to a decrease in spatial resolution, a decrease in contrast sensitivity for low spatial and high temporal frequencies (Merigan and Eskin, 1986) and a 2 log unit decrement in chromatic contrast sensitivity (Merigan, 1989, 1991).

Another, often overlooked, effect of acrylamide poisoning is the disruption of linear responses which may include M as well as P cells (Vidyasagar, 1981). These effects are similar to the disruption of linear summation of colour-opponent ganglion cells which follows administration of the anti-tuberculosis drug Ethambutol (Zrenner and Kruger, 1981; van Dijk and Spekreijse, 1984).

Lesions of primate lateral geniculate nucleus

Studies of the effects of LGN lesions on visual performance are broadly, but not entirely predictable from the anatomical and physiological evidence. For example, ibotenic acid lesions of the magnocellular layers of macaque LGN severely impair contrast sensitivity to a low spatial frequency, achromatic, drifting grating, and raise thresholds for a 1 c/degree drifting grating and flicker detection, whereas contrast sensitivity to a less coarse stationary grating (2 c/degree) is unaffected (Merigan and Maunsell, 1990; Merigan, 1991), but magno-LGN lesions may underestimate the role of this system since some retinal P-alpha cells project outside the LGN. Lesions to the parvocellular laminae of macaque LGN lead to deficits in colour discrimination, high spatial frequency form processing and fine stereopsis (Schiller and Logothetis, 1990; Merigan, 1991; Schiller, 1991). These studies also found that neither magno nor parvo LGN lesions disrupted low spatial frequency shape perception, coarse stereopsis or brightness perception – these functions can be mediated equally well by either system. Moreover, P and M systems cannot be easily segregated by using isoluminant stimuli on the grounds that magnocellular units are silenced at isoluminance (Hubel and Livingstone, 1987; Livingstone and Hubel, 1987, 1988) since Schiller *et al.*, (1991) found that deficits consequent upon lesions to the parvo or magno laminae of LGN were not selectively affected at isoluminance, and that single unit activity is compromised both in the parvo and magno systems. Only low contrast stimuli can silence M-system at isoluminance (Kulikowski, 1991).

The overall picture of the pre-cortical retino-geniculate visual system has been well summarized by Merigan (1991, p. 118): 'with the exception of colour vision, which is dominated by the P-pathway, the P and M pathways differ primarily in the range of spatio-temporal frequencies they provide to the visual cortex. This approach suggests that potentially modular functions of visual cortex such as shape recognition, motion perception, or stereopsis can be done without either the P or the M pathway if the stimuli are chosen to fall within the spatio-temporal range of the remaining pathway'.

Precortical lesions in humans

The dysfunctions which have proven most informative about colour processing are those consequent upon damage to the thin optic nerve fibres (mostly Parvo in origin) and wavelength selective retinal ganglion cells. Two groups involving such optic atrophies have been studied in detail: cases of Diabetes Insipidus Diabetes Mellitus Optic Atrophy and nerve Deafness (DIDMOAD) and several groups of Hereditary Optic Atrophy (HOA). In both DIDMOAD and extreme groups of HOA there is no evidence for functioning (tonic) parvo-cells manifested by the absence of perception of after-images or images with persistent fixation and no colour discrimination shown by the random score on the 100-hue test and only the achromatic (luminosity-like) spectral sensitivity (King-Smith *et al.*, 1980; King-Smith and Kulikowski, 1980; Alvarez and Kulikowski, 1989). In spite of a complete lack of sustained vision, these patients are neither pattern- nor motion blind, which suggests that precortical processing in humans may have the same overlapping characteritics as those found in monkeys (see previous section).

Another group of interest is patients with multiple sclerosis (MS) and optic neuritis (ON) who show very variable degrees of eye-specific dysfunctions, consistent with the damage to either P- or M-systems (or both). This heterogeneity, however, makes it possible to compare the efficacy and correlation between different methods of assessment. In particular, 'objective measures' such as occipitally-recorded Visual Evoked Potentials (VEPs) elicited by 'selective', or preferential stimuli (of low contrast, chromatic or achromatic) correlate with the corresponding (chromatic, or achromatic) psychophysical spectral sensitivity characteristics (Russell *et al.*, 1991). Thus, it seems that VEPs could be a method of choice in non-communicative patients, or monkeys, where rigorous psychophysical observations would be more time consuming (see below and Kulikowski *et al.*, 1988).

Effects of cortical damage on colour vision

Lesions of area V1 – blindsight

Several visual cortical areas contain cells responsive to chromatic stimuli and lesions to any of these areas may be expected to produce deficits in colour vision. Lesions to area V1 are characterized by blind fields (scotomas) within which the subject is not aware of any visual events. However, it is clear that some information about form and colour can reach extra-striate areas directly from subcortical structures (e.g. LGN or pulvinar) and these pathways are implicated in 'blindsight' (see Cowey and Stoerig, 1991 for review).

The area that has been studied most intensively is V4, sometimes referred to as the colour area and it is this area on which the remainder of this review concentrates.

Lesions of area V4

Lesions of cortical visual area V4 in the macaque monkey are of particular interest since it is this area in which Zeki (1983) originally discovered cells that respond to chromatic stimuli not on the basis of the dominant wavelength reflected from a surface, but according to the colour of the surface as perceived by a human observer. The obvious question arises: "Does removal of area V4 produce a deficit specific to colour vision?" Attempts to address this question have found that while monkeys with V4 lesions may take longer than normal monkeys to learn a wavelength discrimination task, their wavelength discrimination thresholds are not elevated appreciably (Dean, 1979; Wild *et al.*, 1985; Butler *et al.*, 1988; Walsh *et al.*, 1993; also recently confirmed by Heywood *et al.*, 1992). It is clear, however, that V4 lesions do produce a deficit in colour constancy (Wild *et al.*, 1985; Butler *et al.*, 1988; Carden *et al.*, 1992; Walsh *et al.*, 1993).

In the most recent and comprehensive demonstration of colour constancy deficits (Walsh *et al.*, 1993), monkeys with V4 lesions were overtrained on a forced-choice wavelength discrimination task which they found very easy to learn. The discriminanda were either 12.5 or 10 Munsell Hue steps apart. When the monkeys were performing reliably at 90% correct or better, the spectral content of the illuminant was changed such that a negative plate (i.e. one for which the monkey did not receive a reward) reflected the same triplet of energies as the previous positive stimulus. The animals with V4 lesions were very much worse than the normal animals under this condition i.e. their colour constancy was impaired. In a control experiment the animals were again overtrained on an easy hue discrimination problem and then required to perform the discrimination when the intensity, but not the spectral content of the illuminant was changed. This condition had no effect on the performance of monkeys with lesions of cortical visual area V4. Figure 1 shows the results of these experiments (see Walsh *et al.*, 1993 for full details).

Visual area V4 provides the major input to the inferotemporal cortex (IT). By comparing the effects of IT lesions and V4 lesions we can add to the picture of the cortical processing of colour. Animals with damage to the inferotemporal cortex are impaired on a wide variety of discrimination tasks and one explanation often given for this is that the animals have difficulty in categorizing visual stimuli (Heywood *et al.*, 1988). However, the only study of IT lesions which has manipulated, as an independent variable, the colour category to which a stimulus belongs (Butter *et al.*, 1965) demonstrated that monkeys with IT cortex lesions show the same discontinuities in wavelength discrimination across the spectrum as do humans (Mullen and Kulikowski, 1990). Thus

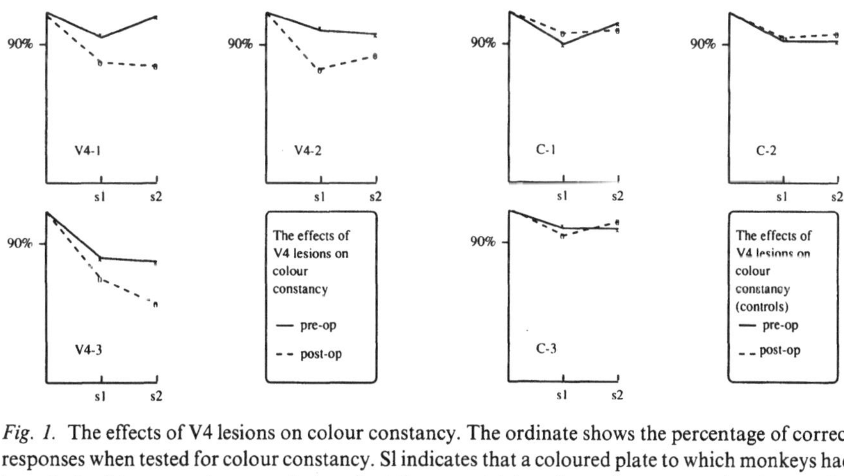

Fig. 1. The effects of V4 lesions on colour constancy. The ordinate shows the percentage of correct responses when tested for colour constancy. Sl indicates that a coloured plate to which monkeys had been trained to respond was made to reflect the same triplet of energies as a previously negative plate. S2 indicates that a negative plate was made to reflect the same triplet of energies as another negative plate (resulting in a larger shift for the positive plate). The origin is at 100% to indicate that all animals had near perfect performance when required to discriminate the plaques under a stable, unchanging illuminant. It is clear from these graphs that all animals with V4 lesions were worse on the constancy task after the operation and all the V4 animals were worse than the unoperated controls after the operation. Data replotted from Walsh *et al.*, (1993).

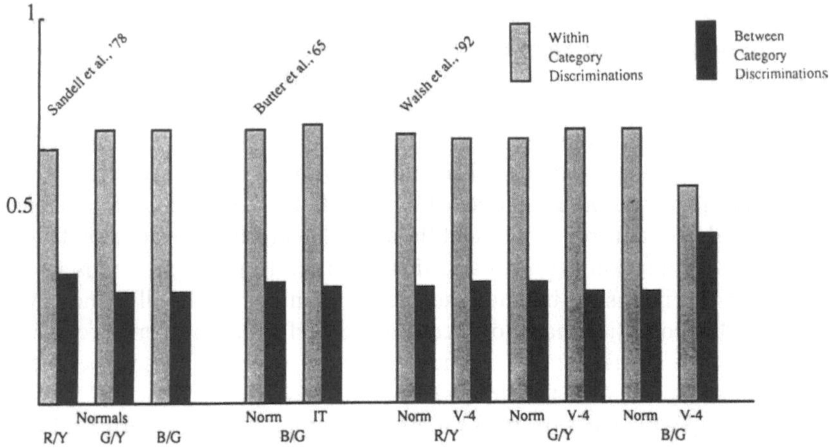

Fig. 2. Colour categorization. The figure shows the proportion of errors made by monkeys on colour categorization tasks. The data are normalized from three studies (Sandell *et al.*, 1978, Butter *et al.*, 1965, and Walsh *et al.*, 1992) and the labels R (red), G (green), B (blue), and Y (yellow) were stimuli with the same dominant wavelength in each study. It is notable that in three studies using extinction, learning, simultaneous, and successive presentations of the coloured stimuli, the animals all made approximately 70% of errors on the within category discriminations and 30% on the between category discriminations. It is also remarkable that the ratios were similar for normal monkeys, monkeys with removal of inferotemporal cortex, and monkeys with V4 lesions. The only anomaly here is the ratio of scores made in the B/G discriminations by V4 monkeys. This is more likely to be due to interference effects (since green had been used with a different contingency in the G/Y discriminations) than to any selective distortion of colour space in this region.

monkeys without inferotemporal cortex do have categorical colour perception. Figure 2 shows a comparison of categorical discrimination by normal monkeys, monkeys with bilateral IT lesions and monkeys with V4 lesions.

If IT lesions do not impair categorical perception then perhaps V4 is the area necessary for the categorization of colours. Thus, as a further test of the categorization hypothesis, Walsh *et al.*, 1992) tested the ability of monkeys with removal of area V4 to discriminate colours which were equally spaced in physical space but belonged to different perceptual colour categories. The stimuli in this study were chosen from a subset of stimuli used by Sandell *et al.*, (1978) who demonstrated that monkeys use the same colour categories as humans. It was found that the monkeys with V4 lesions did partition the spectrum in the same manner as normal monkeys (see Fig. 2).

Following the failure of either V4 of IT lesions to effect impairments of categorical colour perception, the conclusion to which one is driven is that colour categorization occurs at an earlier stage of processing in areas V1 and V2. Three lines of evidence support this view. Some cells in V2 have been shown to respond to a range of chromatic stimuli which are consistent with categorization (Yoshioka *et al.*, 1987, 1988). In particular, cells have been discovered which respond to a combination of hue and luminance. For example a cell may respond to olive but not to green, to maroon but not to red, or to brown but not to yellow. Thus the cells are not driven by the presence of absence of a particular wavelength, since the dominant wavelength used to produce a hue such as yellow can be the same as that used to produce a contrast colour, in this case brown. Clearly, there is a level of chromatic analysis occurring in V2 which may be a suitable precursor for the constancy analysis performed by V4. If this were the case, i.e. categorization precedes constancy, then one must predict that constancy will show category effects. We have begun an analysis of this prediction and do indeed find that constancy is better for colour centroids than for colours at the borders of categories. A similar effect is found for colour memory (Bartleson, 1960). The evidence from global electrophysiological recordings also supports the 'categorization precedes constancy' hypothesis.

Cortical visual evoked potentials (VEPs)

We have recorded VEPs, elicited by isoluminant chromatic gratings and achromatic gratings of low or moderate contrast, from three subject groups: normal human observers, normal macaque monkeys, and macaques with bilateral ablation of area V4. The results of these experiments show that V4 monkeys have normal chromatic VEPs (see Fig. 3). We have described the stimuli and the way in which chromatic VEPs are highly correlated with perceptual thresholds in other publications (Kulikowski, 1991; Kulikowski and Walsh, 1993; Kulikowski *et al.*, in prep; McKeefry and Kulikowski, in press and in this issue).

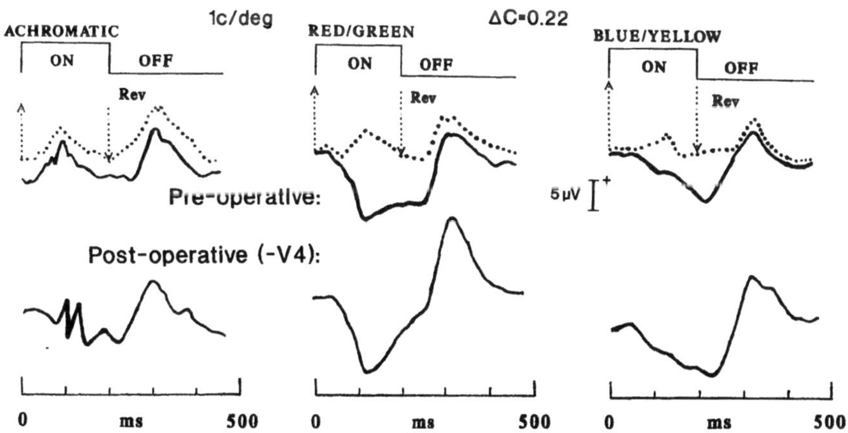

Fig. 3. Visual Evoked Potentials (VEPs) recorded from the rhesus monkey preoperatively (top traces) and after V4 lesions (bottom traces). Grating stimuli were: achromatic i.e. luminance-modulated (left column) and chromatic isoluminant, red/green (middle column) or blue/yellow (right column), presented either on and off (continuous lines) or reversed in contrast (dotted lines). Spatial frequency 1 c/degree, contrast change 0.22.

V2-IT projection

The third line of evidence consistent with the argument above is the discovery of a direct anatomical projection from area V2 to IT (Nakamura *et al.*, 1991). Chromatic information in V2 thus has access to the reward centres of the brain via the inferotemporal cortex. This may explain why lesions of prestriate cortex have often failed to produce deficits in memory tasks (e.g. Cowey and Gross, 1970). However, experiments on the effects of prestriate cortical lesions on colour memory have not been performed using isoluminant colour stimuli.

Comparison of cortical damage in humans and monkeys

How do these lesion effects relate to the colour vision defects seen after cortical damage in humans? King-Smith (1989) reviewed cortical colour deficiencies within two categories: Defects which give rise to 'losses of normal function' and those which give rise to 'abnormal function'. It is clear that the effects of V4 lesions belong to the first of these categories, a loss of normal function, since the spared function, wavelength discrimination and categorization, are quite normal. It is more difficult, however, to select a particular defect observed in humans which matches precisely the deficit of a V4 monkey. Colour perception in humans can clearly be isolated from form and motion perception since cortical colour vision deficits can occur in the absence of other deficits, and colour vision can be spared in cases of severe agnosia (e.g. Abadi and Kulikowski, 1989). There are, however, some interesting differences between monkeys with V4 lesions and human achromatopsiacs. The animals can discriminate very fine differences in hue, achromatopsiacs usually cannot. V4 ablated monkeys are not markedly worse at discriminating colours than they are at discriminating greys, humans achromatopsiacs are. Pure achromatopsia in humans can occur without any deficits in complex form perception, however, monkeys with V4 lesions do have persistent form discrimination deficits (Walsh *et al.*, 1992a). The V4 deficit cannot be defined as a failure to associate colours and objects, since monkeys with V4 lesions are able to perform colour/object conditional discriminations as long as the colour separation is large and the objects are easily discriminable (Wild *et al.*, 1985). The comparison of the effects of lesions in humans and monkeys is hampered by the paucity of experiments carried out with human subjects. For example, we are not aware of any publications in which the colour constancy of a patient has been tested (but see Morland *et al.*, chapter 11 of this volume, for an important exception that came to our notice in proof).

There are several other candidate explanations of the colour constancy deficit without hue discrimination deficits. One possible view is that while colour constancy deficits may not affect simple wavelength differencing they may affect colour memory since one remembers what one perceives – namely colours not wavelengths. Indeed in one study we have observed that monkeys with V4 lesions are slower than normal monkeys on learning a very simple wavelength discrimination task when the stimuli are presented successively (Walsh *et al.*, 1992b). The slowness of learning may also result from lacking a stable standard of colours, for which constancy is needed and indeed present in the animals with effective colour vision from bees and fish to primates.

Acknowledgements

We thank the Wellcome Trust, U.M.I.S.T. Academic Resource & Development Fund and Visual Sciences Fund for funding the projects. The data shown here were obtained in collaboration with S.R. Butler, D. Carden and D. McKeefry.

References

Abadi, R.V. and Kulikowski, J.J. (1989). Seeing without recognizing. In: Kulikowski, J.J., Dickinson, C. and Murray, I.J. (eds), Seeing Contour and Colour: 755–757, Pergamon Press, Oxford.

Alvarez, S.L. and Kulikowski, J.J. (1989). Spatial resolution limits for pattern and movement detection in patients with famillial optic nerve atrophies. In: Kulikowski, J.J., Dickinson, C. and Murray, I.J. (eds), Seeing Contour and Colour: 483–409, Pergamon Press, Oxford.

Bartleson, C.J. (1960). Memory colours of familiar objects. J.Opt. Soc. Am. 50: 73–77.

Butler, S.R., Carden, D., Hilken, H. and Kulikowski, J.J. (1988). Deficit in colour constancy following V4 ablations in rhesus monkeys. J. Physiol. 403: 72P.

Butter, C.M., Mishkin, M. and Rosvold, H.E. (1965). Stimulus generalization in monkeys with inferotemporal and lateral occipital lesions. In: Mostofsky, D.I. (ed), Stimulus Generalization: 119–133, Stanford University Press, Stanford, California.

Burkhalter, A. and Bernardo, K.K. (1989). Organization of corticortical connections in human visual cortex. Proc. Nat. Acad. Sci. USA 86: 1071–1075.

Carden, D., Hilken, H.H. Butler, S.R. and Kulikowski, J.J. (1992). Lesions of primate visual area V4 produce long-lasting deficits to colour constancy. Irish J. Psychol. 13: 455–472.

Cowey, A. and Gross, C.G. (1970). Effects of foveal prestriate and inferotemporal lesions on visual discrimination by rhesus monkeys. Exp. Brain Res. 11: 128–144.

Cowey, A. and Stoerig, P. (1991). The neurobiology of blindsight. TINS 14: 140–145.

Creutzfeldt, O.D., Weber, H., Tanaka, M. and Lee, B.B. (1987). Neuronal representation of spectral and spatial stimulus aspects in foveal and parafoveal area 17 of the awake monkey. Exp. Brain Res. 68: 541–564.

Crook, J.M., Lange-Malecki, B., Lee, B.B. and Valberg, A. (1988). Visual resolution of macaque retinal ganglion cells. J. Physiol. 396: 205–224.

Dean, P. (1979). Visual cortex ablation and thresholds for successively presented stimuli in rhesus monkeys. 2. Hue. Exp. Brain Res. 35: 69–83.

Dow, B.M. (1974). Functional classes of cells and their laminar distribution in monkey visual cortex. J. Neurophysiol. 37: 927–946.

Dow, B.M. (1991). Orientation and color columns in monkey striate cortex. In: Valberg, A., and Lee, B.B. (eds). From Pigments to Perception: 269–274, Plenum Press, New York.

Dow, B.M. and Vautin, R.G. (1987) Horizontal segregation of color information in the middle layers of foveal striate cortex. J. Peurophysiol. 57: 71–739.

Gouras, P. and Kruger, J. (1979). Responses of cells in foveal visual cortex of the monkey to pure color contrast. J. Neurophysiol. 42: 850–860.

Heywood, C.A., Gadotti, A. and Cowey, A. (1992). Cortical area V4 and its role in the perception of color. J. Neurosci. 12: 4056–4065.

Heywood, C.A., Shields, C. and Cowey, A. (1988). The involvement of the temporal lobes in colour discrimination. Exp. Brain Res. 71: 437–441.

Hichs, T.P., Lee, B.B. and Viyasagar, T.R. (1983). The responses of cells in macaque LGN to sinusoidal gratings. J. Physiol. 337: 183–200.

Hockfield, S., Tootell, R.B.H. and Zaremba, S. (1990). Molecular differences among neurons reveal an organization of human visual cortex. Proc. Nat. Acad. Sci. USA. 87: 3027–3031.

Horton, J.C. and Hubel, D.H. (1981). A regular patchy distribution of cytochrome-oxidase staining in primary visual cortex of the macaque monkey. Nature 292: 7620764.

Hubel, D.H. and Livingstone, M.S. (1987). Segregation of form, color and stereopsis in primate area 18. J. Neurosci. 7: 3378–3415.

Hubel, D.H. and Livingstone, M.S. (1990). Color and contrast sensitivity in the lateral geniculate body and primary visual cortex of the macaque monkey. J. Neurosci. 10: 2223–2237.

Kaplan, E., Lee, B.B. and Kulikowski, J.J. (1989). The role of P and M systems (A, B and C). In: Kulikowski, J.J., Dickinson, C. and Murray, I.J. (eds), Seeing Contour and Colour: 222–237, Pergamon Press, Oxford.

Kaplan, E. (1991). The receptive field structure of retinal ganglion cells in cat and monkey. In: Leventhal, A.G. (ed), The Neural Basis of Visual Function, Volume 4 of Vision and Visual Dysfunction, Cronly-Dillon, J.R. (Series ed.). Macmillan Press, London.

King-Smith, P.E., Rosten, J.G., Alvarez, S.L. and Bhargava, S.K. (1980). Human vision without tonic ganglion cells? In: Verriest G. (ed), Colour Vision Deficiencies V: 99–105, Adam Hilger, Bristol.

King-Smith, P.E. and Kulikowski, J.J. (1980). Pattern and movement detection in a patient lacking sustained vision. J. Physiol. 300, 60P.

Kolb, H. and Lipetz, L.E. (1991). The anatomical basis for colour vision in the vertebrate retina. In: Colour Vision, Gouras, P. (ed), Volume 6 of Vision and Visual Dysfunction, Cronly-Dillon, J.R. (ed). Macmillan Press, London.

Kritzer, M.F., Cowey, A. and Somoghy, P. (1992). Patterns of inter- en intralaminar GABAergic connections distinguish striate (V1) and extrastriate (V2, V4) visual cortices and their functionally specialized subdivisions in the rhesus monkey. J. Neuroscience 12: 4545–4564.

Kulikowski, J.J. (1991a). On the nature of VEP, unit responses and psychophysics. In: Valberg, A. and Lee, B.B. (eds), From Pigments to Perception: 197–208, Plenum Press, New York.

Kulikowski, J.J. (1991b). Binocular chromatic rivalry and single vision. Ophth. Physiol. Optics 12: 168–170.

Kulikowski, J.J., Metcalfe, R.A., Murray, J.J. and Russell, M.H.A. (1988). Visual evoked potentials can indicate selective damage to the human visual system. J. Physiol. 403, 73P.

Kulikowski, J.J. Vidyasagar, T.R. and Carden, D. (1989a). Linear/nonlinear analysis of chromatic and achromatic information. In: Kulikowski, J.J., Dickinson, C.M. and Murray, I.J. (eds), Seeing contour and colour: 85–115, Pergamon Press, Oxford.

Kulikowski, J.J., Murray, I.J. and Parry, N.R.A. (1989b). Electrophysiological correlates of chromatic-opponent and achromatic stimulation in man. In: Drum, B. and Verriest, G. (eds), Color Vision Deficiencies IX, Doc. Ophthalmol. Proc. Ser. 62: 145–153, Kluwer Acad. Publ., Dordrecht.

Kulikowski, J.J. and Walsh, V. (1993). Colour vision: Isolating mechanisms in overlapping streams. Prog. in Brain Res. 95: 417–426.

Lee, B.B. (1991a). Spectral sensitivity in primate vision. In: Kulikowski, J.J., Walsh, V. and Murray, I.J. (eds), Limits of Vision: 191–201, Macmillan Press, London.

Lee, B.B. (1991b). On the relation between cellular sensitivity and psychophysical detection. In: Valberg, A. and Lee, B.B. (eds), From Pigments to Perception: 105–115, Plenum Press, New York.

Leventhal, A.G., Rodieck, R.W. and Dreher, B. (1981). Retinal ganglion cell classes in the old world monkey: Morphology and central projections. Science 213: 1139–1142.

Livingstone, M.S. and Hubel, D.H. (1983). Specificity of cortico-cortical connections in the monkey visual system. Nature, 304: 531–534.

Livingstone, M.S. and Hubel, D.H. (1984a). Anatomy and physiology of a color system in the primary visual cortex. J. Neuroscience. 4: 309–356.

Livingstone, M.S. and Hubel, D.H. (1984b). Specificity of intrinsic connections in primate primary visual cortex. J. Neuroscience. 4: 2830-2835.
Livingstone, M.S. and Hubel, D.H. (1987). Psychophysical evidence for separate channels for perception of form, colour, movement, and depth. J. Neuroscience. 7: 3416–3466.
Livingstone, M.S. and Hubel, D.H. (1988). Segregation of form, color, movement and depth: anatomy, physiology, and perception. Science 240: 740–749.
Lund, J.S. (1988). Anatomical organization of macaque monkey striate visual cortex. Ann. Rev. Neurosci. 11: 253–288.
Lund, J.S. (1991). Excitatory and inhibitory circuitry and laminar mapping strategies in the primary visual cortex of the monkey. In: Edelman, G.M., Gall, W.E. and Cowan, W.M. (eds), Signal and sense. Wiley, New York.
Merigan, W.H. (1989). Chromatic and achromatic vision of macaques: role of the P pathway. J. Neurosci. 9: 776–783.
Merigan, W.H. (1991). P and M pathway specialization in the macaque. In: Valberg, A. and Lee, B.B. (eds), from Pigments to Perception: 117–125, Plenum Press, New York.
Merigan, W.H. and Eskin, T.A. (1986). Spatio-temporal vision of macaques with severe loss of Pb retinal ganglion cells. Vision Res. 26: 1751–1761.
Merigan, W.H. and Maunsell, J.H.R. (1990). Macaque vision after magnocellular lateral geniculate lesions. Visual Neuroscience 5: 347: 352.
Michael, C.R. (1987a). Color vision mechanisms in monkey striate cortex: dual-opponent cells with concentric receptive fields. J. Neurophysiol. 41: 572–588.
Michael, C.R. (1987b). Color vision mechanisms in monkey striate cortex: simple cells with dual-opponent color receptive fields. J. Neurophysiol. 41: 1233–1249.
Michael, C.R. (1981). Columnar organization of color cells in monkey's striate cortex. J. Neurophysiol. 46: 587–604.
Michael, C.R. (1989). The origin of double-opponency in the monkey striate cortex. In: Kulikowski, J.J., Dickinson, C.M. and Murray, I.J. (eds), Seeing contour and colour: 59–65, Pergamon Press, Oxford.
Mullen, K.T. and Kulikowski, J.J. (1990). Wavelength discrimination at detection threshold. J. Opt. Soc. Am. A. 7: 733–742.
Reid, R.C. and Shapley, R.M. (1992). Spatial structure of cone inputs to receptive fields in primate L.G.N. Nature 356: 716–717.
Rodieck, R.W. (1991). Which cells code for color? In: Valberg, A. and Lee, B.B. (eds), From Pigments to Perception: 83–93, Plenum Press, New York.
Sandell, J.H., Gross, C.G. and Bornstein, M.H. (1979). Color categories in macaques, J. Comp. Physiol. & Psychol. 93: 626–635.
Schein, S.J. and Desimone, R. (1990). Spectral properties of V4 neurons in the macaque. J. Neurosci. 10: 3369–3389.
Schiller, P.H. (1991). The colour-opponent and broad-band channels of the primate visual system. In: Valberg, A. and Lee, B.B. (eds), From Pigments to Perception: 127–132, Plenum Press, New York.
Schiller, P.H., Logothetis, N.K. (1990). The color-opponent and broad-band channels of the primate visual system. TINS 13: 392–398.
Schiller, P.H., Logothetis, N.K. and Charles, E.R. (1991). Parallel pathways in the visual system: Their role in perception at isoluminance. Neuropsychologia 29: 433–441.
Ts'o, D.Y. (1989). The functional organization and connectivity of color processing. In: Man-Kit Lam, D. and Gilbert, C.C. (eds), Neural mechanisms of visual perception: 87–115, Portfolio Publishing, Texas.
Ts'o, D.Y. and Gilbert, C.D. (1988). The organization of chromatic and spatial interactions in the primate striate cortex. J. Neurosci. 8: 1712–1727.
Vautin, R.M. and Dow, B.M. (1985). Color cell groups in foveal striate cortex of the behaving monkey. J. Neurophysiol. 54: 273–292.
Walsh, V., Carden, D., Butler, S. and Kulikowski, J.J. (1992a). The effects of lesions of area V4 on the visual abilities of monkeys: Shape discrimination. Behav. Brain Res. 50: 115–126.

Walsh, V., Carden, D., Butler, S. and Kulikowski, J.J. (1993). The effects of lesions of area V4 on the visual abilities of monkeys: Hue discrimination and colour constancy. Behav. Brain Res. 53: 51–62.

Walsh, V., Kulikowski, J.J., Butler, S. and Carden D. (1992b). The effects of lesions of area V4 on the visual abilities of monkeys: colour categorization. Behav. Brain Res. 52: 81–89.

Wiesel, T.N. and Hubel, D.H. (1966). Spatial and chromatic interactions in the lateral geniculate body of the rhesus monkey. J. Neurophysiol. 29: 1115–1156.

Wild, H.M., Butler, S.R., Carden, D. and Kulikowski, J.J. (1985). Primate cortical area V4 important for colour constancy but not wavelength discrimination. Nature. 313: 133–135.

Wild, H.M. and Kulikowski, J.J. (1984). Neurotoxic effects of acrylamide on rat retinogeniculate fibres. Behav. Brain Res. 13: 201–207.

Yoshioka, T., Dow, B.M. and Vautin, R.G. (1987). Hue, luminance and hue/luminance cells in foveal V1, V2 and V4 of the behaving monkey. Soc. Neuroscience. Abstr. 13: 177.2.

Yoshioka, T., Dow, B.M. and Vautin, R.G. (1988). Close association of color, orientation and luminance processing in V1, V2 and V4 of the behaving macaque monkey. Soc. Neuroscience Abstr. 14: 457.

Yoshioka, T. and Lund, J.S. (1990). Substrates for interaction of visual channels within area 17 of monkey visual cortex. Soc. Neurosci. Abstr. 16: 707.

Zeki, S.M. (1983). Colour coding in the cerebral cortex: The responses of wavelength-selective and colour-coded cells in monkey visual cortex to changes in wavelength composition. Neuroscience. 9: 767–781.

Zeki, S.M. and Shipp, S. (1988). The functional logic of cortical connections. Nature 335: 311–317.

Zeki, S.M. and Shipp, S. (1989). Modular connections between areas V2 and V4 of macaque monkey visual cortex. Eur. J. Neurosci. 1: 494–506.

Zrenner, E. and Kruger, C.J. (1981). Ethambutol mainly affects the function of red/green opponent neurons. Doc. Ophthalmol. Proc. Ser. 27: 13–25.

Authors' addresses:
Vincent Walsh
Experimental Psychology Dept.
University of Oxford
Oxford, U.K.

Janus J. Kulikowski
Visual Sciences Laboratory
U.M.I.S.T.
Manchester M60 1QD, U.K.

13. Macular and perimacular colour vision in homonymous hemianopsias

DANIELE DENIS, FRANÇOIS DEVIN and JEAN-LOUIS VOLA
(Marseille, France)

Abstract

In a previous study, one of us demonstrated weak colour deficiencies in Homonymous Hemianopsias (HH). In this paper, the Wald-Marré method was utilized to elicit colour mechanisms in HH. The blue mechanism was missing most of the time in the central as well as in the blind areas. However, some weak responses of the Green and Red mechanisms were obtained in the blind areas. This paper presents the results of our investigations.

Introduction

The purpose of this study was to record Homonymous Hemianopsia (HH) colour mechanism defects in the central as well as in the blind areas. Physiologists Moreland (1955), and Moreland and Cruz (1959) studied peripheral colour sensitivity. Clinicians Kitahara *et al.*, (1982) and Vola *et al.*, (1985) recorded colour mechanisms with Stiles' technique. Marré (1971) using Wald's colour adaptation theory succeeded in eliciting in central retina the three colour mechanisms. Thanks to the Wald-Marré technique, the use of spectral lights resulted in the display of the three mechanisms. It also suggested the possibility of recording their central and peripheral sensitivity. A previous study with pigmentary tests by Vola *et al.*, (1989) showed weak defects, especially tritanopia. These weak defects, compared to the severity of dyschromatopsias in prechiasmatic lesions (bitemporal Hemianopsias) are surprising. We tried to find out if residual colour sensitivity could be elicited in central and blind areas of HH visual fields in order to determine whether special pathways operate in these cases. Stoerig and Cowey (1989) found wavelength sensitivity in some visual fields defects. Our findings corroborate their results.

From our clinical point of view, colour vision mechanisms should now be recorded not only in central but also in pericentral retina at least 15° from the fovea on different meridians, in cases of prevalent pericentral pathology, like

B. Drum (ed.), Colour Vision Deficiencies XII, pp. 109–115.
© 1995 *Kluwer Academic Publishers, Dordrecht.*

glaucoma (Devin *et al.*, 1992) in which the Bjerrum area is impaired early on, or like retinal detachment, optic neuropathy and hemianopsias.

Method

Wald-Marré technique

A modified Tübingen perimeter with halogen sources was used.
Enlargement of the fixation point to 11° gave a circular background, the colour of which was changed with Kodak filters. We used a strongly illuminated yellow background (10,000 td) to record the Blue or short wavelength sensitivity (SWS) mechanism, a magenta background (500 td) to record the Green or medium wavelength sensitivity (MWS) mechanism and a cyan background (1000 td) for the Red or long wavelength sensitivity (LWS) mechanism.

A square test 1° of different wavelengths from 400 to 650 nm (range of 50 nm) supplied in the apparatus was superimposed with flashes of 500 ms on the background's centre. A scale of neutral filters expressed in optical density (stepsize of 0.2 LU) was used to record increment thresholds. Neutral filters were increased and decreased by 0.2 LU. After 3 trials a steady response was obtained.

This technique is easy to set up in the peripheral areas on one of more meridians. Shifting fixation was carried out by adding a red light source to the perimeter on the opposite side of the blind area at 15° from the centre and on the 60° meridian in order to set the blind area on the centre of the perimeter on which the background and the test were projected.

Material

Twenty-three cases of Homonymous Hemianopsias were examined, five cases were retained:

1. Diffuse temporo-parieto-occipital vascular lesions, including:
   - one brain temporo-parietal malformation.
   - two carotido-stenosis.
   - one coronography complication.
2. localised lesions, including:
   - one occipital tumour.
   - one occipital ischemia.

Results

As we found in a previous paper (Vola *et al.*, 1989), colour defects in retrochiasmatic lesions are less severe than in prechiasmatic lesions, especially in case of occipital impairments. The central Blue (SWS) mechanism was sometimes normal or missing (Fig. 1), the Green (MWS) and the Red (LWS) mechanisms were more or less weakened (Fig. 2, 3).

It is worth noting that the responses to pseudo-isochromatic plates such as HRR were found to be always normal but the Desaturated D-15 often contained errors.

We report in this paper responses in the central and in the blind areas of three out of the five cases.

The first case was a left brain malformation with temporal hemilobe atrophy. The central Blue mechanism was preserved as well as Green and Red mechanisms. At 15° from the fovea on the 60° meridian in the blind area, the Blue mechanism was missing, but the Green and Red mechanisms could be elicited weakly.

In the second case, left diffuse temporal vascular lesions: in the blind area at 15° from the fovea were found: a missing Blue mechanism (Fig. 4), a weak Green sensitivity mechanism in both eyes (Fig. 5), and a normal Red mechanism (Fig. 6), in right eye, diminished in the left eye.

The third case was a left occipital ischemia with quadranopsias. A weak response of the Red mechanism and no response of the Green or the Blue mechanisms were found in the blind areas.

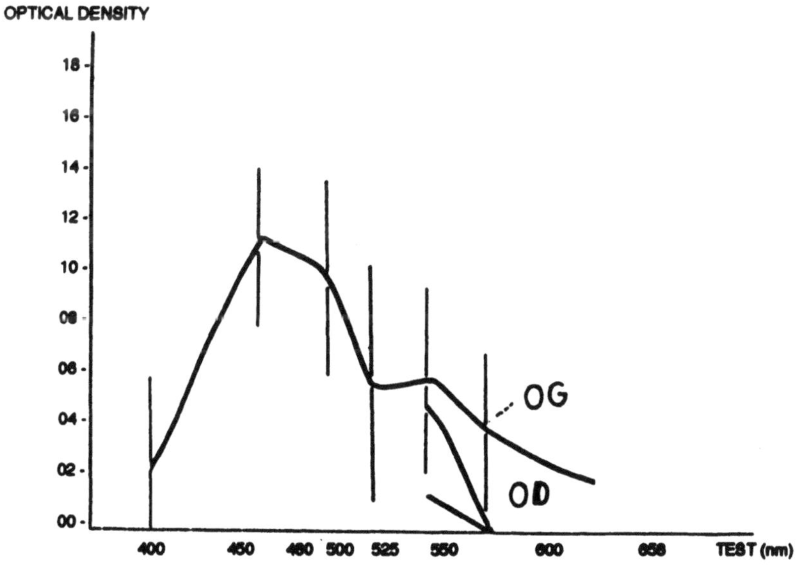

Fig. 1. Central Blue mechanism sensitivity. Abscissa : wavelengths 400 nm to 650 nm. Ordinate: 1° square test spot illumination expressed in optical density filters superimposed on 10,000 td 11° circular yellow background. Solid line : Normal curve with 2 standard deviations. No response recorded; OD: right eye, OG: left eye.

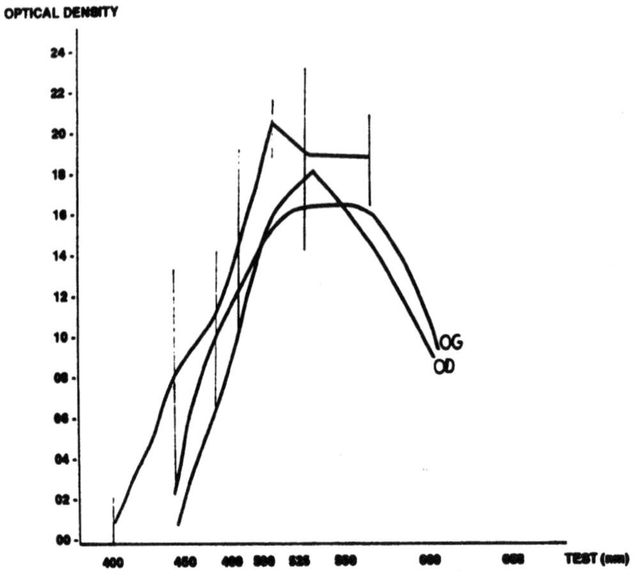

Fig. 2. Central Green mechanism sensitivity, 5,000 td magenta background. For coordinates see Fig. 1. OD, OG: subnormal recorded responses.

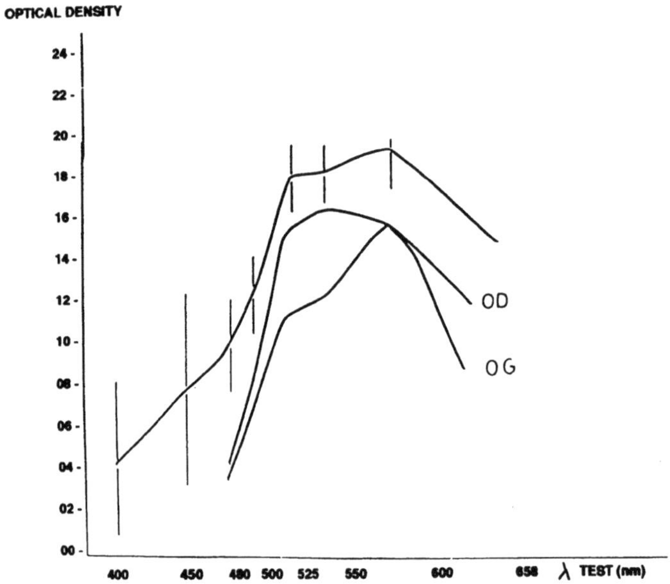

Fig. 3. Central Red mechanism sensitivity, 1000 td cyan background. For coordinates see Fig. 1. OD, OG: recorded responses: decreasing sensitivity.

Discussion

Residual colour mechanisms could be demonstrated in some cases in the blind areas. The Blue mechanism (Fig. 4) was always missing though it was recorded in the centre of the visual field (case 1). The Green mechanism (Fig. 5) was detected but always impaired. The Red (Fig. 6) mechanisms appeared to be the more resistant. These results are in line with the work of Stoerig and Cowey (1989) who found a reduced spectral wavelength sensitivity in a blind area of a visual field: some curves of spectral sensitivity (Fig. 2 of their paper) showed a decrease at 450 nm and a better sensitivity at other wavelengths as we found. The Wald-Marré technique provides evidence for cone contributions in the process. We think that the Stiles two-colour method could detect rod intervention due to Pi 0 and that the ganglion cells still functional due to Pi 3 (work in preparation). It should be noted that our patients did not show evidence of 'blind sight' as they were aware of the stimulus when it was chromatic. In the other cases we examined central colour vision. Central defects were found to be more severe in right HH than in left HH. In the case of a successful carotido stenosis operation, an improvement of central colour defects was displayed. No difference in colour deficiency was found in Hemianopsias with or without macular sparing.

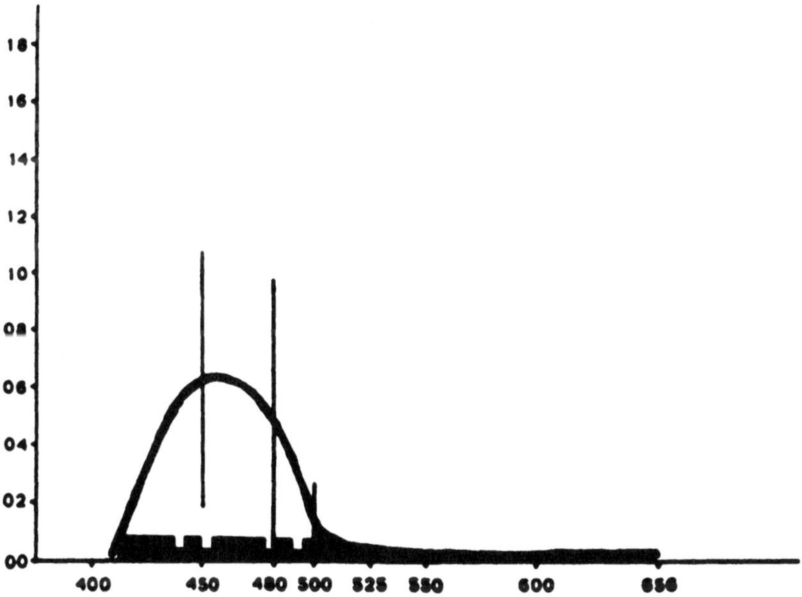

Fig. 4. Paracentral Blue mechanism in the blind area. Solid line : normal sensitivity. Dashed line: recorded response: Missing sensitivity. For coordinates see Fig. 1.

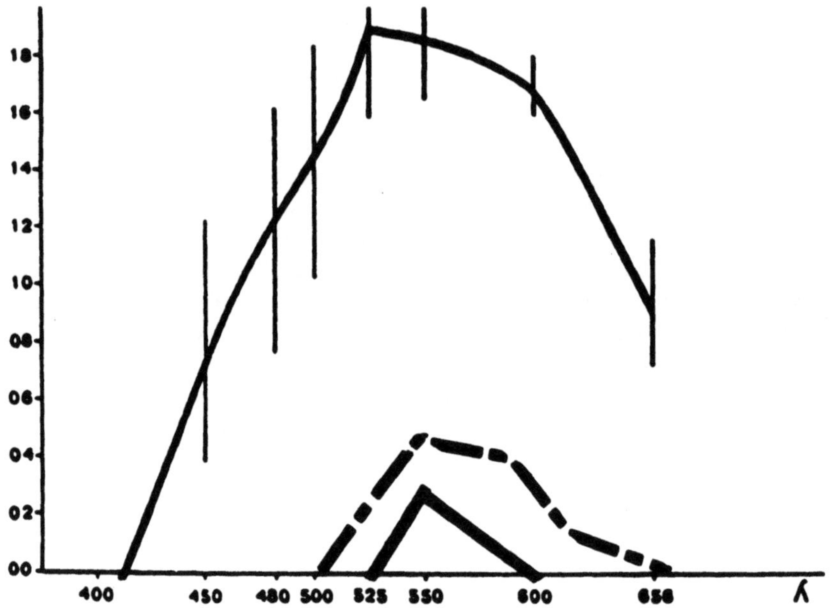

Fig. 5. Paracentral Green mechanism. Smooth line: normal sensitivity. Abscissa : wavelengths. Ordinate: optical density of test's neutral filters (LU). Solid and dashed curves. OD and OG: recorded responses.

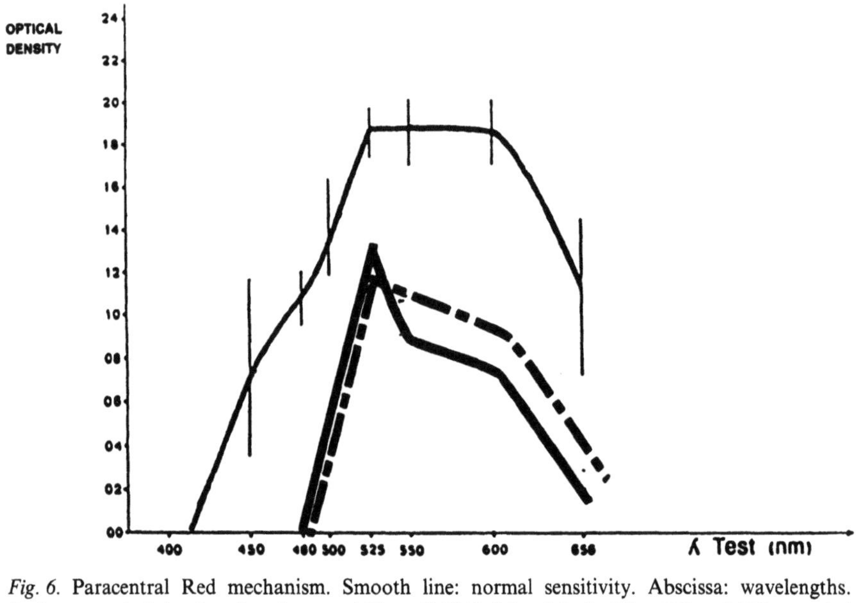

Fig. 6. Paracentral Red mechanism. Smooth line: normal sensitivity. Abscissa: wavelengths. Ordinate: optical density of test's neutral filters (LU) Solid and dashed line OD and OG recorded responses.

Conclusions

The study showed that somc HH blind areas are not as blind as most authors have asserted. Residual colour mechanisms suggest the possibility of other pathways. New studies are necessary to give a better interpretation of these findings.

References

Devin, F., Denis, D. and Vola, J.L. (1992). Vision colorée paramaculaire et glaucome. Ophthalmologie 6: 418–420.

Kitahara, R. Tamaku, R. Nogi, J. Kandabu, A. and Matsusari, J. (1982). Extrafoveal Stiles' π mechanisms. Doc. Ophthalmol. Proc. Ser. 35: 397–403.

Marré, M. (1972). Clinical examination of the three colour vision mechanisms in acquired colour vision defects. In: Verriest, G. (ed.), Acquired Colour Vision Deficiencies, Mod. Probl. Ophthalmol. 11: 224–227, Karger, Basel.

Moreland, J.D. (1955). The Perception of Colour by Extrafoveal and Peripheral Vision. Ph.D. Thesis. University of London.

Moreland, J.D. Cruz, A.C. (1958). Colour perception with the peripheral retina. Optica Acta 6: 117–151.

Stoerig, P. and Cowey, A. (1989). Wavelength sensitivity in blindsight. Nature 342: 916–917.

Vola, J.L., Riss, M., Jayle, G.E., Gosset, A. and Tassy, A. (1989). Acquired deficiency of colour vision in lateral Homonymous Hemianopsias. In: Verriest, G. (ed.), Acquired Colour Vision Deficiencies, Mod. Probl. Ophthalmol. 11: 150–159, Karger, Basel.

Vola, J.L., Gastaud, P. and Gondois, B. (1985). Consideration on Stiles' π mechanism in glaucoma close Bjerrum area. Doc. Ophthalmol. Proc. Ser. 42: 269–272.

Corresponding author;
Dr. Jean Vola
5 La Pelissiére
12 Boulevard de GABES
13008 MARSEILLE, FRANCE.

PART TWO

Genetics and congenital red/green colour deficiencies

14. Correlation between Rayleigh match range in protans and deutans and the difference in λ_{max} between hybrid and normal pigments

SAMIR S. DEEB, JORIS WINDERICKX and ARNO G. MOTULSKY
(Seattle, WA, U.S.A.)

Abstract

Certain protans have gene arrays comprised of L-M hybrid genes and one or more normal M-pigment genes, and certain deutans have arrays containing M-L hybrids in addition to normal L pigment genes. We tested the hypothesis that chromatic discrimination depends upon the difference between spectra of the normal and hybrid pigments in the array. We suggested more severe abnormalities in color vision when normal and hybrid pigments had very similar spectra.

Sequences of the normal and hybrid pigments were determined in 19 protan and 19 deutan male subjects, and the wavelength of maximum absorption (λ_{max} of each pigment was inferred from the recent data of Merbs and Nathans (1992) who determined λ_{max} in vitro on a variety of hybrid and normal pigments. Eight of 9 protanopes had spectrally idential L-M hybrid and normal M pigments. Only six of 10 protanomalous subjects had pigments that differed in λ_{max}. However, the four protanomalous subjects with identical pigments had a markedly wider mean match range (41.5 units) than that (9.3 units) of the 6 subjects who had different pigments. Spectrally identical pigments were found in the arrays of the two deuteranopes as well as in 7 of the 17 deuteranomalous subjects. Thus, the correspondence between match range and the difference in spectra between hybrid and normal pigments was strong in protans and weak in deutans, and the majority of exceptions involved subjects who had a narrower match range than predicted from spectral differences.

Introduction

Correlation between color vision phenotype and the molecular arrangement of the X-linked genes that encode the long wave-sensitive (L) and middle wave-sensitive (M) pigments has been determined in over 120 color-defective males (Nathans *et al.*, 1986; Deeb *et al.*, 1992; Zhang *et al.*, 1992). The methods involved Nagel anomaloscopy to assess the phenotype and Southern blot

B. Drum (ed.), Colour Vision Deficiencies XII, pp. 119–125.
© 1995 *Kluwer Academic Publishers, Dordrecht.*

analysis to determine the structure of the gene arrays. The results of these studies could assign the color deficient subjects into either the protan or deutan series. Within each of these series, there was little correlation between severity of the defect and the type of gene array they possessed (see Fig. 1). In individuals with normal color vision, the L and M pigment genes form head-to-tail tandem arrays comprised of one L pigment gene 5′ upstream of one or more M pigment genes (Feil *et al.*, 1990). In protan subjects, the L pigment gene was replaced by an L-M hybrid gene which is found either by itself or together with one or more normal M pigment genes (Fig. 1). In deutans, the M pigment gene was either deleted or replaced by a M-L hybrid gene. In addition to the normal L and M-L hybrid genes, many deutans (the majority being deuteranomalous) have normal M pigment genes (Fig. 2).

Recent advances in rapid DNA sequencing techniques allowed us to determine accurately the amino acid sequence, and the ability to express and reconstitute the visual pigments in vitro (Merbs and Nathans, 1992 a, b) permits an inference of the absorption characteristics of the various types of hybrid pigments associated with color vision defects. We hypothesized that at least some of the apparent discrepancies between genotype and phenotype might be resolved once the absorption characteristics of the pigments carried by each individual had been defined. To test this hypothesis, we selected protans and deutans whose gene arrays contained one hybrid and at least one normal pigment gene. A milder color vision deficiency (anomalous trichromacy) would be predicted if the hybrid and normal genes encode pigments with different wavelengths of maximum absorption (λ_{max}), allowing better color discrimination. A more severe defect (dichromacy) might result if the two pigments had identical λ_{max} thereby precluding use of both pigments for color perception.

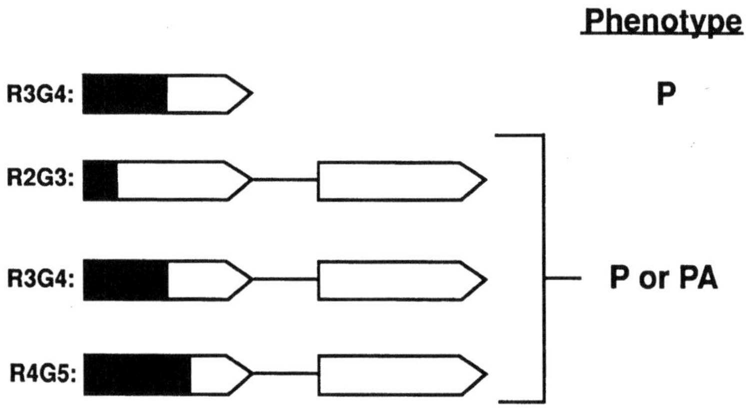

Fig. 1. Gene arrays in protans. Solid and open regions of arrows represent L- and M-derived gene sequences, respectively. Lines connecting arrows represent intergenic regions. Diagram not drawn to scale. R3G4 indicates that the first three exons of the hybrid gene are derived from the L pigment gene and the remainder (3) starting from exon 4 derived from the M pigment gene. P = protanope, PA = protanomalous.

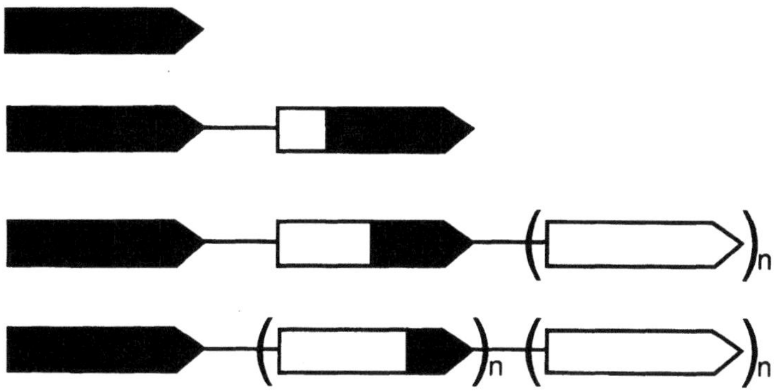

Fig. 2. Gene arrays in deutans. Solid and open regions of arrows represent L- and M-derived gene sequences, respectively. Lines connecting arrows represent intergenic regions. Diagram not drawn to scale. Carriers of any of the patterns shown may be either deuteranopic or deuteranomalous, but in the majority of cases, those with more than one gene in the array test as deuteranomalous. n = one or more copies per array.

Subjects and methods

The subjects were young male Caucasian volunteers recruited through advertisements specifically asking for color-defective individuals. A signed consent form describing the procedures involved was obtained prior to psychophysical testing and collection of peripheral blood samples for DNA isolation. Subjects were tested in each eye by a Nagel model 1 anomaloscope which provided a 2-degree bipartite color-matching field of about 50-td mean retinal illuminance. Details of the protocol used to determine the match range and midpoint were given elsewhere (Deeb *et al.,* 1992). Subjects were assigned to four categories on the basis of the anomalscopic examination: protanope (P), protanomalous (PA), deuteranope (D), and deuteranomalous (DA).

The coding sequences of the normal and hybrid genes in the array were determined by amplification using the polymerase chain reaction followed by single strand conformation polymorphism analysis and direct sequencing according to procedures described in detail elsewhere (Winderickx *et al.,* 1993a). The λ_{max} of hybrid genes was inferred from spectrophotometric measurements made on a variety of in vitro expressed hybrid pigments by Merbs and Nathans (1992b).

Results

Nineteen protan subjects (9 protanopic and 10 protanomalous) who had a normal M pigment gene in addition to the L-M hybrid gene in the X-linked gene array were selected for this study. The sequence of the hybrid and normal pigments in each array were determined and the λ_{max}inferred as described above. Our hypothesis predicts that the gene array of a protanope would encode two pigments of identical λ_{max}, and that of a protanomalous subject would encode pigments that differ in λ_{max}. The results given in Fig. 3a show that, with one exception, protanopes had gene arrays that encoded hybrid and normal pigments of identical λ_{max}. There were more exceptions among the protanomalous subjects. A λ_{max} difference of 3-6 nm between the normal and hybrid pigments was found in only 6 of 10 protanomalous subjects (Fig. 3b). However, the 4 protanomalous subjects whose hybrid and normal pigments were identical had a markedly wider mean match range (41.5 units) than those

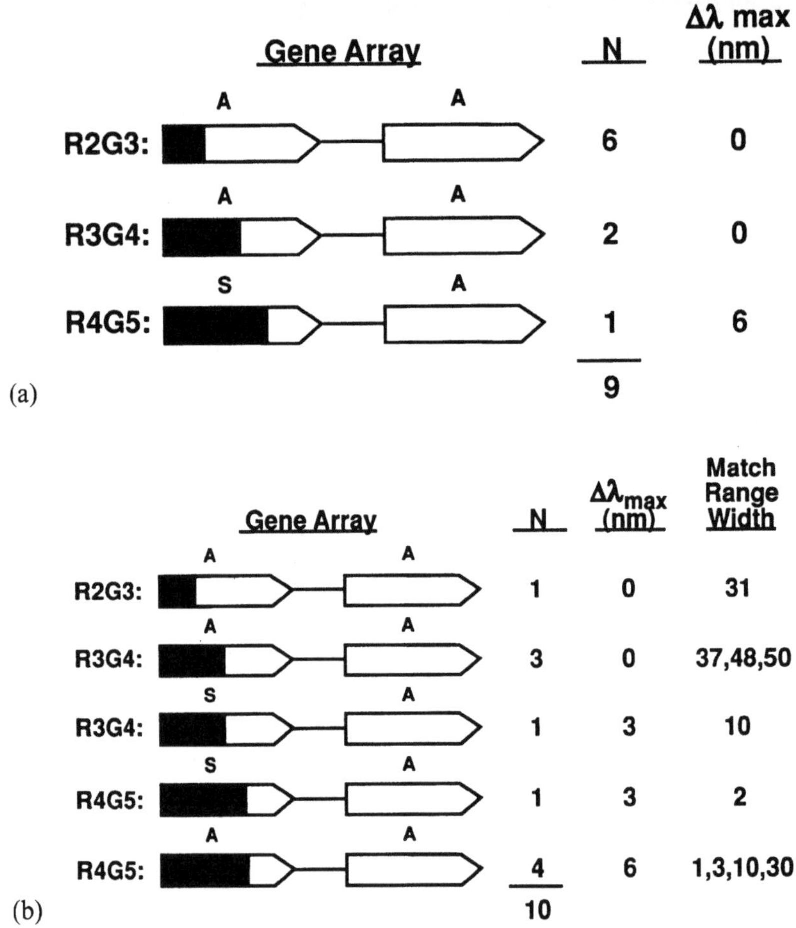

Fig. 3. Correspondence between Rayleigh match range and the difference in λ_{max} between normal and hybrid pigment genes in protans. Refer to legend to Fig. 1 for designations of gene arrays. Fig. 3a. Gene arrays of protanopic subjects. Fig. 3b. Gene arrays of protanomalous subjects. The λ_{max} values were inferred from the data of Merbs and Nathans (1992b). N = number of subjects. The match range is given in units on the red-green-mixture scale, with protanopes having a match range of 73 units. A and S denote the presence of alanine or serine at position 180.

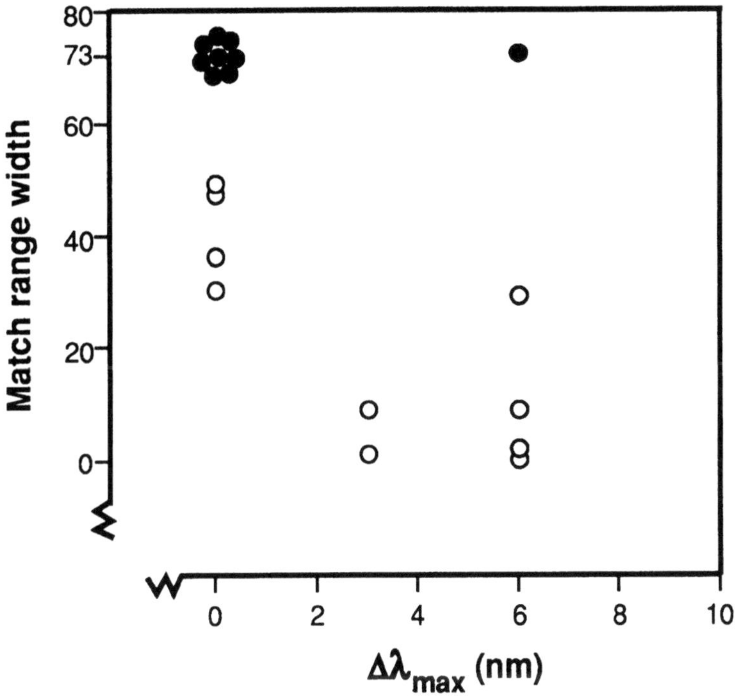

Fig. 4. Rayleigh match range as a function of difference in λ_{max} betwen normal M and L-M hybrid pigments encoded by gene arrays of protans. Protanopes are indicated by solid circles and protanomalous subjects indicated by open circles.

6 who had different pigments (9.3 units). These results are graphically represented in Fig.4. It is interesting to note that in the majority of cases, the presence of serine or alanine at the polymorphic position 180 of the red opsin gene sequence (Winderickx *et al.*, 1992a) determined whether the hybrid and normal pigments differed in absorption characteristics. The serine and alanine-containing forms of the red pigment were shown to differ by 5 nm in λ_{max} (Merbs and Nathans, 1992a). These results indicate good but not perfect correspondence between the amino acid sequence of the pigments in an array and the severity of the color vision defect.

The correspondence between Rayleigh match range and sequence of hybrid and normal opsin genes was studied in 19 deutan subjects (2 deuteranopic and 17 deteranomalous). A plot of Rayleigh match range width as a function of difference in the inferred λ_{max} between the L and M-L pigments in an array is shown in Fig. 5. As expected, the two deuteranopes had normal L and M-L hybrid pigments that were inferred to be identical in λ_{max}. However, 7 of the deuteranomalous subjects had identical pigments in their arrays and 10 had pigments that differed by 4-12 nm. The match range width of the 7 deuteranomalous subjects whose pigments had the same λ_{max}, and who would

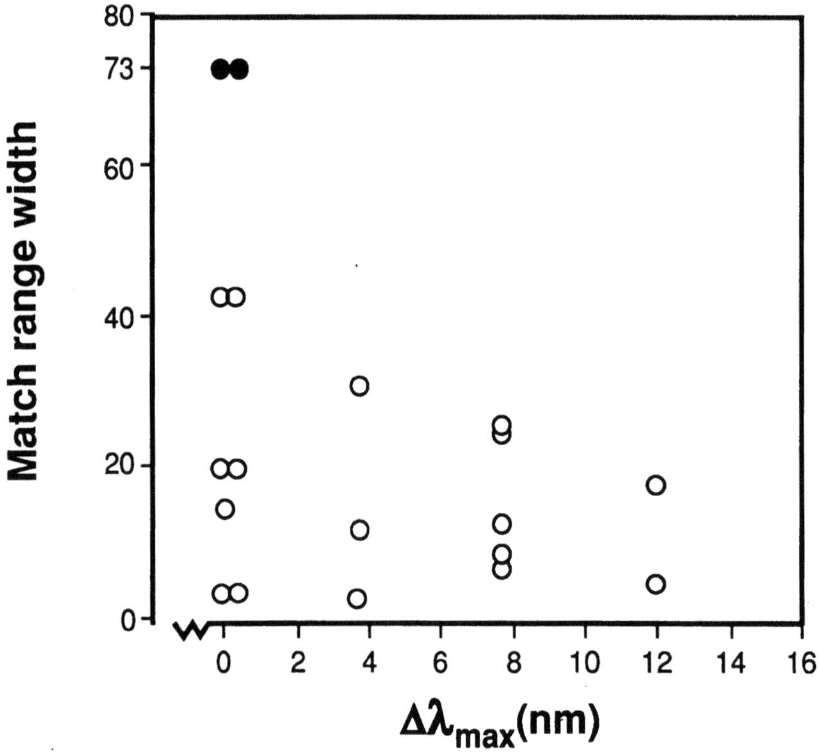

Fig. 5. Rayleigh match range as a function of difference in λ_{max} between normal L and M-L hybrid pigments encoded by gene arrays of deutan subjects. Deuteranopes are designated by solid circles and deuteranomalous subjects by open circles.

have been predicted to test as deuteranopes, did not differ from that of the 10 subjects who had different pigments in their arrays. As in the protan series, the serine/alanine polymorphism at position 180 contributed significantly to the difference in λ_{max} between the normal and hybrid pigments of the array.

Discussion

A better correspondence between genotype and phenotype was achieved by determining the sequence of the hybrid and normal pigment genes than by Southern blot analysis only. Protans showed a much stronger correspondence than did deutans between the severity of the color vision defect and the inferred difference in the λ_{max} of the hybrid and normal pigments. The existence of one or more M pigment genes in addition to the L and M-L hybrid genes in some deutans would not have been expected to contribute to the color vision phenotype since these additional M genes are unlikely to be expressed in the retina (Winderickx *et al.,* 1992b). In the majority of exceptions observed, the subject tested better than was predicted from the spectral characteristics of the pigments. It is likely that some of the remaining discrepancies may be resolved by refinement of the instrumentation and methodologies of assessment of the severity of color vision defects. Alternatively, two factors unrelated to the absorption maxima of the pigments may have contributed to the observed lack of strict correspondence between genotype and phenotype: pigment optical density and efficiency of signal transduction by the photopigment. Sequence variation in the visual pigments could contribute to differences in these properties.

Acknowledgements

The authors wish to thank Laurie Battisti for technical assistance. This research was supported by National Institutes of Health Grant EY 08395.

References

Deeb, S.S., Lindsey, D.T., Hibiya, Y., Sanocki, E., Winderickx, J., Teller, D.Y. and Motulsky, A.G. (1992). Genotype-phenotype relationships in human red/green color vision defects: Molecular and psychophysical studies. Am. J. Hum. Genet. 51: 678–700.

Feil, R., Aubourg, P., Heilig, R. and Mandel, J.L. (1990). A 195-kb cosmid walk encompassing the human Xq28 color vision pigment genes. Genomics 6: 367–373.

Merbs, S.L., and Nathans, J. (1992a). Absorption spectra of human cone pigments. Nature 356: 433–435.

Merbs, S.L. and Nathans, J. (1992b). Absorption spectra of the hybrid pigments responsible for anomalous color vision. Science 258: 464–466.

Nathans, J., Piantanida, T.P., Eddy, R.L., Shows, T.B. and Hogness, D.S. (1986). Molecular genetics of inherited variation in human color vision. Science 232: 203–210.

Winderickx, J., Lindsey, D.T., Sanocki, E., Teller, D.Y., Motulsky, A.G. and Deeb, S.S. (1992a). Polymorphism in red photopigment underlies variation in colour matching. Nature 356: 431–433.

Winderickx, J., Battisti, L., Motulsky, A.G. and Deeb, S.S. (1992b). Selective expression of the human X-linked green pigment genes. Proc. Natl. Acad. Sci. U.S.A. 89: 9710–9714.

Winderickx, J., Battisti, L., Hibiya, Y., Motulsky, A.G. and Deeb, S.S. (1993). Haplotype diversity in the human red and green opsin genes: evidence for frequent exchange in exon 3. Hum. Mol. Genet. 2: 1413–1421.

Zhang, Q., Mao, W., Ma,q., Zeng, R., Wu, L., W. and Chen, Y. (1992). Molecular basis of congenital color vision defects in Chinese patients. Jpn. J. Ophthalmology 36: 479–487.

Samir S. Deeb,
Genetics Department SK-50m
University of Washington,
Seattle, WA 98195 U.S.A.

15. Neural network models for normal and dichromatic color vision

SHIRO USUI and SHIGEKI NAKAUCHI
(Toyohashi, Japan)

Abstract

Neural network models for normal and dichromatic color vision were constructed by a top-down approach rather than an experimental bottom-up approach. The back-propagation learning technique was used to train a three-layer neural network model to realize the nonlinear mapping from cone space to perceived color space. The dichromatic color vision models of the three types of cone loss were represented by the constraint of a loss of the respective cone type. The characteristics of the trained models were analyzed by using spectrally pure light as test color stimuli and compared with available psychophysical data. Each model quantitatively reproduces the available psychophysical evidence on dichromatic vision, such as spectral response property, hue representation of spectrally pure light, and wavelength discrimination without requiring the adjustment of the heuristic parameters characteristic of bottom-up models.

Introduction

A typical form of dichromatic color blindness is a consequence of the loss of one of the three types of cones. The nature of color blindness has been studied by constructing bottom-up color vision models (e.g. Ingling and Tsou, 1977; Ingling, 1977; Guth *et al.*, 1980). These bottom-up models take experimental data as the starting point and are constructed using parameters to fit the data.

This paper proposes neural network models for normal and dichromatic color vision by a top-down approach, that is, by utilizing the back-propagation learning technique (Rumelhart *et al.*, 1986) to realize a nonlinear mapping from cone space defined by responses of cones, to perceived color space defined by one lightness and two chromatic responses without the need for heuristic parameters.

In order to construct the dichromatic color vision models, we introduce the hypothesis that "dichromatic color vision has the best chromatic discrimination ability under the structural constraint of only two cone types instead of three".

B. Drum (ed.), Colour Vision Deficiencies XII, pp. 127–134.
© 1995 *Kluwer Academic Publishers, Dordrecht.*

The dichromatic models also realize a mapping between two-dimensional cone space and perceived color space, although the structural constraint restricts their abilities.

After training was completed, the network performance was analyzed using spectrally pure light as test color stimuli and was compared with psychophysical evidence.

Model structure and learning data

Figure 1 shows a schematic diagram of a three-layer neural network model (1b) and training data (1a and 1c). The model consists of three (for normal color vision model) or two (for dichromatic color vision model) input units, three hidden units and three output units. Each hidden and output unit has a sigmoidal input-output function. The bias unit has a constant output value of 1 and controls the dynamic range of the output units. The input units correspond to the three types of cones (S, M and L) in the retina and have the broadband spectral sensitivities derived from Smith-Pokorny (1975). The output units represent the three opponent color coordinates: lightness, redness-greenness and yellowness-blueness.

Thus, the network models are trained to perform the mapping from cone space to perceived color space. For training, we used as input a set of data consisting of 1569 triplets of cone responses to Munsell color chips under the CIE illuminant C as shown in Fig. 1a, and opponent color coordinates of input color chips which were transformed from the original cylindrical Munsell system into the orthogonal system and normalized as shown in Fig. 1c. Connection weights were initialized randomly for each training.

Protanopia, deuteranopia and tritanopia were represented by a loss of one of the L, M or S cone units, respectively. That is, each dichromatic model has only two cone units in the input layer. For training the dichromatic color vision models, the same data set for training the normal color vision model was used. Of course, the structural constraint (loss of one type of cone units) restricts their performance.

Analysis of the model performance

After training was completed, we anlayzed the model performance to spectrally pure light stimuli, which are often used as color stimuli to measure the characteristics of color vision in human subjects, such as the spectral response property or wavelength discrimination.

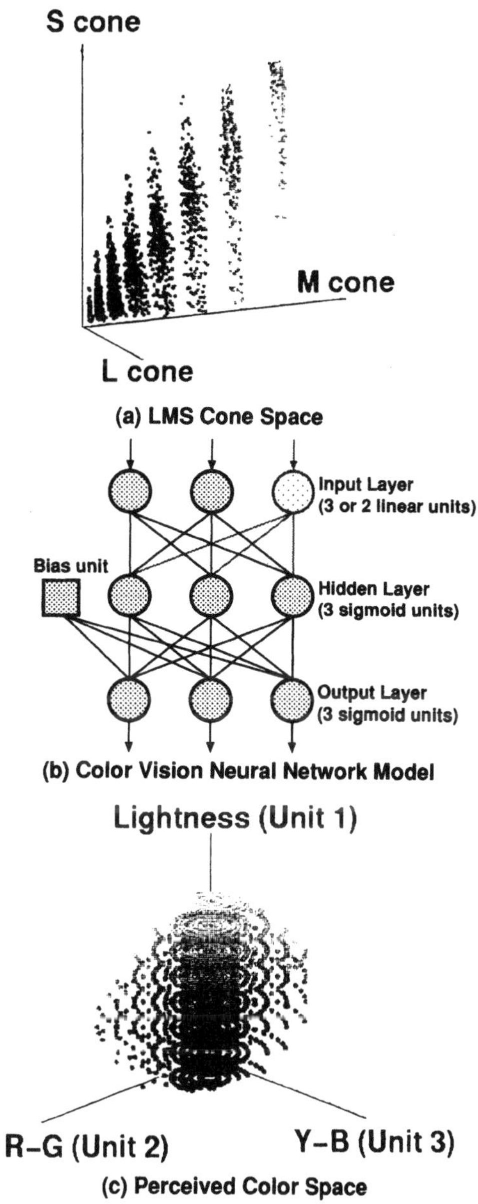

Fig. 1. Color vision neural network model. Model structure (b) and training data (a and c). Each layer consists of three (or two for dichromatic color vision models) cone units, three hidden units and three output units. The set of input-teaching data for training the models consists of (a) triplets of cone responses to 1569 Munsell color chips under the illuminant C and (c) Color opponent coordinates of input color chips, which represent lightness, redness-greenness and blueness-yellowness.

Spectral response and hue representation

Figure 2 shows spectral responses of each output unit and hue angle θ (λ) of the normal and the dichromatic color vision models. In the case of the normal color vision, shown in Fig. 2a, spectral responses of unit 1 (light line), 2 (dark line) and 3 (gray line) are mono-, tri- and bi-phasic, respectively. These characteristics are similar to the luminous efficiency curve and color opponent responses. Hue

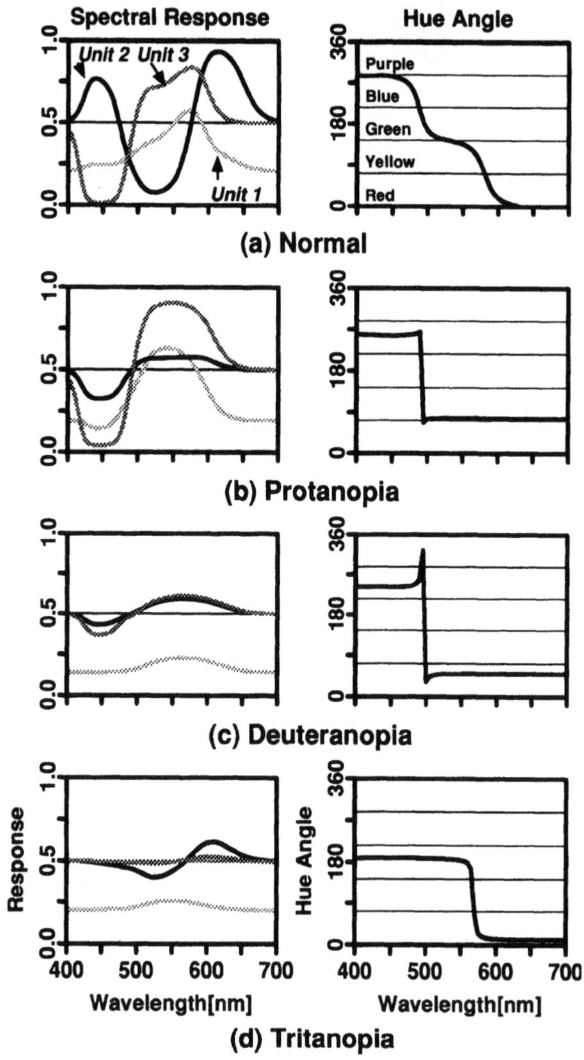

Fig. 2. Spectral response (left-side) and hue angle (right-side) for the spectrum of the normal and the dichromatic color vision models. Light, dark and gray lines show response curves of unit 1, 2, and 3, respectively. (a) normal, (b) protanopia, (c) deuteranopia and (d) tritanopia.

angle θ (λ) for the spectral light stimuli is given by response-ratio of unit 2 and 3 and accounts well for the color appearance such that there is a human perception of purple, despite the fact that "purple" does not appear in the physical spectrum.

It is well-known that in the case of dichromatic color blindness, the spectrum can evoke only two distinguishable hue sensations, yellow-blue for protanopia and deuteranopia, and green-red for tritanopia dichromat. The spectral responses and hue angle of the dichromatic color vision models shown in Figs. 2b and 2c agree with these observations on dichromatic human subjects. The model also correctly predicts that perceived hues for protanopia and deuteranopia dichromats are slightly different.

Wavelength discrimination

Wavelength discrimination has been measured by psychophysical experiments to evaluate the way the visual system performs. For normal color vision, there are two minima as shown in Fig. 3a where the discrimination is best, and discrimination is relatively poorer both at the midspectral region and at the spectral extremes (Wright and Pitt, 1934).

To calculate the wavelength discrimination function for the dichromatic color vision, we define the jnd (just noticeable difference) for wavelength discrimination $\Delta\lambda$ as follows:

$$\Delta\lambda = \left| \frac{d\theta}{d\lambda} \right|^{-1}$$

The wavelength discrimination function of the normal color vision model (Fig. 3b) shows a similar general trend; the best discrimination is at 480 nm and 580 nm.

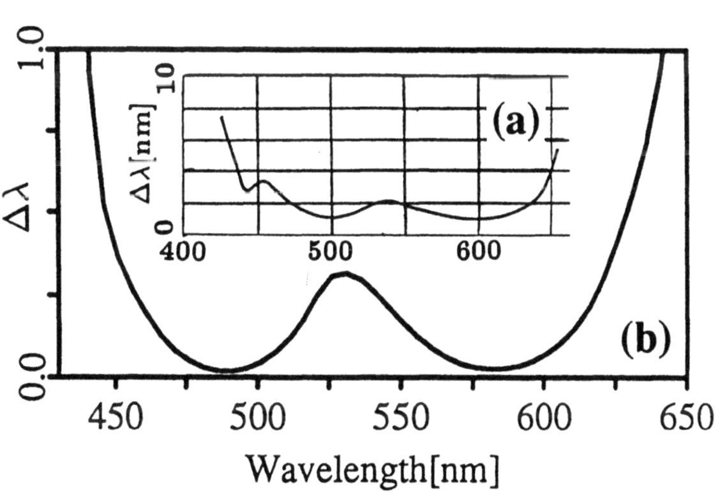

Fig. 3. Wavelength discrimination of normal color vision. (a)Psychophysical experiment (Wright and Pitt, 1934). (b) Prediction by the proposed normal color vision model.

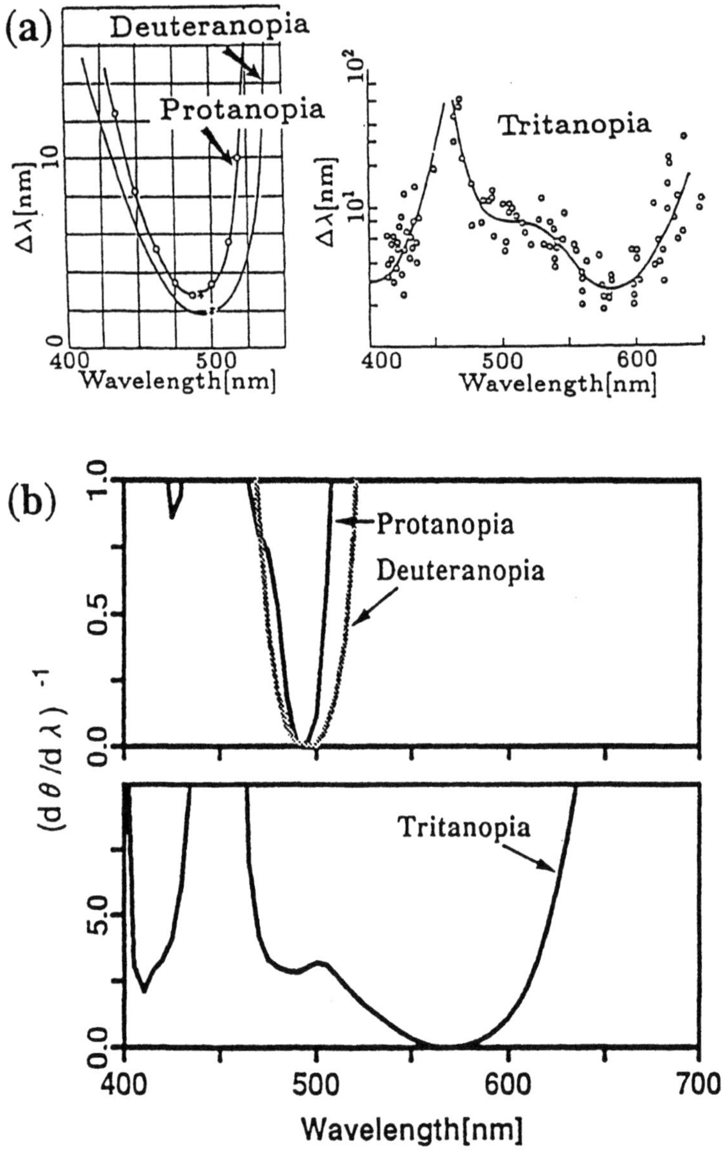

Fig. 4. Wavelength discrimination of dichromatic color vision. (a)Psychophysical experiments for protanopia, deuteranopia (Pitt, 1935) and tritanopia (Wright, 1952). (b) Prediction by the proposed dichromatic color vision models.

The wavelength discrimination function for each dichromatic color vision was also measured (Pitt, 1935; Wright, 1952) as shown in Fig. 4a. As shown in Fig. 4b, each dichromatic model closely fits the experimental wavelength discrimination data.

Prediction of color appearance for dichromats

Our models can be used to generate color images that a dichromat will perceive as the original image, and therefore may suggest to the trichromat what a dichromat is seeing (Fig. 5). To obtain these images, RGB values of the CRT display for the original image were transformed to LMS cone responses, and these LMS cone responses were used as input to the dichromatic color vision models. Each of the three dichromatic images was then created using the color opponent coordinates output by each of the dichromatic color vision models.

For dichromats, the normal color image can only evoke a limited range of distinguishable hue sensations. The image for protanopia has only bluish and yellowish colors. The image for deuteranopia is almost similar to that of protanopia except for the yellow in this image, which is slightly shifted toward the red. Blue and yellow in the image for tritanopia shift to greenish and reddish, respectively. These results correspond well to the hue representation for the spectrum of the dichromats as shown in Figs. 2b, c and d. It is of interest to note that the dichromatic images are not displeasing, and quite colorful to the trichromatic observer.

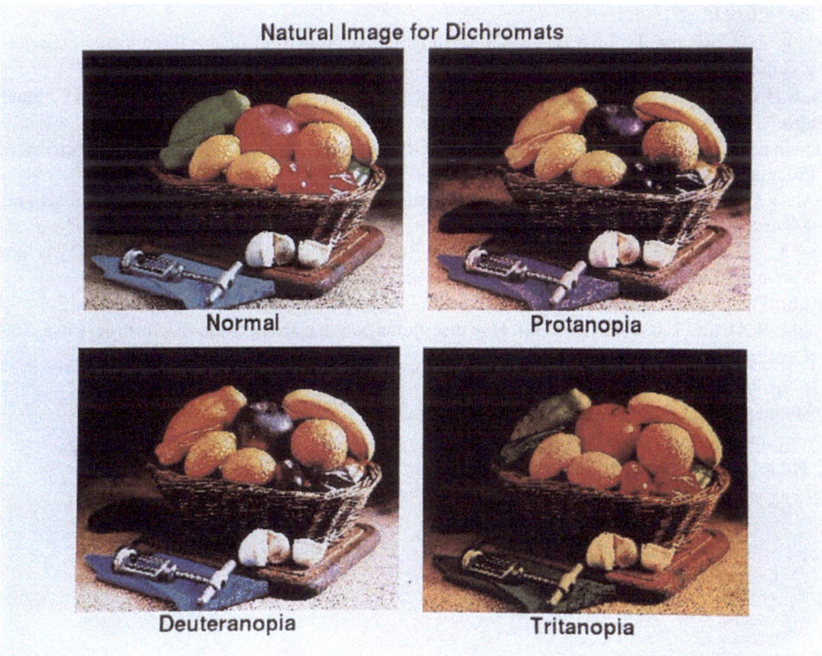

Fig. 5. Prediction of the appearance of a full color image to dichromatic observers as generated by our dichromatic color vision models. The normal color image (upper left) can only evoke a few hue sensations for each dichromat as expected by the hue representations shown in Fig. 2.

Conclusion

Neural network models for normal and dichromatic color vision, which realize the mapping from cone space to perceived color space, were constructed. It is shown that the properties of the trained models provide a quantitative resemblance to the properties of natural color vision, such as hue representation and wavelength discrimination for each of the three dichromatic color vision models. These results suggest that the hypothesis underlying the construction of the dichromatic color vision models is acceptable and further stress our previous suggestion that colors are represented in the visual system by lightness and redness-greenness and blueness-yellowness (Usui *et al.,* 1992). Studies that evaluate the perception of dichromatic observers of synthetic color images produced by our models are highly desirable as they may serve to establish the predictive abilities of our models.

References

Guth, S.L., Massof, R.W. and Benzschawel, T (1980). Vector model for normal and dichromatic color vision. J. Opt. Soc. Am. 70(2): 197–212.

Ingling, Jr. C.R. (1977). The spectral sensitivity of the opponent-color channels. Vision Res. 17: 1083–1089.

Ingling, Jr. C.R. and Tsou, B.H. (1977). Orthogonal combination of the three visual channels. Vision Res. 17: 1075–1082.

Pitt, F.H.G. (1935). Characteristics of dichromatic vision. Great Britain Medical Research Council, Special Report Series, 200: 1–58.

Rumelhart, D.E., McClelland, J.L. and the PDP Research Group (1986). Parallel Distributed Processing. MIT Press, Cambridge, MA.

Smith, V.C. and Pokorny, J. (1975). Spectral sensitivities of the fovea cone photopigments between 400 and 500 nm. Vision Res. 15: 161–171.

Usui, S., Nakauchi, S. and Nakano, M. (1992). Reconstruction of Munsell color space by a five-layer neural network. J. Opt. Soc. Am. A. 9(4): 516–520.

Wright, W.D. (1952). The characteristics of tritanopia. J. Opt. Soc. Am. 42: 509–520.

Wright, W.D. and Pitt, F.H.G. (1934). Hue discrimination in normal color vision. Proc. Phys. Soc. (London) 46: 459–473.

Department of Information & Computer Sciences
Toyohashi University of Technology
1-1 Hibarigaoka Tempaku, Toyohashi 441, Japan

16. Deuteranopia under conditions of a large field

A. ORAZEM and H. SCHEIBNER
(Düsseldorf, Germany)

Abstract

By means of a visual tristimulus colorimeter according to Guild-Bechstein, two types of missing colours and two types of alychne traces were measured by a male small-field deuteranope under the conditions of a 10° visual field. The one type is characterized by a chromaticity region, where the deuteranopic subject could manipulate perfect dichromatic colour matches, the other type is characterized by a chromaticity region where he could not. The nature of this dichotomy and the luminance level prevailing hinted at the interpretation that the emerging extension of deuteranopic colour vision was a deuteranomalous trichomatic vision. By means of three perceptual criteria beyond the colour match, i.e. 'neither blue nor yellow', 'neither green nor red' and 'heterochromatically equally bright', the elements of deuteranopic and deuteranomalous opponent-colour theory were measured, and constitutive relations were derived. They give a quantitative description of the deuteranope's colour vision under large field viewing conditions.

Introduction: aims and methods

As Nagel (1905, 1907) already noticed, some dichromats show signs of trichromatic vision if the visual field is markedly larger than foveal. There are similar reports by, among others, Jaeger *et al.*, (1952, 1987, 1991), Smith and Pokorny (1977), Nagy (1980, 1982) and Breton and Cowan (1981).

By means of a visual tristimulus colorimeter according to Guild-Bechstein (Beck and Richter, 1958), colour vision of a male deuteranope was investigated on a 10° diameter visual field. This choice of field allowed to use the CIE 1964 Supplementary Standard Colorimetric System (DIN 5033, Pokorny and Smith, 1986). An important tool in the psychophysical investigation of normal and defective colour vision is the colour match, in particular in the restricted form of the Rayleigh match. One purpose of this investigation was to include more than one perceptual criterion. Such an extension beyond the colour match will

B. Drum (ed.), Colour Vision Deficiencies XII, pp. 135–147.
© 1995 *Kluwer Academic Publishers, Dordrecht.*

lead to a richer description of colour vision and permit suggestions of the underlying receptoral mechanisms (Scheibner, 1990, 1993).

The luminance of the visual field was between 50 cd/m^2 and 110 cd/m^2 and an exit pupil diameter of 1.6 mm yielded retinal illuminances between 100 and 220 trolands. The luminance was measured by means of a visual luminance meter according to Bechstein (manufacturer: Schmidt and Haensch, Berlin).

Colorimetric reference system

The optical output of the visual colorimeter approximated the reference system according to Wright (1946): The instrumental primaries were represented by approximately monochromatic radiations of wavelength 460 nm (blue, B), 530 nm (green, G), and 650 nm (red, R). Any colour may, therefore, be represented as an additive mixture in the form

$$\mathbf{C} = \mathrm{B} \cdot \mathbf{B} \quad + \quad \mathrm{G} \cdot \mathbf{G} \quad + \quad \mathrm{R} \cdot \mathbf{R}$$

where B, G, R are so-called tristimulus values and **B**, **G**, **R** are basis vectors called 'primaries'.

The blue-green normalisation of primaries was 1:1, whereas the green-red normalisation was 1:0.4. By this measure, the chromaticity loci of the deutan missing colours to be determined took on a convenient finite distance from the instrumental primary **G**.

*Determination of the deuteranopic missing colour **D** and a 'deuteranomalous' quasi-missing colour $\mathbf{S}_d$ (Fig. 1)*

Figure 1 shows a colour triangle BGR assumed to be rectangular and pertaining to our instrumental primaries **B, G, R**. The deuteranopic subject was presented, on the right-hand side of the bipartite visual field, one of five monochromatic test radiations indicated by their wavelengths (from 475 nm to 580 nm). The task was to match such an initial stimulus to a superposition of the three primary components, the 'green' component G being fixed, but the 'red' and 'blue' components R and B being allowed to be freely varied. This procedure of 'dichromatic colour match' within a three-dimensional colour space has been described by several authors (Nuberg and Yustova, 1958; Scheibner, 1968, 1976). In Fig. 1, the filled circles denote chromaticity loci of colour mixtures which perfectly matched the initial spectral stimuli, whereas the open circles indicate loci for which the match was feasible only in an incomplete way, i.e. the perceptual criterion 'indistinguishably equal' was not strictly fulfilled. As can be seen, loci of mixtures matching the same initial monochromatic stimulus form so-called confusion lines, which intersect within a small area. Thus, for complete dichromatic colour matches, there resulted the median deuteranopic missing colour $\overline{\mathbf{D}}$, for incomplete, non-dichromatic colour matches, there resulted the median quasi-missing colour $\overline{\mathbf{S}_d}$. To the latter we have assigned, in anticipation of the discussion, the attribute 'deuteranomalous'.

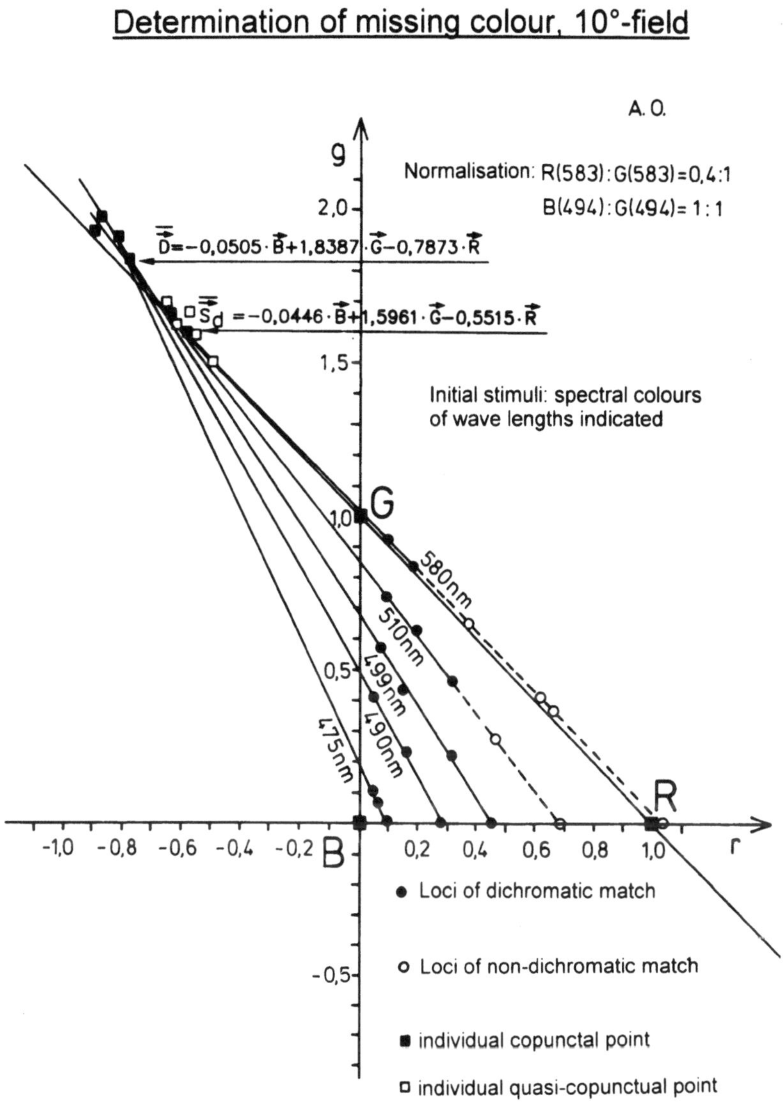

Fig. 1. Colour triangle BGR pertaining to the instrumental primaries **B**, **G**, **R**. The circles designate colour mixtures which matched the spectral initial stimuli characterized by the wavelengths that are written on the pertaining confusion lines. Filled circles designate complete matches, open circles designate incomplete matches. To avoid overcrowding, only some loci of individual missing colours are shown as filled and open squares.

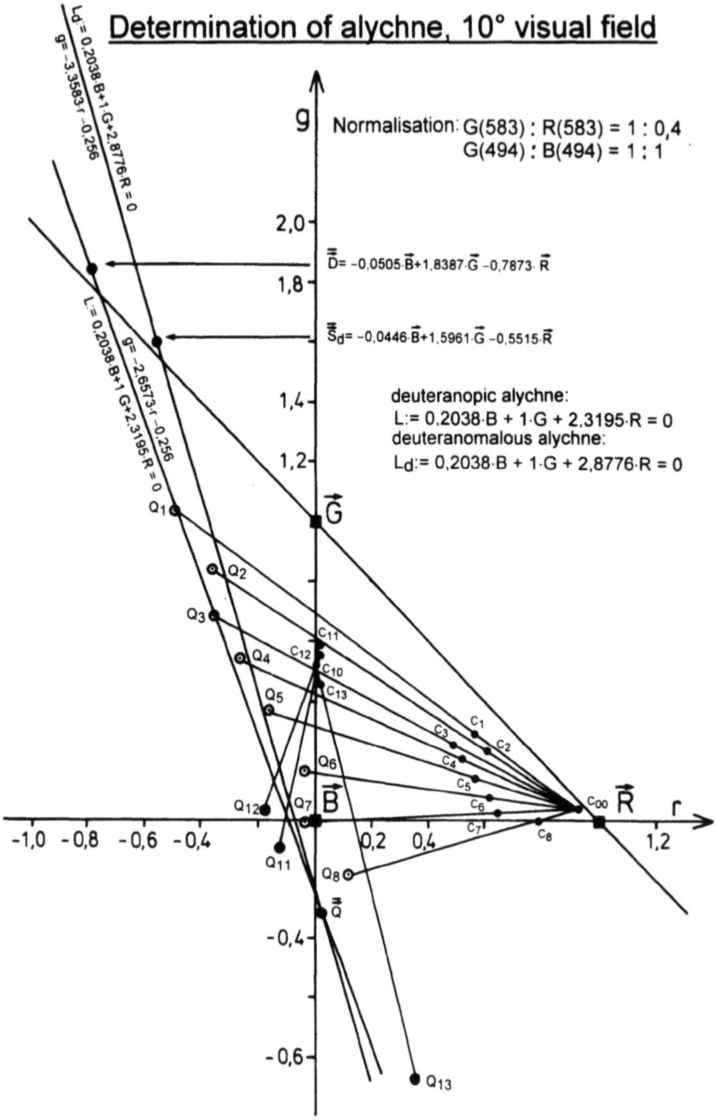

Fig. 2. Instrumental colour triangle as in Fig. 1. Measurement and construction of a deuteranopic and a 'deuteranomalous' alychne trace; see text. The two traces are straight lines and are expressed both in chromaticity coordinates r, g and (homogeneously) in tristimulus values B, G, R, i.e. as linear forms.

Determination of a deuteranopic and a 'deuteranomalous' alychne

According to Schrödinger (1925), the alychne is a plane of the vectorial colour space, i.e. a two-dimensional subspace, in which colour vectors of zero luminance lie coplanarly. Such zero luminance vectors are calculated as free difference vectors of two colours which have been made heterochromatically equally bright (Scheibner, 1976, 1993b).

In Fig. 2, the filled circles Q_{11}, Q_{12}, Q_{13} are chromaticity loci of luminance-free difference vectors, which have been formed from colours denoted by their chromaticity c_{10}, c_{11}, c_{12}, c_{13} located in the region where complete dichromatic colour matches proved to be possible. The straight line approximating the loci Q_{11}, Q_{12}, Q_{13} must run through the locus of the mean missing colour $\overline{\mathbf{D}}$ since this one is likewise a luminance-free difference vector even of matched colours. The result is the straight line L = 0, the trace of the deuteranopic alychne.

The open circles Q_1, Q_2 ... Q_8 are loci of luminance-free difference vectors, which have been formed from the colours denoted by c_{00}, c_1, c_2 ... c_8 located in the region where complete dichromatic colour matches proved not to be possible. Here, similarly, the locus of the mean quasi-missing colour $\overline{\mathbf{S}}_\mathbf{d}$ is a constitutive element of the fitted straight line, since $\overline{\mathbf{S}}_\mathbf{d}$ resulted from very safe heterochromatic brightness matches. This yields the straight line $L_d = 0$, the trace of an alychne, to which we, again, assign, in anticipation of the discussion, the attribute 'deuteranomalous'.

The deuteranopic opponent-colour vision

With reference to the instrumental colour triangle BGR $_1$a fundamental colour triangle PDT is shown in Fig. 3, where the 'green' fundamental primary **D** has been taken over from Fig. 1, primaries **P** and **T** have been taken from Bruckwilder and Scheibner (1988/89).

Figure 3 also shows the deuteranopic alychne L = 0 taken over from Fig. 2. Additionally, Fig. 3 shows the deuteranopic neutral zone K = 0 which was experimentally determined by the subject applying the perceptual criterion 'neither blue nor yellow'. Both the line L = 0 and the line K = 0 are members of the pencil of straight lines of which the carrier point is the locus D of the deuteranopic missing colour **D.** (Following geometric terminology (e.g. Hopkins and Hails, 1953), we denote by 'pencil' a (plane) set of straight lines that intersect in a common point, the 'copunctal' point).

The linear forms describing the straight lines (as in Fig. 2) provide the linear transformation from the three-dimensional instrumental colour space (B, G, R) onto the two-dimensional opponent-colour space (K, L) (Scheibner, 1990; Knottenberg and Scheibner, 1993). We call K the 'chrominance', more exactly, the 'blue-yellow chrominance', L the 'luminance'. The transformation equation is shown in Fig. 4 (upper matrix equation). It transforms the three instrumental colour matching functions $\bar{b}(\lambda)$, $\bar{g}(\lambda)$, $\bar{r}(\lambda)$ into the two opponent spectral functions $\bar{k}(\lambda)$, $\bar{l}(\lambda)$. The resulting curves are shown in Fig. 4. The elements of

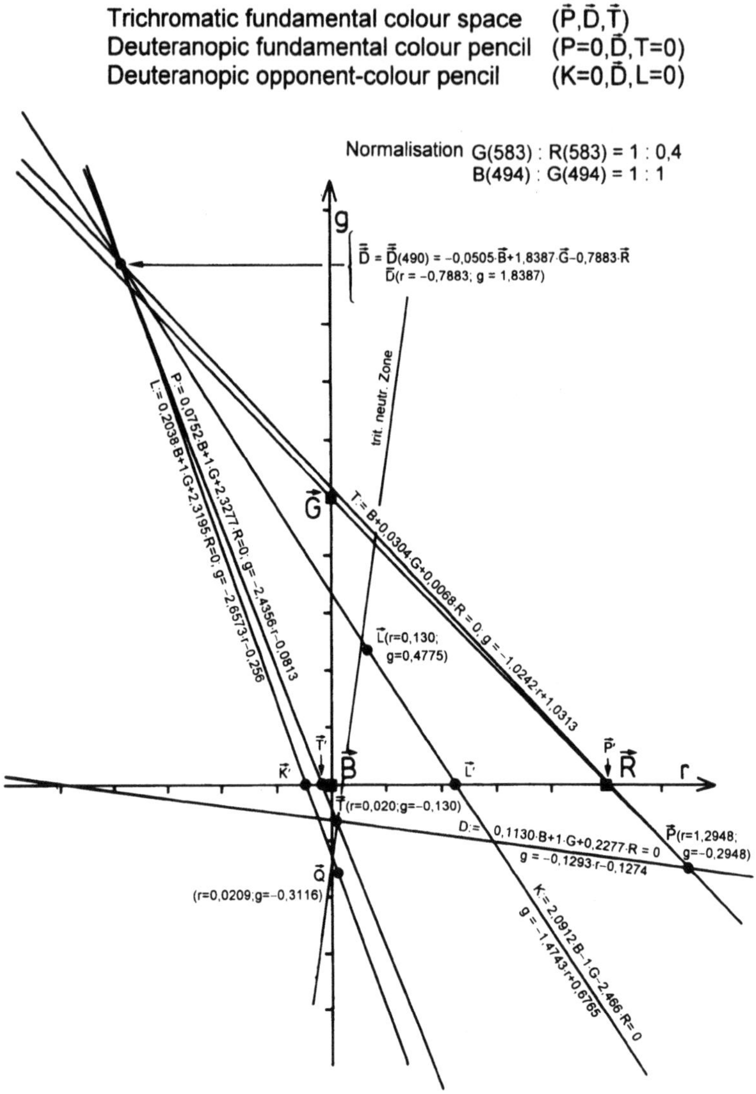

Fig. 3. Instrumental colour triangle BGR as in Fig. 1. The deuteranopic missing colour **D** (Scheibner, 1968) is the carrier point of two dichromatic 'pencils' (see e.g. Hopkins and Hails, 1953): a) one with the referential straight lines K = 0 and L = 0, b) one with the referential straight lines P = 0 and T= 0. The latter pencil is supplemented by the line D = 0 to a trichromatic fundamental colour triangle PDT.

the two matrix rows are in agreement with the coefficients of the linear forms K = 0 and L = 0 of Fig. 3 up to a normalization factor.

By matrix concatenation, the connection between the deuteranopic fundamental spectral functions $\bar{p}(\lambda)$, $\bar{t}(\lambda)$ and the deuteranopic opponent

Deuteranopic opponent spectral functions, 10°-field

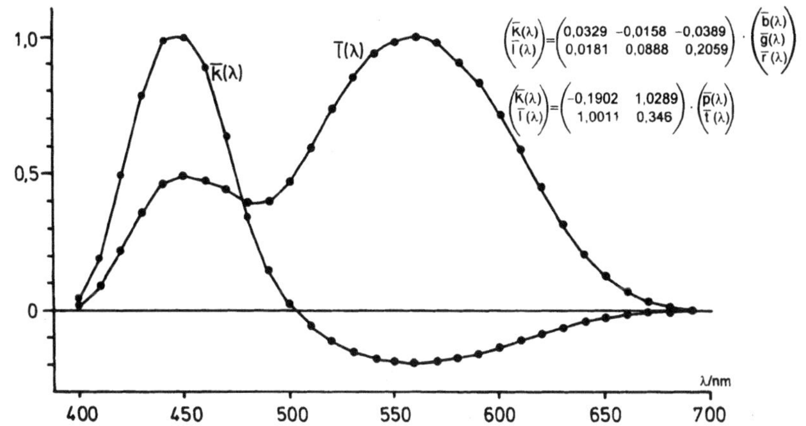

Fig. 4. Opponent spectral functions of the deuteranopic subject for 10° visual field; $\bar{b}(\lambda)$, $\bar{g}(\lambda)$, $\bar{r}(\lambda)$ are trichromatic instrumental colour matching functions (Wright, 1946); $\bar{k}(\lambda)$ is the spectral blue-yellow chrominance function, $\bar{l}(\lambda)$ is the spetral luminance function; $\bar{p}(\lambda)$, $\bar{t}(\lambda)$ are spectral fundamental functions, where $\bar{p}(\lambda)$ may be interpreted as the spectral sensitivity of the 'red' retinal cones, $\bar{t}(\lambda)$ as the spectral sensitivity of the 'blue' retinal cones.

spectral functions $\bar{k}(\lambda)$, $\bar{l}(\lambda)$ can be established. This transfer equation, also shown in Fig. 4 (lower matrix equation), represents the essence of the deuteranopic opponent-colour description.

Choice of the deuteranomalous fundamental colour space

In Fig. 5, similar to Fig. 3, the chromaticity locus of the deuteranomalous quasi-missing colour $\mathbf{S_d}$ of Fig. 1 (and Fig. 2) has been made the copunctal point of the pertaining pencil. Together with the two other fundamental primaries **P** and T of Fig. 3, a 'deuteranomalous' fundamental colour triangle PS_dT is fixed. Its triangle sides are, again, characterized in Fig. 5 by, among others, homogeneous linear forms. If we compare Fig. 5 with Fig. 3 we may recognize that only the left triangle side $P_d = 0$ has changed its coefficients. Similar to Fig. 3, the three linear forms supply the mapping equations from the instrumental colour space (B, G, R) onto the 'deuteranomalous' fundamental colour space (P_d, D_d, T_d), where P_d is identical with P, D_d is identical with S_d, and T_d is identical with T. This transfer equation is shown in Fig. 5, too. The choice of the usual protanopic missing colour as 'red' deuteranomalous fundamental primary **P** has consequences that are discussed later.

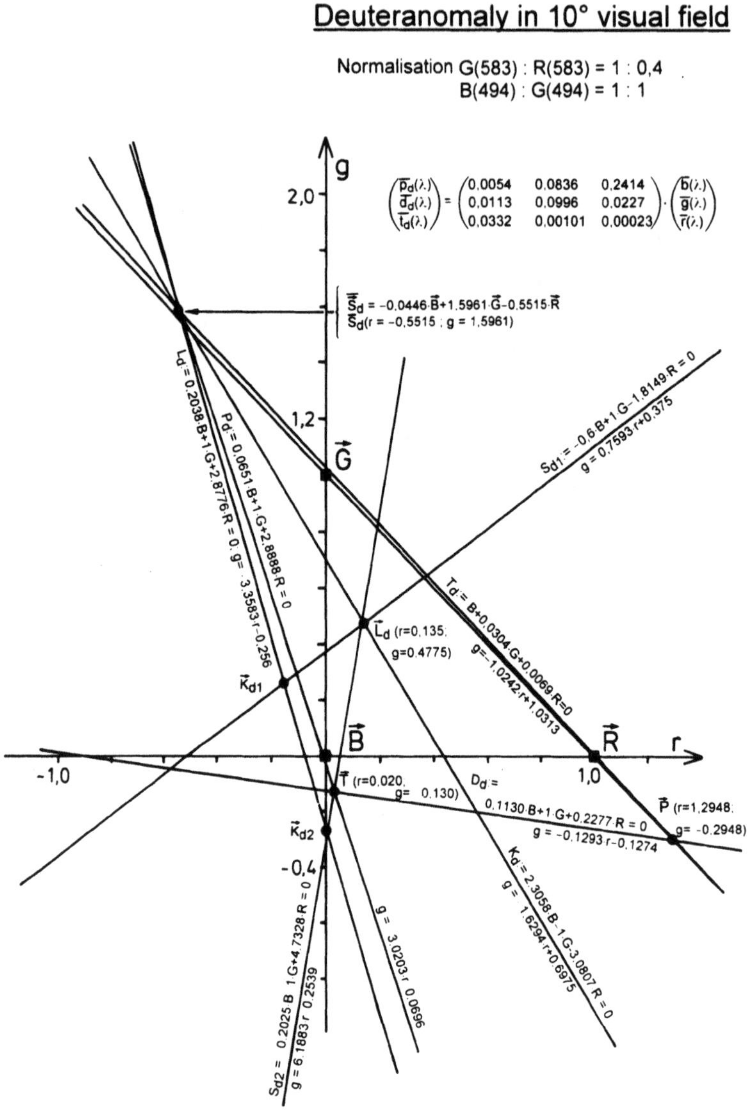

Fig. 5. Fundamental colour triangle P S_dT with reference to the instrumental colour triangle BGR. Compared to Fig. 3, the copunctal point D of Fig. 3 has been changed to the copunctal point S_d of Fig. 1 and, therefore, the left fundamental triangle side has changed from P = 0 to P_d = 0. Additionally, two deuteranomalous opponent colour triangles K_{d1} S_d L_{d2} and K_d $S_d L_d$ are shown, for which the luminance-relevant triangle side, i.e. the alychne trace, has undergone a change from L = 0 to L_d = 0 according to Fig. 2. For further details see text.

The deuteranomalous opponent-colour vision

In Fig. 5, also the 'deuteranomalous' alychne trace $L_d = 0$ is shown. It has been taken over from Fig. 2. In order to be able to construct a deuteranomalous opponent-colour system we need, moreover, a locus on which the perceptual criterion 'neither blue nor yellow' holds, and we need a locus on which the perceptual criterion 'neither green nor red' holds.

The first locus is given through the neutral zone. Experimentally, we could not distinguish between a deuteranomalous and a deuteranopic neutral zone. However, in order to have a consistent deuteranomalous opponent-colour system, we made the deuteranomalous neutral zone a member of the pencil carried by the copunctal point S_d. The neutral zones of Figs. 3 and 5 thus differ slightly. Particularly in the region of real chromaticities, however, the difference does not exceed the experimental scatter.

Weak green-red vision was exhibited by the subject on the 'yellow' side of the neutral zone; cf. the region of open circles in Fig. 1. Indeed, the deuteranopic subject could manipulate the perceptual criterion 'neither green nor red' on the large visual field only in an insecure way. For that purpose, he gauged binary mixtures that obeyed the criterion. Their loci were approximated through the straight line $S_{d1} = 0$ (Fig. 5). The problem arose how to continue this line on the 'blue' side of the neutral zone. The following experiment gave an answer: the subject was presented the short wavelength spectrum in sequence with the help of an interference filter wedge installed in the colorimeter. The result was that the subject could not see any red in the form of a violet sensation as far as the realizable spectrum end. From that we decided that a theoretical tritanopic neutral zone be a continuation of the straight line $S_{d1} = 0$. This tritanopic neutral zone is drawn in Fig. 5 and denoted by $S_{d2} = 0$. It connects the locus of $\lambda = 570$ nm with the short wavelength end point of the spectral chromaticity locus.

Due to the piecewise linearisation of the curve-shaped locus at which the perceptual criterion 'neither green nor red' held, there are two opponent-colour triangles $K_{d1} L_dS_d$ and $K_{d2}L_dS_d$ in Fig. 5. The linear forms describing their triangle sides, again, supply the mapping equations from the instrumental colour space (B, G, R) onto the opponent-colour space (K_d, S_d, L_d) as shown in Fig. 6. The piecewise linearisation appears in the form of two matrix rows so that the matrix possesses four rows. The resulting opponent spectral curves are shown in Fig. 6, the dashed curve branches indicating the non-valid continuations.

The physiologically relevant transfer of the deuteranomalous fundamental spectral functions $\bar{p}_d(\lambda)$, $\bar{d}_d(\lambda)$, $\bar{t}_d(\lambda)$ (Fig. 5) into the deuteranomalous opponent spectral functions $\bar{k}_d(\lambda)$, $\bar{s}_{d1}(\lambda)$, $\bar{s}_{d2}(\lambda)$, $\bar{l}_d(\lambda)$ is also shown in Fig. 6.

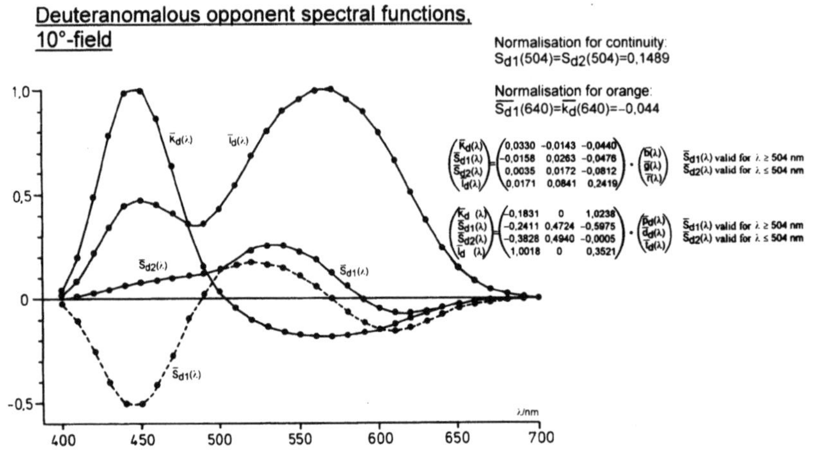

Fig. 6. Opponent spectral functions effective in the orange-red region as indicated by open circles in Fig. 1. The notation is analogous to Figs. 4 and 5. Additionally to the two functions shown in Fig. 4, a spectral green-red chrominance function emerges, which is composed of two pieces $\bar{s}_{d1}$ and $\bar{s}_{d2}$. Invalid curve pieces are dashed. For further details see text.

Discussion

There are controversial reports on residual green-red discrmination in dichromats, e.g. by Scheibner and Boynton (1968) and Wooten and Wald (1973). The present contribution attempts to quantify the emergence of a green-red-vision out of deuteranopia under the conditions of a large visual field, at photopic luminance levels.

For luminance levels of about 10 cd/m^2 and lower, retinal rods are said to be effective and induce a chroma 'red' in protanopes and deuteranopes for a 8° visual field according to Smith and Pokorny (1977). Nagy (1980, 1982), Breton and Cowan (1981) and Scheufens and Scheibner (1984, 1987) essentially confirm these findings. Additionally, Breton and Cowan report on an anomalous receptor type that induces a kind of trichromatic vision for higher luminance levels. This would be the case for levels higher than about 100 troland.

Our measurements were made at a level, between 100 and 220 trolands. Therefore, our data lend themselves to the interpretation that a third, altered cone pigment causes an emerging deuteranomalous trichromatic vision. Our evaluation using the perceptoric tools of the opponent-colour theory permits to separate deuteranopic and deuteranomalous traits by presupposing only the scheme of simple linear opponent-colour vision (cf. e.g. Knottenberg and Scheibner, 1991, 1993). Beyond that, any predictions based on hypothetical assumptions are not necessary.

The large field deuteranopia, Figs. 3 and 4, shows an unconspicuous chrominance curve $\bar{k}(\lambda)$, whereas the luminance curve $\bar{l}(\lambda)$, i.e. the spectral brightness sensitivity, shows a short wavelength side maximum. As in Fig. 3 the straight line L = 0 and the 'blue' fundamental primary **T** are not incident, this contribution results from the 'blue' cones. For a strict foveal visual field, however, this contribution is zero (Kröger-Paulus and Scheibner, 1980; Orazem and Scheibner, 1992).

The large field deuteranomaly – here derived from a residual green-red vision of the deuteranope – shows a weak 'alteration' of the long wavelength curve $\bar{p}_d(\lambda)$ towards longer wavelengths compared to large field deuteranopia, Fig. 5. However, the middle wavelength curve $\bar{d}(\lambda)$ of normal trichromacy and that of large field deuteranomaly are identical, if we assume a linear connection between normal large field trichromacy and deuteranomalous large field trichromacy, because the triangle sides D = 0 (Fig. 3) and $D_d = 0$ (Fig. 5) are identical. An alternative could be to use an altered function $\bar{d}_d(\lambda)$ from the beginning, e.g. the values provided by DeMarco *et al.*, (1992). Experimental evidence for this is the site of the straight line $S_{d1} = 0$ in Fig. 5. But such a modification would possibly require to change the chromaticity locus of the 'red' fundamental primary **P** in Fig. 5 (Scheibner, 1993a), a possibility that we have not worked out so far. The curve $\bar{k}_d(\lambda)$ in Fig. 6 shows great similarity to that of large field deuteranopia (Fig. 4). A new phenomenon – compared to large field deuteranopia, Figs. 3 and 4, – is the emergence of the green-red-chrominance curve $\bar{s}_d(\lambda)$ (Figs. 5 and 6). Its short wavelength branch $\bar{s}_{d2}(\lambda)$ shows a missing violet sensation, its long wavelength branch $\bar{s}_{d1}(\lambda)$ shows a spectral yellow with a zero crossing at about 590 nm – Donders' shift rule (Donders, 1881).

At $\lambda = 640$ nm (Fig. 6), the ordinate of the curve $\bar{k}_d(\lambda)$ has been made equal to that of the curve $\bar{s}_{d1}(\lambda)$. This 'orange' normalisation rests on the following experimental finding: at about 640 nm, the subject saw orange 'as red as yellow'.

In summary, it is a remarkable feature of deuteranopic large field colour vision, that, compared to small field deuteranopia, the chroma 'red' reappears in the long wavelength spectrum, but does not in the short wavelength spectrum. We consider this a postreceptoral phenomenon, which illuminates the 'pivotal role' (Scheibner, 1990) of the 'blue' cones in the process of creating colour sensations.

Acknowledgements

We thank Joel Pokorny for critical comments and helpful suggestions, Gisela Voos and Claudia Wittrock for technical assistance.

References

Beck, J. and Richter, M. (1958) Neukonstruktion des Dreifarbenmeßgerätes nach Guild-Bechstein, Die Farbe 7: 141–152.

Breton, M.E. and Cowan, W.B. (1981) Deuteranomalous color matching in the deuteranopic eye. J. Opt. Soc. Am. 71: 1220–1223.

Bruckwilder, R. and Scheibner, H. (1988/89) Spektralwerte des Protanopen, insbesondere ihre Bestimmung bei minimaler Buntsättigung. Die Farbe 35/36: 215–258.

Deutsche Normen DIN 5033 (1972), Blatt 2, Beuth-Verlag, Berlin.

DeMarco, P., Pokorny, J. and Smith, V.C. (1992) Full-spectrum cone sensitivity functions for X-chromosome-linked anomalous trichromats. J. Opt. Soc. Am. A. 9: 1465–1476.

Donders, F.C. (1881) Über Farbensysteme. Graefe's Archiv 27 (1): 155–223.

Hopkins, E.J. and Hails, J.S. (1953). An introduction to plane projective geometry. Clarendon Press, Oxford.

Jaeger, W. and Kroker, K. (1952), Über das Verhalten der Protanopen und Deuteranopen bei großen Reizflächen. Klin. Monatsbl. Augenheilk. 121: 445–449.

Jaeger, W. and Krastel, H. (1987), Normal and defective colour vision in large field. Jpn J. Ophthalmol. 31: 20–40.

Jaeger, W., Krastel, H. and Marat, G. (1991) Large field spectral matches in dichromats. In: Drum, B., Moreland, J.D. and Serra, A. (eds.), Colour Vision Deficiencies X, Doc. Ophtalmol. Proc. Ser. 54: 14–19, Kluwer Acad. Publ., Dordrecht.

Knottenberg, Th. and Scheibner, H. (1991) Approaching the Abney effect within linear opponent-colour vision. In: Drum, B., Moreland, J.D. and Serra, A. (eds.,) Colour Vision Deficiencies X, Doc. Ophtalmol. Proc. Ser. 54: 315–322, Kluwer Acad. Publ., Dordrecht.

Knottenberg, Th. and Scheibner, H. (1993) Berücksichtigung des Abney-Effekts im Rahmen der linearen Gegenfarbentheorie. Ophthalmologe 90: 155–160.

Kröger-Paulus, A. and Scheibner, H. (1980) Variation of deuteranopic blind fundamentals and the visual cone pigment erythrolabe. Ophthalmic Research 12: 177–183.

Nagel, W.A. (1905) Dichromatische Fovea, trichromatische Peripherie. Z. Psychol. Physiol. Sinnesorg. 39: 93–101.

Nagel, W.A. (1907) Neue Erfahrungen über das Farbensehen der Dichromaten auf großen Feld. Z. Sinnesphysiologie 41: 319–337.

Nagy, A.L. (1980) Large field substitution Rayleigh matches of dichromats. J. Opt. Soc. Am. 70: 778–784.

Nagy, A.L. (1982) Homogeneity of large-field color matches in congential red-green color deficients. J. Opt. Soc. Am. 72: 571–577.

Nuberg, N.D. and Yustova, E.N. (1958) Researches on dichromatic vision and the spectral sensitivity of the receptors of trichromats. In: National Physical Laboratory (ed.), Visual Problems of Colours, Vol. II: 475–486, Her Majesty's Stationery Office, London.

Orazem, A. and Scheibner, H. (1992) Gegenfarbensehen eines Deuteranopen. Pflügers Archiv, European Journal of Physiology 420, Supplement no. 1, R48.

Pokorny, J. and Smith, V.C. (1986) Colorimetry and color discrimination. In: Boff, K.R.,Kaufmann, L. and Thomas, J.P. (eds.), Handbook of perception and human performance, Vol. I, Chap. 8. Wiley, New York.

Scheibner, H. (1968) Dichromasie als Homomorphismus der Trichromasie. Optica Acta 15: 339–349.

Scheibner, H. (1976) Missing Colours (Fehlfarben) of Deuteranopic and Extreme Deuteranomalous Observers. In: Verriest, G. (ed.), Colour Vision Deficiencies III, Mod. Probl. Ophthalmol. 17: 21–26, Karger, Basel.

Scheibner, H. (1990) Aspects of anomalous trichromatic opponent-color vision. In: Ohta, Y. (ed.), Color Vision Deficiencies: 45–60, Kugler and Ghedini, Amsterdam.

Scheibner, H. (1993a) Interconnections between fundamental-color spaces and opponent-color spaces. In: Optical Society of America (ed.), 1993 Technical Digest Series, Vol. 16: 84.

Scheibner, H. (1993b) Transformation of luminance coefficients. J. Opt. Soc. Am. A. 10: 1392–1395.

Scheibner, H.M.O. and Boynton, R.M. (1968) Residual red-green discrimination in dichromats. J. Opt. Soc. Am. 58: 1151–1158.

Scheufens, P. and Scheibner, H. (1984) Mesopic deuteranopic vision with a large observation field. In: Verriest, G. (ed.), Colour Vision Deficiencies VII, Doc. Ophtalmol. Proc. Ser. 39: 311–318, Dr. W.Junk, The Hague.

Scheufens, P. and Scheibner, H. (1987) Mesopic deuteranopic vision: two-dimensional versus three-dimensional, In: Verriest, G. (ed.) Colour Vision Deficiencies VIII, Doc. Ophtalmol. Proc. Ser. 46: 319–325, Nijhoff/Junk, Dordrecht.

Schrödinger, E. (1925) Über das Verhältnis der Vierfarben- zur Dreifarbentheorie. Sitzungsb. Akad. Wiss. Wien IIa/134: 471–490.

Smith, V.C. and Pokorny, J. (1977) Large-field trichromacy in protanopes and deuteranopes. J. Opt. Soc. Am. 67: 213–220.

Wooten, B.R. and Wald, G. (1973) Color-vision mechanisms in peripheral retinas of normal and dichromatic observers. J. Gen. Physiol. 61: 125–145.

Wright, W.D. (1946) Researches on Normal and Defective Colour Vision. Kimpton, London.

Physiologisches Institut II
Universität Düsseldorf
Moorenstr. 5
D-4000 Düsseldorf
Germany

17. The red-green chromatic system in X-chromosome-linked anomalous trichromats

ERIKO MIYAHARA, JOEL POKORNY and VIVIANNE C. SMITH
(Chicago, IL, U.S.A.)

Abstract

We compared the spectral sensitivity of a protanomalous (PA), a deuteranomalous (DA), and a normal (N) trichromatic observer as measured by increment detection thresholds and hue identification thresholds on a white pedestal. Detection sensitivity functions for wavelengths greater than 500 nm showed two peaks and a notch around 570 nm (PA and N) or 580 nm (DA). The peaks of the function for the two anomalous trichromats were much less sensitive and the depth of the notch was much shallower than the peaks and notch of the normal trichromat. Hue detection sensitivity functions clearly separated into the 'red' and 'green' region of the spectrum for each of the observers. Both detection and hue identification data were well fitted by assuming that there was an inhibitory interaction between long-wavelength-sensitive (LWS) and middle-wavelength-sensitive (MWS) cones.

Introduction

Increment threshold spectral sensitivity functions for normal trichromats using a relatively large and long test stimulus on a white background showed three peaks at 450, 530 and 610 nm with local minima near 480 and 570 nm (Sperling and Harwerth, 1971; King-Smith and Carden, 1976; Kranda and King-Smith, 1979; Klingaman *et al.*, 1980; Krastel *et al.*, 1991). The short wavelength peak has been modelled as due to short-wavelength-sensitive (SWS) cone activity, the middle-wavelength peak due to a differential (MWS-LWS) cone signal and the long wavelength peak due to a differential (LWS-MWS) cone signal (Sperling and Harwerth, 1971). Previous increment threshold spectral sensitivities of dichromats showed only one broad peak in the middle to long wavelength region (Verriest and Uvijls, 1977; Klingaman *et al.*, 1980; Foster *et al.*, 1989).

In this study, we measured increment thresholds for anomalous trichromats with a methodology designed to facilitate the test detection by the red-green chromatic system of color defectives. First, the test stimuli were presented on a

B. Drum (ed.), Colour Vision Deficiencies XII, pp. 149–157.
© 1995 *Kluwer Academic Publishers, Dordrecht.*

steady, spatially coextensive white pedestal to favor the detection of the test stimuli by the chromatic mechanism (Foster and Snelgar, 1983; Snelgar *et al.*, 1987; Ueno *et al.*, 1985; Ueno and Swanson, 1989). Second, the temporal presentation included gradual onset and offset. Thus our paradigm minimized both spatial and temporal stimulus transients. A third methodological variable we considered was field size. We included a large test stimulus that included parafoveal retina to maximize the possibility of chromatic detection. In dichromats and extreme anomalous trichromats, chromatic discrimination in the long-wavelength region is improved by the use of large fields (Smith and Pokorny, 1977; Nagy, 1980, 1982; Breton and Cowan, 1981). In color normal observers, an increase in field size from 2° to 10° or larger improves discrimination (Brown, 1952; Stiles and Burch, 1959; Wyszecki and Stiles, 1982) by a factor of about two. We considered that anomalous trichromats are also likely to show greater chromatic sensitivity when a larger test stimulus is used than when a smaller stimulus is used.

Methods

Apparatus and stimuli

Three channels of a four-channel Maxwellian optical view system (Yeh *et al.*, 1993) controlled by a Macintosh II computer were used. The test stimulus was a circular field with 2° or 10° diameter and presented with a temporal profile of one trough-to-trough period of a 1 Hz raised cosine. The test was either chromatic (dominant wavelengths of 458, 480, 500, 510, 529, 541, 559, 571, 580, 590, 610, 639 and 659 nm) light or 4600 K white light. A steady 4600 K white background of 200 td and 19° in diameter and a steady 4600 K white pedestal

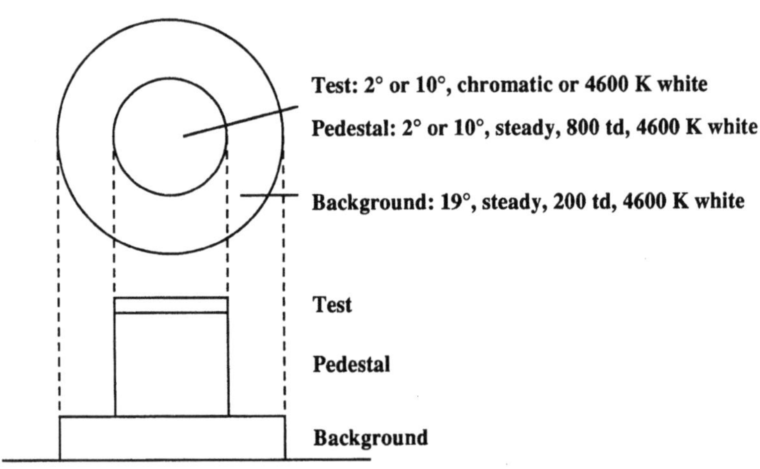

Fig. 1. Spatial configuration of stimulus.

of 800 td were presented concentrically. The test stimulus when present was spatially superimposed upon the pedestal as shown in Fig. 1.

Observers

Observers were either laboratory personnel or individuals who responded to an advertisement asking for 'red-green color defective people' on campus. In this report, we show data for two of the color defective observers: a PA and a DA, and from a color normal observer. All observers were tested for color vision using Ishihara pseudoisochromatic plates, Standard Pseudoisochromatic Plate test II, Farnsworth-Munsell 100 hue test and Neitz OT anomaloscope. The error score of the 100 hue test of the protanomalous (PA) observer (JK) was 76, his anomaloscope matching range was 55 to 65, and the matching midpoint was 60. The error score of the 100 hue test of the deuteranomalous (DA) observer (TJ) was 218, his anomaloscope matching range was 10 to 24, and the midpoint was 17. The error score of the 100 hue test of the normal (N) observer (EM) was 8; the midpoint of the anomaloscope setting was 41 and no other settings were accepted. We have data for four other deuteranomalous observers, but the results of TJ were representative of all five deuteranomalous trichromats.

Detection threshold experiment

The observer adapted to the steady pedestal and background for two minutes. Then a detection threshold for the test stimulus was measured by two randomly interleaved staircases. A three-alternative forced-choice method was used for each trial. The observer pressed a button to initiate a trial. Upon the release of the button, three 1 sec intervals were identified by 0.5 sec beeps. One of the three temporal intervals contained a test stimulus and the two others were blank. The observer then pressed one of the three buttons to indicate which interval contained the stimulus. Two short beeps let the observer know that the response had been recorded and the device was ready for the next trial. After two consecutive responses, the luminance of the test stimulus was decreased. After an incorrect response, the luminance of the test light was increased. On the first trial of each staircase, a stimulus with the maximum amplitude was presented. The stimulus amplitude was initially decremented by 0.3 log unit. After each reversal, the step size was decreased by a factor of 1.6 and after the third reversal the step was 0.08 log unit, and remained so throughout the rest of the staircase. The mean amplitude of four reversal points with a 0.08 log unit step was taken as the threshold for each staircase and the mean of the two staircases was taken to represent the threshold. Then the filter wheel was rotated to the next randomly chosen wavelength. If an incorrect response occurred four times at the maximum amplitude stimulus, the tracking was terminated and recorded as no threshold measurable. Two thresholds were estimated for each stimulus and the mean of the two was taken as the final threshold.

Hue identification experiment

Methods and conditions were the same as for the detection threshold experiment except for the following: The dominant wavelengths of the chromatic stimuli used ranged from 500 to 659 nm, giving 11 stimuli. The stimulus was a single presentation of one sec of a 1 Hz raised cosine wave. The observer's task was to press one of the two buttons on the switch box. In one session, either a 'green' response or a 'red' response was assigned to be the 'correct' response. The observer's task was to press button 1 to indicate that the light was greenish and to press button 2 for anything else in a 'green' threshold session. The observer was instructed to press button 1 to indicate that the light was reddish and to press button 2 to indicate for anything else in a 'red' threshold session. Two short beeps let the observer know that the response had been recorded and the device was ready for the next trial. One session was run for each 'red' or 'green' staircase giving thresholds for the eleven wavelengths. Two thresholds for the energy required for correct hue identification were estimated for each stimulus and the mean of the two thresholds was taken as the final chromatic threshold.

Results

The data are expressed as relative sensitivities on an energy basis. Figure 2 shows relative spectral sensitivity for detection (closed circles) and hue identification (open triangles: 'green', open squares: 'red') for observer EM (N). The 'green' (or 'red') sensitivities plotted at minus infinity mean that the observer responded as 'non-green' (or 'non-red') four times after the maximum amplitude stimulus presentation for these test wavelengths. The spectral sensitivity for detection shows peaks near 529 and 610 nm and notches near 480 and 571 nm. The spectral loci of these peaks and notches are consistent with previous studies (Sperling and Harwerth, 1971; King-Smith and Carden, 1976; Kranda and King-Smith, 1979; Klingaman *et al.,* 1980; Foster and Snelgar, 1983; Snelgar *et al.,* 1987; Krastel *et al.,* 1991). We focus on the data for wavelengths above 500 nm where the red-green chromatic system is more prominent. We called the part of the data left of the 571 nm notch the 'M lobe' and the part of the data right of the 571 nm notch the 'L lobe'. The theoretical transition point from the M lobe to the L lobe can be predicted by the calculation described in Pokorny and Smith (1977) for a dichromatic neutral point. Briefly, the expected transition point wavelength is that wavelength for which the pigment sensitivity ratio is identical to the pigment sensitivity ratio for the corresponding white adaptation light, 4600 K in this study. We used LWS and MWS cone sensitivity for N, MWS and anomalous LWS cone sensitivity for PA, and LWS and anomalous MWS cone sensitivity for DA. These sensitivities are tabulated in De Marco *et al.,* (1992). The predicted transition point is 571 nm for N, 563 nm for PA and 578 nm for DA (rounded to the nearest nanometer).

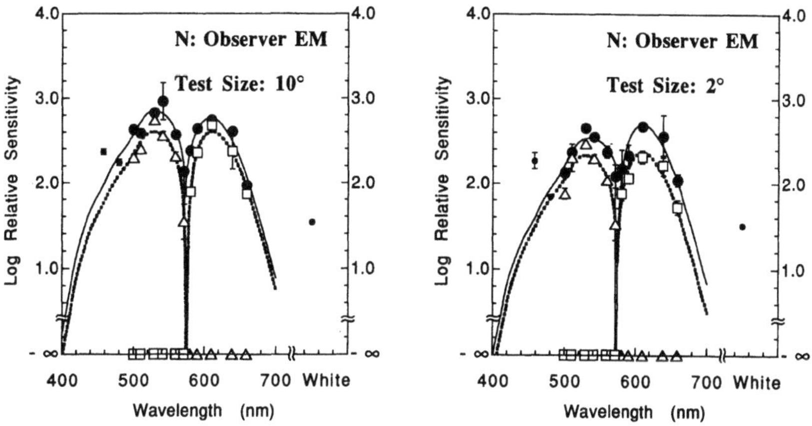

Fig. 2. Detection sensitivity and hue identification sensitivity as a function of wavelength for observer EM (N). Solid circles are detection sensitivity, open triangles are 'green' sensitivity and open squares are 'red' sensitivity. Error bars show the ranges of the two thresholds at that wavelength. Solid lines are fits to detection sensitivity and dotted lines are fits to hue identification sensitivity. The fitting equations are described in text. For the hue identification data points plotted on the abscissa, the observer gave the 'wrong' response four times after the presentation of the maximum amplitude stimulus and the experiment was terminated, therefore the sensitivity was not measurable. Detection sensitivities of 458 mm, 480 nm, and white test stimulus are plotted with smaller symbols because we focus on chromatic stimulus whose dominant wavelength is 500 nm and longer.

According to the model of Sperling and Harwerth (1971), detection sensitivity and hue identification sensitivity data are predicted by a subtractive signal of the LWS and MWS cones. For a normal observer, N:

$$S_\lambda = k_1 * M_\lambda - k_2 * L_\lambda,\ 500\text{ nm} \leq \lambda \leq 570\text{ nm} \quad (1)$$
$$S_\lambda = k_3 * L_\lambda - k_4 * M_\lambda,\ 572\text{ nm} \leq \lambda \leq 659\text{ nm} \quad (2)$$

where S_λ is the test sensitivity as a function of test wavelength, L_λ and M_λ are Smith and Pokorny MWS and LWS cone sensitivities (Smith and Pokorny, 1975) at wavelength λ nm as tabulated in De Marco *et al.*, (1992) and k_1 through k_4 are weighting factors for LWS and MWS cone sensitivities. The test stimulus may be detected by the achromatic system at or near the transition point. Thus, we excluded one data point closest to the predicted transition point (571 nm) when we fit equations (1) and (2) to the detection threshold data. The solid lines in Fig. 2 are fits to detection sensitivity data using equations (1) and (2). For hue identification data, equation (1) is used to fit 'green' sensitivities and equation (2) is used to fit 'red' sensitivities. The dotted lines in Fig. 2 are upper envelopes of the red and green lobes. To get the best fit, we allowed the k_i to vary to minimize the mean of squared residuals of the data and prediction. As can be seen from Fig. 2, both detection and hue identification data are well fitted with the two equations (e.g. MSR = 0.0051 for the L lobe and 0.0092 for the M lobe for 10° detection sensitivity). Table 1 shows values of k_i for detection and hue identification data.

Figure 3 shows the spectral sensitivity data for JK (PA). The arrow on the right side of the figures is sensitivity to the white test of observer EM (N) for comparison. The detection sensitivity of JK is about 1.0 log unit lower than EM (N) at the peaks while the sensitivity for white is only 0.2–0.3 log unit lower. Detection data show a notch at 571 nm, rather than at 559 nm which is closest to the predicted transition point 563 nm and the notch is much shallower compared to EM. The solid and dotted lines are fits to the detection and the hue identification sensitivities by the following equations:

$$S_\lambda = k_1 * M_\lambda - k_2 * L'_\lambda,\ 500 \text{ nm} \leq \lambda \leq 562 \text{ nm} \quad (3)$$
$$S_\lambda = k_3 * L'_\lambda - k_4 * M_\lambda,\ 564 \text{ nm} \leq \lambda \leq 659 \text{ nm} \quad (4)$$

where L'_λ is the sensitivity of the anomalous LWS cone (DeMarco *et al.*, 1992) at wavelength λ nm and other symbols are as in equations (1)–(2). The equations describe the sensitivity data well (e.g. MSR = 0.0128 for the L lobe and 0.0059 for the M lobe for 10° detection sensitivity). The reduced sensitivity is reflected in the lower values of the fitting parameters in Table 1 compared with those for EM.

Figure 4 shows the spectral sensitivity data for TJ (DA). His peak detection sensitivity is about 1.0 log unit lower than that of EM (N) and the sensitivity for the white is 0.7–0.8 log unit lower. The detection sensitivity of TJ shows a shallow notch at 580 nm, close to the predicted transition point 578 nm. The solid and dotted lines in Fig. 4 are fits to the detection and the hue identification sensitivity data by the following equations:

$$S_\lambda = k_1 * M'_\lambda - k_2 * L_\lambda,\ 500 \text{ nm} \leq \lambda \leq 577 \text{ nm} \quad (5)$$
$$S_\lambda = k_3 * L_\lambda - k_4 * M'_\lambda,\ 579 \text{ nm} \leq \lambda \leq 659 \text{ nm} \quad (6)$$

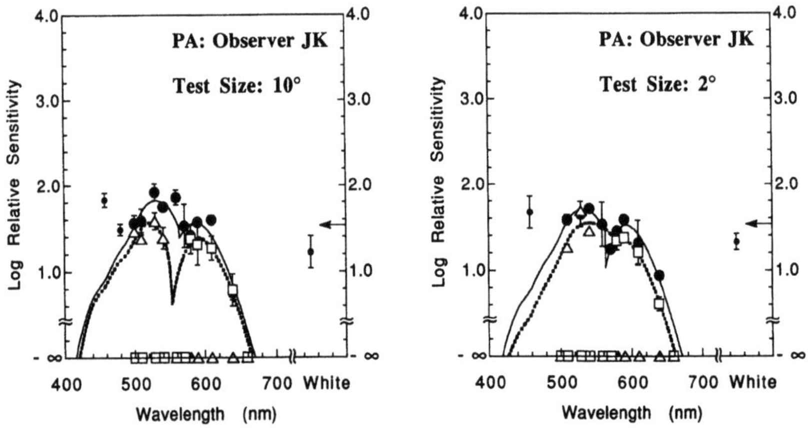

Fig. 3. Detection sensitivity and hue identification sensitivity as a function of wavelength for observer JK (PA). Symbols and lines are the same as Fig. 2. The arrow at the right side of the figure indicates the sensitivity to the white test stimulus for observer EM. The apparatus did not provide sufficient energy to measure detection thresholds of 659 nm light for 10° test stimulus and of 480, 500, and 659 nm light for 2° test stimulus for this observer.

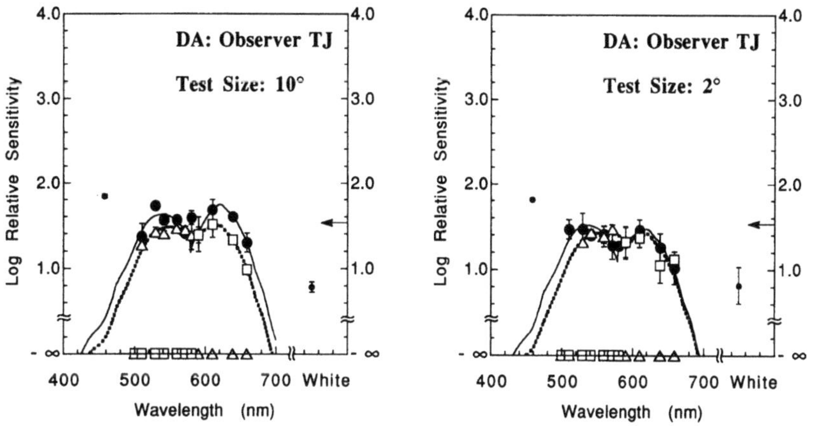

Fig. 4. Detection sensitivity and hue identification sensitivity as a function of wavelength for observer TJ (DA). Symbols, lines and arrows are the same as Fig. 3. The Apparatus did not provide sufficient energy to measure detection thresholds of 480 and 500 nm light for both test stimulus sizes for this observer.

where M'_λ is the sensitivity of the anomalous MWS cone (De Marco *et al.*, 1992) at wavelength λ nm and other symbols are as in equations (1)–(2). Observer TJ identified 580 nm stimulus as being 'green' for a 10° test stimulus. The data are well fitted with equations (5) and (6) (e.g. MSR = 0.0001 for the L lobe and 0.0046 for the M lobe for 10° detection sensitivity). As for the PA observer, the reduced sensitivity is reflected in the lower values of the fitting parameters in Table 1 compared with those for EM.

Discussion

The two X-chromosome-linked anomalous trichromats JK (PA) and TJ (DA) show two peaks and a notch for detection sensitivities, implying red-green color opponency. But the peaks and notch are much less distinct than those measured for the color normal observer EM, implying a loss of chromatic discrimination. The shallowness of the notch may have resulted from our relatively coarse spectral sampling in the wavelength region of the notch but is more likely due to the comparatively weak chromatic sensitivity. The hue identification data show a clear separation between the 'red' and 'green' lobes for all the observers. Anomalous trichromats can tell 'red' from 'green' reliably for these near threshold conditions. Other studies have reported hue naming by anomalous trichromats above threshold (Smith *et al.*, 1973) and existence of cancellation colors (Romeskie, 1978) in the red-green region.

All three observers showed a tendency to increased chromatic sensitivity with the 10° field. However, contrary to our expectation, the observers did not show a deeper notch at the transition point with a 10° test compared to a 2° test. Thus it appears that suppression of the achromatic spatial and temporal transients is more important in revealing chromatic sensitivity in anomalous trichromats than use of large fields. Nagy *et al.*, (1985) have noted the roles of field size, spatial and temporal transient vary in improving chromatic discrimination varies among anomalous trichromats and dichromats.

Acknowledgements

This research was supported in part by NEI research grant EY00901 (J. Pokorny).

References

Breton, M. and Cowan, W. (1981). Deuteranomalous color matching in the deuteranopic eye. J. Opt. Soc. Am. 71: 1220–1223.

Brown, W.R.J. (1952). The effect of field size and chromatic surroundings on color discrimination. J. Opt. Soc. Am. 42: 837–844.

DeMarco, P., Pokorny, J. and Smith, V.C. (1992). Full-spectrum cone sensitivity functions for X-chromosome-linked anomalous trichromats. J. Opt. Soc. Am. A9: 1465–1476.

Foster, D.H. and Snelgar, R.S. (1983)Test and field spectral sensitivities of colour mechanisms obtained on small white backgrounds: action of unitary opponent-colour processes? Vision Res. 23: 787–797.

King-Smith, P.E. and Carden, D. (1976). Luminance and opponent-color contributions to visual detection and adaptation and to temporal and spatial integration. J. Opt. Soc. Am. 66: 709–717.

Klingaman, R.L., Zrenner, E. and Baier, M. (1980). Increment flicker and hue spectral sensitivity functions in normals and dichromats: the effect of flicker rate. In: Verriest, G. (ed.), Color Vision Deficiencies V: 240–243, Adam Hilger Ltd., Bristol.

Kranda, K. and King-Smith, P.E. (1979). Detection of coloured stimuli by independent linear systems. Vision Res. 19: 733–745.

Krastel, H., Jaeger, W., Zimmerman, S., Heckmann, B. and Krystek, M. (1991). Systematics of human photopic spectral sensitivity. In: Drum, D., Moreland, J.D. and Serra, D. (eds.), Colour Vision Deficiencies X, Doc. Ophtalmol. Proc. Ser. 54: 323–339, Kluwer Acad. Publ., Dordrecht.

Nagy, A.L. (1980). Large-field substitution Rayleigh matches of dichromats. J. Opt. Soc. Am. 70: 778–784.

Nagy, A.L. (1982). Homogeneity of large-field color matches in congenital red-green color deficients. J. Opt. Soc. Am. 72: 571–577.

Nagy, A.L., Purl, K.F. and Houston, J.S. (1985). Cone mechanisms underlying the color discrimination of deutan color deficients. Vision Res. 25: 661–669.

Pokorny, J. and Smith, V.C. (1977). Evaluation of single pigment shift model of anomalous trichromacy. J. Opt. Soc. Am. 67: 1196–1209.

Romeskie, M.(1978). Chromatic opponent-response functions of anomalous trichromats. Vision Res. 18: 1521–1532.

Smith, V.C., Pokorny, J. and Swartley, R. (1973). Continuous hue estimation of brief flashes by deuteranomalous observers. Am. J. Psychol. 86: 115–131.

Smith, V.C. and Pokorny, J. (1975). Spectral sensitivity of the foveal cone photopigments between 400 and 500 nm. Vision Res. 15: 161–171.

Smith, V.C. and Pokorny, J. (1977). Large-field trichromacy in protanopes and deuteranopes. J. Opt. Soc. Am. 67: 213–220.

Snelgar, R.S., Foster, D.H. and Scase, M.O. (1987). Isolation of opponent-colour mechanisms at increment threshold. Vision Res. 27: 1017–1027.

Sperling, H.G. and Harwerth, R.S. (1971). Red-green cone interaction in the increment-threshold spectral sensitivity of primates. Science 172: 180–184.

Stiles, W.S. and Burch, J.M. (1959). NPL-colour-matching investigation: Final report. Optica Acta 6: 1–26.

Ueno, T., Pokorny, J. and Smith, V.C. (1985). Reaction times to chromatic stimuli. Vision Res. 25: 1623–1627.

Ueno, T. and Swanson, W.H. (1989). Response pooling between chromatic and luminance systems. Vision Res. 29: 325–333.

Verriest, G. and Uvijls, A. (1977). Central and peripheral increment thresholds for white and spectral lights on a white background in different kinds of congenitally defective colour vision. Atti della Fondazione Giorgio Ronchi 32: 213–254.

Wyszecki, G. and Stiles, W.S. (1982). Color Science: Concepts and Methods, Quantitative Data and Formulae (2nd ed.). Wiley, New York.

Yeh, T., Pokorny, J. and Smith, V.C. (1993). Chromatic discrimination with variation in chromaticity and luminance: data and theory. Vision Res. 33: 1835–1845.

Visual Sciences Center
The University of Chicago
939 East 57th Street
Chicago, IL 60637 U.S.A.

18. The spectral sensitivity characteristics of congenital red-green color vision deficiencies

KENJI KITAHARA, AKIKO KUBO and ATSUSHI KANDATSU
(Tokyo, Japan)

Abstract

The spectral sensitivity in congenital red-green color vision deficiencies was measured using intense 430 nm and 700 nm backgrounds. In protanopia and protanomaly, the spectral sensitivity curves in the middle and long wavelength regions on both backgrounds coincided with the spectral sensitivities of the middle-wavelength (M) cones. In deuteranopia and six out of the nine observes with deuteranomaly, the spectral sensitivity curves in the middle and long wavelength regions on both backgrounds coincided with the spectral sensitivities of the long-wavelength (L) cones. Three of the nine observers with deuteranomaly showed spectral sensitivity curves in the middle and long wavelength regions that showed slight but significant differences on the 430 nm background as compared to the 700 nm background.
As a result, it is felt that protanomalous subjects have abnormal pigment whose spectral characteristics are similar to those of M cones, while some of the deuteranomalous subjects have abnormal pigment whose spectral characteristics are different from those of L cones.

Introduction

One of us (Kandatsu, 1993) previously reported that the spectral sensitivities measured on a monochromatic background can be directly compared with the Stiles' Π-mechanisms. He found that the pattern of the spectral sensitivity curve on a 430 nm monochromatic background with a retinal illuminance of 530 photopic trolands was similar to that of the Stiles Π 5 mechansim (L cones). The pattern of the spectral sensitivity curve on a 700 nm monochromatic background with a retinal illuminance of 530 photopic trlands showed two peaks. The peak in the short wavelength region coincided with that of the short-wavelength (S) cones and the peak in the middle and long wavelength regions coincided with the Stiles Π 4 mechanism (M cones).
Applying this method we (Kandatsu and Kitahara, 1993) investigated the

B. Drum (ed.), Colour Vision Deficiencies XII, pp. 159–162.
© 1995 *Kluwer Academic Publishers, Dordrecht.*

spectral sensitivity characteristics of deutan type of Pigmentfarbenanomalie and found the responses of both M and L cones.

In this paper we investigated whether the spectral sensitivities in dichromats are different from anomalous trichromats when they are measured on intense 430 nm and 700 nm backgrounds.

Method

A three-channel Maxwellian view optical system was used in this experiment. A 1° diameter circular test light was superimposed in the center of an 8° circular background field. Narrow-band (6 to 10 nm half-band width) interference filters were used for both the test and background lights. The test lights were exposed for 200 msec every second. The test sensitivities were measured both on 430 nm and on 700 nm backgrounds with a retinal illuminance of 530 photopic trolands for color defective observers, three with protanopia, four with protanomaly, one with deuteranopia and nine with deuteranomaly. The type of red-green color vision defect was diagnosed with the Nagel model I anomaloscope. All dichromats failed the panel D-15 test, while the results showed a normal pattern for all anomalous trichromats except one observer with deuteranomaly.

Results

In all observers with protanopia and protanomaly, the spectral sensitivity curves in the middle and long wavelength regions on both 430 nm and 700 nm backgrounds coincided with the spectral sensitivity of the M cones. Fig. 1a shows the representative sensitivity curves on a 430 nm background (solid line) and a 700 nm background (open circles) for protanopic observer R.I. The curves are shifted vertically to agree in the long wavelength range. Fig. 1b shows the representative sensitivity curves of protanomalous observer K.O.

In deuteranopia and six of the nine observers with deuteranomaly, the spectral sensitivity curves in the middle and long wavelength regions on both backgrounds coincided with the spectral sensitivity of the L cones. Three of the nine observers with deuteranomaly including one who failed the panel D-15 test, showed spectral sensitivity curves in the middle and long wavelength regions that were slightly but significantly different on the 430 nm background than on the 700 nm background. Fig. 2a shows the sensitivity curves of deuteranopic observer H.K. on a 430 nm and a 700 nm background plotted the same as in Fig. 1. The curves are normalized to agree in the long wavelength region. Fig. 2b shows the sensitivity curves of representative observer Y.N. whose sensitivity curves in the middle and long wavelength regions were different on both backgrounds.

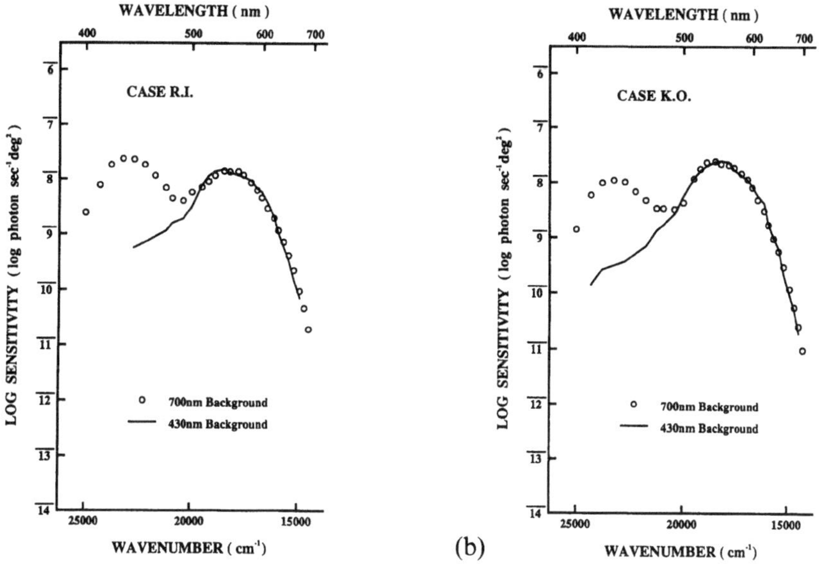

(a) (b)

Fig. 1. (a) The representative sensitivity curves on a 430 nm background (solid line) and a 700 nm background (open circles) for protanopic observer R.I. The curves are shifted vertically to agree in the long wavelength range. (b) The representative sensitivity curve for protanomalous observer K.O.

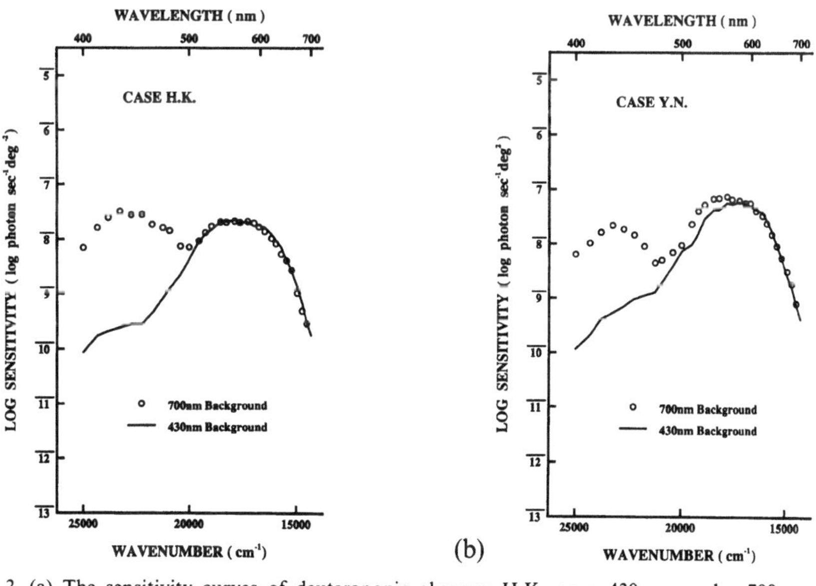

(a) (b)

Fig. 2. (a) The sensitivity curves of deuteranopic observer H.K. on a 430 nm and a 700 nm background plotted the same as in Fig. 1. The curves are normalized to agree in the long wavelength region. (b) The sensitivity curves of representative observer Y.N. whose sensitivity curves in the middle and long wavelength regions were different on both backgrounds.

Discussion

Three out of 9 observers with deuteranomaly showed spectral sensitivity curves in the middle and long wavelength regions that were different on the 430 nm background than on the 700 nm background. These differences are due to the contribution of cones containing the deuteranomalous anomalous pigment. These results also showed that the degree of the difference in the spectral sensitivity curves did not relate to the color discrimination ability. However, in all observers with protanomaly the spectral sensitivity curves in the middle and long wavelength regions on a 430 nm background coincided with the spectral sensitivity on a 700 nm background. That is, the response of the cones containing anomalous photopigment could not be detected. As a result, it is felt that protanomalous subjects have abnormal pigment whose spectral characteristics are very similar to those of M cones, while some of the deuteranomalous subjects have abnormal pigment whose spectral characteristics are slightly different from those of L cones. Another reason may be the fact that the L/M cone sensitivity ratio is approximately 16.8/1 at 700 nm (Smith and Pokorny, 1975). Therefore, the conditions of the adaptation are not sufficient in this experiment for selective adaptation in protanomaly. Further study is necessary.

References

Kandatsu, A. (1992). Spectral sensitivity on monochromatic backgrounds. J. Jpn. Ophthalmol. Soc. 97: 190–195.

Kandatsu, A. and Kitahara, K. (1993). The visual characteristics of a case of Pigmentfarbenanomalie. In: Drum, B. (ed.), Color Vision Deficiencies XI, Doc. Ophthalmol. Proc. Ser. 56: 113–117, Kluwer Acad. Publ., Dordrecht

Smith, V.C. and Pokorny, J. (1975). Spectral sensitivity of the foveal cone photopigments between 400 nm and 500 nm. Vision Res. 15: 161–171.

Department of Ophthalmology
The Jikei University School of Medicine
3–25–8 Nishi-Shinbashi
Minato-ku, Tokyo 105, Japan.

PART THREE

Basics of colour vision

19. Dual bases in dichromatic color space

KENNETH KNOBLAUCH
(New York, NY, U.S.A.)

Abstract

The set of directions in color mixture space to which each of the visual mechanisms subserving discrimination is most sensitive is shown to be mathematically dual to the set of directions that stimulates each of the mechanisms in isolation. This result implies that mechanisms can be characterized equally well by either set of directions. By using a simultaneous detection/identification task, the directions in color space were determined to which a protanopic observer was most sensitive identifying the color appearance of a Gaussian-modulated ($\sigma = 0.1$ sec), 1 degree disk. Despite a small response bias in the identification judgements, the mechanisms estimated as mediating the identification responses accounted reasonably well for the detection data.

Mechanisms and bases in color space

While classic approaches in color vision have emphasized the analysis of sensitivity to narrow-band increments from various parts of the spectrum (e.g. Stiles, 1978), a growing number of studies are being devoted to analysis of arbitrary stimulus modulations in a color mixture space (e.g. Cole *et al.*, 1993; Derrington *et al.*, 1984; King-Smith *et al.*, 1983; Krauskopf *et al.*, 1982; Poirson and Wandell, 1990; Stromeyer *et al.*, 1983). In the former case, a mechanism is characterized by its spectral sensitivity along the wavelength axis. In the latter case, a mechanism is characterized by its set of contrast or modulation thresholds in the color mixture space.

Assuming that a mechanism is linear, these thresholds lie on a line in 2-D mixture space, on a plane in 3-D mixture space and on an $n - 1$ hyperplane in n-D mixture space. The orientation of such a linear surface is determined by its normal vector, u_i, (also called the gradient vector). The equation of the surface can be written as the dot product between u_i and the vectors w representing color modulation depths that satisfy

B. Drum (ed.), Colour Vision Deficiencies XII, pp. 165–176.
© 1995 *Kluwer Academic Publishers, Dordrecht.*

$$u_i \cdot w = c \qquad (1)$$

where c is a constant. In the determination of the effect of a given modulation in color space on a mechanism, this gradient vector plays the same role as a spectral sensitivity does in the wavelength domain. Formally,

$$\phi_i(w) = \int_{400}^{700} S_i(\lambda) \sum_j w_j E_j(\lambda)\, d\lambda = u_i \cdot w = \sum_j u_{ij} w_j \qquad (2)$$

where ϕ_i is the excitation of a mechanism with spectral sensitivity S_i (λ) or equivalently with gradient vector u_i, and w_j are the components of a vector w, representing the direction of modulation of a mixture of lights, E_j (λ) (e.g. the primaries, R, G and B, of a TV display) in the color mixture space. The excitation of the mechanism by the modulation of the stimulus in the direction w is given by the orthogonal projection of w along the direction u_i. The gradient direction can also be viewed as the direction of modulation to which the mechanism is most sensitive, i.e. which produces the greatest change in excitation, ϕ_i, per modulation depth.

A set of n such vectors, $\{u_i\}$, arising from n independent visual mechanisms forms a basis of n-dimensional color mixture space. Associated with the basis $\{u_i\}$ is another basis that will be designated $\{v_i\}$. The elements of this basis are the modulation directions formed by the intersection of the constant response surfaces of each set of all of the mechanisms but one. Since by construction the modulations of all of the mechanisms but one are constant in these directions, each of these directions modulates one mechanism in isolation. With suitable normalization of the lengths of the $\{u_i\}$, the relation between the two bases can be summarized as

$$\phi_i(v_j) = u_i \cdot v_j = \begin{cases} 1 & \text{if } i = j \\ 0 & \text{if } i \neq j. \end{cases} \qquad (3)$$

which defines the $\{v_i\}$ to be the *reciprocal* or *dual* basis of the $\{u_i\}$ (Lang, 1972).

Note that if we make each u_i into a row of a matrix, U, and each v_i into a column of a second matrix, V, then

$$U \cdot V = I \text{ or } U^{-1} = V, \qquad (4)$$

where I is the identity matrix with 1's on the diagonal and 0's elsewhere. Importantly, this result is valid in the case of n mechanisms in an n-dimensional mixture space. Also, note that since multiplication by the matrix U takes each isolation direction into a distinct element of the usual basis of n-dimensional space, that U maps each *stimulus-modulation* coordinate vector into a *mechanism-modulation* coordinate vector. Thus, a given set of visual mechanisms is characterized by two bases – a sensitivity or gradient basis and an isolation basis. Only one needs to be determined, however, in order to specify the other. $\{v_i\} = \{u_i\}$ only if the basis is an orthogonal basis.

Strategies for deriving mechanism bases

Many experiments aimed at elucidating the fundamental mechanisms of color discrimination have been cast in terms of the above framework. When several mechanisms are operational, however, the threshold contour is not a linear surface, but the envelope of threshold surfaces of each of the substrate discrimination mechanisms. Given that such mechanisms combine their signals independently and given the presence of inherent measurement error, thresholds in color space usually fall along an elliptical- or ellipsoidal-like contour (depending on the dimension of the space and the number of mechanisms). Under these circumstances, the underlying mechanisms cannot be resolved and are not unambiguously associated with the principal axes of the measured threshold surface (Poirson *et al.*, 1990; Knoblauch and Maloney, 1992).

An approach that has met with some success has been to determine isolation directions as those directions along which repetitive modulation habituates the response of a single mechanism, leaving the thresholds along all other isolation directions unaffected (Krauskopf *et al.*, 1982, 1986). The mechanisms thus revealed seem consistent with color discrimination thresholds, but their relationship to color appearance data is not well understood.

One can also inquire as to how well color appearance judgements account for discrimination data. For example, if we assume that the underlying mechanisms signal the elemental hues that the observer sees at a suprathreshold level, then we can, in principle, determine null directions for a given mechanism based on its perceptual organization, e.g. a modulation direction that signals neither red nor green (Thornton and Pugh, 1983). Such directions are members of a constant criterion surface for a given mechanism and, thus, should be useful for defining elements of a mechanism isolation basis (e.g. equilibrium red-green hues presented at equal luminance should isolate a yellow-blue system). This approach will fare poorly under conditions in which the threshold contour is elongated along the direction in which one is searching for a null, in which case there will be a large region of uncertainty in setting the criterion. The approach may not be viable if the perceptual organization of the mechanism does not lend itself to making a null judgement.

A fourth strategy related to this latter one assumes that not only can the observer set his modulation thresholds with reasonable precision but that near threshold the observer can report hue and brightness judgements accurately. If this is the case and if these judgements do reflect the magnitude of activity in the substrate discrimination mechanisms, then perhaps by evaluating the thresholds for these percepts, we can determine threshold surfaces for each perceptual criterion from which the mechanism bases could be derived as described above.

The success of this fourth strategy depends on the assumption that identification thresholds will lie on a well-defined threshold surface of the mechanism probed and will not depend on the interaction of other mechanisms. This idea was tested initially by measuring detection and identification thresholds in a dichromatic observer. Since such an observer's visual system is limited to two degrees of freedom, the analyses are simplified and the number of conditions to be tested is reduced. If this procedure does not work with such an observer, then there would be little hope of applying it to the more complicated case of a trichromat.

Methods

The observer was the author, a protanope, 40 years of age at the time of testing, with 20/20 uncorrected visual acuity. His right eye was used in all testing.

The stimulus was a 1 degree disk presented in a dark surround. It consisted of a mixture of the three primaries of a television display (Electrohome), under computer control (PC-XT, Data Translation 12-bit D/A converters, linearized with software lookup tables) with CIE chromaticity coordinates of the guns: B (0.161, 0.072), G (0.255, 0.610), R (0.606, 0.356). The luminances of the guns were modulated as a Gaussian pulse ($\sigma = 100$ msec, duration 660 msec) from a pedestal value of chromaticity (0.32, 0.32) and luminance 100 cd/m^2. All modulations were constrained to lie in a plane in color mixture space in which one axis corresponded to the luminance of the B primary and the other was a fixed combination of the R and G primaries (R:G luminance ratio of 1:2.7).

Psychometric functions were generated over several sessions using the method of constant stimuli and two-alternative forced choice with feedback. Each interval was preceded by a 250 msec tone and the intervals were separated by a 500 msec interval. With each trial, the observer responded initially as to which of two intervals the stimulus was presented followed by a judgement as to the appearance of the stimulus as yellow or blue in one set of sessions and black or white (or light or dark) in a separate set. Four directions were tested within a session, and these were chosen to counterbalance the pairing of yellow-blue with white-black sensations.

Five modulation levels were tested for each direction in a session, and each level was repeated ten times, with order randomized in each block of twenty. Thus, 200 stimuli were presented in each session, lasting about twenty minutes. Five sessions were run for each of the six sets of four directtions, giving 250 judgements per psychometric function and 6000 stimulus presentations in total.

Results

Thresholds for detection and identification were estimated from Weibull functions fitted individually to the data from each criterion and color direction, using a maximum likelihood technique (Watson, 1979). The threshold was defined as the modulation level that yields a probability of 0.816. In several instances, the proportion of stimuli identified as a specific hue was higher than the proportion correctly detected. Such a paradoxical reversal may represent a response bias. This interpretation was reinforced when the identification proportions recorded in the complementary direction fell below chance. In general, the thresholds for identification varied independently of the positions of the detection psychometric functions.

Figure 1 shows the estimated thresholds from each of the detection psychometric functions plotted as unfilled circles in the color mixture plane of the experiment. Eight of the directions in Fig. 1 were replicated. The maximum variation between replications is only 36.1% or 0.134 log units, well within the expected variability for threshold measurement. Also, there are no apparent asymmetries in the thresholds in complementary directions.

The threshold contour (solid line) was fitted to the detection data using a fitting procedure devised by Poirson (1991) that generalizes the maximum likelihood procedure to data collected in more than one color direction and that

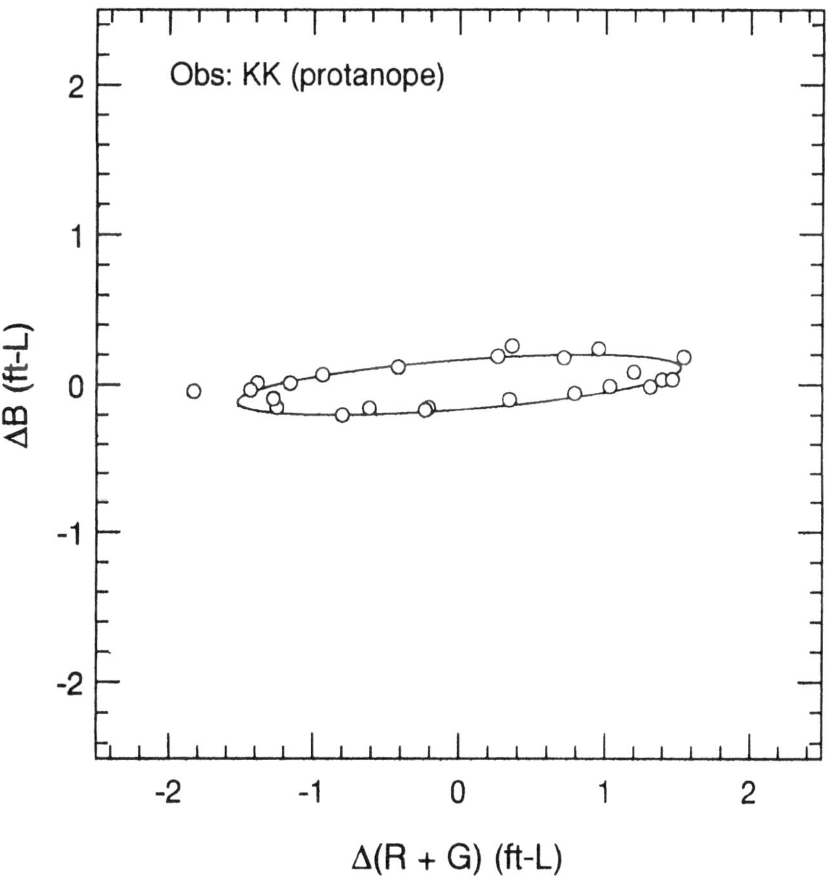

Fig. 1. Detection thresholds estimated from individual psychometric functions, plotted in color mixture space. Solid curve is the estimated detection contour from the quadratic fit to all of the detection data.

uses all of the percent correct data. Conceptually, this procedure seeks a quadratic transformation that assigns equal percent correct to equal Euclidean length. Visually, using two dimensional detection data, the data are fit to a circular bowl whose sides have the form of a Weibull function, with threshold parameter $\alpha = 1.0$ and steepness parameter β. Four parameters are required for this fit, three to describe the quadratic transform and one that gives the steepness parameter for the sides of the bowl. In the transformed space, the fitted threshold contour is a circle that is the cross-section through the bowl for $\alpha = 1.0$. The parameters of the quadratic transform give the threshold ellipse shown in Fig. 1 with $\beta = 1.95$.

The thresholds estimated from individual hue-identification psychometric functions are plotted in Fig. 2 as filled and unfilled circles and triangles. The

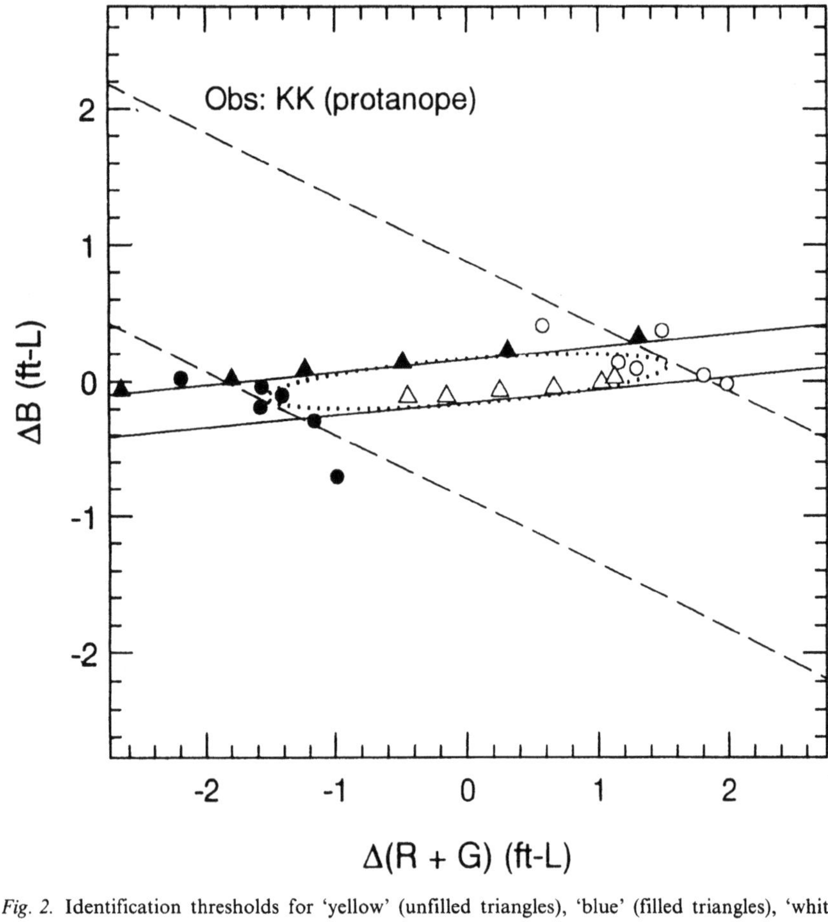

Fig. 2. Identification thresholds for 'yellow' (unfilled triangles), 'blue' (filled triangles), 'white' (unfilled circles) and 'black' (filled circles) response criteria, estimated from individual psychometric functions. Parallel lines represent threshold-identification contours for yellow-blue (solid lines) and white-black (dashed lines) estimated from quadratic fits to all of the identification data for a particular opponent-pair criterion.

dotted ellipse is the threshold contour from the quadratic model from Fig. 1. The parallel solid and dashed lines were obtained by fitting the quadratic model separately to the yellow-blue and white-black sets of psychometric functions. In the case of the white-black data, one of the dimensions of the fitted ellipse became effectively infinite, resulting in parallel lines (dashed lines). In the case of the yellow-blue data, the quadratic fit was not singular, but the ratio of the principal axes of the predicted ellipse was nearly 30:1. The solid pair of parallel lines represents the prediction obtained from the singular value decomposition (Press *et al.,* 1992) of the linear transform matrix, using the threshold contour based only on the largest singular value. The steepness parameters from these fits were slightly lower than that for detection ($\beta_{y-bl} = 1.88$; $\beta_{w-bk} = 1.55$).

The yellow-blue identification contours closely follow the major axis of the discrimination ellipse, although they are not exactly parallel to it. Enveloping the detection contour in a parallelogram with interior angles of 30° and 150°, the two sets of identification contours are not orthogonal. Thus, the white-black identification contours deviate in direction considerably from the minor axis of the detection ellipse. The gradient directions for the identification contours are: $u_{y-bl} = (-0.592, 6.334)$ and u_{w-bk} (0.541, 1.141). Applying Equation 4, the isolation directions are $v_{y-bl} = (-0.278, 0.132)$ and $v_{w-bk} = (1.544, 0.144)$.

At first glance, the identification data conform to expectation. The 'black' and 'white' identification thresholds (filled and unfilled circles, respectively), though noisy, fall along parallel lines with negative slope. The 'yellow' and 'blue' data (unfilled and filled triangles, respectively) follow positively sloped linear contours with the 'blue' data continuing along the line even outside the detection contour. The yellow-blue data show an asymmetry, however, in that 'blue' identification data are slightly outside the linear contour and 'yellow' identification data are all inside. This asymmetry reflects the response bias indicated earlier.

Figure 3 shows the detection thresholds re-expressed in terms of their projections along the gradient directions of the identification contours (equation 1 or 2). If the identification contours represent the threshold contours for the detection data as well, then these data are expected to lie on a unit circle (solid curve) in this space. The mapping is not unreasonable given variability of the data.

Discussion

Mullen and Kulikowski (1990) measured the minimum wavelength difference between two stimuli producing identification as accurate as detection performance. Proceeding in this fashion across the spectrum, they found evidence for four fixed boundaries at locations consistent with transitions between post-receptoral opponent channels. In a related experiment, Krauskopf *et al.,* (1986) studied the relation between detection and discrimination for pairs of lights that fell along or between their estimates of

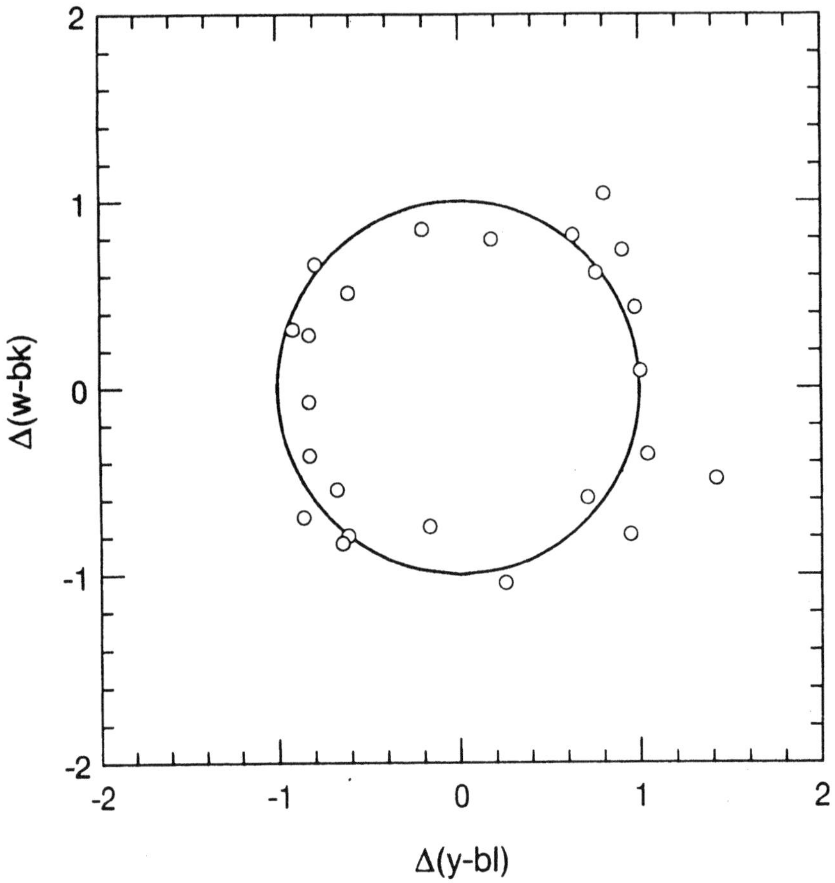

Fig. 3. Detection thresholds from Fig. 1 replotted in terms of their projections along the gradients of the identification mechanisms estimated in Fig. 2.

mechanism-isolation directions. Contrary to the previous citation, they found no difference in performance between these two sets of conditions, suggesting that their habituation-isolated mechanisms assume no special status with respect to the signalling of color appearance. The reasons for the disagreement between these two studies are unclear, although they differed considerably in their experimental conditions.

Pugh and colleagues (Thornton and Pugh, 1983a, b; Friedman *et al.,* 1985) demonstrated that red-green and yellow-blue equilibrium contours are parallel to detection contours under conditions in which detection is likely to be mediated by a single mechanism, thus, linking detection and appearance data to specific opponent mechanisms at threshold. Subsequently, Calkins *et al.,* (1992) showed that the locus of stimuli indiscriminable from a long-wavelength, 'reddish' light produced identical spectral sensitivities up to 0.7 log units above threshold. Recall that spectral sensitivities are equivalent to threshold surfaces in the mixture domain, so their demonstration tracks the constant criterion surfaces of a mechanism to suprathreshold levels.

By measuring identification thresholds for opponent hues, I have attempted to characterize the mechanisms subserving color appearance. Interestingly, the identification contour could be followed, even when its threshold was greater than that of a second mechanism. Another advantage to this technique is that the mechanism subserving white-black discrimination could be probed under the same spatial and temporal conditions as the chromatic mechanism. An equilibrium criterion like hue cancellation (e.g. neither white nor black) is not possible for white-black with light mixtures. Psychophysical estimates of the characteristics of this system typically come from spatial or temporal comparisons (e.g. minimum border or minimum flicker judgements) and may not be equivalent to estimates of chromatic systems obtained under different spatio-temporal conditions if cone inputs to chromatic and achromatic pathways vary independently with spatial and/or temporal frequency.

In this study, it has been assumed that at threshold the fundamental mechanisms are linear. It follows that the threshold contours of an isolated mechanism in color mixture space are linear, also. The identification criteria that were used in this experiment arise from a specific version of opponent-colors theory (Hurvich and Jameson, 1957). To the extent that the measured identification contours based on these criteria are linear, the data are consistent with this theory. An alternative hypothesis would be that identification judgements are based on the activity in several channels, each with different spectral sensitivities. Assuming that the signals from such channels are combined independently, this hypothesis would predict curved identification contours (much like the detection contour which arises from the combination of independent mechanisms with different spectral sensitivities). Additional data would be required to distinguish strongly between these hypotheses. The hypothesis that more than two mechanisms contribute to these data has also not been tested. Given that two seem to account for the data, it is not clear what advantages would be gained in considering more in the current experimental context.

Since Le Grand's analysis (1949) of MacAdam's data (1942), there have been several attempts to account for color discrimination data by mechanisms aligned along the principal axes of the discrimination ellipses (Boynton and Kambe, 1980; Guth, 1991; Rodieck, 1973). A difference between these axes and the ones derived from appearance judgements such as hue cancellation (Hurvich and Jameson, 1957) has been interpreted as evidence of an incompatibility between data arising from detection and appearance judgements (Guth, 1991; Krauskopf *et al.*, 1982). Poirson *et al.*, (1990) demonstrated that theoretically in the case of elliptical (or ellipsoidal) discrimination contours, the mechanism axes need not be aligned with the principal axes, but could be related by an arbitrary rotation in the mechanism space. I have shown here that axes derived from appearance judgements are not aligned with the principal axes of the discrimination contour but they nevertheless provide a reasonable account of discrimination data.

The gradient directions of the S- and M-cones (based on the Smith and Pokorny fundamentals, 1975) in the coordinate system of the primary lights used in this study are: $u_s = (0.0172, 0.9998)$ and $u_M = (0.597, 0.802)$. The white-black gradient direction, $u_{w\text{-}bk}$, does not coincide with the M-cone gradient direction but makes an angle of 11.3° with it. This value is close to 0°, and it remains to be tested whether measurement error excludes that value. An angle different from 0° would imply a difference between the M-cone and white-black spectral sensitivities, suggesting the presence of input from S-cones to the white-black system. Similarly, the yellow-blue gradient direction makes an angle of 6.3° with the S-cone gradient direction, indicating input from the M-cones to identification along the yellow-blue contour.

Finally, a problem in interpreting these data arises because of the presence of a response bias in the identification performance. This bias appears most obviously in the yellow-blue data for which identification of 'yellow' is uniformly better than detection and 'blue' worse. One method of dealing with this bias would be to use a criterion-free measure, such as d' from signal detection theory. The application of this approach to the detection data is straightforward and requires no further discussion.

The identification data for a single direction might be analyzed as a classification experiment using a cumulative d' measure (see MacMillan and Creelman, 1991). The difficulty is that for the identification task there is no correspondence function as exists for a detection task. In other words, for detection, there is a correct answer on each trial because the stimulus occurs in either the first or second interval. Consequently, the false positive rate is simply computed from the data. No such simple assignment is possible with the identification data. As a result, cumulative d' measures for a given direction are relative and only comparable within a direction. One possible solution to this difficulty would be to include catch trials in the experiment for which the stimulus is not presented in either interval. There is no correct answer for the detection judgement, but the identification judgements would give a baseline response rate from which an estimate of response bias might be computed.

In summary, the data show a close correspondence between threshold contours arising from detection and identification criteria. With recognition of the need for better control of potential response biases, identification measures hold promise for characterizing color mechanisms when inferences from detection data are not well determined. Since the derived mechanism axes are not aligned with the principal axes of the detection ellipse and since they are not orthogonal, the distinction between the dual bases of color space (i.e. in gradient and isolation directions) is critical for properly characterizing mechanisms in color mixture space.

Acknowledgement

I wish to thank Laurence T. Maloney, Allen B. Poirson and John S. Werner for helpful discussions and comments. This research was conducted at the Lighthouse Research Institute in New York.

References

Boynton, R.M. and Kambe, N. (1980). Chromatic difference steps of moderate size measured along theoretically critical axes. Color Res. App. 5: 13–23.

Calkins, D.J., Thornton, J.E. and Pugh, E.N. Jr. (1992). Monochromatism determined at a long-wavelength/middle-wavelength cone-antagonistic locus. Vision Res. 32: 2349–2367.

Cole, G.R., Hine, T. and McIlhagga, W. (1993). Detection mechanisms in L-, M- and S-cone contrast space. J. Opt. Soc. Am. A 10: 38–51.

Derrington, A.M., Krauskopf, J and Lennie, P. (1984). Chromatic mechanisms in lateral geniculate nucleus of macaque. J. Physio. 357: 241–265.

Friedman, L.J., Thornton, J.E. and Pugh, E.N. Jr. (1985). Cone antagonism along visual pathways of red/green dichromats. Vision Res. 25: 1647–1654.

Guth, S.L. (1991). Model for color vision and light adaptation. J. Opt. Soc. Am. A, 9: 976–993.

Hurvich, L.M. and Jameson, D. (1957). An opponent-process theory of color vision. Psych. Rev. 64: 384–404.

King-Smith, P.E. Chioran, G.M., Sellers, K.L. and Alvarez, S.L. (1983). Normal and deficient colour discrimination analysed by colour television. In: Mollon, J.D. and Sharpe, L.T. (eds.). Colour Vision: Physiology and Psychophysics: 167–172, Academic, London.

Knoblauch, K. and Maloney, L.T. (1992). Indeterminacy of a privileged coordinate frame based on discrimination measures: Analysis and Experimental Tests. Advances in Color Vision Technical Digest 1992 (Optical Society of America, Washington, D.C.) 4: 198–200.

Krauskopf, J., Williams, D.R. and Heeley, D.W. (1982). Cardinal directions of color space. Vision Res. 22: 1122–1131.

Krauskopf, J., Williams, D.R., Mandler, M.B. and Brown, A.M. (1986). Higher order color mechanisms. Vision Res. 26: 23–32.

Lang, S. (1972) Linear Algebra, 2nd ed., Addison-Wesley, Reading.

Le Grand, Y. (1949). Les seuils différentiels de couleurs dans la théorie de Young. Revue d'Optique 28: 261–278.

MacAdam, D.A. (1942). Visual sensitivities to color differences in daylight. J. Opt. Soc. Am. 32: 247–274.

Macmillan, N.A. and Creelman, C.D. (1991). Detection Theory: A User's Guide. Cambridge University Press, Cambridge.

Mullen, K.T. and Kulikowski, J.J. (1990). Wavelength discrimination at detection threshold. J. Opt. Soc. Am. A 7: 733–742.

Poirson, A.B. (1991). Appearance and detection of colored patterns. Ph. D. dissertation. Stanford University (University Microfilms, Ann Arbor).

Poirson, A.B. and Wandell, B.A. (1990). The ellipsoidal representation of spectral sensitivity. Vision Res. 30: 647–652.

Poirson, A.B., Wandell, B.A., Varner, D.C. and Brainard, D.H. (1990). Surface characterization of color thresholds. J. Opt. Soc. Am. A 7: 783–789.

Press, W.H., Teukolsky, S.A., Vetterling, W.T. and Flannery, B.P. (1992). Numerical Recipes in C: The Art of Scientific Computing, 2nd ed., Cambridge University Press, Cambridge.

Rodieck, R. (1973). The Vertebrate Retina: Principles of Structure and Function. W.H. Freeman, San Francisco.

Smith, V.C. and Pokorny, J.H. (1975). Spectral sensitivity of the foveal cone pigments between 400 and 500 nm. Vision Res. 15: 161–171.
Stiles, W.S. (1978). Mechanisms of Colour Vision. Selected Papers of W.S. Stiles with A New Introductory Essay. Academic, London.
Stromeyer, C.F. III, Kronauer, R.E. and Cole, G.R. (1983). Adaptive mechanisms controlling sensitivity to red-green chromatic flashes. In: Mollon, J.D. and Sharpe, L.T. (eds.), Colour Vision: Physiology and Psychophysics: 313–330, Academic, London.
Thornton, J.E. and Pugh, E.N. Jr. (1983a). Red/green color opponency at detection threshold. Science 219: 191–193.
Thornton, J.E. and Pugh, E.N. Jr. (1983b). Relationship of opponent-colours cancellation measures to cone-antagonistic signals deduced from increment threshold data. In: Mollon, J.D. and Sharpe, L.T. (eds.). Colour Vision: Physiology and Psychophysics: 361–373, Academic, London.
Watson, A.B. (1979). Probability summation over time. Vision Res. 19: 515–522.

INSERM Unité 371
Cerveau et Vision
18 avenue du Doyen Lépine
69500 Bron
France

20. Tritan pairs estimated by modulation photometry of red, green and blue lights

BARRY B. LEE and TSAIYAO YEH
(Göttingen, Germany)

Abstract

We wished to determine if, in flicker photometry, residual flicker at equal luminance is minimal for a tritanopic pair of lights. Modulation photometry was performed using red, green and blue lights, in phase or counterphase and in different proportions. Modulation thresholds reached a maximum when modulation of the three lights was in the equiluminance plane. At equal luminance, the rise in threshold was most pronounced with chromatic modulation close to a tritanopic confusion line. When the data were transformed into a cone excitation space, they were consistent with detection usually being dependent on a middle-to-long wavelength cone difference signal. This is analogous to residual distinctness of chromatic borders in the minimally distinct border task, which is also dependent on a middle-to-long wavelength cone difference signal, probably arising from a non-linearity in the magnocellular pathway. However, some observers appeared to be able to detect flicker at 20 Hz using their short-wavelength sensitive cones.

Introduction

In the minimally distinct border task (MDB), the relative radiance of two colored fields is adjusted until the border between them is minimally distinct. This occurs at equal luminance. Residual distinctness at equal luminance depends on the chromaticities across the border (Tansley and Boynton, 1976; Valberg and Tansley, 1977). If they fall along a tritanopic confusion line, the border is very indistinct and tends to 'melt'. A melting border can be used as a criterion to determine tritanopic pairs for individual observers (e.g. Nakano and Kaiser, 1992). There is strong evidence that the magnocellular (MC-)pathway forms the physiological substrate of MDB (Kaiser *et al.*, 1990; Valberg *et al.*, 1992). The MC-pathway also gives a residual, excitatory response to equal-luminance borders which closely matches residual distinctness reported psychophysically. This residual response is due to a non-linearity of M- and L-

B. Drum (ed.), Colour Vision Deficiencies XII, pp. 177–184.
© 1995 *Kluwer Academic Publishers, Dordrecht.*

cone summation and is absent with tritanopic pairs of lights. It has also been described with chromatic modulation (Lee *et al.*, 1989) when it is seen as a response at twice the modulation frequency.

We wished to determine tritanopic pairs for different observers so as to be able to compare psychophysical temporal modulation sensitivities to responsivities of ganglion cells with short (S-) wavelength-sensitive cone input (Yeh *et al.*, 1995). However, our Maxwellian view system did not lend itself easily to construction of colored borders. The results presented here derive from an attempt to find a temporal analogue of the melting border criterion.

There is evidence of intrustion of a luminance mechanism into detection of red-green chromatic modulation at high frequencies (e.g. Swanson *et al.*, 1987b). If the frequency-doubled response in the MC-pathway is the physiological substrate, then we may expect the intrusion to be least for tritan pairs. To test if this were the case, we systematically explored observers' modulation sensitivities at and around equal luminance with different proportions of red, green and blue lights, as in modulation photometry (Pokorny *et al.*, 1989). In a given condition, relative modulation depths of the three lights (and mean chromaticity) was kept constant. An observer set thresholds by adjusting the yoked modulation depths of the three lights. In modulation photometry, thresholds are highest at equal luminance. We predicted that thresholds at equal luminance should be dependent on the chromaticities used and highest close to the predicted tritan pair. This was indeed the case.

Methods

We used a Maxwellian view system with light-emitting diodes (LEDs) as light sources (Swanson *et al.*, 1987b). Light from red (638 nm), green (554 nm) and blue (480 nm) LEDs was combined with conventional Maxwellian optics. Light from the red and green diodes was of high colorimetric purity (>98%), that from the blue LED had a purity of 80.9%. The observer viewed the stimuli through a 2 mm artificial pupil. A one-degree stimulus field was surrounded by a 4.7 degree annulus with one-tenth the retinal illuminance of the stimulus field and matching chromaticity co-ordinates. The annulus was used to suppress stray light effects. Mean retinal illuminance of the stimulus was 700 td (Westheimer, 1966). This consisted of 257 td from the red LED, 241 td from the green LED, and 202 td from the blue LED, as measured with a Photo Research (Burbank, CA) spectrophotometer focused on a diffusing surface placing a few mm behind the plane of the pupil. The CIE (1931) chromaticity co-ordinates were (x, y) = (0.53, 0.34).

It is convenient to visualize the stimulus in a MacLeod-Boynton (MacLeod and Boynton, 1979) cone excitation space (Fig. 1), in which vertical lines represent changes in S/(L+M) cone excitation (and tritan confusion axes) and horizontal lines changes in L/(L+M) cone excitation. The three filled symbols

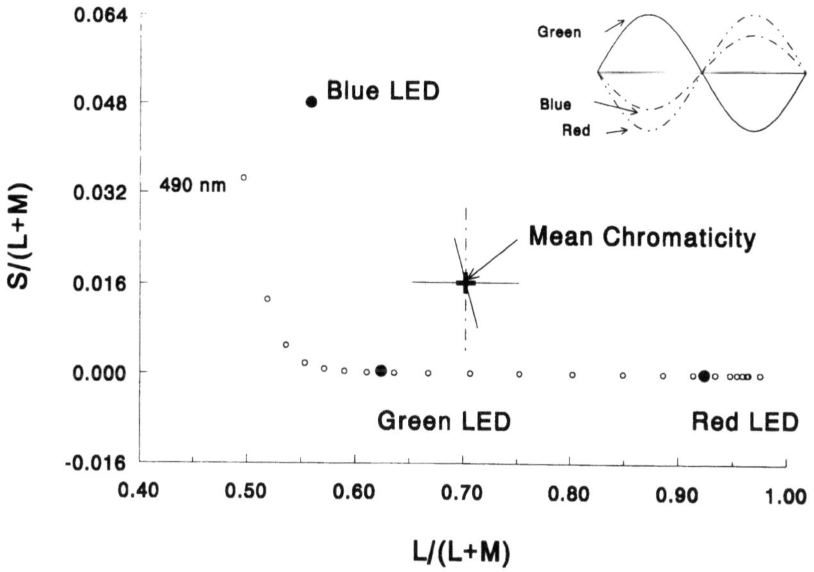

Fig. 1. The stimuli presented in cone excitation space. Open circles represent spectrum loci at 10 nm intervals. Closed circles represent chromaticities of the three diodes and the cross is the mean chromaticity. We usually modulated the red and blue diodes in phase, in counterphase with the green diodes, as in the insert. If the proportion of red light in the red/blue mixtures is varied, the near-horizontal solid line segment represents modulation with a red:blue proportion of 1:0 and the other solid line segment represents with a red:blue proportion of 0:1. Changing green light modulation amplitude would tilt the modulation vector away from the equiluminance plane. The dashed vertical line represents the modulation along the tritanopic confusion line for a standard observer.

represent the chromaticities of the three diodes. The cross is the mean chromaticity. The solid-line segments through the cross represent modulation of the red against the green diode (the almost horizontal segment) and the blue against the green diode. It can be seen that some combination of the red and blue against the green should yield modulation along a tritanopic confusion line, indicated by the dashed vertical line segment.

We usually modulated the red and blue diodes in phase, in counterphase with the green, as shown in the inset in Fig. 1. We varied proportions of red to blue. Some proportion should be tritanopically confused with the green at equal luminance. We also varied the relative modulation amplitude of the green to the other two, to span equal luminance. With modulation of all three diodes yoked together, observers set thresholds by the method of adjustment. Data shown are means of three estimates. Two color-normals (T.Y., D.L.) and two dichromats (B.L., a deuteranope; C.W., a protanope) served as observers. We tested frequencies of 10 and 20 Hz. Similar results were obtained, but the higher frequency yielded more reliable data for the trichromatic observers.

Results

Heterochromatic modulation photometry (HMP) is a photometric method in which thresholds to counterphase-modulated lights are measured as a function of the ratio of their modulation depths (Pokorny, *et al.,* 1989). Thresholds are maximal when the lights are of equal luminance, when, in conventional flicker photometry, the sensation of flicker is minimized. We performed modulation photometry with various proportions of red to blue modulated against green. Threshold data are shown in Fig. 2 for two trichromatic subjects. For each curve, the red and blue lights were mixed in different proportions, as indicated. The abscissa represents the luminance, log (G/(R+B)), ratio for the CIE (1931) standard observer. The ordinate plots thresholds as the summed troland modulation contrasts of the three lights at threshold. In the foremost curve only the red and green diodes are modulated, and threshold rises toward a modest peak close to equal luminance, when log(G/(R+B)) is zero. This condition corresponds to the horizontal line segment in Fig. 1, far from the tritan confusion line. With an increasing blue component, the threshold rise becomes sharper at equal luminance as the modulation vector in Fig. 1 rotates toward the vertical, tritan axis. The predicted tritan pair lies, for a standard observer, between the second and third curves from the back.

Variation between subjects in data in Fig. 2 might be expected to depend on two factors. The first of these is inter-individually variability in M/L-cone ratio (e.g. Lutze *et al.,* 1990), which is expected to be most apparent in a shift in the luminance ratio of the peak when only the red and green diodes are modulated (the foremost curves in Fig. 2). The second is inter-individual variability in

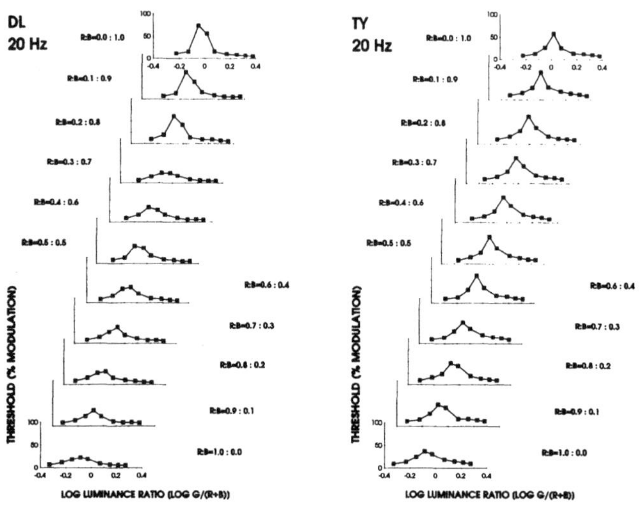

Fig. 2. Modulation thresholds for two color normal observers at 20 Hz plotted as a function of luminance ratios of the green light to the red/blue light, normalized for the CIE (1931) observer. Each curve represents data for different proportions of red:blue light.

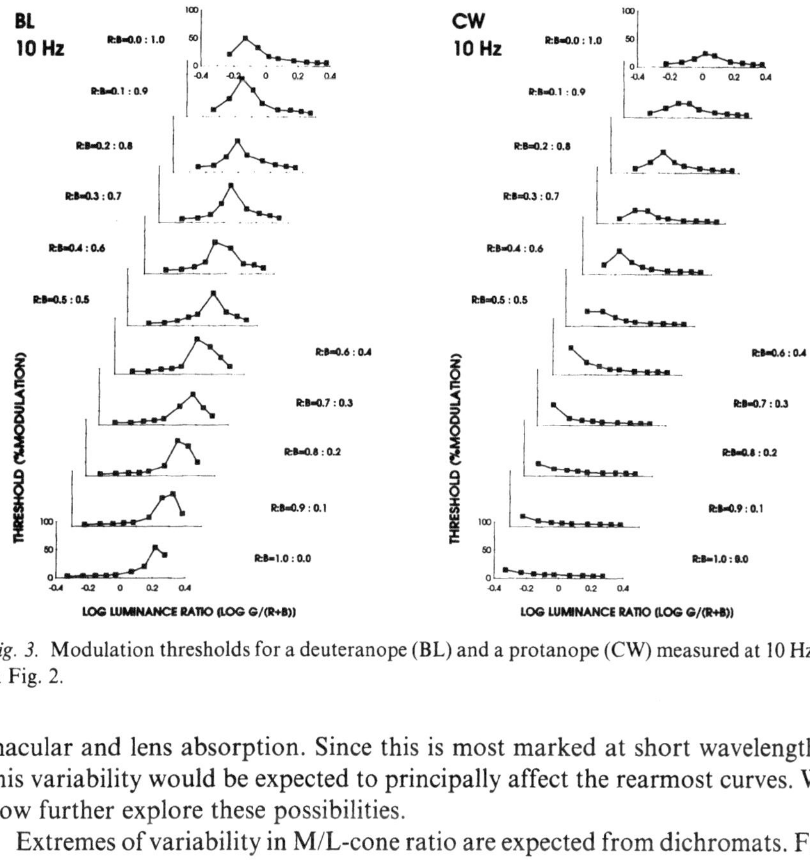

Fig. 3. Modulation thresholds for a deuteranope (BL) and a protanope (CW) measured at 10 Hz as in Fig. 2.

macular and lens absorption. Since this is most marked at short wavelengths, this variability would be expected to principally affect the rearmost curves. We now further explore these possibilities.

Extremes of variability in M/L-cone ratio are expected from dichromats. Fig. 3 shows threshold data from two dichromatic subjects, one of the authors (B.L.), a deuteranope, and a protanope (C.W.). Since dichromats only have the M- or L-cone, threshold peaks are expected to be similar under all conditions. This is seen in Fig. 3 to be the case. For the protanope, the range of red-green ratios explored in the foremost curves was not sufficient to encompass the threshold peak representing the flicker minimum. The progression of the peaks is in opposite directions for the two dichromats, as expected. Close to the predicted tritan modulation condition, the peak is similar in location for the two observers.

One purpose of the experiments described here was to test if a temporal analogue of the melting border might serve to define tritanopic pairs for individual observers. Although the data in Fig. 2 show that peak thresholds are maximal close to the predicted tritanopic confusion line, the data do not permit precise determination of the maximum. In particular, it would be useful to explore sensitivities at other angles than those available between the two line segments in Fig. 1. This could be achieved by modulating the green and red diodes in counterphase with the blue, and the green and blue diodes in counterphase with the red. We did not explore different luminance ratios for these other conditions, but measured thresholds at individuals' equal luminance points for the different combinations, since from data in Fig. 2, it was possible to determine the sensation luminance (Kaiser, 1988) for each observer for the three lights. We could then provide chromatic modulation at any desired angle within the cone excitation space in Fig. 1. Observers were required to set modulation thresholds under each condition.

Fig. 4a, b show threshold data for the two observers of Fig. 2. Modulation thresholds have been plotted against the ratio R/(R+B). A ratio of near 0.2 would be a tritanopic confusion line for a standard observer. For observer D.L., there was a wide region about this value in which the observer could not detect the modulation. However, observer T.Y. could detect the modulation throughout the range.

Fig. 4c, d show the same data replotted as detection contours in the MacLeod-Boynton cone excitation space. Contours for both observers are

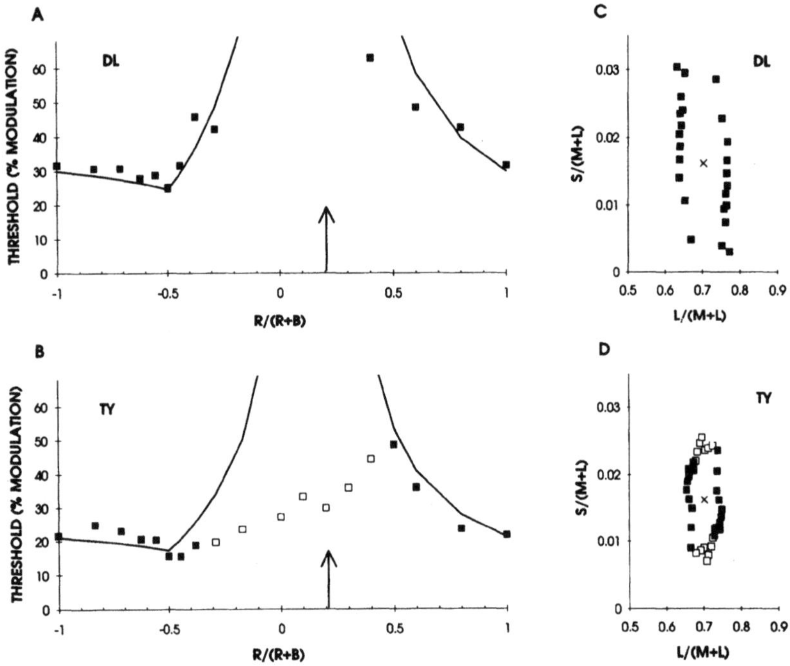

Fig. 4. Equiluminance 20 Hz modulation thresholds for two color normal observers. a, b. Thresholds plotted as a function of proportion of red modulation to the sum of red and blue modulation depths. Negative values represent red and blue diodes modulated in counterphase. The arrow indicates the R/(R+B) ratio required for modulation along a tritanopic confusion line for the Judd observer. The solid line is a template which should be followed if detection is based on a |M-L| cone signal. Deviant points for observer T.Y. are plotted as open symbols. c, d. Data replotted in the MacLeod-Boynton cone excitation diagram. The cross is the mean chromaticity for the standard observer. The deviant points in b for T.Y. are again plotted as open symbols, and their loci suggest detection through S-cone modulation.

elongated parallel to the S-cone axis, and most data points fall on approximately vertical straight line segments, especially for observer D.L. This indicates that L-cone modulation determines threshold to the equal-luminance modulation along these line segments.

We have fitted the data in Fig. 4a, b with a template derived from the L-cone modulation under the different conditions. As expected, the template satisfactorily fits data from observer D.L. The template fails to fit data from T.Y., and deviant points are drawn as open symbols in both Fig. 4b and d. We interpret these thresholds as reflecting detection through S-cone modulation by this subject. In Fig. 4d, the open symbols are seen to lie approximately parallel to the L-cone axis, consistent with detection through the S-cone. Detection of S-cone modulation at this frequency has been reported by others (Nakano and Kaiser, 1992; Stockman *et al.*, 1991).

Discussion

Our data indicate that psychophysical sensitivity to chromatic modulation at high frequency tends to be minimal along a tritanopic confusion line. Under most conditions, detection of such stimuli appears dependent on the L-cone modulation at equal luminance (Fig. 4c, d). This is consistent with the hypothesis that the MC-pathway then underlies detection, for the amplitude of the MC-pathway non-linear response at equal luminance is linearly related to the |M-L| cone difference signal, which is a linear function of L-cone modulation (Lee *et al.*, 1989, Fig. 3). This correlation between psychophysics and physiology is analogous to the correlation between psychophysical residual distinctness and the residual MC-pathway response in MDB (Kaiser *et al.*, 1990). It should be noted that although a response in the MC-pathway to equal luminance modulation may also be revealed on changing the relative phase of the light stimuli, this phase shift has become very small at 20 Hz, the frequency used here (Lindsey *et al.*, 1986; Smith *et al.*, 1992; Swanson *et al.*, 1987a).

However, if residual flicker can still be detected close to a tritan line, as with observer T.Y., this indicates that in some subjects S-cone modulation may produce a sensation of flicker at high temporal frequencies, as suggested by Nakano *et al.*, (1992). In this respect, residual distinctness of the minimally distinct border and residual flicker at equal luminance do not appear to behave in a strictly similar manner. Although the residual response in the magnocellular pathway may underlie detection at equal luminance through an |L-M| cone signal, we have not seen a residual response in this pathway on S-cone modulation. The physiological locus and substrate of detection of high-frequency S-cone modulation thus remains unclear.

Acknowledgements

We thank David Lee and Christian Wehrhahn for their co-operation in these experiments. Tsaiyao Yeh was partially supported by DFG Grant II B 7 – Le 524/8 – 1.

References

Kaiser, P.K. (1988). Sensation Luminance: a new name to distinguish CIE luminance from Luminance dependent on an individual's spectral sensitivity. Vision Res. 28: 455–456.

Kaiser, P.K., Lee, B.B., Martin, P.R. and Valberg, A. (1990). The physiological basis of the minimally distinct border demonstrated in the ganglion cells of the macaque retina. J. Physiol. (Lond.). 422: 153–183.

Lee, B.B., Martin, P.R. and Valberg, A. (1989). Nonlinear summation of M- and L-cone inputs to phasic retinal ganglion cells of the macaque. J. Neuroscience. 9: 1433–1442.

Lindsey, D.T., Pokorny, J. and Smith, V.C. (1986). Phase-dependent sensitivity to heterochromatic flicker. J. Opt. Soc. Am. A 3: 921–927.

Lutze, M., Cox, N.J., Smith, V.C. and Pokorny, J. (1990). Genetic studies of variation in Rayleigh and photometric matches in normal trichromats. Vision Res. 30: 149–162.

MacLeon, D.I.A. and Boynton, R.M. (1979). Chromaticity diagram showing cone excitation by stimuli of equal luminance. J. Opt. Soc. Am. 69: 1183–1185.

Nakano, Y. and Kaiser, P.K. (1992). Color fusion and flicker fusion frequencies using tritanopic pairs. Vision Res. 32: 1417–1424.

Pokorny, J., Smith, V.C. and Lutze, M. (1989). Heterochromatic modulation photometry. J. Opt. Soc. Am. A. 6: 1618–1623.

Smith, V.C., Lee, B.B., Pokorny, J., Martin, P.R. and Valberg, A. (1992). Responses of macaque ganglion cells to the relative phase of heterochromatically modulated lights. J. Physiol. (Lond.). 458: 191–221.

Stockman, A., MacLeod, D.I.A. and DePriest, D.D. (1991). The temporal properties of the human short-wave photoreceptors and their associated pathways. Vision Res. 31: 189–208.

Swanson, W.H., Pokorny, J. and Smith, V.C. (1987a). Effects of temporal frequency on phase-dependent sensitivity to heterochromatic flicker. J. Opt. Soc. Am. A 4: 2266–2273.

Swanson, W.H., Ueno, T., Smith, V.C. and Pokorny, J. (1987b). Temporal modulation sensitivity and pulse detection thresholds for chromatic and luminance perturbations. J. Opt. Soc. Am. A 4: 1992–2005.

Tansley, B.W. and Boynton, R.M. (1976). A line, not a space, represents visual distinctness of borders formed by different colors. Science. 191: 954–957.

Valberg, A., Lee, B.B., Kaiser, P.K. and Kremers, J. (1992). Responses of macaque ganglion cells to movement of chromatic borders. J. Physiol. (Lond.). 458: 479–602.

Valberg, A. and Tansley, B.W. (1977). Tritanoptic purity-difference function to describe the properties of minimally distinct borders. J. Opt. Soc. Am. 67: 1330–1336.

Westheimer, G. (1966). The Maxwellian view. Vision Res. 6: 669–682.

Yeh, T., Lee, B.B. and Kremers, J. (1995). The temporal response of ganglion cells of the macaque retina to cone-specific modulation. J. Opt. Soc. Am. A / In Press.

Dr. Barry B. Lee
Department of Neurobiology
Max Planck Institute for Biophysical Chemistry
37077 Göttingen, Germany

21. Interocular difference in Rayleigh matches of color normals

STEVEN K. SHEVELL and JI CHANG HE
(Chicago, IL, U.S.A.)

Abstract

The Rayleigh match typically is used to assess whether an individual has normal or red/green-defective color vision. There is substantial variation, however, in the Rayleigh matches of color-normal observers, and even differences in the matches from the left eye and right eye of a single observer. We confirmed this interocular difference in two independent samples of normals, and measured additional color matches to seek the cause of the difference. Color matches that varied the wavelength in the mixture field or the field size (i) fail to provide any evidence of an interocular difference in the wavelength of peak sensitivity of photopigment (λ_{max}) and (ii) suggest the difference often is due to effective optical density of photopigment. We conclude that an interocular difference in Rayleigh-match midpoint (but not range) is common, and due to effective optical density and/or pre-receptoral spectrally selective filtering.

Introduction

Metameric color matching is a sensitive measure of individual differences in color vision. Normal color vision is trichromatic, which requires by definition three primary lights to match any arbitrary spectral power distribution. Two primaries, however, are sufficient to match a stimulus restricted to wavelengths above 540 nm because only two classes of cone, the M cone and L cone, are appreciably stimulated in this wavelength range. Thus a normal trichromat can set an admixture of 545 and 670 nm lights to match 589 nm. This is the Rayleigh equation, which is used widely to assess abnormal color vision and, more recently, to quantify variation among normal trichromats.

The Rayleigh-match measurement is the proportion of 670 nm light in the admixture that matches 589 nm. A normal value is about 0.50 when the levels of 545 and 670 nm lights are equated for L-cone quantal absorption (the deutan mode—Mitchell and Rushton, 1971; Pokorny *et al.*, 1973). Variation among normals is considerable, both in the range of proportions accepted as a match

B. Drum (ed.), Colour Vision Deficiencies XII, pp. 185–191.
© 1995 *Kluwer Academic Publishers, Dordrecht.*

and in the midpoint of the range. For example, one normal trichromat may accept as a match any proportion in the range 0.52–0.56, while another accepts only proportions 0.47–0.49. Simple deuteranomalous (protanomalous) trichromats set matches well below (above) normal values.

Individual differences in Rayleigh matches of normals are attributed, in part, to genetic differences among people, which result in slightly different cone photopigments (Neitz and Jacobs, 1986, 1990; Sanocki *et al.*, 1993; Sanocki *et al.*, 1994; Winderickx *et al.*, 1992). We consider here whether slightly different cone pigments might account also for the frequently-observed difference in Rayleigh matches between the left eye and right eye of a single observer (Humanski and Shevell, 1991; Palacz *et al.*, 1972). We first report results from two new samples, which show that an interocular difference in match-midpoint is common: a statistically reliable difference was found (i) in 12 of 38 males (31%), and (ii) in 9 of 31 observers in an independent sample of 15 males and 16 females (29%). Next, an additional color match was measured for the second sample (620 nm light replaced 670 nm light in the admixture). The admixture ratio (I_{620}/I_{545}) set in this match divided by the admixture ratio in the usual Rayleigh match (I_{670}/I_{545}) is a value particularly sensitive to the wavelength of peak sensitivity (λ_{max}) of L-cone photopigment (He and Shevell, 1994). Variation in the λ_{max} of photopigment is attributed to visual pigment genes on the X-chromosome (Nathans *et al.*, 1986; Neitz and Jacobs, 1990; Neitz *et al.*, 1991). The results show no evidence of a difference between the two eyes of a single observer with respect to the λ_{max} of photopigment.

Methods

Color matches were measured with a three-channel, computer-controlled Maxwellian-view system using LEDs as light sources (similar to the instrument described by Pokorny *et al.*, 1989). Stimuli were presented in either a 2° or a 7° bipartite field: a mixture of 550 plus 670 nm light in the left hemifield, and 590 nm light in the right hemifield (the results below give the exact wavelengths, which can be slightly different owing to the particular interference filters used). The retinal illuminance of a standard-normal match was 57 td. On each trial the bipartite field came on steadily until the observer responded, via a button press, that the mixture appeared ‘greener than’, ‘redder than’, or ‘the same as’ the 590 nm field. The observer could vary only the radiance of the 590 nm light while viewing the bipartite field. After a response, the mixture ratio for the next trial was set by the computer according to a double interleaved staircase method designed to determine independently the two ends of the match range (for details see He and Shevell, 1994).

Each observer participated in four sessions on separate days. The first day was considered practice. Results reported here are from the following three days. Statistical analyses are based on the three replicated measurements of each end point of the match range, so that interocular differences in midpoint are assessed with respect to variability of the measurements used to compute each midpoint. Note that match range is statistically independent of the midpoint valuc and its cstimatcd standard error.

Observers were volunteers from the university community paid hourly to participate in the study.

Results

Experiment 1

Rayleigh matches of 38 males are shown in Fig. 1. In this experiment the field size was 2 deg. The mixture hemifield was a combination of 551 and 667 nm light, and the monochromatic hemifield was 588 nm. The abscissa is the proportion of 667 nm light in the mixture field, $I_{667}/(I_{667} + I_{551})$. The ordinate is the subject number. Independent measurements are shown for the left eye (solid circles) and right eye (open triangles). The line through each plotted symbol is the average match range, estimated from the mean of the replicated measurements of each end point; the symbol is the midpoint of the range. Asterisks along the ordinate indicate observers with a statistically reliable difference between the left-eye

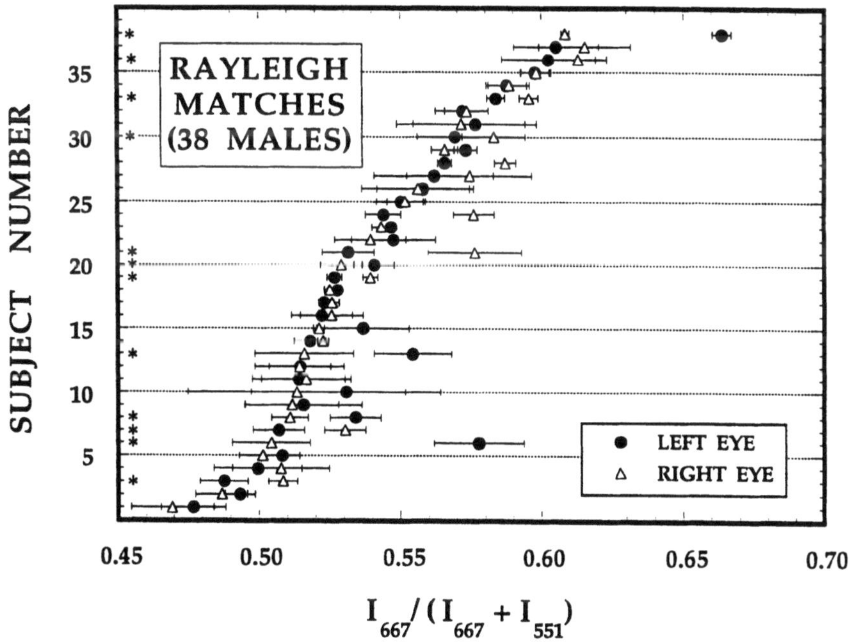

Fig. 1. Rayleigh match measurements (midpoint and range) from 38 male observers. Asterisks indicate observers with a statistically reliable difference between eyes in Rayleigh-match midpoint ($p<0.05$).

midpoint and right-eye midpoint ($p<0.05$). Twelve of the 38 observers (31%) reveal a reliable interocular difference. Rayleigh match range, on the other hand, never showed a reliable interocular difference (see Humanski and Shevell, 1991, for a statistical procedure that simultaneously assesses interocular differences in match midpoint and match range). Overall, a difference between eyes is common for match midpoint but not match range.

Rayleigh matches from the same subjects were measured also using a 7° field. Matches with a 7° field typically require a higher proportion of 667 nm light in the mixture than at 2° because of lower optical density of photopigment at 7° (Pokorny and Smith, 1976). Of the 12 observers with a reliable interocular difference in midpoint at 2°, only 3 of them (#7, #13, and #36) had a statistically reliable difference between eyes with a 7° field. This suggests the difference between an observer's two eyes in Rayleigh-match midpoint often is due to a difference in effective optical density of photopigment.

Experiment 2

An interocular difference in match midpoint and not match range was confirmed in an independent sample of 31 observers (15 males and 16 females). These measurements, with a 2° field, are shown in Fig. 2. Results from males

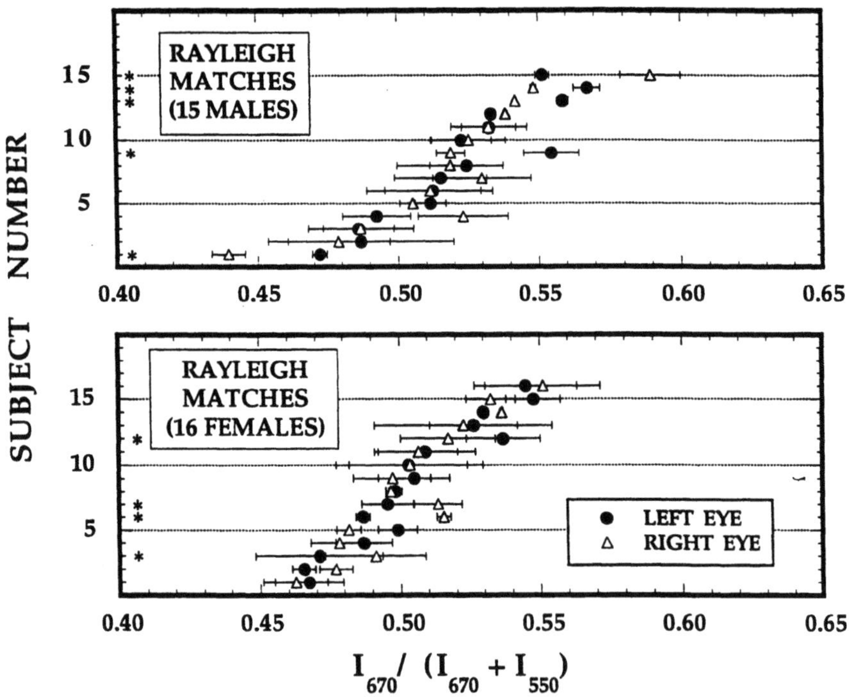

Fig. 2. Rayleigh match measurements (midpoint and range) from 15 male and 16 female observers. Asterisks as in Fig. 1.

(females) are in the upper (lower) panel. The wavelengths in this experiment were 550 and 670 nm in the mixture hemifield, and 590 nm in the monochromatic hemifield. Asterisks denote observers with a statistically reliable interocular difference in match midpoint. The proportion of observers showing a difference is nearly the same as in the first experiment: 29% (9 of 31 observers). As before, there is no evidence of a difference between eyes in match range.

These observers also performed a second color match, with 622 nm light replacing 670 nm in the mixture field. The proportion of 622 nm light in the match was measured as before, with the match midpoint used to compute I_{622}/I_{550}. This value, divided by the analogous proportion from the match using 670 nm, is relatively insensitive to variation in photopigment optical density and pre-receptoral spectrally selective filtering but varies considerably with differences of even a few nanometers in L-cone-pigment λ_{max} (He and Shevell, 1994). This measure, which we call K (= $[I_{622}/I_{550}]/[I_{670}/I_{550}]$), was used to assess whether the interocular difference in match midpoint might be due to an interocular difference in the λ_{max} of L-cone photopigment. Relative values of K are plotted in Fig. 3 (males in upper panel, females in lower

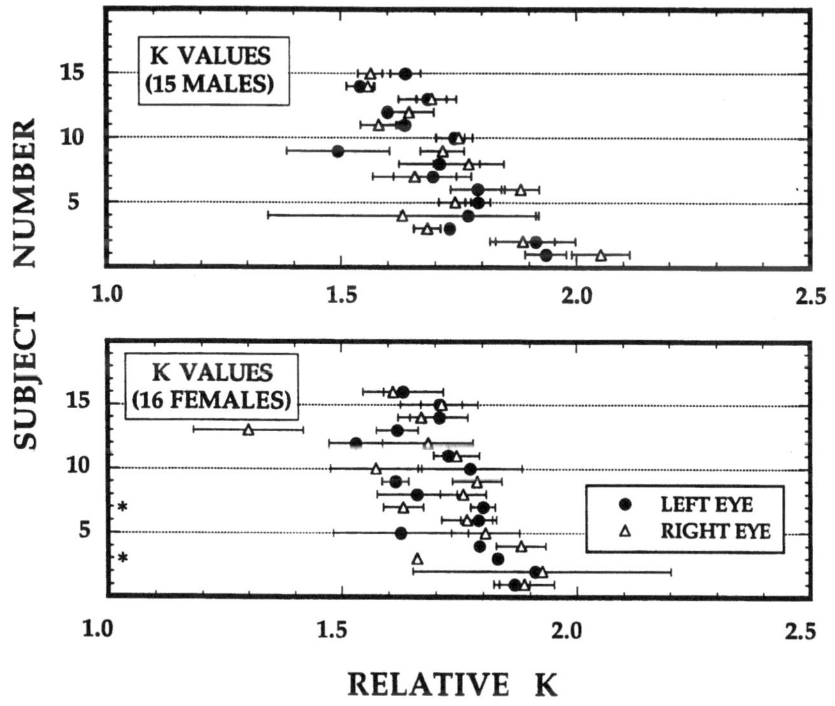

Fig. 3. The ratio relative K (= $[_{622}/I_{550}]/[I_{670}/I_{550}] \times$ const.) of two Rayleigh-type match-measurements (see text), from the same observers as in Fig. 2. Each plotted symbol is the mean value of relative K from 3 replications on separate days. Error bars are ± 1 standard error of the mean. Asterisks indicate observers with a statistically reliable difference between eyes in the mean value of relative K ($p < 0.05$).

panel)*. Independent measurements from the left-eye and right-eye are plotted as solid circles and open triangles, respectively. The horizontal line through each point is ±1 standard error of the mean, based on replicated measurements of K on each of 3 days.

Only two of the 31 observers show a statistically significant interocular difference in the value of relative K, using a 5% ($p < 0.05$) criterion for rejecting by chance the null hypothesis of no difference between eyes. As there were 31 observers, this is very close to the expected number of spuriously significant observations (0.05 x 31 = 1.55). There is, therefore, no evidence for a difference between an individual's two eyes with respect to the wavelength of peak sensitivity of photopigment (λ_{max}).

Discussion

These measurements confirm that color matches often differ in the left-eye and right-eye of a single observer. The interocular difference is always one of match midpoint, which strongly suggests a mechanism affecting the relation between incident light and the quantal match in each eye. Match width, on the other hand, which is reflective of discriminative ability, never was found to be reliably different in the two eyes. These results confirm an earlier report with 20 observers (Humanski and Shevell, 1991), which found a highly reliable interocular difference in midpoint but not range.

A quantal match is affected by three factors: the wavelength of peak sensitivity (λ_{max}) of photopigment, effective optical density of photopigment, and pre-receptoral spectrally selective filtering. The ratio of two Rayleigh-type match-measurements is sensitive primarily to a small difference in λ_{max}. Figure 3 shows, for 31 subjects, values of this ratio, which differ between the two eyes of an observer at only the level of chance. Thus there is no evidence of an interocular difference in λ_{max}. The ratio K is a highly reliable measure of an *interobserver* difference of a few nanometers in L-cone λ_{max} (Sanocki *et al.*, 1994) so the lack of a difference between eyes in Fig. 3 strongly suggests that variation in λ_{max} is not a major factor.

Of the 12 observers with an interocular difference measured with a 2° field (Fig. 1), only 3 had an interocular difference also with a 7° field. As photopigment optical density is lower at 7° than 2°, this suggests the interocular difference is mediated in part by effective optical density. Effective optical density can be influenced by the actual optical density of photopigment within cones and by receptor alignment. Pre-receptoral filtering is also a likely contributor to differences between the two eyes.

Finally, we note that the ratio of two Rayleigh-type measurements (K) may

* Relative K is the value of K in absolute units (He and Shevell, 1994) multiplied by a fixed scaling constant. The value of the scaling constant is inconsequential here.

be useful for distinguising variation in λ_{max} from other factors that affect color vision, in studies of both interocular and interobserver differences.

Acknowledgements

Supported by National Eye Institute grant EY-04802.

References

He, J.C. and Shevell, S.K. (1994). Individual differences in cone photopigments of normal trichromats measured by dual Rayleigh-type color matches. Vision Res. 34: 367–376.

Humanski, R.A. and Shevell, S.K. (1991). Factors contributing to differences in Rayleigh matches of normal trichromats. In: Drum, B., Moreland, J.D. and Serra, A. (eds.), Colour Vision Deficiencies X, Doc. Ophthalmol. Proc. Ser. 54: 273–283. Kluwer Acad. Publ., Dordrecht.

Mitchell, D.E. and Rushton, W.A.H. (1971). The red/green pigments of normal vision. Vision Res. 11: 1045–1056.

Nathans, J., Thomas, D. and Hogness, D.S. (1986). Molecular genetics of human color vision: The genes encoding blue, green and red pigments. Science 232: 193–202.

Neitz, J. and Jacobs, G.H. (1986). Polymorphism of the long-wavelength cone in normal human colour vision. Nature 323: 623–625.

Neitz, J. and Jacobs, G.H. (1990). Polymorphism in normal human color vision and its mechanism. Vision Res. 30: 621–636.

Neitz, M., Neitz, J. and Jacobs, G.H. (1991). Spectral tuning of pigments underlying red-green color vision. Science 252: 971–973.

Palacz, O., Glinksi, J. and Kuprianowicz, W. (1972). The problem of practical value of Ishihara's test and anomaloscopic examination on the basis of the results of examinations of right and left eye in normal subjects. Klinika Oczna, Acta Ophthalmologica Polonica 42: 81–85.

Pokorny, J. and Smith, V.C. (1976). Effect of field size on red-green color mixture equations. J. Opt. Soc. Am. 66: 705–708.

Pokorny, J., Smith, V.C. and Katz, I. (1973). Derivation of the photopigment absorption spectra in anomalous trichromats. J. Opot. Soc. Am. 63: 232–237.

Pokorny, J., Smith, V.C. and Lutze, M. (1989). A computer-controlled briefcase anomaloscope. In: Drum, B. and Verriest, G. (eds.), Colour Vision Deficiencies IX, Doc. Ophthalmol. Proc. Ser. 52: 515–522, Kluwer Acad. Publ., Dordrecht.

Sanocki, D., Lindsey, D.T., Winderickx, J., Teller, D.Y., Deeb, S.S. and Motulsky, A.G. (1993). Serine/alanine amino acid polymorphism of the L and M cone pigments: Effects on Rayleigh matches among deuteranopes, protanopes and color normal observers. Vision Res. 33: 2139–2152.

Sanocki, E., Shevell, S.K. and Winderickx, J. (1994). Serine/alanine amino acid polymorphism of the L-cone photopigment assessed by dual Rayleigh-type color matches. Vision Res. 34: 377–382.

Winderickx, J., Lindsey, D.T., Sanocki, E., Teller, D.Y., Motulsky, A.G. and Deeb, S.S. (1992). A ser/ala polymorphism in the red photopigment underlies variation in colour matching among colour-normal individuals. Nature 356: 431–433.

Visual Sciences Center
The University of Chicago
939 East 57th Street
Chicago, IL 60637
U.S.A.

22. Dichoptic color mixing

M.L.F. DE MATTIELLO, A.R. BIONDINI and H. SALINAS
(Buenos Aires, Argentina)

Abstract

Chromatic vision is mediated by three types of cones. Although 'green' and 'red' cones are treated similarly by the visual system, the 'blue' cones are different in several aspects, for example in their distribution and number in the retina, in their lack of spatial antagonism, in their poor spatial and temporal responses, in their fast loss of adaptation and in their different dichoptic resolutions.

The present work studied dichoptic mixtures of short-wavelengths between 430 and 520 nm with central fixation. These mixtures produced a marked chromatoestereopsis which was not observed when working with red and green stimuli. The zones in which this phenomenon is more pronounced were determined. Finally, a complementary study with eccentric fixation resulted in the disappearance of this phenomenon. The results obtained support the proposal that the chromatic processing of information for this task is principally carried out at the retinal level.

Introduction

The binocular mixture of chromatic lights has been studied since Hering and Helmholz (see Dunlap, 1944; Das, 1953; De Weert and Levelt, 1976, for a summary on this subject). The conditions required to obtain dichoptic color mixing are: a) small stimuli of equal size, b) stimuli of similar density and chromaticity, and c) 'vision unity' or fusion. If any of these requirements are not met, alternation and/or chromatostereopsis of the stimuli is observed.

In previous work (Mattiello and Salinas, 1992), we obtained reasonable dichoptic mixtures using middle and long wavelengths, but when analyzing short wavelengths, recurrent problems were encountered. The aim of this work was to analyze the influence of short wavelengths in these types of mixtures.

B. Drum (ed.), Colour Vision Deficiencies XII, pp. 193–196.
© 1995 *Kluwer Academic Publishers, Dordrecht.*

Method

Apparatus

A complementary optic channel (W3) was added to a monocular Wright colorimeter (Wright, 1946). This consisted of a Barr & Stround interference filter, a diaphragm to regulate the size of the images and a lens to adjust the chromatic fusion. The colors from the interference filters – eight hues between 482 and 650 nm (see Table 1) – were metamerically matched with the colorimeter's channel W2, using wavelengths of 460, 580 and 650 nm as primary colors. The four blues from the colorimeter were 430, 460, 480 and 510 nm, with luminances of 2.59, 5.13, 6.40 and 13.01 cd/m^2 respectively. The luminances of the hues were not equalized, as this would have produced a strong inequality in saturation and consequently no perceptibly equivalent stimuli would have been obtained.

Table 1. Wavelength and luminance of the eight stimuli generated with the interference filters (W3).

Wavelength (nm)	650	619	598	577	556	525	504	482
Luminance (cd/m^2)	21.62	31.89	29.82	43.91	62.98	18.91	11.65	2.61

Observers

Two observers with normal color vision participated in the experiment. They were both emmetropic, orthophoric and had good stereo acuity (20′ or less).

Procedure

Four sets of functions were generated, using one of the blues selected for each of them. For the monocular mixture (mM), one blue was chosen and mixed with the other hues from the interference filters. The resulting color was then equalized with colors coming from W1. The stimulus from W1 was considered as a reference and was presented in periods of 5 sec every 30 sec. Once this first operation was completed, the pairs of colors were separated: one was reproduced through channel W2 and the other through channel W3. Binocular fusion was then adjusted.

The observer's task was to match the dichoptic mixture (dM) from W2 and W3, with the comparison stimulus represented by mM (from W1) by adjusting the colorimeters's controls. Both 50′ visual angle stimuli were vertically aligned and placed 1° apart. Each experimental situation was carried out twice, alternating the stimuli in order to compensate for the effects of ocular dominance.

Results and discussion

With central vision, as already previously observed (Mattiello and Salinas, 1992), the dichoptic mixture of the four selected blues with the remaining hues, resulted in different grades of alternation and chromatostereopsis. These phenomena slowly disappeared as the wavelengths from W3 approached the short wavelengths from W2. For 430 and 460 nm, the alternation and stereopsis disappeared when these hues were mixed with another wavelength under 500 nm; for 480 nm, when mixed with another wavelength under 555 nm, and for 510 nm, when mixed with another wavelength under 597 nm. Starting from these values, the dM could be readily made. However, when the stimuli were displaced to 1°40′ eccentricity over the horizontal axis, all the dichoptic mixtures could be obtained with the higher density of the blues cones (Ronchi, 1992).

Figure 1 shows the average dichoptic and monocular matches made by the two subjects for the four sets of data. It can be seen that component G requires more luminosity in the dM than components R and B. Consequently, the summation of the luminosities remains relatively constant (Fig. 2).

Fig. 1 also shows the points, named S1 and S2 by Wright (see Fig. 72, Wright, 1946), which correspond to the crossing of the spectral coefficient curves R-G and B-G. In our study the former corresponds to 590 nm approximately and the latter to 480 nm. It will also be observed that the difference between the maximum points of the G and R curves are similar to those obtained by König (Wright, 1946).

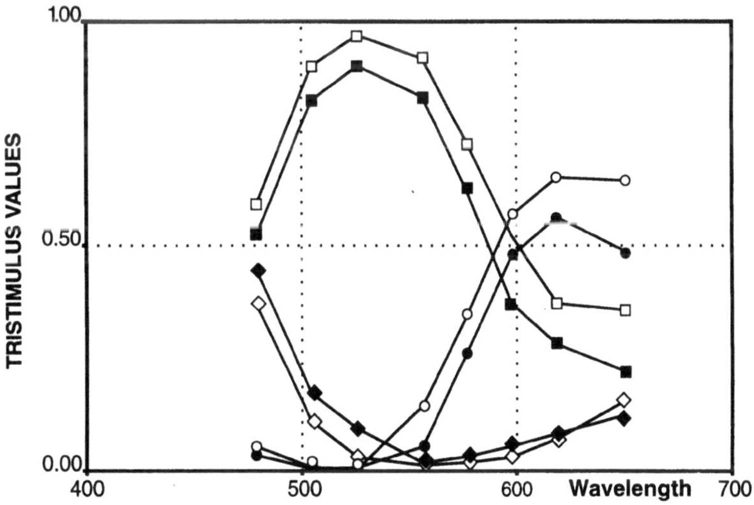

Fig. 1. Mean dichoptic (empty symbols) and monocular (filled symbols) color matching functions using a 50′ matching field at 1°40′ eccentricity. Each spectral coefficient curve is represented by a different symbol: blue ◇, green □ and red ○. Each symbol represents the average of the four data sets each with four estimations per observer.

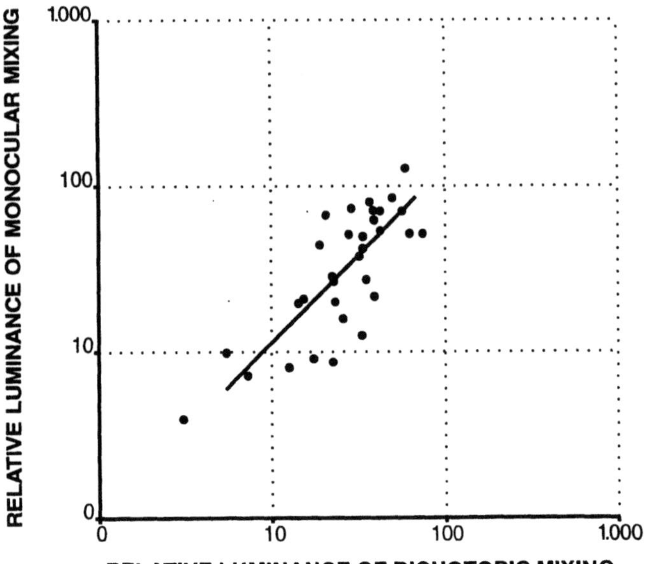

Fig. 2. Relation between the luminance of the dichoptic and monocular mixtures corresponding to 32 combinations of hues used in the experiment. Each symbol represents the average of four estimations per observer.

The measurements taken further confirm the dM and show that dichoptic fusion for short wavelengths can be obtained at 1°40′ eccentricity. In addition, the difference between the fundamental mono- and dichoptic components is small, indicating that the processing of chromatic information for this task may be carried out at the retinal level.

References

Das, S.R. (1953). Binocular Color Mixture. Thesis. University of Illinois.

De Weert, C.M.M. and Levelt, W.J.M. (1976). Comparison of normal and dichoptic colour mixing. Vision Res. 16: 59–70.

Dunlap, K. (1944). Alleged binocular color mixing. Am. J. Psychol. 57: 559–563.

Mattiello, M. and Salinas, H. (1992). Actas de la Tercera Reunion Nacional de Optica: 531–532. Barcelona.

Ronchi, L. (1992). Visual responses in the short-wavelength range. A bibliographical review. ATTI della Fondazione Giorgio Ronchi XLVII, no. 6.

Wright, W.D. (1946). Researches on Normal and Defective Colour Vision. Henry Kimpton, London.

Maria L.F. de Mattiello
PRIVIS-CONICET
Paraguay 2155 1er piso
(1121) Buenos Aires
Argentina

23. Absence of lightness constancy as a deficit of monochromatic vision

C. von CAMPENHAUSEN and H.A. TAUSCH
(Mainz, Germany)

Abstract

Dark-adapted human subjects arranged a collection of colored papers in a sequence according to their lightness. The sequence of the papers (which appeared gray to the subjects) varied under different illuminant spectra. These variations could be predicted quantitatively from the radiometric data and human scotopic spectral sensitivity. By substitution of natural daylight spectra in the computation it was shown that monochromats cannot perceive lightness reliably under the changing illuminant spectra of our planet. Animals need some degree of lightness constancy to detect figures against a background reliably. This can be achieved only with the help of additional light receptors of different spectral sensitivity.

Introduction

In this communication a problem of monochromatic vision is described which is caused by the natural variation of the illuminant spectrum. Consider an animal with only one type of light receptor cell, for instance a rod-monochromat. Suppose this animal eats fruit that is green when unripe and red when ripe. To the monochromatic animal everything will appear achromatic. The ripe and unripe fruit can only be discriminated by their lightness. But lightness will be unreliable for the animal because of the diurnal shift in the illuminant spectrum. At noontime the green fruit will appear lighter than the red fruit, in the evening darker. So the monochromatic animal cannot recognize the ripe fruit reliably. If the unripe fruit is poisonous, the animal will become extinct.

To avoid this end to the story, the animal should at least be a dichromat, i.e. capable of recording the shift of power in the illuminant spectrum. With this information it could regulate its spectral sensitivity and achieve some degree of lightness constancy to become independent of the illuminant spectrum.

We pose the question: is the situation for monochromats really so dangerous

B. Drum (ed.), Colour Vision Deficiencies XII, pp. 197–202.
© 1995 *Kluwer Academic Publishers, Dordrecht.*

on our planet? The problem cannot be settled by calculations alone since there are too many visual parameters that might be helpful for recognition and discrimination of natural objects under changing natural illumination. For this reason we developed an experiment in which monochromats failed in a visual task because of changes in the illuminant spectrum under controlled experimental conditions (von Campenhausen, 1986). This experiment was improved so that the empirical results of the laboratory experiment could be confirmed quantitatively by calculations. Then, natural illuminant spectra were substituted in the calculation to show that visual failure also has to be expected under natural conditions.

Methods and results

Our subjects were normal trichromatic humans, wearing tightly-fitting goggles with gray filters of extremely low transmittance which rendered them rod-monochromats. In pilot experiments under natural daylight conditions we found that some papers appeared lighter to our monochromats than others at noon, but darker in the evening. Unfortunately observations of this type can hardly be repeated, since daylight is unreliable, changing in unpredictable ways, so a laboratory experiment had to be done first.

The light sources in the laboratory were blue and yellow fluorescent lamps which could be dimmed separately. Their spectra are shown at the top of Figure 1. The lower part of Fig. 1 shows the color loci of the 30 colored papers used in the experiment. It can be seen that the papers represented a wide range of colors and different excitation purities. These papers were arranged by the subjects in a sequence according to lightness.

Figure 2 shows the averaged sequence of the colored papers from seven subjects under blue and yellow illumination along the abscissa and the ordinate, respectively. Each paper is identified by an individual number. If the sequence had been the same under both illuminations, all points should lie on the diagonal. The deviations from the diagonal indicate that the lightness of the papers changes considerably depending on the illuminant spectrum. The yellow paper no. 4, for instance, is at the tenth position in the lightness sequence in blue light and at position three in yellow light. Under our laboratory conditions the lack of lightness constancy must be regarded as a major deficit of monochromatic vision.

Calculations showed that our subjects responded exactly as one would expect on the basis of their spectral sensitivity and the spectra of the light reflected by the colored papers. This is demonstrated in Figure 3. The ordinate shows the calculated sequence and the abscissa gives the empirical sequence found under a specific light, here combined (B+Y). There is a close coincidence of the empirical and theoretical sequences.

Therefore we can also calculate the expected monochromatic performance under natural light. We substituted in our computation morning and evening

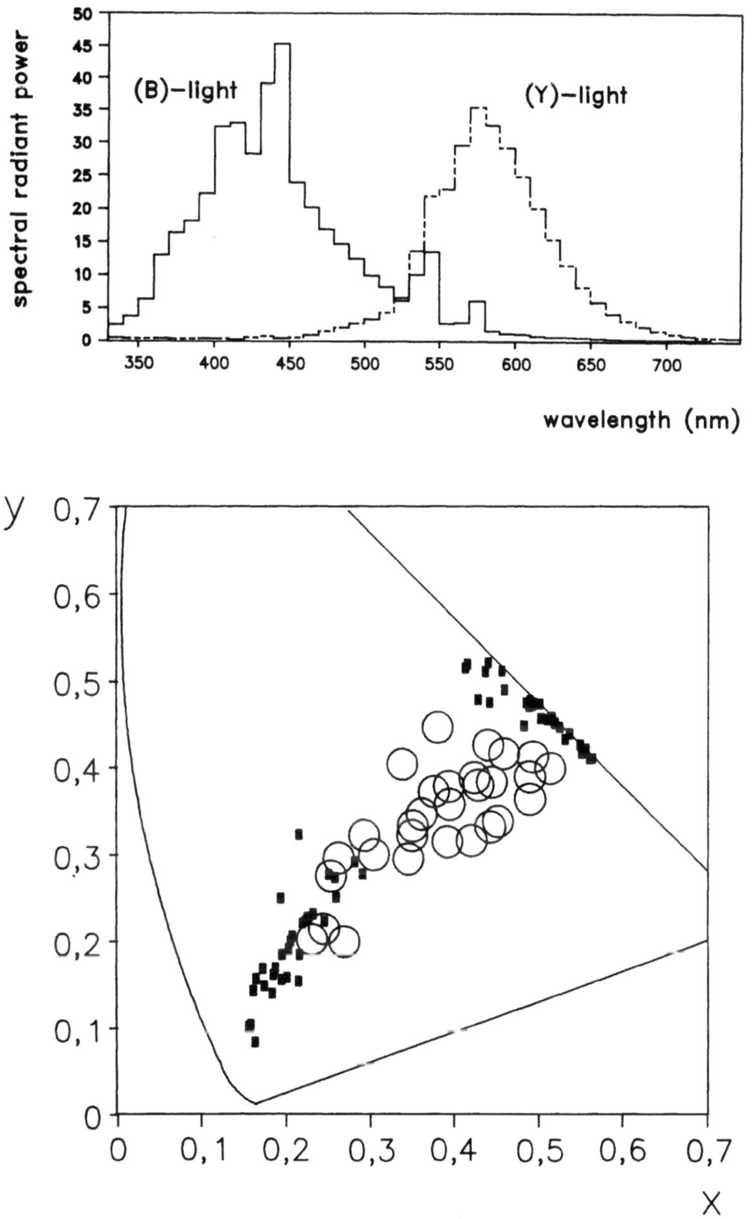

Fig. 1. Illumination spectra of the blue (B) and yellow (Y) fluorescent lamps (top). Color loci of 30 colored papers are shown in the figure at bottom. Color loci under combined (B+Y) illumination are indicated by open circles (*o*), those for blue (B) and yellow(Y) illumination by filled rectangles (■).

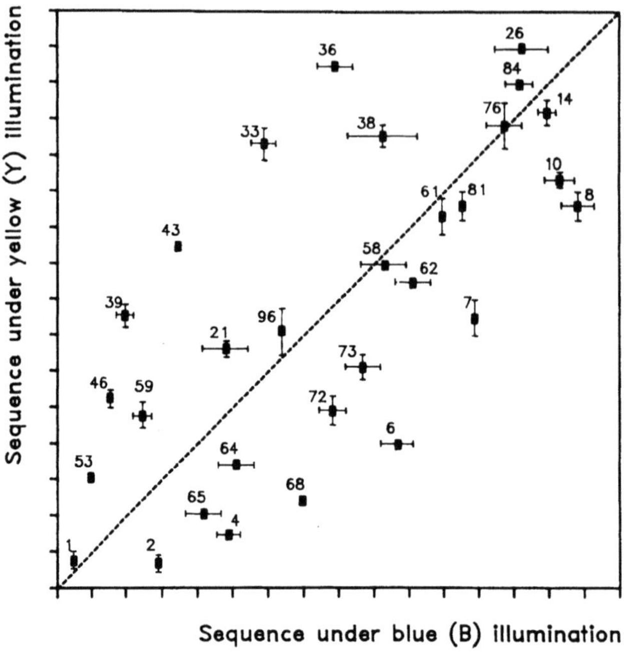

Fig. 2. Lightness sequences of colored papers under yellow (ordinate) and blue (abscissa) illumination arranged by 7 monochromatic subjects. Papers are marked by numbers. Vertical and horizontal bars are the standard deviations of the positions in the sequence.

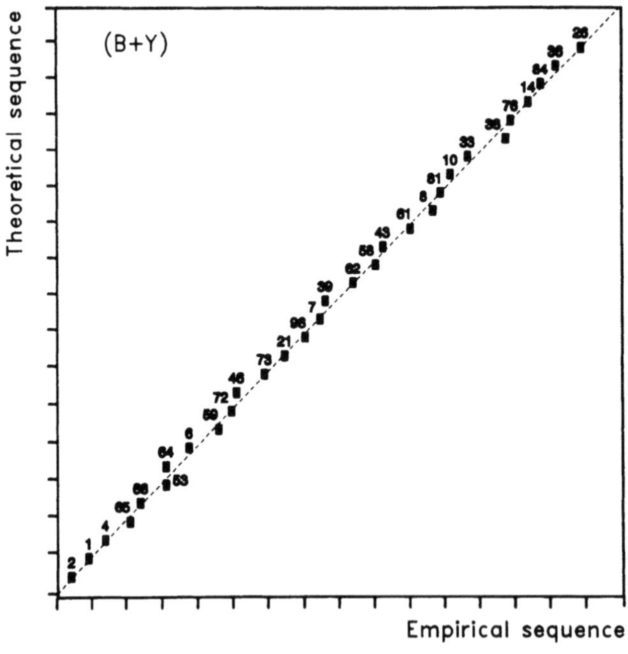

Fig. 3. Comparison of a theoretical versus an empirical sequence under white (B+Y) light.

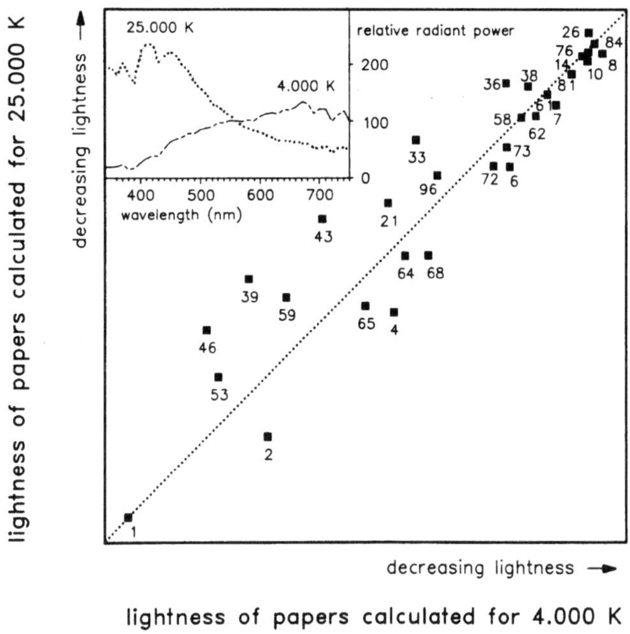

Fig. 4. Calculated lightness sequences of monochromats under different daylights, the relative spectral power spectra of which are shown in the inset.

light spectra, as shown in the inset of Figure 4. These relative spectral power distributions were calculated with the method of Judd, MacAdam, and Wyszecki (1964) as described in Wyszecki and Stiles (1982, pp. 145–146). The calculated sequences under these illuminations, Fig. 4, again show that the points deviate considerably from the diagonal. The yellow paper no. 4 is at position 10 in the lightness sequence under noon light but at position 5 under evening light. This means that the lightness of our colored papers as perceived by the monochromat changes considerably with time of day.

Conclusions

These results demonstrate that lightness cannot be perceived reliably by monochromats under natural conditions on our planet. Objects may appear lighter than, darker than, or equal to the background depending on the varying natural illuminant spectrum. Therefore, the most important visual task, the recognition of objects against a background, cannot be established reliably by monochromats under natural conditions.

Monochromatic eyes are sufficient to detect contours against the sky, the direction of a light source or darkening due to a shadow cast on the eye. Monochromatic fish are found in the deep sea (Lythgoe, 1972) where light comes from bioluminescence rather than from the variable sun. But in biotopes illuminated by the sun, monochromatic animals are rare.

Reliability of perception is more important than discrimination of many colors. Therefore additional receptors may have evolved initially to solve the constancy problem rather than to provide color vision. One may speculate on the existence of animals with more than one receptor type providing some degree of constancy but no color discrimination. Animals of this type have not yet been found, but may have survived as living fossils.

We used papers of uniform color in this study. Krinov (1947) recorded spectral reflectance functions of natural formations like forests, crops and soils. In a recent theoretical investigation, Dannemiller (1993) calculated sequences for the photon catches of the three human cone types from these scenes under different illuminations. This work, which is of interest for the lightness constancy in cone vision, is mentioned here because of the similarity in the treatment of the data.

Our work will be continued with natural objects under natural variations of illuminant spectra, to investigate empirically the limitations of monochromatic vision under biologically-relevant circumstances.

References

von Campenhausen, C. (1986). Photoreceptors, lightness constancy and color vision. Naturwissenschaften 76: 82–83.

Dannemiller, J.L. (1993). Rank orderings of photoreceptor photon catches from natural objects are nearly illuminant-invariant. Vision Res. 33: 131–140.

Krinov, E.L. (1947). Spectral reflectance properties of natural formations. National Research Council of Canada, Ottawa, TT-439.

Lythgoe, J.N. (1972). The adaptation of visual pigments to the photic environment. In: Dartnall, H.J.A. (ed.), Handbook of Sensory Physiology, Vol. VII/1, Photochemistry of Vision: 566–603, Springer-Verlag, Berlin and New York.

Wyszecki, G. and Stiles, W.S. (1982). Color Science, Concepts and Methods, Quantitative Data and Formulas (2nd ed). John Wiley and Sons, New York.

C. v. Campenhausen and H.A. Tausch
Johannes Gutenberg-Universität
Institut für Zoologie,
Abteilung III
55099 Mainz
Germany

24. Foveal densitometry and color matching in oligocone trichromacy

J.E.E. KEUNEN[1], S.R.S. DE BRABANDERE[2], and A.T.A. LIEM[1]
([1]Utrecht and [2]Rotterdam, The Netherlands)

Abstract

Oligocone trichromacy (a subtype of incomplete achromatopsia) is a congenital and stationary cone dysfunction with trichromatism. It is characterized by reduced visual acuity and day blindness, a reduced amplitude of the photopic electroretinogram, a normal fundus and normal color vision.

It is hypothesized that these patients might have a reduced number of normal functioning cones with preservation of the three cone types. As far as we know, this is the first report in which this hypothesis is tested by screening the foveal cone photopigment in four patients with oligocone trichromacy. Foveal densitometry was performed (a technique measuring the foveal cone photopigment density difference and its time constant of photopigment regeneration) and color matching; increment threshold spectral sensitivities were also measured. A reduced density difference of the foveal cone photopigment with a normal time constant of photopigment regeneration was found in all patients. The color matching data were normal. Increment threshold spectral sensitivity was tested in two patients; the results were normal.

We conclude that foveal cone photopigment screening in oligocone trichromacy confirms *in vivo* the hypothesis of a reduced number of cones (decreased density differences) with otherwise normal functioning cones.

Introduction

In cone dysfunctions one or more of the three cone types are deficient or absent. Patients generally have a congenital color vision defect which remains stationary during life. Several cone dysfunctions have been described in terms of color vision impairment: anomalous trichromatism, dichromatism, monochromatism and achromatism. Complete achromats have no functioning cones at all. Incomplete achromats have a residual color vision. François *et al.* (1955, 1959) published eight different combinations of rods and residual cones in

B. Drum (ed.), Colour Vision Deficiencies XII, pp. 203–210.
© 1995 *Kluwer Academic Publishers, Dordrecht.*

incomplete achromatopsia. An improved classification was published by Pokorny *et al.* (1982).

In 1973 van Lith described four patients with a congenital and stationary dysfunction of the cone system in which the color vision is unaffected. He introduced the name oligocone trichromacy, which can be classified as a subtype of incomplete achromatopsia in which all three cone types are preserved (Pokorny *et al.*, 1982). Oligocone trichromacy includes a decreased visual acuity with a striking improvement in dim light (day blindness) and a reduced photopic electroretinogram (ERG). The visual acuity is determined by the remaining cones in the fovea. The fundus, color vision, scotopic ERG and electro-oculogram (EOG) are normal. Patients may have nystagmus. The inheritance pattern of oligocone trichromacy is presumably autosomal recessive (van Lith, 1973) and the prevalence seems low. Similar cases were described by François and Verriest (1960), Goodman *et al.* (1963), Krill (1968), Neuhann *et al.* (1978) and Pokorny *et al.* (1982).

The pathophysiology of oligocone trichromacy can be derived from the name: the hypothesis is that patients might have a reduced number of normal functioning cones (oligocone), whereas each of the three cone types has a normal function (trichromacy).

As far as we know, this is the first report in which this hypothesis is tested by screening the foveal cone photopigment. In four patients foveal densitometry, color matching and increment threshold spectral sensitivity measurements were performed. Our results confirm the hypothesis that patients with oligocone trichromacy do have a reduced number of normal functioning cones.

Materials and methods

Patients

We investigated four patients with oligocone trichromacy after informed consent was obtained. Routine color vision tests included the Ishihara (edition 1970), the Standard Pseudoisochromatic Plates 1 and 2, the Farnsworth F2 Tritan Plate, the Panel D-15 test (saturated and desaturated) and anomaloscope testing.

Patient #1 is a white male born in 1952. Visual acuity had been decreased as long as he could remember without progression and and he was easily dazzled by light. The family history was negative. Best-corrected visual acuity was 0.25 in both eyes under standard conditions. Under mesopic light level, however, visual acuity improved to 0.6. The refraction anomaly was +1.50-0.50x170 in the right eye and +1.50-2.00x170 in the left eye. Slit-lamp examination and funduscopy were normal. Visual field testing (Goldmann) and dark adaptation were normal. The photopic ERG was almost absent, the scotopic ERG was normal.

Patient #2 is a white female patient born in 1967. She is a sister of patient #3.

Both were the only affected family members; consanguinity between the parents was ruled out. Visual acuity had been decreased as long as she could remember and was without progression. Best-corrected visual acuity was 0.2 for the right eye and 0.3 for the left eye. Under mesopic conditions visual acuity improved to 0.4 in both eyes. The refraction anomaly was +3.50-0.50x20 in the right eye and +0.50-1.50x180 in the left eye. Slit-lamp examination and funduscopy were normal, as was visual field testing with a Goldmann perimeter. The amplitudes of the photopic ERG were seriously decreased, the scotopic ERG was normal.

Patient #3 (sister of patient #2) was a white female born in 1973. Visual acuity had been reduced as long as she could remember, without any progression; she complained seriously of day blindness. Best-corrected visual acuity was 0.4 in both eyes; under mesopic conditions best-corrected visual acuity improved to 1.0. The refraction anomaly was +2.00-1.50x180 right eye and +0.50-1.50x180 left eye. Slit-lamp examination and funduscopy were normal, as was visual field testing (Goldmann). The amplitudes of the photopic ERG were seriously decreased, the scotopic ERG was normal.

Patient #4 was a white female, born in 1979. She had a stationary decreased visual acuity as long as she could remember. She complained also of day blindness. The family history was negative. In 1985 an active toxoplasmosis retinochoroiditis in the macular area of the left eye caused a considerable drop in visual acuity. Best-corrected visual acuity in 1992 was 0.5 in the right eye and 0.1 in the left eye under standard conditions. At mesopic light level the visual acuity of the right eye improved to 0.8, the visual acuity of the left eye remained 0.1. The refractive error in both eyes was –1.00D. Slit-lamp examination and fundus were normal. Multiple stimulus static perimetry of the right eye was normal, but revealed in the left eye a dense central scotoma. The photopic ERG amplitudes were seriously reduced, the scotopic responses were subnormal. Fluorescein fundus angiography in the right eye was normal.

Foveal densitometry

The Utrecht fundus reflection densitometer used in this study allows direct measurements of changes in optical reflection between fully dark-adapted and fully bleached states of the retina. Such measurements provide information concerning the in situ photopigment density of the tested area. Detailed technical information has been described previously (van Norren and van der Kraats, 1989). In brief, the instrument includes a single 30 W halogen lamp, which serves as the light source for the bleaching and measuring beams. To avoid possible photochemical light damage, a yellow filter in the bleaching beam eliminates wavelengths shorter than 500 nm. In our study, a fixation aid was positioned so that all measurements were taken at the fovea. To obtain a reasonable signal, the diameter of the measuring field (that is, what is viewed by the photomultiplier) is 2.4°. The measuring sequence included a two-minute bleaching exposure at 5.43 log trolands, followed by a dark period of photopigment regeneration. When regeneration had stabilized for at least two minutes (after an approximate 10–15 minute period in the dark-adapted condition), a second bleaching exposure of three minutes was included to ensure that no substantial head movement had occurred during the run. Data were discarded if the second (bleached) part of the measurement differed by more than 0.05 log unit from the initial bleached part of the measurement. The following two values were calculated from the data: the density difference and the time constant of photopigment regeneration. The density difference is the logarithmic difference of the measuring-light reflectance of a specific wavelength and the reference-light reflectance (wavelength, 730 nm), from fully bleached to fully dark-adapted states. The 730 nm wavelength is chosen as a reference, because this wavelength is not absorbed appreciably by the visual pigments. After completion of a measurement, the density difference is available for 15 wavelengths. In case of foveal fixation the density difference spectrum peaks at 551 nm; this wavelength is used to express the results of densitometry in our study. The regeneration data were expressed as a function of time and a time constant of photopigment regeneration was calculated by the available curve-fitting program. This was obtained from data of the density trace at ten-second intervals for the first four minutes after extinction of the bleaching light. The logarithm of the difference between these values and the mean value in the fully dark-adapted condition was then plotted against time and a straight line was fitted through the data points (Smith *et al.,* 1983).

Normative data for foveal densitometry were obtained from a previous study (Liem *et al.*, 1992). The mean foveal density difference at 551 nm of a control group of 27 healthy young subjects was 0.38 ± 0.12 (2 SD). The mean time constant of photopigment regeneration was 103 ± 46 seconds (2 SD).

Anomaloscope

The Rayleigh equation was measured with a Nagel type I anomaloscope, using a standard technique (Pokorny *et al.*, 1979). Midmatching points of the patients were assesssed at a constant yellow scale setting of 12.5. The normal matching range on our anomaloscope includes red-green scale numbers 37 through 45.

Increment Threshold Spectral Sensitivity

We studied increment threshold spectral sensitivity in two patients on a 1000 td white background, using a portable instrument described previously by Went and de Vries (1976).

Results

Clinical, densitometric and psychophysical data of all patients are given in Table 1. Although both eyes were investigated, only the data of the eye with the best-corrected visual acuity are presented.

Table 1. Clinical, Densitometric and Psychophysical data of one eye in four Patients with Oligicone Trichromacy.

Patient # Age (yrs), Sex	VA photopic	VA mesopic	Density Difference[a] at 551 nm	Time Constant[b]	Mean Rayleigh Match	ITSS
1,40, M	0.25	0.6	0.10	NM	39	normal
2,26, F	0.3	0.4	0.14	77	41	normal
3,19, F	0.4	1.0	0.24	98	42	NA
4,13, F	0.5	0.8	0.12	92	40	NA

VA best-corrected visual acuity
ITSS increment threshold spectral sensitivity
NM not measurable
NA not available
[a] foveal cone density difference, normal mean 0.38 ± 0.12 (2 SD)
[b] time constant of photopigment regeneration, normal mean 103 ± 46 seconds (2 SD)

An example of a foveal cone densitometric measurement at 551 nm in a healthy young subject is given in Figure 1, top left. The foveal density difference is 0.40 and the time constant of photopigment regeneration is 135 seconds. Plus signs represent the number of counts as a function of time for the reference wavelength at 730 nm. The stability of this trace is used for checking the reliability of the measurement. Crosses represent the number of counts at 419 nm. Because at short wavelengths the reflection factor of the eye is theoretically small, we chose to display the count numbers at this wavelength as a sensitive indicator for specular reflections at layers anterior to the photoreceptors. The open squares represent density, which is the logarithm of the ratio of the number of counts at the deep-red wavelength (730 nm) and the counts at 551 nm. At the 551 nm wavelength, maximal density differences between fully bleached and fully dark-adapted retina were found. The results of the corresponding density difference spectrum, obtained from all 15 wavelengths, are presented in Fig. 1, top right.

The foveal cone density difference in all patients ranged from 0.10 to 0.24, which were lower than normal values. The time constant of foveal cone photopigment regeneration was normal in three patients, and not measurable in patient #1 due to noisy data. An example of the density trace in patient #1 is given in Fig. 1, bottom left. The foveal density difference was 0.10 (obviously reduced compared to normal) and the time constant of photopigment regeneration was not measurable due to a too low foveal density difference. The corresponding density difference spectrum, obtained from 15 wavelengths, is

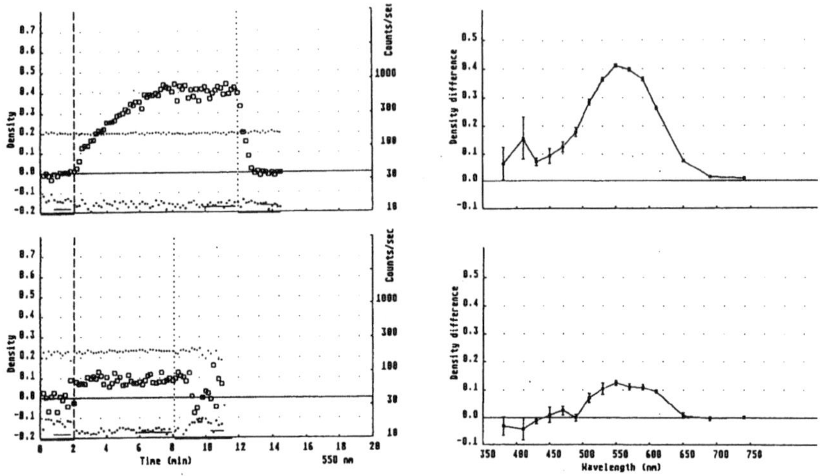

Fig. 1. Top left, Foveal cone densitometry in a healthy 28-year-old white female. Open squares represent density, plus signs represent reflected light at 730 nm, crosses represent reflected light at 419 nm. Light off is indicated by a vertical dashed line, light on by a vertical dotted line. Top right, The corresponding density difference spectrum. The highest density difference is at 551 nm. Bottom left, Foveal cone densitometry in patient #1 with oligocone trichromacy, and bottom right, the corresponding density difference spectrum. A reduced density difference is determined, and the density difference spectrum is flattened.

given in Fig. 1, bottom right. The shape of the density difference spectrum in this patient was flattened compared to the shape of the normal density difference spectrum.

Results of color vision tests were normal in all four patients. The color matches demonstrated also normal values in all patients (see Table 1).

Increment spectral sensitivity curves were obtained in patients #1 and #2. The results compared well with normal data, showing the typical three humped curve.

Discussion

It was the purpose of this study to test the hypothesis that oligocone trichromacy patients have a reduced number of normal functioning cones. A screening of the foveal cone photopigment was used to investigate the hyposthesis: foveal densitometry, color matching and increment threshold spectral sensitivity measurements were performed. As far as we know, foveal densitometry and increment threshold spectral sensitivity never have been investigated before in oligocone trichromacy.

We found decreased density differences of the foveal cone photopigment in all four patients, and normal time constants of foveal cone photopigment regeneration in three patients (in patient #1 the time constant of photopigment regeneration was not measurable due to a too low density difference). The density difference spectrum was flattened in all four patients. However, the Rayleigh matches, the color vision tests and the increment threshold spectral sensitivity curves (measured in two patients) were normal.

Possible causes for a reduced foveal cone photopigment density difference include: (1) loss of foveal cone photoreceptors, (2) shortening of photoreceptor outer segments, (3) metabolic dysfunction of the retinal pigment epithelium-photoreceptor complex, (4) disorientation of photoreceptors, (5) stray light, or any combination of these factors (Liem *et al.*, 1992). Although foveal densitometry does not differentiate between these factors (because the data reflect the optical density of all cones in the measuring area), the time constant of photopigment regeneration, the color matching data and the increment threshold spectral sensitivity data make interpretation of the foveal cone density differences possible.

Shortening of the photoreceptor outer segments (2) is unlikely as in this case pseudoprotanomaly on the anomaloscope would have been expected. The Nagel anomaloscope is sensitive to cone pigment density changes occuring in individual cones, such as shortened outer segments (Alpern, 1979). Our color vision measurements show a normal Raleigh match midpoint in all oligocone trichromacy patients.

The normal time constants of cone photopigment regeneration are not compatible with a metabolic dysfunction of the photoreceptors (3). In photoreceptor disorientation (4), we would expect pseudoprotanomaly on the anomaloscope. Also the normal fundus makes an impairment at the level of the foveal cone photoreceptor outer segment-retinal pigment epithelium complex unlikely. Fluorescein fundus angiography in patient #4 was normal. In this way, photoreceptor disorientation is unlikely.

Stray light (5) can be excluded as the fundus looks normal in patients with oligocone trichromacy.

Only one factor remains as explanation for the decreased density differences in patients with oligocone trichromacy: a loss of foveal cone photoreceptors (1). The lacking correlation between the Rayleigh matches and foveal cone density differences, together with normal time constants of photopigment regeneration make a reduction in the number of foveal cones likely. This explains also the reduced amplitudes of the photopic ERG.

The three cone types remain preserved as can be derived from the increment threshold spectral sensitivity and color vision tests; the remaining cones have a normal function. The remarkable increase of visual acuity when changing from photopic to mesopic conditions, which was noticed in all four patients, might point to disturbances of cone-rod interactions at a post-receptoral level as suggested by Schneck *et al.* (1993) in other forms of incomplete achromatopsia.

Based on foveal densitometry, color matching and increment threshold spectral sensivity, we confirmed that patients with oligocone trichromacy do have a decreased number of otherwise normal functioning cones.

References

Alpern, M. (1979). Lack of uniformity in colour matching. J. Physiol. 288: 85–105

François, J., Verriest, G. and De Rouck, A. (1955). L'achromatopsie congénitale. Doc. Ophthalmol. 9: 338–424.

François, J. and Verriest, G. (1959). Contribution a l'étude des dyschromatopsies congénitales à symptomes intermédiaires entre ceux des systèmes dichromatiques classiques et ceux de l'achromature. Ann. Oculist. 192: 8–112.

François, J. and Verriest, G. (1960). Trois nouvelles observations d'achromatopsie congénitale atypique. Ann. Oculist. 193: 123–137.

Goodman, G., Ripps, H. and Siegel, I.M. (1963). Cone dysfunction syndromes. Arch. Ophthalmol. 70: 214–231.

Krill, A.E. (1968). The electroretinogram in congenital color vision defects. In: The clinical value of electroretinography, Proc. 5th ISCERG Symposium Ghent 1966:205-214, Karger, Basel.

Liem, A.T.A., Keunen, J.E.E. and van Norren, D. (1992). Foveal densitometry in adult-onset diffuse drusen. Am. J. Ophthalmol. 114: 149-157.

van Lith, G.H.M. (1973). General cone dysfunction without achromatopsia. In: Pearlman, J.T. (ed.), 10th ISCERG Symposium, Doc. Ophthalmol. Proc. Ser. 2: 175–180, Dr. W. Junk, The Hague.

Neuhann, T., Krastel, H. and Jaeger, W. (1978). Differential diagnosis of typical and atypical congenital achromatopsia. Graefes Arch. Ophthalmol. 209: 19–28.

Van Norren, D. and van de Kraats, J. (1989). Retinal densitometer with the size of a fundus camera. Vision Res. 29: 369–374.

Pokorny, J., Smith, V.C., Verriest, G. and Pinckers, A.J.L.G. (1979). Congenital and Acquired Color Defects: 119–120. Grune and Stratton, New York.

Pokorny, J., Smith, V.C., Pinckers, A.J.L.G. and Cozijnen, M. (1982). Classification of complete and incomplete autosomal recessive achromatopsia. Graefes Arch. Ophthalmol. 219: 121–130.

Schneck, M., Haegerstrom-Portnoy, G. and Verdon, W. (1993). Interactions between rod and long-wavelength cones in incomplete achromatopsia. In: Zrenner, E. (ed.), Abstract Book, 12th IRGCVD Symposium: 38, Tübingen.

Smith, V.C., Pokorny, J. and van Norren, D. (1983). Densitometric measurement of human cone photopigment kinetics. Vision Res. 23: 517–524.

Went L.N. and de Vries-de Mol E.C. (1976). Genetics of colour vision. In: Verriest, G. (ed.), Colour Vision Deficiencies III, Mod. Probl. Ophthalmol. 17: 96-107, Karger, Basel.

J.E.E. Keunen
F.C. Donders Institute of Ophthalmology
P.O. Box 85500
3508 GA Utrecht
The Netherlands

25. Study on color mechanism dependency in the off effect

MOMOYO KONO and SHINICHI YAMADE
(Ohtsu, Shiga, Japan)

Abstract

The present study examines whether the off effect, the phenomenon of temporal edge emphasis, exhibits any type of wavelength selectivity as in the monochromatic Mach band which is the spatial edge emphasis. At first, off effect measurements were taken using a monochromatic background and a monochromatic test light. The off effect was observed in all combinations of test (red, green or blue) and background (red, green or blue) colors. However, we were unable to demonstrate wavelength selectivity. In the second, the on-off effect like overshoot of increment threshold on the chromatic change of background field was studied. The overshoot was observed in all combinations of test (red, yellow, green or blue) and background (red-green or yellow-blue). However, we were unable to demonstrate any color mechanism specificity. The degree of the overshoot seemed to depend solely on the chromatic difference of the background field. These results suggest that the off effect, a function of temporal specificity, has no relationship to color mechanism.

Introduction

The off effect (Crawford, 1947) is interpreted as a phenomenon of temporal edge emphasis which may have a relationship to the receptive field. Wavelength selectivity (Mitsuboshi, 1983; Green and Fast, 1971) has been verified in the monochromatic Mach band, which is a phenomenon of spatial edge emphasis with the overshoot in the increment threshold found on the brighter side of the monochromatic luminance-step boundary. This suggests color mechanism specificity in the lateral inhibitory interactions. On the other hand, Cole *et al.*, (1993) reported that there was virtually no effect for equiluminant Craik-Cornsweet stimuli (Cornsweet, 1970), and refuted the existence of color mechanism specificity in lateral inhibitory interactions.

The present study examined whether the off effect exhibits any type of wavelength selectivity as in the Mach band.

B. Drum (ed.), Colour Vision Deficiencies XII, pp. 211–219.
© 1995 *Kluwer Academic Publishers, Dordrecht.*

Experiment 1

At first, off effect measurements were taken using a monochromatic background and a monochromatic test light.

A two channel Maxwellian view system and interference filters were used. Figure 1 shows the position of the backgrounds, test fields and the time course. The background field is a 4.9° circle, exposed for 2 sec on and 2 sec off. Its intensity is approximately 1000 td. The test light at the center of the background is a 1.1° circle which is exposed for 10 msec. Three wavelength variations were used as test (red 648 nm, green 552.5 nm, and blue 477 nm) and background lights (red 663 nm, green 544 nm and blue 479 nm). The thresholds were determined at several time points on an average of 3 to 5 trials for each point. The adjustment method was used.

Figure 2 shows the results using a red background light. Both subjects, MK (left) and SY (right), have normal color vision. The points on the graph indicate the log relative threshold of the test stimulus as a function of the time from the exposure of the background light to the start of the test light. The upper line indicates the results using a red test light, the middle line a green test light, and the lower line a blue test light. These lines are separated to avoid overlapping. The overshoot in the increment threshold was observed on the brighter side of the temporal monochromatic luminance-step boundary in all test colors. If wavelength selectivity was seen, the off effect using the red test light should be greater than in the other two colors because the background light was red, but neither subject demonstrated such selectivity.

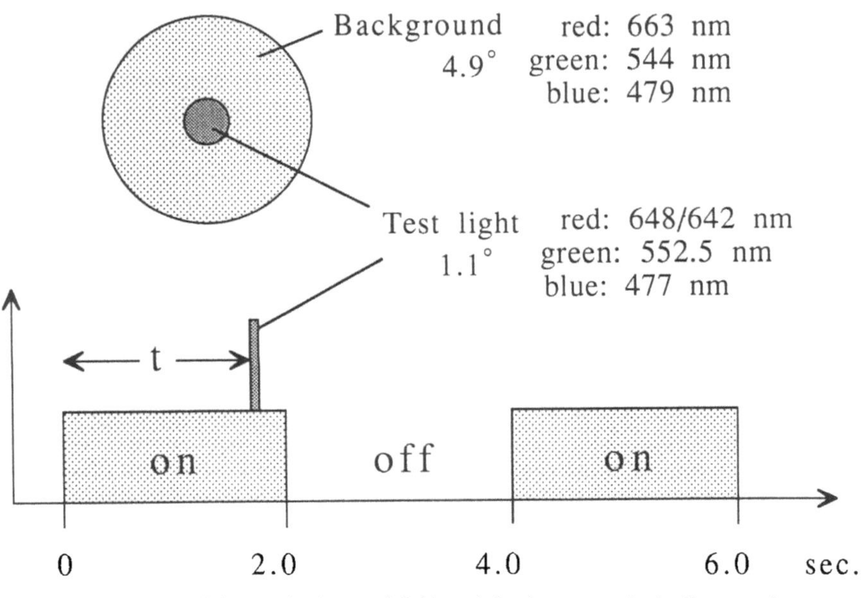

Fig. 1. Position of the test, background fields and the time course in the first experiment.

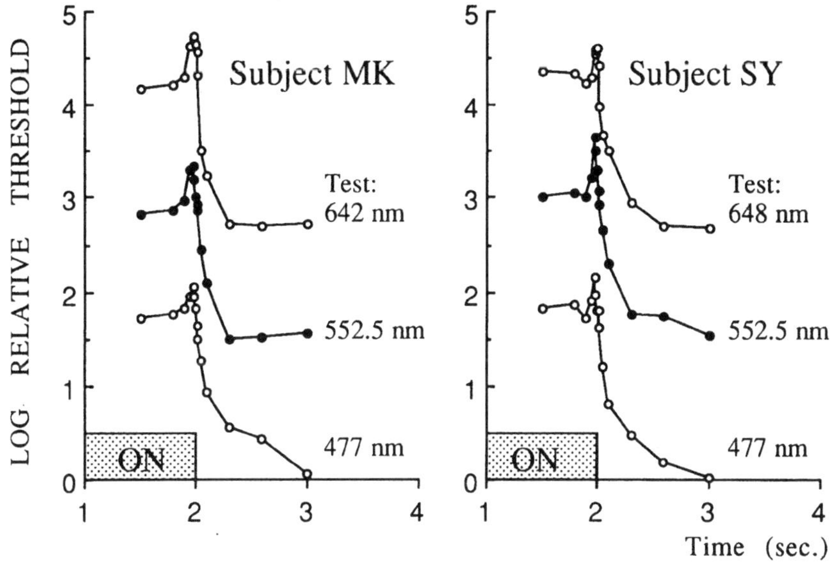

Fig. 2. Off effect using red (663 nm) background.

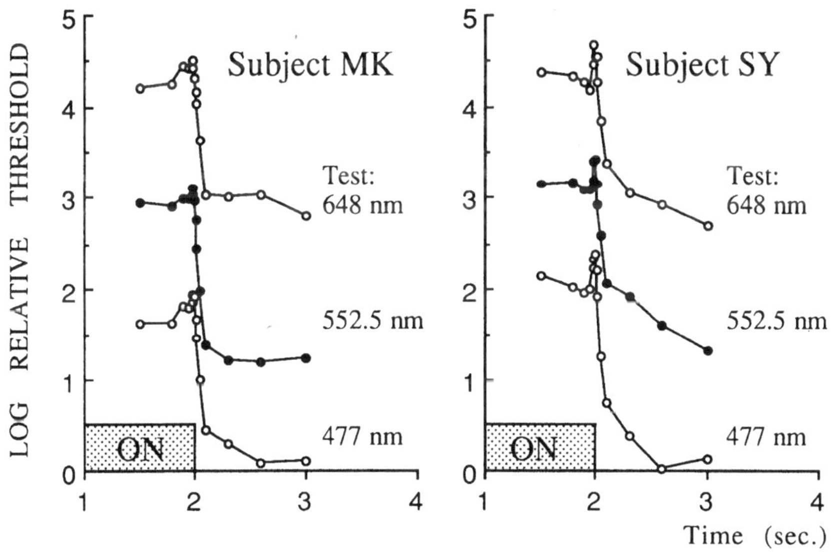

Fig. 3. Off effect using green (544 nm) background.

Figure 3 shows the results using a green background. The off effect was observed in all three test colors, but those using the green test light were not greater than the other two.

Figure 4 shows the results using a blue background light. The off effect was

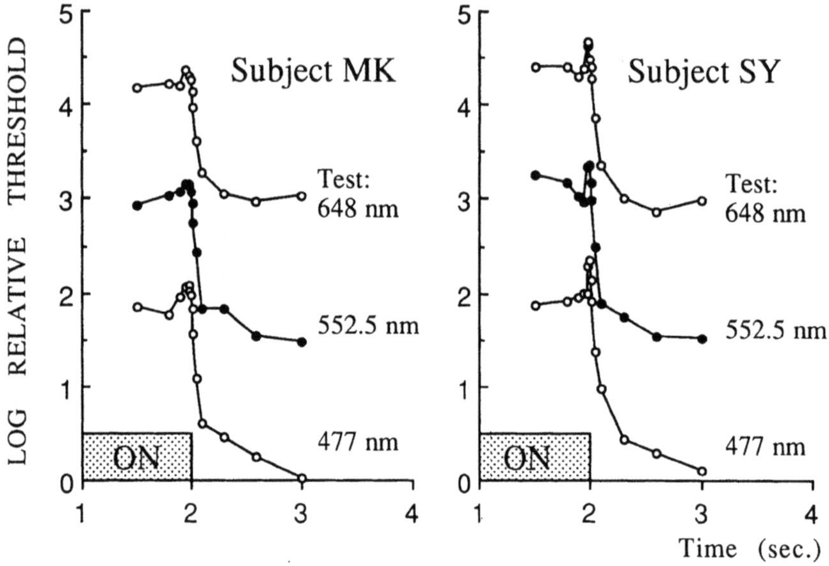

Fig. 4. Off effect using blue (479 nm) background.

observed in all three test colors, but those using the blue test light were not greater than the other two.

As shown in Figs. 2, 3 and 4, the off effect using a monochromatic background and a monochromatic test light did not demonstrate wavelength selectivity. There are two possible explanations; that the off effect cannot be interpreted as having a relationship to the receptive field, or that our methods were not applicable to demonstrate wavelength selectivity (King-Smith and Carden, 1976; King-Smith and Kranda, 1981; Foster and Snelgar, 1983). For example, changes of luminance in the background field existed in this experiment, so we did another experiment using only chromatic changes of background light.

Experiment 2

In the second experiment the increment threshold with a chromatic change of background, which showed an on-off effect like overshoot, was studied. A three channel Maxwellian view system was used in this experiment. As shown in Fig. 5, the background and the test fields were the same, as in the first experiment. The sequence of the background colors were red-green, as shown in Fig. 5, or yellow-blue. These background colors were matched in brightness using the flicker method before we began the experiment. A yellow 575 nm test light was added to the other three test lights. T is the time from the exposure of the red background light.

Figure 6 shows the results of subject SY. The lines on the left show the results

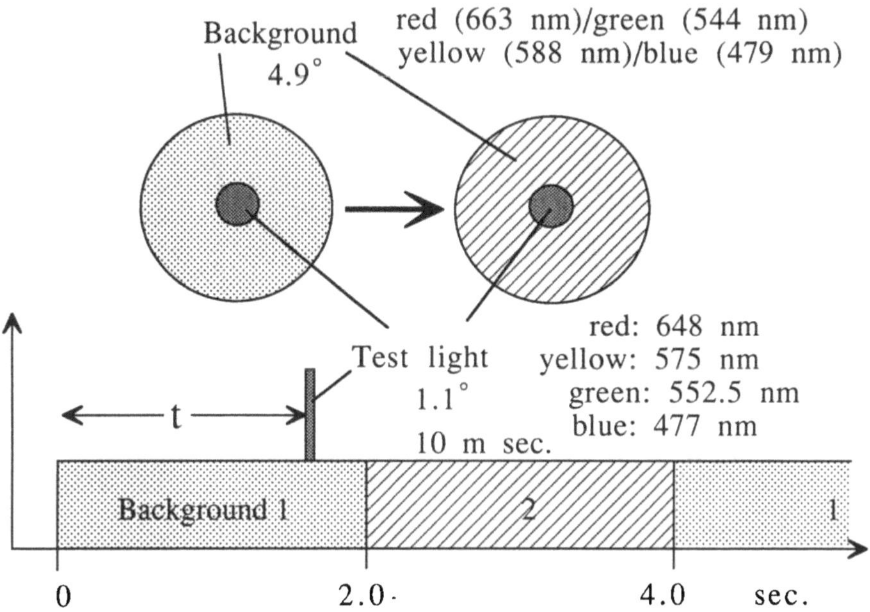

Fig. 5. Position of the test, background fields and the time course in the second experiment.

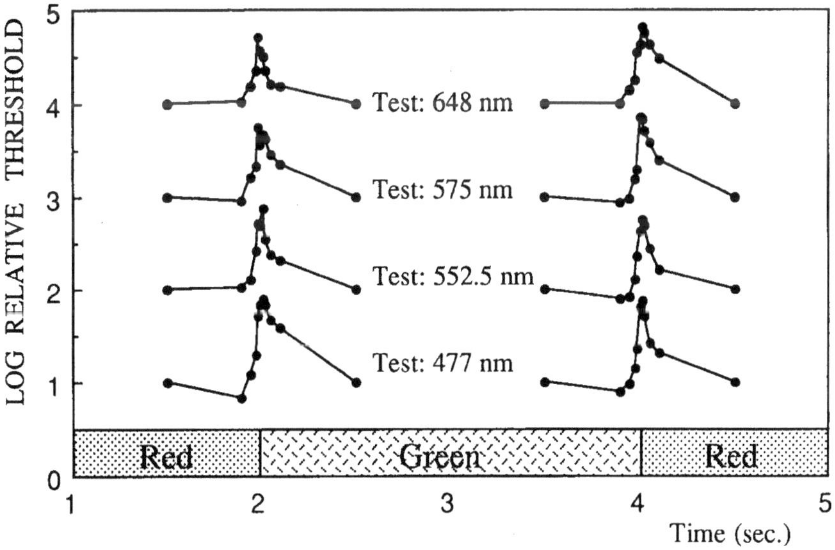

Fig. 6. Overshoots using red (663 nm)-green (544 nm) background: subject SY.

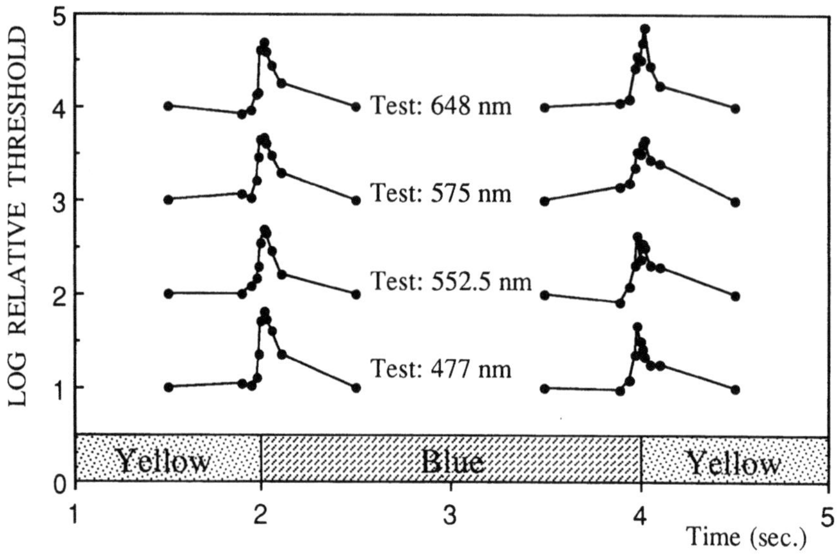

Fig. 7. Overshoots using yellow (588 nm)-blue (479 nm) background: subject SY.

when the background light changed from red to green, and the lines on the right those when the background light changed from green to red. The upper line indicates the results using a red test light, the next line a yellow, the next line a green, and the lowest line a blue test light. The on-off effect like overshoot of the increment threshold on the chromatic change of the background was observed in all colors. If this phenomenon had some relationship with color mechanism, then the overshoot observed using the red or green test lights should be somewhat different from the other test light colors because the background light was red-green. But neither showed any remarkable differences.

Figure 7 shows the results using yellow-to-blue and blue-to-yellow backgrounds. Overshoot was observed for all test lights, but those using yellow and blue test lights were not so remarkably different from the other test colors.

Experiment 3

We then examined the overshoot in relation to the degree of chromatic difference in the background field. One background color was fixed at 544 nm (green), and the other background color was changed from 460 nm (blue) to 640 nm (red) in 20 nm steps. The wavelength of the test light was 552.5 nm (green).

Figure 8 shows the results of the subject SY (normal color vision). The log relative thresholds of the on-off effect like overshoot are plotted as a function of the wavelength of the changing background color. This result indicates that the degree of the overshoot seemed to depend on the chromatic difference of the two background lights.

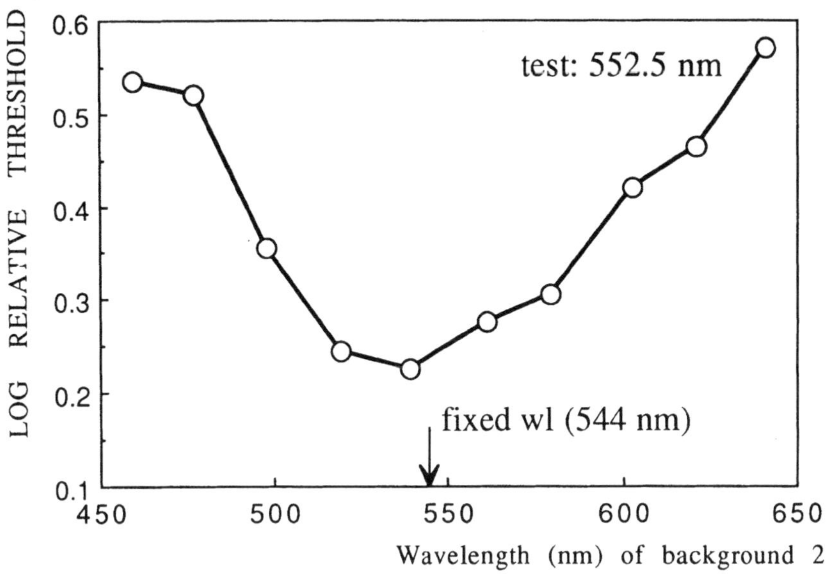

Fig. 8. Degree of overshoot vs. chromatic difference of background lights: subject SY.

Discussion

The present study examined whether the off effect, a phenomenon of temporal edge emphasis, exhibited any type of wavelength selectivity as in the Mach band. However, we were unable to demonstrate any wavelength selectivity in the off effect using a monochromatic background and a monochromatic test light, nor in the on-off effect like overshoot of the increment threshold on the chromatic change of the background. These results suggest that the off effect, a function of temporal specificity, has no relationship to color mechanism, although the blue wavelength used in experiments 1 and 2 (477 and 479 nm) were not short enough to estimate S-cone mechanism. We thought that it might have been because this phenomenon is controlled mainly by the Y-cell system and more by the central neural system.

The third experiment showed that the degree of overshoot on the chromatic change of the background field seemed to depend on the chromatic difference of the two background colors. So, this test could be used to study temporal color discrimination. For example, it may be of interest to test this phenomenon in color vision deficiencies.

Figure 9 shows the result of a deuteranomalous subject, indicating that he had relatively poor discrimination ability between green and the longer wavelength areas. This was easily predicted by his color abnormality.

In experiments 2 and 3, two background colors were matched in brightness using the flicker photometric method, but the possibility of some artifacts could

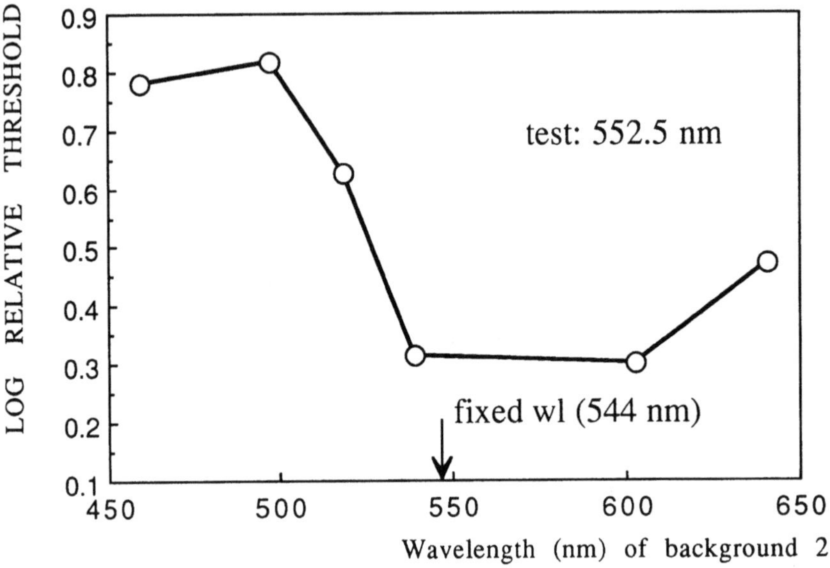

Fig. 9. Degree of overshoot vs. background chromatic difference: subject DA.

not be excluded. Some temporal artifacts, positive or negative flashes, may have existed at the time point when the background colors changed. Some amount of overshoot seen in experiments 2 and 3 might possibly be caused by these artifacts. The minimum value of the overshoot in Fig. 8 (about 0.2 log unit) may be artifactual, but dependency on the chromatic difference also seen in Fig. 8 means that these overshoots were caused mainly by the chromatic change of the background field.

Conclusion

The off effect using a monochromatic background and a monochromatic test light and the on-off effect like overshoot on the chromatic change of the background light were examined. (1) Overshoots were observd in all combinations of the test and background lights in both experiments, but neither were able to demonstrate any color mechanism specificity between the test and background colors. (2) The degree of overshoot on the chromatic change of the background seemed to depend on the chromatic difference of the two alternating background lights. This result indicates that this test may be useful in the study of temporal color discrimination.

References

Cole, G.C., Hine, T. and Scott, J. (1993). Relative contribution of luminance and chromaticity to the Craik-Cornsweet effect. In: Drum, B. (ed.) Colour Vision Deficiencies XI. Ophthalmol. Proc. Ser. 56: 51–57, Kluwer Acad. Publ., Dordrecht.

Cornsweet, T. (1970). Visual Perception, Academic Press, New York.

Crawford, B.H. (1947). Visual adaptation in relation to brief conditioning stimuli. Proc. Roy. Soc. 134B: 283–300.

Green, D.G. and Fast, M.B. (1971). On the Appearance of Mach Bands in Gradients of Varying Color. Vision Res. 11: 1147–1155.

Foster, D.H. and Snelgar, R.S. (1983). Test and field spectral sensitivity of colour mechanisms obtained on small white backgrounds: Action of unitary opponent-color processes?. Vision Res. 23: 787–797.

King-Smith, P.E. and Carden, D. (1976). Luminance and opponent-color contributions to visual detection and adaptation and to temporal and spatial integration. J. Opt. Soc. Am. 66: 709–717.

King-Smith, P.E. and Kranda, K. (1981). Photopic adaptation in the red-green spectral range. Vision Res. 21: 565–572.

Mitsuboshi, M. (1983). Sensitivity at the Edge of a Monochromatic Luminance-Step and the Mechanism of Color Vision. In: Mollen, J.D. and Sharpe, L.T. (eds.), Colour Vision Physiology and Psychophysics: 455–463, Academic Press, London.

Momoyo Kono, M.D.
Department of Ophthalmology
Shiga University of Medical Science
Ohtsu, Shiga 520–21, Japan

26. Psychological factors and defective colour vision

J. PÉREZ-CARPINELL, M.C. IBAÑEZ VALLEJO and
J.V. DIAZ ESTEVE
(València, Spain)

Abstract

In this study, involving students between 14 and 16 years old, the psychological dimensions of cognition and personality affected by red-green colour vision deficiencies were investigated. From the results of a factorial analysis it appears that the influence of colour vision abnormality on the psychological factors that we tested is always minor. Colour vision defect is correlated with such cognitive variables as spatial aptitude and visual difference perception, and with such personality variables as praxemia and submission. The weighting of the cognition and personality factors linked to colour vision abnormality, however, varies considerably and does not in any case contribute more than 20% to the total variance.

Introduction

This paper investigates how the psychological variables of cognition and personality influence colour vision testing in normal and anomalous trichromat subjects. The results extend those reported by Lakowski (1970) for a population of normal trichromats.

Method

The studied population consists of a total of 312 secondary school students of ages ranging between 14 and 16 years.

The colour vision testing of the subjects was realized using the Ishihara test and the Davico anomaloscope (Capilla and Aguilar, 1993).

As a measure of personality variables each student performed the 16PF test (Catell, 1972). This test is based on the evaluation of sixteen primary dimensions or factors which are functionally independent and are argued to be psychologically significant, and four broader secondary factors, obtained from

B. Drum (ed.), Colour Vision Deficiencies XII, pp. 221–226.
© 1995 *Kluwer Academic Publishers, Dordrecht.*

the primary factors. The factors included in the test are: Sizothymia-Affectothymia [A]; Lower scholastic mental capacity – Higher scholastic mental capacity [B]; Lower ego strength – Higher ego strength [C]; Submissiveness – Dominance [E]; Desurgency – Surgency [F]; Weaker superego strength – Stronger superego strength [G]; Threctia – Parmia [H]; Harria – Premsia [J]; Alaxia – Protension [L]; Praxemia – Autia [M]; Artlessness – Shrewdness [N]; Untroubled adequacy – Guilt proneness [O]; Conservatism of temperament – Radicalism [Q_1]; Group adherence – Self sufficiency [Q_2]; Low integration – High strength of self-sentiment [Q_3]; Low ergic tension – High ergic tension [Q_4]; Personal adjustment – Anxiety [Q_I]; Introversion-Extroversion [Q_{II}]; Paternia (affectiveness) – Cortertia (cortical alertness) [Q_{III}] and Submission-Independence [Q_{IV}].

The intellectual capacity of the subjects was operationally defined by the BADYG test (Yuste Hernanz, 1989), which attempts to measure intelligence and other special mental abilities. This test includes the following eleven subtests: General Mental Maturity [G.M.M.]; General Intelligence Verbal [G.I.V.]; General Intelligence non-Verbal [G.I.nV.]; Verbal Mental Ability [V.M.A.]; Numerical Aptitude [N.Ap.]; Reading Comprehension [R.C.]; Memory [M]; Non-verbal Mental Ability [nV.M.A.]; Logic Reasoning [L.R.]; Spatial Aptitude [S.Ap.] and Visual Differences Perception [At.].

Results and conclusions

Only sixteen subjects (5.7% of the males) had defective red-green colour vision; these results are similar to those found by other authors. None of the 29 females tested had any colour vision defect.

The sample of 16 subjects with defective red-green colour vision and 20 subjects with normal colour vision were selected from the total population studied. Colour vision tests were administered individually, while cognitive and personality tests were done in groups. Our results show that according to the Ishihara test (15th edition), 7 subjects had a strong deutan defect, 4 had a strong protan defect and 5 subjects had undefined red-green deficiencies; whereas according to the anomaloscope, 9 subjects were diagnosed as deutans (8 simple deuteranomalous and 1 deviant) and 7 were protanomalous (6 simple and 1 extreme).

The degree of association among all the personality and colour vision variables was tested in order to find correlations between the 21 different variables (1 for colour vision defect, Cvd, and 20 for personality factors, 16 primary and 4 secondary) by applying the point-biserial of the correlation coefficient r_{pb} defined as:

$$r_{pb} = (M_p - M_q/\sigma_t)\sqrt{pq}$$

since the discriminatory variable is really dichotomous in the sense that the anwer generated by the observer tested can only be pass or fail. The calculations and factorial analysis were done using the Statview 512$^+$ computer programme (Valero *et al.*, 1991), which provided the matrix of correlations between variables, principle components and varimax rotated factor loadings. Table 1 shows the correlation matrix among the 21 variables. For a significant correlation between Cvd and the other personality variables, the correlation coefficient has to be equal to or greater than 0.335 for a gl = 33 and a df = 0.005. From the data in the table it can be seen that the Cvd variable correlates positively with the variables M – Praxemia (autistic imagination), Q_1 (Personal adjustment anxiety) and Q_{IV} (Submission); but it relates negatively to the variables A (Affectothymia) and C (Higher ego strength). Values close to significance occur positively with the variable Q_1 (Conservatism of temperament) and negatively with the variables Q_2 (Self-sufficiency) and Q_{II} (Extroversion). A possible (but speculative) interpretation of these correlations is that subjects with defective colour vision tend towards greater autistic imagination, greater anxiety and a greater sense of independence, and that they also tend to be affectively (weaker), have a low ego strength and show a possible tendency towards radicalism and introversion.

Factorial analysis of all the data, principal component procedure, the method of root extraction by defect and finally a transformation of vectorial values using the Orthatran/Varimax method with Statview 512$^+$ revealed seven factors. Factor III (the one linked to Cvd variable) is heavily saturated at the Cvd variable and also at A (Affectothyma), M (Praxemia) and Q_2 (Self-sufficiency). The intercorrelation of factors indicates that the Cvd variable tends to relate to higher scholastic mental capacity (variable B), self-sufficiency (variable Q_2) and independence (variable Q_{IV}). The loading of the personality factors, obtained after factorial analysis show that only factor III is saturated (.817) in the Cvd variable, contributing slightly to the total variance, i.e. the proportion of influence of the colour vision defect on the personality factors does not appear to be more than 16% of the total variance of the behaviour proper to each personality.

With the aim of finding the possible influence that colour vision defective factor has on the elements of intellectual capacity in a subject, Statview 512$^+$ was used again to calculate the point-biserial of the correlation coefficient r_{pb}. The analysis involved 12 variables (1 for colour vision defect, Cvd, and 11 for cognitive factors). The following table, Table 2, shows the correlation matrix among the 12 variables. The data in the table show that the Cvd variable only correlates significantly with the variable S.Ap. (Spatial Aptitude). With the variables At. (Visual Difference Perception), G.M.M. (General Mental Maturity) and G.I.nV. (General Intelligence non-Verbal) the correlation occurs at values close to the limit value (0.335); with the N.Ap. (Numerical Aptitude) it does so negatively. This might be expected if one argues that these variables are somehow related to perception whilst the others are purely intellectual factors. In the same way as before, factorial analysis, principal component procedure, defective root extraction and a transformation of the vectorial values by the Orthotran/Varimax method produced four factors. Factor 3

Table 1. Correlation matrix among the different variables (colour vision and personality factors).

	Cvd	A	B	C	E	F	G	H	J	L	M	N	O	Q_1	Q_2	Q_3	Q_4	Q_I	Q_{II}	Q_{III}	Q_{IV}
Cvd	1																				
A	–.437	1																			
B	.271	.016	1																		
C	–.425	.234	.114	1																	
E	.144	–.109	–.174	–.048	1																
F	.033	.49	.099	.180	–.011	1															
G	.021	–.066	.430	–.072	–.076	–.305	1														
H	–.035	.417	.148	.556	–.117	.483	–.010	1													
J	.072	–.369	–.19	–.114	.453	–.603	.113	–.283	1												
L	.049	.243	.296	.156	.138	.534	–.231	.510	–.250	1											
M	.506	–.338	.335	–.053	.107	.048	.234	.099	.373	–.016	1										
N	–.163	.567	.531	.183	–.167	.371	.081	.288	–.168	.338	.143	1									
O	.247	–.255	.410	–.386	–.075	.186	.323	–.240	–.179	.083	.207	.039	1								
Q_1	.318	–.140	–.246	–.426	.198	.326	–.512	–.270	–.276	.111	–.121	–.287	.129	1							
Q_2	–.323	.218	.068	.051	.106	–.131	–.234	–.238	.114	–.185	–.284	.061	–.298	.151	1						
Q_3	–.163	–.000	.016	.635	.027	–.008	.222	.516	.116	.000	–.058	–.115	–.154	–.405	–.007	1					
Q_4	.125	–.079	–.096	–.654	–.119	–.130	–.316	–.612	.078	–.027	–.197	–.124	.245	.478	.265	–.674	1				
Q_I	.366	–.125	.099	–.852	.053	.012	–.090	–.497	.058	.179	.048	.040	.559	.415	–.019	–.657	.775	1			
Q_{II}	–.231	.724	–.059	.340	.075	.590	.180	.722	–.235	.331	–.001	.349	–.088	–.282	–.261	.344	–.502	–.320	1		
Q_{III}	–.090	.254	.583	.142	–.255	–.162	.721	.217	.122	–.068	.173	.561	.035	–.599	.114	.344	–.305	–.172	.246	1	
Q_{IV}	.427	–.056	.407	–.117	.423	.389	–.231	.048	–.152	.608	.186	.157	.188	.603	.138	–.217	.155	.334	–.106	–.185	1

Table 2. Correlation matrix among the different variables (colour vision and intellectual capacity factors).

	Cvd	GMM	GIV	GInV	VMA	NAp	RC	M	nVMA	LR	SAp	At
Cvd	1											
GMM	.258	1										
GIV	–.058	.658	1									
GInV	.217	.706	.329	1								
VMA	.069	.624	.827	.411	1							
NAp	–.277	.511	.604	.382	.422	1						
RC	.008	.455	.720	.054	.512	.176	1					
M	.063	.480	.596	.389	.476	.512	.572	1				
nVMA	.180	.616	.359	.812	.397	.346	.196	.404	1			
LR	.095	.636	.333	.846	.534	.335	–.001	.311	.539	1		
SAp	.351	.471	–.017	.772	.061	.166	–.155	.208	.505	.564	1	
At	.207	.138	–.024	.266	.090	–.063	–.049	–.306	.299	.224	1	1

(the one linked to Cvd) is heavily saturated at the Cvd variable and lightly saturated at the S.Ap variable (Spatial Aptitude); the loading of the cognitive factors obtained after factorial analysis shows that only Factor 3 is heavily saturated (.949) at the variable Cvd which is linked to possible red-green defect in the subject. Nevertheless, it contributes only 20% to the total variance.

All these results are along the lines of those obtained by authors such as Lakowski (1970) who, as stated in our Introduction, found in a population of normal trichromats, a tendency for certain color vision tests to be associated with cognitive or personality elements, although the loadings were very slight.

Our results suggest (but only very weakly) that people with red-green colour vision deficiencies tend to be emotionally stable, mature, calm, face reality and show greater dependence (than normals) in decision making. Further, they may be generally passive and carried along by the group and may be affected in intellectual factors such as Spatial Aptitude and Difference Perception (but this is highly speculative). A study that used tests specifically to investigate spatial aptitute and difference perception might show a greater influence of these factors in subjects with defective color vision.

References

Capilla, P. and Aguilar, P. (1993). Avico 'Color vision analyzer': a new apparatus for different studies on human color vision. Clin. Vision Sci. 8: 201–206.

Catell, R.B. (1972). 16 PF, form A: Personality questionaire published by TEA Ediciones S.A. Madrid, 1985. Spanish adaptation of Manual for the 16 PF (Sixteen Personality factors) by R.B. Catell. Institute for Personality and Ability Testing, Illinois, U.S.A.

Lakowski, R. (1970). Psychological variables in colour vision testing. In: Richter, M. (ed.), Proceedings of the 1st International Colour Congress, Stockholm 1969: 79–87, Musterschmidt Verlag, Gottingen.

Valero, P., Pitarque, A. and Sanmartin, J. (1991). Prácticas de estadística con ordenador. Cristobal Serrano, Valencia.

Yuste Hernanz, C. (1989). Batería de Aptitudes Diferenciales y Generales (BADYG), Ciencias de la Educación Preescolar y Especial, Madrid.

J. Pérez-Carpinell and M.C. Ibañez Vallejo
Dept. d'Optica, Facultat de Física
Universitat de València
E-46100 Burjassot, Spain.

J.V. Diaz Esteve
Dept. de Metodologia
Facultat de Psicologia
Universitat de València
E-46010 València, Spain.

27. Color discrimination under mesopic conditions in cats and humans

MARK F. TRITSCH
(Mainz, Germany)

Abstract

The ability to discriminate colors under various levels of mesopic illumination was studied in three cats using colored papers as stimuli and with a computer controlled training schedule. Discrimination based on brightness differences was prevented using a new method in which illuminant color was varied. The luminance threshold for color discrimination in cats was found to be 0.5 cd/m^2, close to the cone threshold reported by others from ganglion cell recordings. The same experiment was carried out with human subjects and revealed a luminance threshold of 0.00042 cd/m^2 for color discrimination.

Introduction

The domestic cat is a model species for investigating primitive color vision (Meyer and Anderson, 1965; Brown *et al.*, 1973). It has a predominantly rod retina, but in the visual streak cones reach a density of 40,000/mm^2 while rods fall to 200,000/mm^2 (Williams *et al.*, 1993). Thus, cats can engage in both daytime and nighttime activities. Although color vision is precluded at night under purely scotopic conditions, it has yet to be shown that cats can discriminate colors under the mesopic conditions characteristic of twilight. According to one theory (Hammond, 1978), cat color vision is based on a comparison of the outputs of the rods and the predominant cone type (λ_{max} = 550 nm). However, color discrimination has been demonstrated at high photopic levels (Bonaventure, 1964), and antagonistic input from 550 nm and 450 nm cones as found in some ganglion cells in the cat retina (Crocker *et al.*, 1980) would be sufficient to explain all the color discrimination results obtained so far.

In experiments already reported (Tritsch, 1993), domestic cats quickly learned to discriminate between surface colors when the color of the illuminating light was varied. Since the colored surfaces reflected different wavebands of light, this variation changed the brightness of the surfaces relative

B. Drum (ed.), Colour Vision Deficiencies XII, pp. 227–234.
© 1995 *Kluwer Academic Publishers, Dordrecht.*

to one another. As a result, the color rewarded as the positive stimulus was in some trials brighter and in others darker than the unrewarded one, and in some trials they were equally bright. This made it impossible for the cats to use the brightness cue reliably for discrimination. They responded by switching to the color cue. That work was carried out at photopic levels. In the experiment described below, the effect of reducing overall illumination to mesopic and scotopic levels was investigated.

Method

Three cats were trained to choose a blue (rewarded) rather than an orange (unrewarded) colored paper (10 cm x 11 cm) in a 2-alternative forced-choice procedure. Diffuse illumination of the experimental room was provided by blue and yellow fluorescent tubes, the light from which could be presented mixed in various ratios. Training took place in a wire cage in the middle of the experimental room; at the front end of the cage and separated from each other by a partition were two hanging doors, to which were attached the two papers. The cats were rewarded with food from an automatic dispenser when they pushed their heads through the door with the blue paper. Olfactory cues were controlled for by frequent changes of stimulus papers and providing a food dispenser behind each door. The cats learned to initiate each trial by first going to the back of the cage, where they received a tiny additional morsel of food. Each cat participated in 100 trials per day. The experimental procedure and the ratio of blue to yellow illumination were computer controlled.

During each experiment, the doors with the stimulus papers were randomly switched in position from trial to trial. Independently from this, the lighting mixture was also varied from trial to trial using nine different ratios of blue and yellow illumination (9/1, 8/2, ..., 1/9) presented in random order. These mixtures of illuminant light were made equally bright for the cats (see below). Finally, by shielding the fluorescent lights, the overall level of illumination was reduced from the maximum value of 43 cd/m^2 employed in the original experiments (measured in terms of luminance of the orange paper at a 5/5 mixture of blue and yellow lighting), to either 3, 1.26, 0.42, 0.125, or 0.038 cd/m^2 for each experiment. At each illumination level, the cats were trained on the discrimination task until the performance stabilized. The results given below are based on the final two sessions (i.e. 20 trials at each of the nine illuminant color ratios) at each illumination level for each cat. In additional experiments with 2 human subjects, illumination was further reduced in three steps to a minimum of 1.19 x 10^{-4} cd/m^2. Each subject was tested in 100 trials with varying illuminant color ratio at each illumination level.

Spectroradiometric data were collected for the illuminants and all colored papers (Figure 1). The photopic lightness factor for the combination of a given paper and illumination was calculated by weighting the light reflected from the paper with the cat photopic spectral sensitivity function peaking at 550 nm

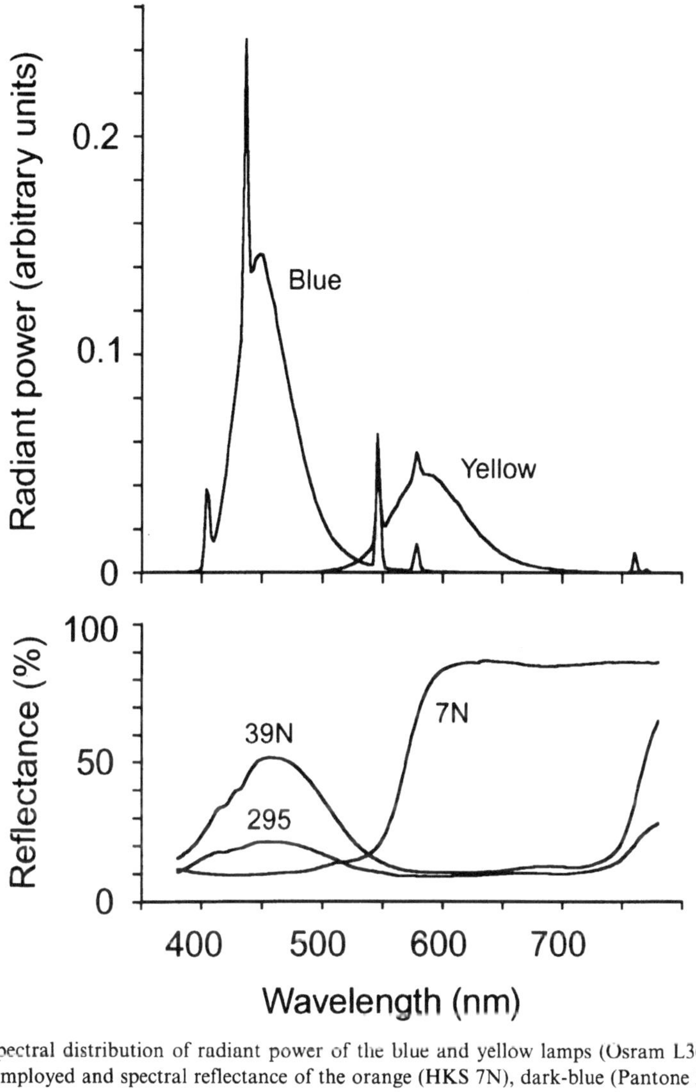

Fig 1 Spectral distribution of radiant power of the blue and yellow lamps (Osram L36/67 and L36/62) employed and spectral reflectance of the orange (HKS 7N), dark-blue (Pantone 295) and blue (HKS 39N) papers.

measured by Gunter (1954) using a behavioral method similar to the one used here. (Gunter's lens absorption correction was not used.) This function was also used to set all lighting mixtures to equal brightness. The scotopic lightness factor was calculated in the same way but using a Dartnall nomogram peaking at 503 nm as spectral sensitivity function. These lightness factors are plotted against the ratio of blue and yellow illumination in the top (photopic) and bottom (scotopic) panels of Fig. 2, and show the lighting mixture at which the colored papers should appear equally light to the cat.

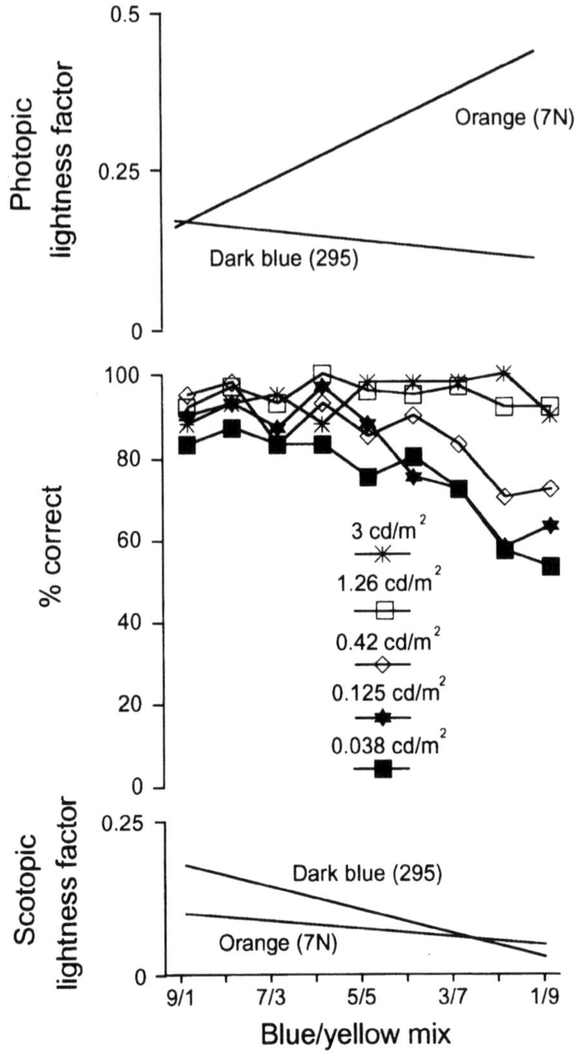

Fig. 2. Middle panel: Color discrimination performance (% correct) when the mixture of blue and yellow lighting is varied (the abscissa is shown on the bottom panel). The results from three cats are averaged and plotted for five different levels of illumination between 3 and 0.038 cd/m^2 (each data point is based on 60 trials). Top and bottom panels: Variation of the photopic and scotopic lightness factors of the orange and the dark-blue papers under the different mixtures of blue and yellow light used in the experiment.

In order to investigate the extent of color discrimination under mesopic conditions it is necessary to have a pair of colored papers which can be made to appear equally light for the cats under both photopic and scotopic conditions when different mixtures of blue and yellow illumination are used for the two conditions. HKS 7N orange and Pantone 295 dark blue satisfied this requirement. Under photopic conditions (Fig. 2 top panel) the two papers should appear equally light at a lighting mixture of 9/1, but still be discriminable by color. Under scotopic conditions, however, thc absence of color discrimination should cause the cats to confuse the papers at a lighting mixture of about 2/8, their point of equal lightness for rod vision (Fig. 2 bottom panel). When the level of illumination is reduced in steps from photopic through the mesopic, therefore, a point should be reached at which cone activation is no longer sufficient to support color discrimination, and at this point the cats should begin to confuse the colored papers at or around a lighting mixture of 2/8.

Results

The experiments confirmed the predictions made for the cats' discrimination. In the middle panel of Fig. 2 the mean discrimination performance of three animals is plotted against lighting mixture for each of 5 levels of illumination. At lighting levels above 1 cd/m^2 (measured as described above), all the cats were able to successfully discriminate by color at all lighting mixtures (>90% correct). Below 1 cd/m^2 the colored papers were confused at lighting mixtures 3/7, 2/8 and 1/9. In Fig. 3 the average discrimination performance at these three lighting mixtures is plotted against the luminance level (log luminance of the orange card at 5/5), and can be seen to fall off steeply below 1 cd/m^2. Taking a discrimination performance of 75% as criterion, the luminance threshold for color discrimination by the cat under the conditions of this experiment is 0.5 cd/m^2. The same plot also shows the cats' discrimination performance at the maximum level of illumination (42 cd/m^2) using a different, brighter blue paper.

In order to provide a comparison with the cat data, two human observers performed the same discrimination in the same apparatus with the same colored papers. The sizes of the blue and yellow units in the lighting mixtures were adjusted at each illumination level so that the papers appeared equally light to the observers at a mixture of 5/5. Therefore, when illumination fell below the level necessary for conc activation, discrimination performance fell for all lighting mixtures, since observers could not reliably use a lightness cue. In the experiment each observer placed his head where the cat normally made its choice and was equipped with a stick with which to open the appropriate door. The results are shown in Fig. 3. The humans confused the papers below 0.001 cd/m^2, at luminances at which the central scotoma became noticeable. Taking the same 75% criterion as for the cat, the luminance threshold for human color discrimination is 0.00042 cd/m^2. Thus, the threshold for colour vision in cats lies 3 log units higher than in man.

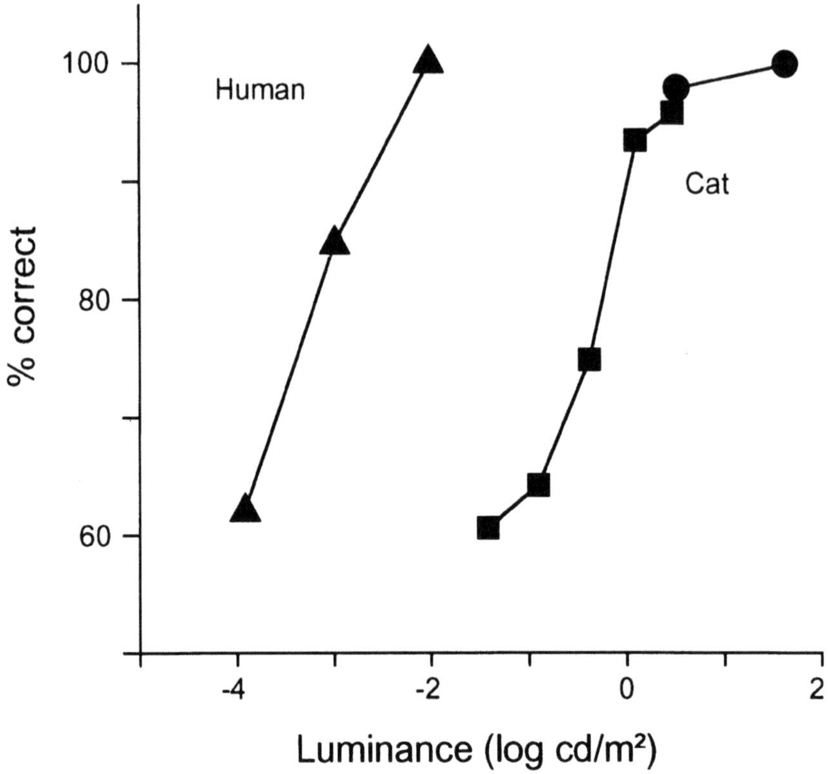

Fig. 3. Color discrimination performance under varying levels of illumination (shown on the abscissa as luminance of the orange paper under a 5/5 lighting mixture).
■ : Mean correct choices of the cats for the illumination mixtures 3/7, 2/8 and 1/9 from Fig. 2. ●: Mean correct choices of the same cats for all illumination mixtures at the highest levels of illumination and with the bright-blue paper. ▲: mean correct choices of the dark-blue paper by two human observers; each data point is the average of 200 trials.

Discussion

Recordings of cone signals in ganglion cells in retina, optic tract and LGN of the cat show a lower limit for the mesopic range of between –1 and –0.5 log cd/m^2, the lower value being obtained with dilated pupils (Daw and Pearlman, 1969; Hammond and James, 1971; Enroth-Cugell *et al.*, 1977; Ahmed *et al.*, 1977; Chan *et al.*, 1992). This corresponds well with the data presented here, although differences in experimental conditions and particularly in lighting spectra make a direct comparison difficult. In behavioral measurements of cat spectral sensitivity for large increments on a white background, Loop *et al.* (1987) found that on backgrounds of 3 cd/m^2 or less the stimuli were detected through the rods. Presumably cone activity is still present on this background but is exceeded by that of the rods. Thus the 0.5 cd/m^2 threshold for color discrimination corresponds to the known lower limit of mesopic vision in the cat.

In humans the low end of the human mesopic range may be a little below the lowest luminance (0.001 cd/m^2) for which Kokoschka (1972) measured luminous efficiency functions deviating from the standard scotopic $V\lambda'$ function. This corresponds well with the threshold of 0.00042 cd/m^2 for color discrimination. The results in cats and humans suggest that color discrimination fails at the lower end of the mesopic range. This does not explain, however, why the mesopic range ends at a luminance 3 log units higher in cats than in humans. A reason that is plausible is the much lower cone density in cats.

The results do not exclude the possibility that rods play a role in cat color vision. Indeed, it is interesting to note that color vision is fully intact at a background luminance of 3 cd/m^2, against which large-increment detection in cats is mediated by rods (Loop *et al.*, 1987). However, there is no sign that color discrimination fails at higher luminances, where the rods would be expected to saturate (Fig. 3), so it is reasonable to propose that cat color vision is principally based on interaction between cones.

Finally, the results show that cats are indeed able to employ their limited capabilities for color vision under the twilight conditions favored by them for moving about.

Acknowledgements

This work was supported by grant Tr 320/1-1 from the Deutsche Forschungsgemeinschaft

References

Ahmed, B., Hammond, P. and Nothdurft, H.-C. (1977). A reappraisal of the feline mesopic range. J. Physiol. (Lond.) 266: 94P.

Bonaventure, N. (1964). La vision chromatique du Chat. C.R. Acad. Sc. Paris 259: 2012–2015.

Brown, J.L., Shively, F.D., LaMotte, R.H. and Sechzer, J. (1973). Color discrimination in the cat. J. comp. Physiol. Psychol. 84: 534–544.

Chan L.-H, Freeman, A.W. and Cleland, B.G. (1992). The rod-cone shift and its effect on ganglion cells in the cat's retina. Vision Res. 32: 2209–2219.

Crocker, R.A., Ringo, J., Wolbarsht, M.L. and Wagner, H.G. (1980). Cone contributions to cat retinal ganglion cell receptive fields. J. Gen. Physiol. 76: 763–785.

Daw, N.W. and Pearlman, A.L. (1969). Cat colour vision: one cone process or several? J. Physiol. (Lond.) 201: 745–764.

Enroth-Cugell, C., Hertz, B.G. and Lennie, P. (1977). Cone signals in the cat's retina. J. Physiol. (Lond.) 269: 273–296.

Gunter, R. (1954). The spectral sensitivity of light-adapted cats. J. Physiol. (Lond.) 123: 409–415.

Hammond, P. (1978). The neural basis for colour discrimination in the cat. Vision Res. 18: 233–235.

Hammond, P. and James, C.R. (1971). The Purkinje shift in cat: extent of the mesopic range. J. Physiol. (Lond.) 216: 99–109.

Kokoschka, S. (1972). Untersuchungen zur mesopischen Strahlungsbewertung. Die Farbe 21: 39–112.

Loop, M.S., Millican, C.L. and Thomas, S.R. (1987). Photopic spectral sensitivity of the cat. J. Physiol. (Lond.) 382: 537–553.

Meyer, D.R. and Anderson, R.A. (1965). Color discrimination in cats. In: de Reuck, A.V.S. and Knight, J. (eds.), Color Vision: 325–344, Little Brown, Boston.

Tritsch, M.F. (1993). Color choice behavior in cats and the effect of changes in the color of the illuminant. Naturwissenschaft 80: 287–288.

Williams, R.W., Cavada, C. and Reinoso-Suárez, F. (1993). Rapid evolution of the visual system: a cellular assay of the retina and dorsal lateral geniculate nucleus of the Spanish Wildcat and domestic cat. J. Neurosci. 13: 208–228.

Dr. Mark F. Tritsch
Institute of Zoology III
Johannes Gutenberg University
55099 Mainz
Germany

28. Ageing and chromatic contrast sensitivity

Annette Werner, Guntram Schwarz and Walter Paulus
(Freiburg, München and Göttingen, Germany)

Abstract

Chromatic contrast sensitivity (chromatic CS) was measured for 40 subjects aged between 18 and 67 (mean = 37.1). Detection thresholds were determined for an equiluminant sinewave grating for which chroma was modulated along the protanopic axis (0.3–8.1 cpd, 1Hz counterphase modulated). An age-related reduction of chromatic CS was observed for all but the lowest spatial frequencies; i.e. for all spatial frequencies above 2.1 cpd. Age-related changes in spectral properties of the elderly eye were experimentally mimicked but did not reproduce the effects observed in elderly test subjects. Age-related neural changes are discussed.

Introduction

Luminance CS, like other visual performance, e.g. acuity, steropsis, has been shown to decrease with age (e.g. Owsley *et al.*, 1983; Sloane *et al.*, 1988; Tulunay-Keesey *et al.*, 1988). Such changes have been attributed to optical factors, e.g. senile miosis, changes in spectral transmittance of the ocular media, or scattering of the lens, as well as neural changes at the retinal and cortical levels (see Elliot *et al.*, 1990). However, the effect of age on chromatic CS has not yet been investigated, even though a number of studies have demonstrated a general loss in spectral sensitivity (e.g. Werner and Steele, 1988) and a tritanlike worsening in colour discrimination associated with ageing (Ohta and Haruo, 1975). Furthermore, it has been suggested that the parvoceullular pathway may be more affected by ageing than the magnocellular pathway (Elliot *et al.*, 1990; Porciatti *et al.*, 1992) and thus chromatic CS may be more affected by ageing than luminance CS. In order to determine the influence of ageing on chromatic CS, we measured the CS of differently aged observers for heterochromatic, equiluminant sinewave gratings at spatial frequencies between 0.3 and 8.1 cpd.

B. Drum (ed.), Colour Vision Deficiencies XII, pp. 235–241.
© 1995 *Kluwer Academic Publishers, Dordrecht.*

Method

Observers

Observers were 40 healthy subjects between 18 and 67 years with no obvious colour vision deficiencies (tested with Ishihara plates and the Farnsworth-Munsell 100–hue test, and with normal or corrected-to-normal visual actuity. The observers were assigned to one of two groups: younger subjects, aged between 18 and 31 (n = 24, mean = 26.0 ± 4.7); and older subjects, aged between 40 and 67 (n = 16; mean = 53.7 ± 10.6).

Experimental apparatus

All experiments were performed in a darkened room. The stimuli for testing chromatic CS were generated on a computer controlled colour monitor using the Moorfield Vision System (Arden *et al.*, 1988). The colour monitor had a 0.31 dot-pitch resolution and was driven at 90 Hz by a graphic card with a 24-bit intensity variation of the primaries. The test program was provided by the computer-graphic software included in the system. Observers were placed 1.5 m in front of the monitor; their head position was not fixed.

Stimuli

The test-pattern was a vertical sinewave grating of 0.3–8.1 cycles per degree (cpd) square wave counterphase modulated at 1 Hz. It subtended 9.5 x 9.5 degrees of visual angle and was surrounded by a homogeneous background of mean luminance (26.5 cd/m^2) and mean chromaticity, both of which were kept constant throughout the experiments. The test-pattern consisted of equiluminous red and green stripes, whose chromatic contrast was varied by choosing colours from the protanopic confusion line. The chromaticities of the extreme colours of the confusion line were: $x_1 = 0.3471$, $y_1 = 0.4838$; and $x_2 = 0.3902$. In order to obtain equiluminance, the relative luminances of the RGB phosphors were first determined individually for each observer by heterochromatic brightness flicker photometry. Contrast was calculated as a percentage of the euclidian distance in the CIE XYZ colour space between the loci of the extreme colours of the protanopic confusion line.

In a second experiment, the effect of ageing on the spectral transmittance properties of the eye was simulated by filtering the stimuli through a long-wavelength filter (Kodak, Wratten No. 81 EF; Figure 1). The filter was placed in lens holders worn by the observers during the experiments. The effect of using this filter is indicated in Fig. 1 by the combined spectral absorbance (●) of the combined long-wavelength filter and a young lens (from Weale, 1988). The combined spectral absorbance is calculated as the log of the inverse of the product of the spectral transmittances of the filter and a younger lens (data taken from Fig. 1). For wavelengths above 400 nm it matches the spectral absorbance of an older lens (from Weale, 1988).

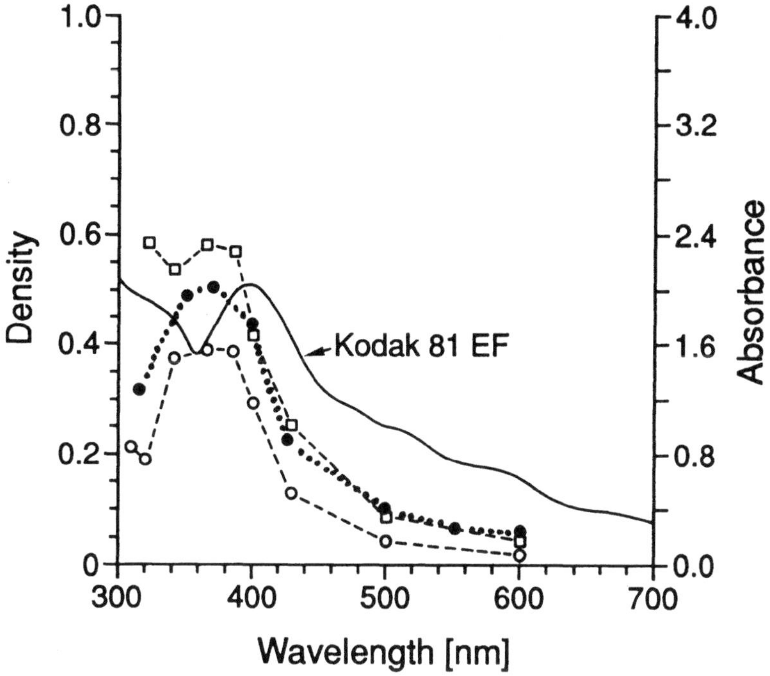

Fig. 1. Age mimicking long-wavelength filter (Kodak Wratten No. 81EF). Functions of spectral absorbance of the lenses of a 13-year-old (o) and a 63-year-old person (□) are marked by dashed lines (according to Fig. 1 in Weale, 1988). The spectral absorbance of the combined long-wavelength filter and younger lens is given by ●. Note that the left ordinate (density) refers to the long-wavelength filter, the right ordinate (absorbance) to the lenses and the combined long-wavelength filter and younger lens.

Experimental procedure

Each test session was preceded by a heterochromatic flicker photometry test. In the contrast experiments, chromatic contrast was determined by means of detection thresholds for the grating, which were measured in a double staircase manner. Threshold was defined as the contrast that allowed a 70% probability of correct answers (Arden *et al.*, 1988). All tests were performed binocularly and for central vision.

Results

The effect of ageing on chromatic CS is demonstrated in Figure 2 where contrast detection thresholds for a 4.1 cpd grating are given as a function of age. The curve has a correlation coefficient of 0.92. In Figure 3, the chromatic CS for spatial frequencies between 0.3 to 8.1 cpd is compared for a group of younger

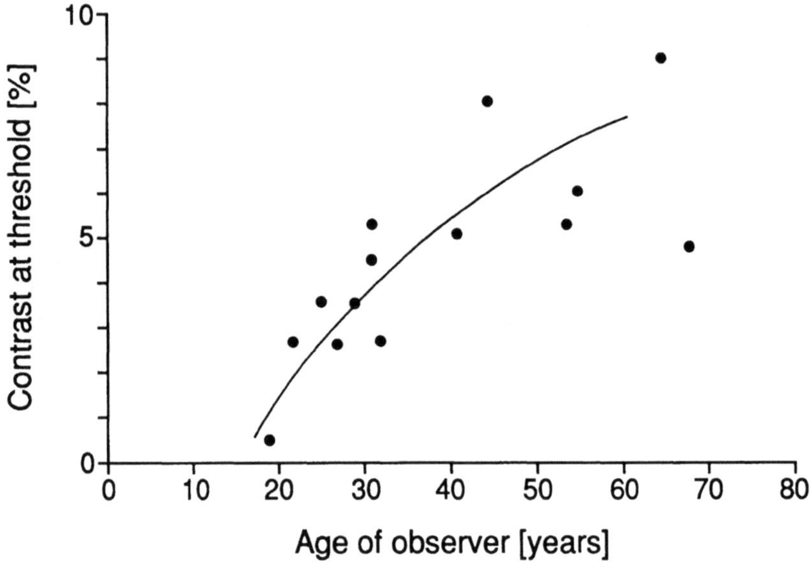

Fig. 2. Detection thresholds for a 4.1 cpd chroma modulated red/green sinewave grating (1 Hz counterphased) as a function of the observers age. Correlation coefficient = 0.92; regression coefficient = 0.67.

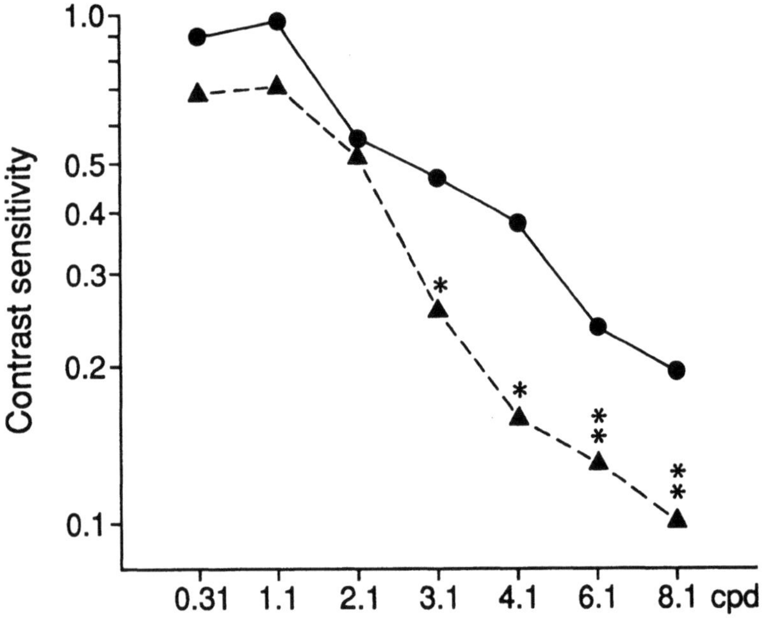

Fig. 3. Chromatic CS (1/threshold) for a chroma modulated red/green sinewave grating as a function of spatial frequency in observers aged between 18 and 31 (●; n = 14) and observers aged between 40 and 67 (▲; n = 8). Significant differences between the groups of older and younger observers are indicated by asterisks (t-test; ★★ for $p < 0.001$ and ★ for $p < 0.01$).

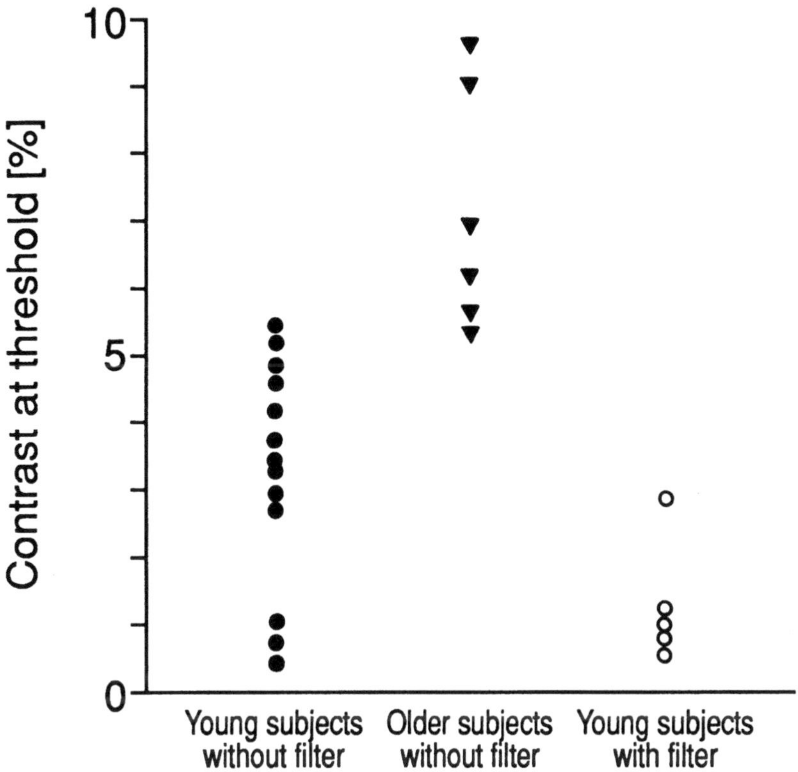

Fig. 4. Individual results for observers younger than 30 years with (o; n = 5) and without (●; n = 15) age mimicking long-wavelength filter. The results of observers older than 40 are also shown (▼; n = 8). Same experiment as described in Figs. 2 and 3.

(aged 18 to 31; ●) and older observers (aged 40 to 67; ▲). The chromatic contrast detection thresholds of the older group are significantly increased for all spatial frequencies higher than 2.1 cpd and the fall off of the function is steeper.

In order to simulate age-related changes in ocular spectral transmittances we had several observers, younger than 28 years, repeat the experiments while viewing the gratings through an age-mimicking longwavelength filter (see Fig. 1). As can be seen in Figure 4, chromatic CS is not reduced by this procedure.

Discussion

What is the possible origin of the observed age related decrease of chromatic CS (Figs. 2 and 3)? Retinal blur, caused by cataract or poor refraction, has been found to be a critical factor influencing luminance CS but not chromatic CS (Seim and Valberg, 1993). Reduced accommodation of the elderly eye cannot account for a reduction of chromatic CS, since the distance between the monitor and the observers was 1.5 m which is beyond the accommodation range. Senile miosis, and the changed spectral properties of the optical media, reduce retinal illumination on average by 3 times (Weale, 1963) and lead to a decline in luminance CS. This is particuarly so for higher spatial frequencies (Kulikowski, 1971; Owsley *et al.*, 1983). The observed reduction in chromatic CS in elderly observers may therefore be related to such changes in retinal illumination. However, our simulation of the aged optical media by means of a longwavelength filter in younger observers did not produce a similar reduction in chromatic CS (Fig. 4) and, therefore, the observed chromatic CS loss cannot be traced back to a decrease in retinal illumination in the aged eye. This is consistent with the observations of Owsley *et al.*, (1983) and Sloane et al., (1988) and suggests that additional neural changes must be taken into account. For example, a random cell loss within the visual pathway, as has been suggested by Weale (1975) for luminance CS loss, may also explain the pronounced chromatic CS loss at higher spatial frequencies. Chromatic CS reduction in the elderly may also be due to changes in transmitter efficiency as has been reported for the dopaminergic and GABA-ergic system (Carlsson and Winblad), 1976; Carlsson, 1987). It is interesting to note that luminance CS in elderly people can be enhanced by the dopamine agonist piribedil (Corbe *et al.*, 1992). Furthermore, impaired colour vision (Price *et al.*, 1992) and a similar selective reduction of luminance CS for medium and higher spatial frequencies, if modulated with 1 Hz temporal frequency (Bodis-Wollner, 1990), has been found in patients with Parkinson's disease. Further investigations should help clarify the relation between the level of dopamine/GABA and luminance and chromatic CS loss in the elderly.

Acknowledgements

The study was supported by a grant from the German Ministry for Research and Technology (BMFT) to Prof. W. Paulus. We thank Mrs. Claudia Frenzel for ophthalmological investigations of the test subjects and help with the experiments and Dr. Lindsay Sharpe and Mirko Whitfield for help with preparation of the manuscript.

References

Arden, G.B., Gündüz, K. and Perry, S. (1988). Colour vision testing with a computer graphics system. Clin. Vision Sci. 2: 303–320.

Bodis-Wollner, I (1990). The visual system in Parkinson's disease. In: Cohen, B. and Bodis-Wollner, I. (eds.), Vision and the brain: 297–316, Raven Press, New York.

Carlsson, A. (1987). Brain neurotransmitters in aging and dementia: similar changes across diagnostic dementia groups. Gerontology 33: 159–167.

Carlsson, A. and Winblad, B. (1976). Influence of age and time interval between death and autopsy on dopamine and 3-methoxytyramine levels in human basal ganglia. J. Neurol. Trans. 38: 271–276.

Corbe, C., Arnauld, F., Brault, Y. and Janiak-Bolzinger, C. (1992). Effect of a dopaminergic agonist, piribedil (Trivastal 50 mg LP), on visual and spatial integration in elderly subjects. J. Neurol. 239, suppl. 1: 22–37.

Elliot, D., Whitaker, D. and MacVeigh, D. (1990). Neural contribution to spatiotempora contrast sensitivity decline in healthy ageing eyes. Vision Res., 30: 541–547.

Kulikowski, J.J. (1971). Some stimulus parameters affecting spatial and temporal resolution of human vision. Vis. Res. 11: 83–93.

Ohta, Y. and Haruo, K. (1975). Colour perception changes with age. Mod. Probl. Ophthalmol. 17: 345–352.

Owsley, C., Sekular, R. and Siemsen, D. (1983). Contrast sensitivity throughout adulthood. Vision Res., 23: 689–699.

Porciatti, V., Burr, D.C., Morrone, C. and Fiorentini, A. (1992). The effect of ageing on the patterns electro-retinogram and visual evoked in humans. Vision Res., 32: 1199–1209.

Price, M., Feldman, R.G., Adalberg, D., Kayne, H. (1992). Abnormalities in color vision and contrast sensitivity in Parkinson's disease. Neurol. 42: 887–890.

Seim, T. and Valberg, A. (1993). Image diffusion in cataract affects chromatic and achromatic contrast perception differently. In: Drum, B. (ed.), Color Vision Deficiencies XI, Doc. Ophthalmol. Proc. Ser. 56: 153–161, Kluwer Acad. Publ., Dordrecht.

Sloane, M., Owsley, C. and Jackson, C.A. (1988). Aging and luminance-adaptation effects on spatial contrast sensitivity. J. Opt. Soc. Am. A, 5: 281–290.

Tulunay-Keesey, U., Hoeve, J. and Terkla-McGrane, C. (1988). Threshold and suprathreshold spatio-temporal responses through adulthood. J. Opt. Soc. Am. A 5: 2191–2203.

Weale, R.A. (1963). The Aging Eye. Lewis, London.

Weale, R.A. (1975). Senile changes in visual acuity. Trans. Ophthalmol. Soc. U.K. 95: 36–38.

Weale, R.A. (1988). Age and the transmittance of the human crystalline lens. J. Physiol. 395: 577–587.

Werner, J. and Steele, V. (1988). Sensitivity of human color mechanisms throughout the life span. J. Opt. Soc. Am. A, 5: 2122–2130.

Annette Werner
Universität Freiburg
Hansastr. 9
79104 Freiburg, Germany

29. Unique-colors in honeybees?

WERNER BACKHAUS
(Berlin, Germany)

Abstract

Neuronal color coding and color choice behavior in ordinary color training experiments, in which a bee is trained to one stimulus, are very well described by the color theory for the honeybee on the basis of the excitations of two types of tonic color opponent coding opponent (see Backhaus, 1991, 1992b, 1993 for review). Under certain conditions, color perception in bees was already suspected of being not always related directly to the excitations of color opponent coding neurons but to the amounts of unique-colors which should be linearly related to the excitations of these neurons (Backhaus, 1988). These unique-colors might be comparable to the six unique-colors (Urfarben, elemental, primary, basic, fundamental, or irreducible colors) blue, yellow, red, green, black and white in humans (Hering, 1878, 1905). Since color vision in honeybees has no 'brightness' dimension, only five different unique-colors, four 'chromatic' and one 'achromatic', were postulated. Furthermore, it was expected that the bee learns in double color training experiments the unique-color which two alternately rewarded colors have in common and to choose the stimuli in unrewarded tests according to the amounts of these unique-colors. The results of the first double color training experiments (Backhaus and Kratzsch, 1993) indeed did not agree with the predictions of the common color theory for the honeybee, and thus could not be explained on the basis of the electrical properties of neurons alone. But the results of these experiments agreed well with the predictions from the color theory extended by the unique-color hypothesis for the bee. It obviously depends on the training method used, whether the color choice behavior of the bee is directly related to the excitations of the color opponent neurons or to the amounts of unique-colors which are steered by the excitations of the color opponent coding neurons. Comparable relations between the amounts of unique-colors, the axes of the color space, and the excitations of color coding neurons are hypothesized to exist in human color vision.

B. Drum (ed.), Colour Vision Deficiencies XII, pp. 243–250.
© 1995 *Kluwer Academic Publishers, Dordrecht.*

Introduction

Our color sensations consist of the six unique-colors (Urfarben, elemental, primary, basic, fundamental, or irreducible colors) blue, yellow, red, green, black, and white (Hering, 1878, 1905). Since color vision in honeybees and in humans is quite similarly organized, it might well be that unique-colors also exist in honeybees.

Unique-colors in humans

In color discrimination and color similarity experiments, human probands are asked to judge the total color difference of test stimuli with respect to a standard (e.g. Sixtl, 1982 for review). In this case, the judgments are well described on the basis of a three-dimensional subjective color space (e.g. Wyszecki and Stiles, 1982 for review). The color space is spanned by three axes (psychophysical scales). Different colors are represented by different loci in the color space. The total color difference between two colors is derived according to the law of Pythagoras (Euclidean metric) from the individual differences in the projections of the two color loci on the three axes. In another type of experiment, humans are asked, for example, to judge the amounts of the unique-colors blue, yellow, red, green, black and white (Hering, 1878, 1905; Hard, 1966). They may also be asked to cancel the chromatic unique-colors (e.g. green and blue) in a color sensation by adjusting the intensities of monochromatic lights which evoke the opponent-colors (e.g. red and yellow) in order to produce an achromatic color sensation (Jameson and Hurvich, 1955). Or they may relate their judgments about the similarity of two colors on the amounts of a specific unique-color (e.g. Sixtl, 1982 for review). The results of both types of experiments with humans can not yet be unequivocally related to the excitations of specific types of color coding neurons (see e.g. De Valois and De Valois, 1993).

Unique-colors in honeybees?

Both color discrimination and color similarity experiments can also be performed with honeybees (Backhaus, 1992 for review). Multi dimensional scaling analysis of the results of multiple choice experiments (Backhaus *et al.*, 1987) showed that the color space of the honeybee is two-dimensional with a city block metric (sum of the absolute differences on the axes). A 'brightness' dimension is not involved in color vision of the bee (Hertz, 1937; Backhaus *et al.*, 1987; Backhaus, 1992; Dittrich, 1992; Brandt *et al.*, 1993). The two scales of the color space of the bee were identified (Backhaus, 1991) to be identical with the excitations of the two types of tonic color opponent coding neurons exclusively recorded (Kien and Menzel, 1977) in the optic lobes of the honeybee. This interpretation was unequivocally possible in the honeybee, because the city block metric fixes the orientation of the axis and does not allow for rotation before interpretation. The physiological models of color vision in honeybees

The Theory of Color Vision and Color Choice Behavior of the Bee

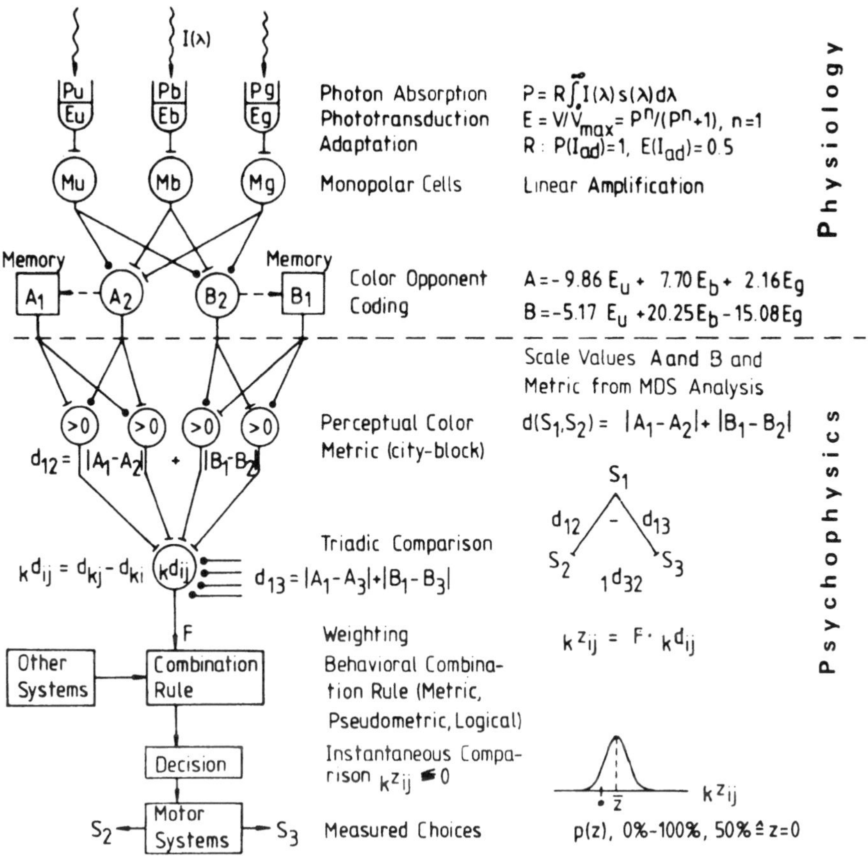

Fig. 1. The theory of color vision and color choice behavior of the honeybee. The upper part (above dashed line) is related to the physiology of the color vision system. The lower part represents the psychological model of color vision. The complete mathematical description (right side) as well as the neronal interpretation (left side) of the theory is presented. I(λ): spectral light intensity distribution; u, b, g: photoreceptor cell types; R: range sensitivity; s(λ): spectral sensitivity; P: absorbed photon flux; E: cell excitation; V: cell potential; M: monopolar cell; A, b: excitations of color opponent coding neurons (circles: interneurons, boxes: memories); d_{ki}: color difference of two stimuli S_k and S_i; ${}_kd_{ij}$: judgment values of stimuli S_i and S_j with respect to the training stimulus S_k; F: experiment type dependent scaling factor; ${}_kd_{ij}$: weighted judgment values; p(z): inverse z-transformation to choice percentages p. Also the psychphysical part of the theory (lower part, right side) allows for a complete interpretation by neurons with simple (linear) properties. ${}_kd_{ij}$: judgment value neuron which realizes the city block metric by adding the results from pairs of (>0)-neurons. In the example, exclusively the two alternatives S_2 and S_3 are shown, and the rewarded stimulus S_1 is not presented to the bee in the tests. The judgment value neuron calculates the difference between the two color differences between the excitation values A_1 and B_1 of the learned color stored in memory and the excitation values A_2, A_3 and B_2, B_3 corresponding to the alternatives. If ${}_kd_{ij}$ is greater than zero at the instant of decision, stimulus S_3 is chosen. If ${}_kd_{ij}$ is less than zero, stimulus S_2 is chosen. (From Backhaus, 1993, Fig. 3).

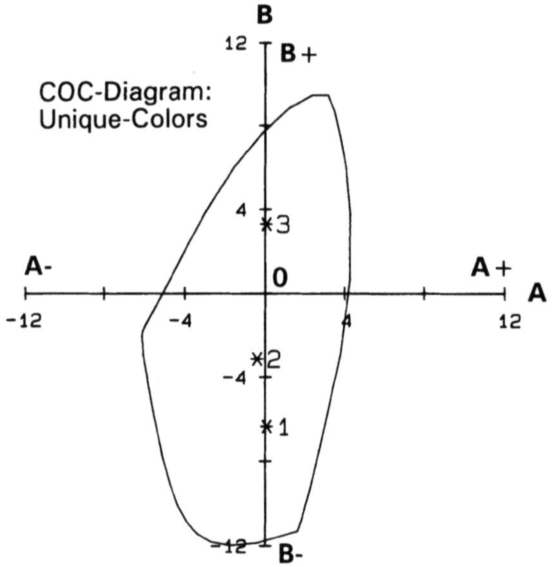

(a)

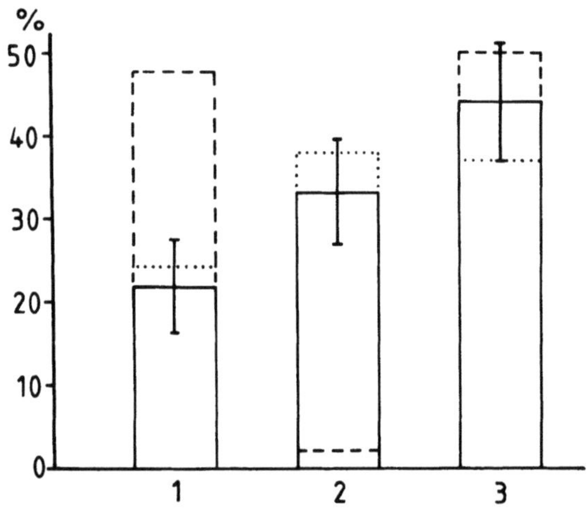

(b)

Fig. 2. Results of a double color training experiment. (a) Color opponent coding (COC) diagram of the honeybee showing the excitation values A and B of the two types of color opponent coding neurons due to monochromatic light (closed spectral curve) and to three color stimuli (1–3). A^+, A^- and B^+, B^-: locations of the four hypothetical (pure) 'chromatic' unique-colors and of a hypothetical (pure) 'achromatic' unique-color (0). Three color stimuli (1–3) were selected which are well discriminated (>94%) by the bee from each other and from the grey background. The stimuli differ only for neuron type B. According to the unique-colors hypothesis for the bee (see text), color

(Backhaus, 1991, 1992, 1993 for review) have been successfully tested in several behavioral and electrophysiological experiments and thus received the status of a theory (see Fig. 1). The two excitation values A and B of the two types of color opponent coding neurons span the color opponent coding (COC) diagram (Fig. 2a) which represents the equispaced subjective color space of the bee.

Color perception in bees was already suspected of being based, under certain conditions, on unique-colors (Backhaus, 1988) comparable to the six unique-colors (see above) in humans. Since bees possess no 'brightness' dimension in color vision (see above), only five different unique-colors, four 'chromatic' (A^+, A^- and B^+, B^-) and one achromatic (0), were postulated (see Fig. 2a). Since the unique-colors diagram of humans is equispaced (e.g. Hurvich and Jameson, 1956), it was assumed in the unique-colors hypothesis for the honeybee that the amounts of the unique-colors are linearly related to the excitations A and B of the two color opponent coding (COC) neurons which span the equispaced COC diagam (see above). In double color training experiments, the bee is expected to learn the unique-colors as a food signal which the two alternately rewarded colors have in common and to choose the stimuli in the tests according to the amounts of these unique-colors. The bee can show its innate flower constancy in this case also during foraging at two very differently colored 'flowers' by choosing the one unique-color which both the colors have in common. In the first double color training experiments (Backhaus and Kratzsch, 1993), three stimuli (1–3) were used which differed, according to the unique-colors hypothesis for the bee, in the amounts of the 'achromatic' unique color (0). The stimuli were very different (>94%) from each other and from the grey background, as checked in usual color discrimination experiments. The bee was trained alternately to two of the stimuli which had only the hypothesized 'achromatic' unique-color (0) in common. In the tests, the two training stimuli (1 and 3) and a stimulus (2), which was not presented during training, appeared indeed to be chosen by the bee according to the amounts of 'achromatic' unique-color (0) in the respective colors. Predictions were derived from the common color theory (Fig. 1) on the basis of the electrical properties of the neurons alone. In this case, the bee was assumed to learn the two colors 1 and 3 independently of each other as in usual

---

1 consists of 47% 0 and 53% B^-, color 2 of 74% 0 and 26% B^-, and color 3 of 72% 0 and 28% B^+. An individual bee was alternately trained to the two stimuli 1 and 3. In the unrewarded tests, all three stimuli were presented. (b) Comparison of the choice percentages measured in the double color training experiment (solid line) and the predictions of the common color theory for the honeybee (see Fig. 1) (dashed line) and of the color theory extended by the unique-color hypothesis for the bee with the assumption that the bee learns the unique-color which both the rewarded colors have in common, i.e. the achromatic unique-color 0 (dotted line). The measured choice percentages do not agree with the predictions of the common color theory on the basis of the electrical properties of neurons alone. Instead, the results agree well (X^2, 5% level) with the predictions from the color theory extended by the unique-colors hypothesis for the bee. The results thus support the hypothesis that unique-colors also exist in the color vision of the honeybee. (From Backhaus and Kratzsch, 1993, Figs. 1 and 2).

color training experiments (no common unique-color learned). According to the common color theory, the bee would choose in this case each of the alternative colors (1–3) according to the smallest of the two color differences (greatest similarity) of the respective alternative color (1, 2, or 3) to the two learned colors (1 and 3) (Fig. 2b, dashed lines). In addition, predictions were derived from the color theory extended by the unique-colors hypothesis for the bee. In this case, the number of choices was assumed to be proportional to the amounts of the 'chromatic' unique-color (0) of the three alternative colors (1–3) (Fig. 2b, dotted lines). The measured choice percentages obviously do not agree with the predictions of the common color theory on the basis of the electrical properties of neurons alone. Instead, the results agree well (χ^2, 5%-level) with the predictions from the color theory extended by the unique-colors hypothesis for the bee. The results thus support the hypothesis that unique-colors also exist in the color vision of the honeybee. Further experiments are in process.

Conclusions and discussion

On the basis of the presented results and because of the similarities between honeybee and human color vision, it is supposed now that also in human color vision the axes of the color space may be represented by the electrical excitations of color coding neurons (see e.g. Valberg and Lee, 1989; De Valois and De Valois, 1993) and that the amounts of unique-colors might be linearly related to the excitations of color opponent coding neurons. Unfortunately, there are still greater mismatches between the structure of the subjective color space of humans obtained by multidimensional scaling (MDS) analysis and subjective color order systems like the Munsell atlas (see e.g. Indow and Ohsumi, 1972). In addition, the many types of color opponent coding neurons in the striate cortex do not allow for an unequivocal neuronal interpretation of the axes of the human color space (see e.g. De Valois and De Valois, 1993). Thus more detailed psychophysical and electrophysiological data as well as more adequate physiological models have to be obtained for testing this hypothesis about human color vision quantitatively.

From an ontological point of view, the unique-colors cannot be identical to the excitations of neurons. We know from introspection that the unique-colors appear to be six different qualities, whereas the excitations of neurons are only temporal changes of the electrical potential which is only one quality. This raises the interesting question of consciousness of the different types of judgments discussed. If color discrimination and similarity judgments of humans would also rely exclusively on the excitations of neurons, as is the case in the honeybee (see above), this would imply that this type of judgment would be unconsciously performed. Only judgments related to unique-colors would presuppose color sensations and thus would be consciously performed. First hints that this might indeed be the case come from blindsight experiments

(Stoerig and Cowey, 1989). Wavelength discrimination functions measured in humans, which lack color sensations only, possess the same form as the wavelength discrimination functions measured in humans with normal color vision.

Acknowledgments

I would like to thank the anonymous referees for helpful comments on the manuscript.

References

Backhaus, W. & Menzel, R. (1987). Color distance derived from a receptor model of color vision in the honeybee. Biol. Cybern. 55: 321–331.

Backhaus, W., Menzel, R. and Kreißl, S. (1987). Multidimensional scaling of color similarity in bees. Biol. Cybern. 56: 293–304.

Backhaus, W. (1988). Opponent color coding and color perception in bees. In: das Gracas de Souza, D. *et al.,* (eds.), Anais da XVIII Reuniao Anual de Psicologia, 123–126, Sociedade de Psicologia de Ribeirao Preto.

Backhaus, W. (1991). Color opponent coding in the visual system of the honeybee. Vision Res. 31: 1381–1397.

Backhaus, W. (1992a). Color vision in honeybees. Neurosci. Biobehav. Rev. 16: 1–12.

Backhaus, W. (1992b). The Bezold-Brücke effect in the color vision system of the honeybee. Vision Res. 32: 1425–1431.

Backhaus, W. (1993). Color vision and color choice behavior of the honeybee. In: Recent Progress of Neurobiology of the Honeybee. Special issue, Apidologie, 24: 309–331.

Backhaus, W. and Kratzsch, D. (1993). Unique-colors in color vision of the honeybee? In: Elsner, N. and Richter, D.W. (eds.), Genes, Brain and Behavior. Proceedings of the 21st Göttingen Neurobiology Conference 39: 830, Thieme, Stuttgart.

Brandt, R., Backhaus, W., Dittrich, M. and Menzel, R. (1993). Simulation of threshold spectral sensitivity according to the color theory for the honeybee. In: Elsner, N. and Richter, (eds.), Genes, Brain and Behavior, Proceedings of the 21st Göttingen Neurobiology Conference 374. Thieme, Stuttgart.

De Valois, R.L. and De Valois, K. (1993) A Multi-Stage Color Model. Vis. Res. 33: 1053–1065.

Dittrich, M. (1992). Vision in the Honeybee: Color Induction. Diss. Freie Universität, Berlin.

Hard, A. (1966). Philosophy of the Hering-Johansson Natural Colour System. Farbe 15: 287.

Hering, E. (1878). Zur Lehre vom Lichtsinn. Carl Gerold's Sohn, Wien.

Hering, E. (1905). Grundzüge einer Theorie vom Lichtsinn, Leipzig. [Outlines of a theory of the light sense (translated by L.M. Hurvich and D. Jameson). Harvard University Press, Cambridge, Mass. (1964].

Hertz, M. (1937). Beitrag zum Lichtsinn und zum Formensinn der Biene. Zeitschr. vergl. Physiol. 24: 413–421.

Hurvich, L.M. and Jameson, D. (1956). Some quantitative aspects of an opponent-colors theory. IV. A psychological color specification system. J. Opt. Soc. Am., 46: 416–421.

Indow, T. and Ohsumi, K. (1972). Multidimensional scaling of sixty Munsell colors by nonmetric procedure. In: Vos, J.J. *et al.,* (eds.), Color Metrics, Proceedings of the Helmholtz Memorial Symposium, 1971. Driebergen, The Netherlands, 124–133, AIC/Holland, c/o Institute for Perception, TNO, Soesterberg.

Jameson, D. and Hurvich, L.M. (1955). Some quantitative aspects of an opponent-colors theory. I. Chromatic responses and spectral saturation. J. Opt. Soc. Am. 45: 546–552.
Kien, J., Menzel. (1977). Chromatic Properties of neurons in the optic lobes of the bee. II. Narrow band and colour opponent neurons. J. Comp. Physiol. A 113: 35–53.
Menzel, R. and Backhaus, W. (1991). Colour vision in insects. In: Gouras, P. (ed.), Vision and Visual Dysfunction, Cronilly-Dillon, J. (ed.), Vol. 6, The Perception of Colour, chap. 14, 262–293, Macmillan, London.
Sixtl, F. (1982). Meßmethoden der Psychologie. Theoretische Grundlagen und Probleme, 2. Auflage, Beltz, Weinheim.
Stoerig, P. and Cowey, A. (1989). Wavelength sensitivity in blindsight. Nature 342: 916–917.
Valberg, A. and Lee, B. (1989). Detection and discrimination of colour, a comparison of physiological and psychophysical data. Physica Scripta 39: 178–186.
Wyszecki, G. and Stiles, W.S. (1982). Color Science. Concepts and Methods, Quantitative Data and Formulae (2nd) ed., Wiley, New York.

Dr. Werner Backhaus
Institut für Neurobiologie
c/o Institut für Biophysik
Freie Universität Berlin
Thielallee 63
14195 Berlin,
Germany.

30. Contribution of two colour opponent mechanisms to Fechner-Benham subjective colours

JEAN LE ROHELLEC and FRANÇOISE VIÉNOT
(Paris, France)

Abstract

Fechner-Benham (FB) subjective colours are assumed to originate from retinal neural interactions. We address the question of whether colours elicited by a rotating FB disk require the involvement of two colour opponent mechanisms. It is hypothesized that short presentations and narrow stripes favour the $L_{opp.}$ and $M_{opp.}$ mechanisms and that long presentations and wide stripes favour the $S_{opp.}$ mechanism. FB disks, differing by width of the stripes (6, 12, 24, 48 min arc) were rotated at 7 Hz and presented for various durations (1, 2, 4, 8, 16 and 32 rotation cycles). Responses of three observers were collected with a hue naming procedure. The results show two time exposure effects and one area effect. Inversion between blue naming and green naming frequently occurs with short exposures (1–2 cycles) and a gradual shift in the perceived hue appears when the exposure time increases to 32 rotation cycles. In some cases, FB subjective colours vary with stripe width. We suggest that the $L_{opp.}$ and $M_{opp.}$ mechanisms can generate various FB subjective colours on their own but that modulation from the $S_{opp.}$ mechanism further modifies the colours.

Introduction

Rotating a Fechner-Benham (FB) disc under white illumination produces characteristic colours on each ring: reddish for pattern p_1, yellowish for pattern p_2, greenish for pattern p_3, bluish for pattern p_4 (see Figure 1). These colours are named PIFCs (Pattern Induced Flicker Colours).

Numerous scientists have studied the temporal and spatial aspects of the phenomenon (Piéron, 1922; von Campenhausen, 1968; Jarvis, 1977; Kozak *et al.*, 1989). They have shown that PIFCs are determined by the phase relationship between the luminous template of the background signal and the luminous template of the arc signal (Fig. 1). Besides, it has been shown that PIFCs depend upon the relative contrast between these signals (Tritsch, 1992).

Von Campenhausen (1969) has clearly established the retinal origin of the

B. Drum (ed.), Colour Vision Deficiencies XII, pp. 251–258.
© 1995 *Kluwer Academic Publishers, Dordrecht.*

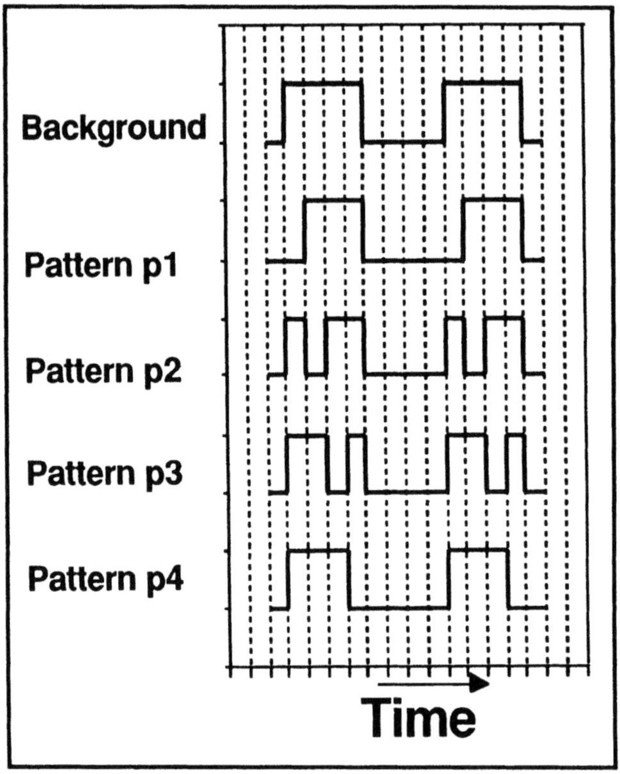

Fig. 1. Temporal patterns of luminance created by the background and by the four stripes of a rotating Fechner-Benham's top.

phenomenon at the post-receptoral level. He has insisted upon the role of lateral interactions in the retina. Recently various retinal sites have been proposed: horizontal cells (von Campenhausen *et al.*, 1992), bipolar cells (Tritsch, 1992), ganglion cells (Zrenner, 1983). Finally, it should be emphasized that unlike stationary colours, the PIFCs are not related to the maximum absorbance of photoreceptors but to the wiring in the retina. In this paper, we intend to investigate the neuronal coding of PIFCs.

De Valois and De Valois (1993) have recently proposed a model that can serve as a framework. In this model, $L_{opp.}$ and $M_{opp.}$ pathways on the one hand, and the $S_{opp.}$ pathway on the other hand stay segregated until the cortex. Signals from $S_{opp.}$ serve to modulate various combinations of $L_{opp.}$ and $M_{opp.}$ signals and constitute the opponent perceptual systems Red-Green and Yellow-Blue.

Furthermore, $L_{opp.}$ and $M_{opp.}$ mechanisms on the one hand and the $S_{opp.}$ mechanism on the other hand, have different spatio-temporal characteristics. We have addressed these different mechanisms by modifying the exposure time and the thickness of the arcs of an FB disc. We expect to favour $L_{opp.}$ and $M_{opp.}$

activity with short exposure times and narrow arcs or to favour $S_{opp.}$ activity with long exposure times and wide arcs.

Experiment

We examined Fechner-Benham colours which arose from modified Fechner-Benham discs with three single stripes selected out of the four angular positions (p_1, p_2, p_3, p_4). All possible combinations of three stripes were used (Figure 2). The thickness of the three stripes was selected among 4 possibilities: 6, 12, 24, 48 min. of arc. Occurrence of every stripe, exposure time, and thickness was equally probable. Speed (7 cycles per second) and average luminance (100 cd.m^{-2}, 2900 K tungsten light) were kept constant during the presentation.

Each disk was successively presented to the observer during 1, 2, 4, 8, 16 and 32 rotation cycles (143, 286, 572, 1144, 2288, and 4576 ms). All combinations were presented in each session. Measurements were repeated during 5 sessions. Data related to exposure time result from 60 trials for each pattern (one stripe, all widths) per subject and data related to stripe width result from 90 trials for each pattern (one stripe, all exposure times) per subject.

Task

The observer did not know which stripe was missing. Observers were asked to give a verbal description of each of the three patterns of the disk in terms of unique hues (red, yellow, green, blue) plus orange and purple, or a combination of two of these hues. The terms orange and purple were allowed since they frequently arose in the verbal description given by the subjects in preliminary experiments.

Control experiment

The same subjects participated in a conventional colour naming experiment. A monochromatic test (vertical bar of 1.5° x 10 min., 15 cd.m^{-2}) was superimposed

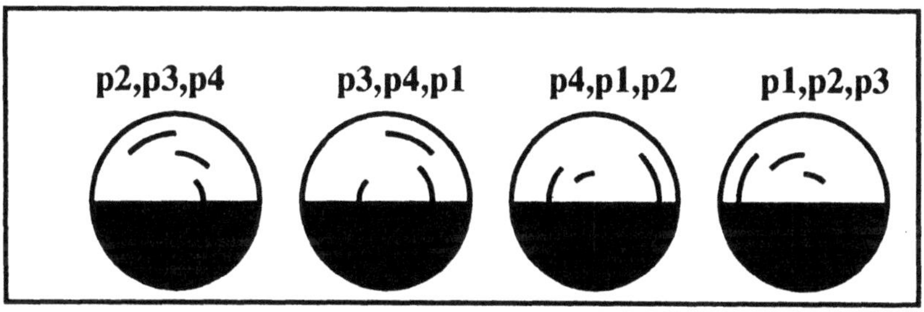

Fig. 2. Modified Fechner-Benham disks used in this experiment. One stripe out of 4 is missing on each disk. Each stripe is equally presented during the experiment. All stripes have the same thickness on one disk. Each disk is replicated for various thicknesses.

on a steady white background (20 cd.m^{-2}), during 143 ms. Twenty-five wavelengths between 410 nm and 650 nm, at 10 nm step, were selected. Measurements were repeated in 5 sessions. Unique hues, plus orange and purple and any simple combination of these terms were allowed, as in the main experiment.

Subjects

Four subjects participated in the experiment: Two authors JLR (male 28 years), FV (female, 45 years) and two naive subjects HS (female 20 years), and RP (male 22 years). All subjects had normal colour vision (Farnsworth-Munsell 100 hue) and normal or corrected to normal acuity.

Results

The results show firstly that FB subjective colours change as exposure time increases, and secondly that FB subjective colours vary with stripe width.

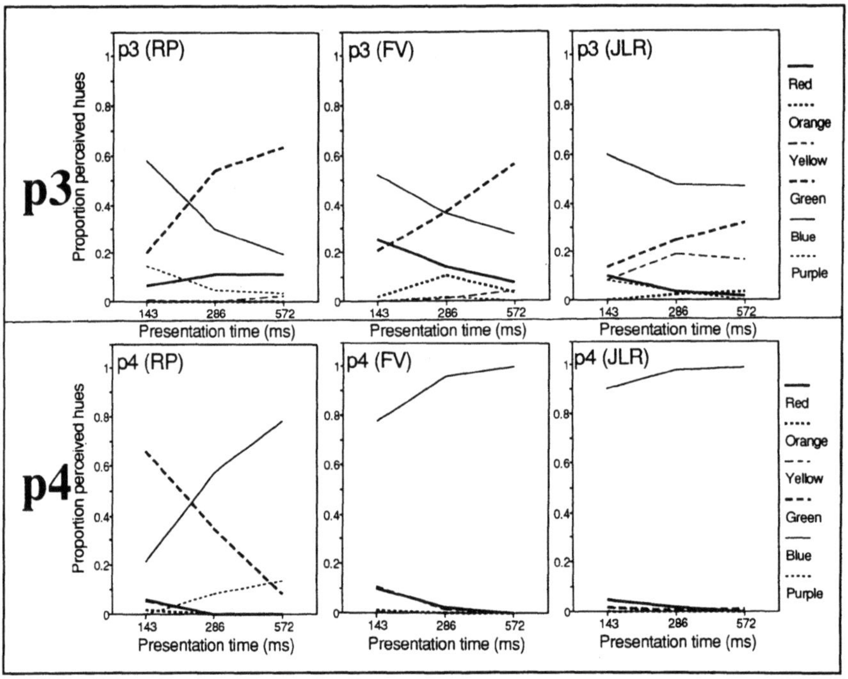

Fig. 3. Short exposure change in the proportion of perceived hues of pattern p_3 and pattern p_4 during the first 572 ms, for 3 subjects. Top: Pattern p_3. With one cycle presentation (143 ms), the stripe is mainly seen *blue*. With four cycles presentation (572 ms), the stripe is mainly seen *green* for 2 subjects RP and FV. The change does not occur for subject JLR so markedly. Bottom: Pattern p_4. An inversion occurs between green and blue naming for subject RP. The two other subjects always report it *blue*.

Changes do not occur with all patterns and can occur with different patterns for different subjects. (Results of subject HS which exhibited a strong intra-observer variability were excluded from the analysis.)

Despite inter-observer variability, the result show two effects related to exposure time and one effect related to the spatial content of the stimulus.

1. Short exposure effect (presentation limited to one or two cycles): it appears on patterns p_3 and p_4 which are usually described as green and blue. When only one or two rotation cycles are presented, misnaming, confusion or inversion of blue naming and green naming are observed on patterns p_3 and p_4 (Fig. 3).
2. Long exposure effect: a change with the number of presentation cycles. It appears mainly on pattern p_1 which is usually described as red. When the exposure time increases to 4576 ms, a gradual shift in the perceived hue of the stripe is noticed on pattern p_1 (Fig. 4).
3. Area effect: With some patterns, the proportion of perceived hues given by one or another observer is modified as the arc width increases (Fig. 5).

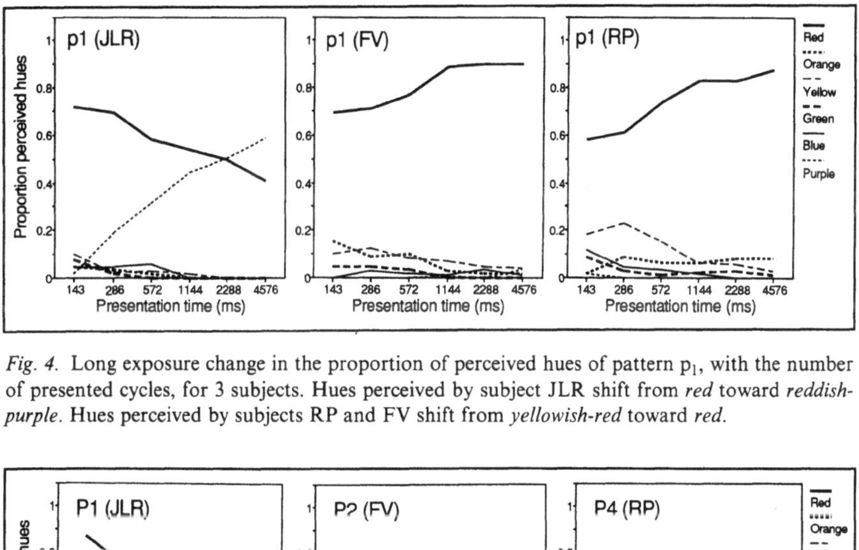

Fig. 4. Long exposure change in the proportion of perceived hues of pattern p_1, with the number of presented cycles, for 3 subjects. Hues perceived by subject JLR shift from *red* toward *reddish-purple*. Hues perceived by subjects RP and FV shift from *yellowish-red* toward *red*.

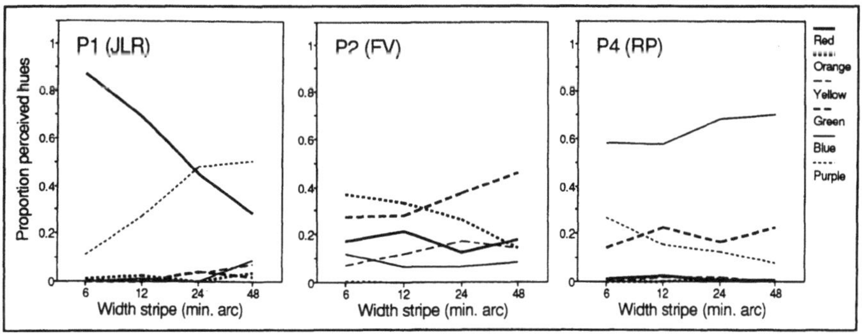

Fig. 5. Area effect. Some cases of change in the proportion of perceived hues when the width of the stripe increases. Pattern and subject identification as indicated in the inset. From left to right: Change from *red* to *purple* on pattern p_1 (subject JLR). Change from *orange* to *green* on pattern p_2 (subject FV). Change from *purple-blue* to *greenish-blue* on pattern p_4 (subject RP).

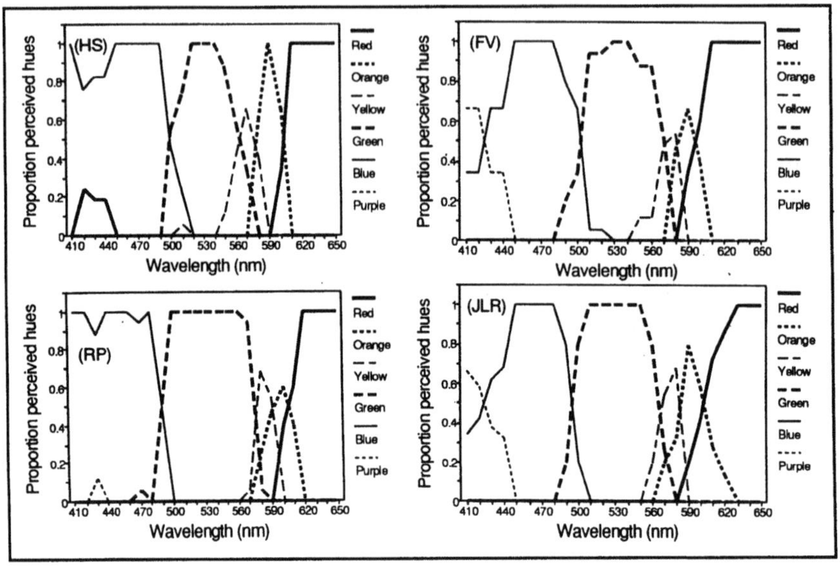

Fig. 6. Control experiment. Results of colour-naming task on a vertical bar (1.5° x 10 min., 143 ms, 15 cd.m^{-2}) superimposed on a steady white background (20 cd.m^{-2}). Organisation of colour names slightly changes between subjects.

Colour naming

With the conventional colour naming task, we have found differences between subjects: various hues do not cover exactly the same range of wavelengths (Fig. 6). Subject HS was able to report subtle changes of colour, whereas subject RP reported only abrupt changes.

Discussion and conclusion

We have reported that increasing the exposure time of a rotating Benham's top produces a short exposure effect and a long exposure effect. In some cases, increasing the stripe width produces a change in dominant hue.

These changes with exposure time or with stripe width indicate that several mechanisms with different spatio-temporal characteristics are involved in eliciting PIFCs. It is known that $L_{opp.}$ and $M_{opp.}$ mechanisms and $S_{opp.}$ mechanisms differ in their spatio-temporal characteristics. We propose these differences to explain the observed changes.

The short exposure effect consists of a confusion between blue naming and green naming that resembles in some respects a tritan confusion of colour. In their model, De Valois and De Valois (1993) hypothesize that the $S_{opp.}$ mechanism has a modulatory role on the $L_{opp.}$ and $M_{opp.}$ systems to separate the green and the blue signals. Several psychophysical findings suggest that the

$S_{opp.}$ mechanism is slow relative to $L_{opp.}$ and $M_{opp.}$ mechanisms. Therefore, the confusion of green and blue colours at short exposure times can be explained by the sluggishness of the $S_{opp.}$ mechanism. However, when several rotation cycles are presented, the cooperation of $L_{opp.}$ and $M_{opp.}$, and $S_{opp.}$ produces the classical gamut of colours.

When the width of the stripes increases, we notice in some cases a shift of dominant hue, like the one from red to purple on pattern p_1 (observer JLR), or the one from orange to green on pattern p_2 (observer FV), or the one from purple-blue to greenish blue on pattern p_4 (observer RP). This could be explained by the assumption that the various mechanisms involved in eliciting PIFCs have different spatial integration characteristics. $S_{opp.}$ mechanism draws input from S cones which are parsimoniously and regularly distributed. It integrates over a larger area than $L_{opp.}$ and $M_{opp.}$. We hypothesize that when the stripe widens, the probability of tapping the $S_{opp.}$ pathway increases and the cooperation of $L_{opp.}$ and $M_{opp.}$ with $S_{opp.}$ modifies the full gamut of colours.

As far as the long exposure effect is concerned, several stages of colour processing with a longer integration time might be involved. While peripheral stages might be sufficient for colour discrimination, more central stages yield colour identification. We suggest that changes in hue of pattern p_1 might be related to such a colour identification process.

To summarize, temporal and spatial effects indicate that several mechanisms are active in producing PIFCs. Modulation of $L_{opp.}$ and $M_{opp.}$ signals by $S_{opp.}$ signals, as proposed by De Valois and De Valois (1993) provides a suitable model to explain these spatio-temporal effects.

References

Both, R. and Von Campenhausen, C. (1978). Sensitivity of a sensory process to short time delays: a study in pattern-induced flicker colours (PIFCs). Biological Cybernetics 30: 63–74.

Von Campenhausen, C. (1968). Uber die Farben der Benhamschen Scheibe. Zeitschrift für Vergleichende Physiologie 60: 351–374.

Von Campenhausen, C., Hofstetter, J., Schramme. and Tritsch, M.F. (1962). Color induction via non-opponent lateral interactions in the human retina. Vision Res. 32: 913–923.

Von Campenhausen, C. (1973). Detection of short time delays between photic stimuli by means of pattern-induced flicker colors (PIFCs). Vision Res. 13: 2261–2272.

Von Campenhausen, C. (1968). Uber den Ursprungsort von musterinduzierten Flickerfarben im visuellen System des Menschen. Zeitschrift für Vergleichende Physiologie 61: 355–360.

De Valois, R.L. and De Valois, K.K. (1993). A Multi-Stage Color Model. Vision Res. 33: 1053–1065.

Jarvis, J.R. (1977). On Fechner-Benham subjective colour. Vision Res. 17: 445–452.

Kozak, W.M., Reitboeck, H.J. and Meno, F. (1989). Subjective color sensations elicited by moving patterns: effect of luminance. In: Kulikowsky, J.J. and Cronilly-Dillon, J.R. (eds.), Seeing Contour and Colour: 294–310, Pergamon Press, Oxford.

Piéron, H. (1922). Le mécanisme d'apparition des couleurs subjectives de Fechner-Benham. l'Année Psychologique 23: 1–49.

Rosenblum, K., Anderson, M.L. and Purple, R.L. (1981). Normal and color defective perception of Fechner-Benham colors: implications for color vision theory. Vision Res. 21: 1483–1490.

Tritsch, M.F. (1992). Fourier analysis of the stimili for pattern-induced flicker colors. Vision Res. 32: 1461–1470.

Zrenner, E. (1983). Neurophysiological Aspects of Color Vision in Primates: 106-109, Springer-Verlag, Berlin.

Jean Le Rohellec and Françoise Viénot
CNRS UPR 257
Laboratoire de Physique Appliquée
Muséum National d'Histoire Naturelle
43, Rue Cuvier
75005 Paris, France.

31. The spectral sensitivity of the acuity criterion: Effect of nonlinear summation of isolated parvocellular receptive field centers

VINCENT A. BILLOCK
(Fort Rucker, AL, U.S.A.)

Abstract

Spectral sensitivity for the detection of acuity targets is additive and indistinguishable from spectral sensitivity to flicker (Pokorny *et al.*, 1968; Ingling *et al.*, 1992). Acuity is believed to be mediated by color opponent neurons with Type I receptive fields. The spectral sensitivity of V_λ (flicker) is approximately 5R+3G while the spectral sensitivity of the r-g channel is 2R-3G. Ingling and Martinez (1983) showed that the opponent channel is additive for luminance gratings and should therefore have a spectral sensitivity of 2R+3G, which is distinguishable from V_λ. Ingling and Tsou (1987) showed that high frequency gratings (and other acuity targets) do not modulate surrounds of r-g Type I cells; therefore, if only centers are active and if centers are present in the same proportions as cones, the spectral sensitivity of acuity may reflect the spectral sensitivity of the pooled, isolated centers, and not the spectral sensitivity of the r-g channel. This argument did not take into account the nonlinear effects of probability summation among pooled isolated centers. Modelling indicates that nonlinear summation does not grossly affect the shape of the resulting spectral sensitivity function. In addition, if isolated centers are allowed to retain their channel weights after being stripped of modulated surrounds, an opponent channel of 2R:3G actually gives a better approximation to V_λ (5R:3G) than does a channel of 5R.3G after nonlinear pooling. These results illustrate the difficulty of working backward from spectral sensitivity to channel mechanisms.

Introduction

It is curious that luminance as measured by an acuity criterion tapping the parvocellular system strongly resembles luminance as determined by methods tapping the magnocellular system (e.g. CFF or HFP). This is odd for two reasons, one of which involves additivity, the other which involves spectral sensitivity.

B. Drum (ed.), Colour Vision Deficiencies XII, pp. 259–265.
© 1995 *Kluwer Academic Publishers, Dordrecht.*

First, the parvocellular system is known to be color opponent (subtractive), not additive as the channel supporting luminosity must be. This contradiction was addressed by Ingling and Martinez (1983), who made it clear that an additive luminance signal (R+G) can be carried by a cell whose color signal is R-G. Considering only spatial and chromatic interactions, the total signal is

(R-G)(Center + Surround)/2	+	(R+G)Center − Surround)/2	(1)
{chromatic, opponent & lowpass}		{achromatic, additive & bandpass}	

Secondly, the CIE V_λ function, which represents luminosity for the standard human observer, implies a weighting of R-cones to G-cones of about 5R:3G (13R:8G to be more exact), while the r-g opponent system has a weighting of 2R:3G, as determined by the fact that the r-g system has a crosspoint near 580 nm. (All calculations assume Smith and Pokorny cone sensitivity functions (Wyzecki and Stiles, 1982), each normalized to a maximum sensitivity of 1.0). There are various ways to view these weightings. For the r-g cells the 2R:3G weighting probably reflects a difference in potency between centers and surrounds. This difference could result from either a difference in the number of cones in center and surround and/or a difference in gain between center and surround mechanisms. (Note that if the first term in Equation 1 is set to 2R-3G, the second term must be 2R+3G).

Criteria that must tap the magnocellular system – large flickering targets – produce a 5R:3G spectral sensitivity. Magno cells have large enough receptive fields that both center and surround should have cone ratios that reflect the ratio of cones in the retina. Thus, the 5R+3G luminosity function produced by flickering targets suggests that R-cones are about twice as numerous on the retina as G-cones – a notion with much support in the literature (Walraven, 1974; Cicerone and Nerger, 1989; Vimal *et al.*, 1989). In other words: *flicker criteria count cones, but acuity critera (which stimulate small receptive fields) should not.* The problem with this reasonable argument is that spectral sensitivity for acuity is indistinguishable from spectral sensitivity for flicker (Ingling and Tsou, 1987; Ingling *et al.*, 1992). Ingling *et al.* (1992) attempted directly to compare flicker and grating sensitivity. They presented 10 c/deg, monochromatic gratings on identical wavelength monochromatic fields that were flicker matched to a standard background. Ingling *et al.* (1992) found that the Weber contrast of gratings at threshold was constant (in flicker units relative to the monochromatic field) regardless of wavelength. Such a result would not be expected from a comparison of the expected spectral sensitivities of the acuity (parvo – 2R:3G) and flicker (magno – 5R:3G) channels (see Fig. 1). Figure 1 would predict that less contrast would be required at short wavelengths and more contrast would be required at long wavelengths. Since Ingling *et al.* (1992) found no such wavelength dependence, this would imply that the parvocellular system does not mediate detection of acuity targets. However, since there is strong evidence the parvocellular system does mediate detection of high spatial frequencies and acuity targets (Lynch *et al.*, 1992; Schiller *et al.*, 1990), there must be an alternate explanation.

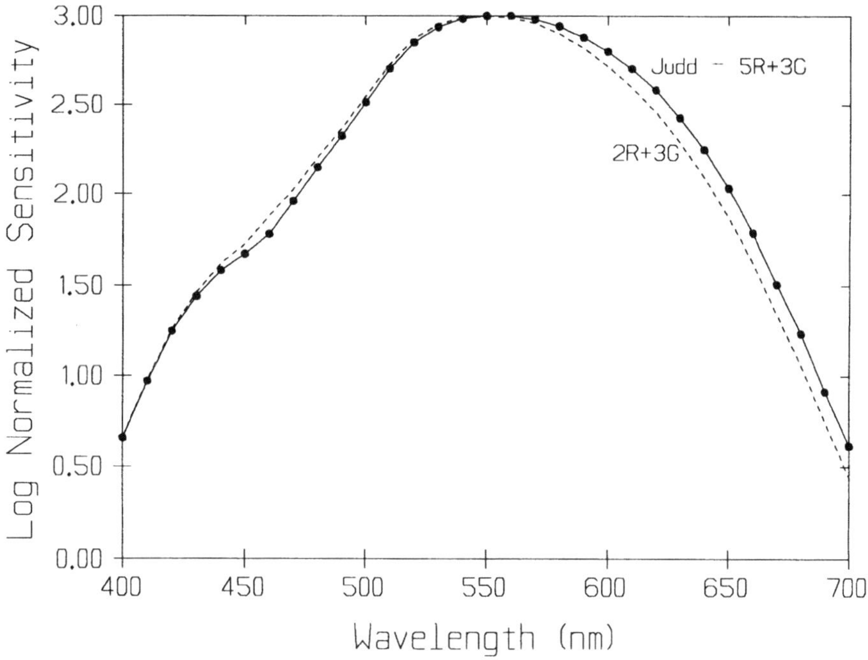

Fig. 1. The Judd modified V_λ (points) has a spectral sensitivity of 5R+3G (13R+8G to be exact – solid line). This is distinguishably different (.15 – .18 log units over much of the long wavelength spectrum) than the 2R+3G spectral sensitivity (580 nm crosspoint – broken line) expected for criteria that excite the parvo channel. In general, the parvo channel – with a greater contribution from medium wavelength sensitive cones – should be more sensitive than the flicker channel to short-wavelength stimuli and less sensitive to long-wavelength stimuli. Yet experiments comparing the spectral sensitivity for flicker and acuity lead to indistinguishable results (Ingling *et al.*, 1992)

Ingling and Tsou (1987) noted that the single cone centers of parvo cells have much higher spatial frequency cutoffs (by a factor of 3–4) than the surrounds. Therefore, moderate to high frequency acuity targets are resolved only by centers and if the ratio of R-cone centers to G-cone centers is the same as the ratio between R-cones and G-cones, then acuity 'counts cones', as does flicker. Only for moderate spatial frequencies at high contrasts (conditions that modulate surrounds) would acuity targets produce a 2R:3G spectral sensitivity. Although Ingling and Tsou's ingenious argument removes the contradiction between flicker and acuity spectral sensitivity, it rests on an unsupported assumption: that somehow and somewhere R-cone and G-cone center responses of Type I cells are added linearly. However, since the R-cone and G-cone driven receptive field centers remain separate until reaching the cortex (similar to the separate on- and off-center cells which constitute functionally separate systems in the early visual system) simple linear addition is suspect. Even on reaching the cortex, the signals may not be combined immediately. Color and luminance appear to be extracted by cortical lowpass and bandpass filtering operations on geniculate units of similar spectral sensitivity. Thus, it seems likely that R-cone center acuity signals and G-cone center acuity signals are carried by separate channels until rather late stages of cortical processing. Moreover, units beyond the earliest stages of cortical processing show evidence of nonlinearities (Heeger, 1992), making linear combination unlikely. This situation is reminiscent of spatial frequency channels, which contribute to detection in a nonlinear fashion that is well modelled by probability summation. Probability summation is also a plausible method for combining spectral sensitivity information. This study examines the consequences of relaxing the linear addition assumption.

Calculations

Most vision researchers use the following approximation to probability summation suggested by Quick (1974):

$$\text{Response} \propto (A^n + B^n + C^n + \ldots)^{1/n} \quad (2)$$

Where n is generally the slope of the frequency of seeing curve for the stimulus and A, B, C ... are the signals of individual channels. Models similar to this have been used to describe interactions of chromatic mechanisms; some typical values are $n = 2$ (vector models of color space – Guth *et al.*, 1980; Ingling and Tsou, 1977), $n = 2.5$ to 3.5 under various conditions (Cowan *et al.*, 1984), and $n = 4$ (Kranda and King-Smith, 1979). If n is large (for all practical purposes $n > 10$), probability summation is equivalent to peak detection (Graham, 1989). This study explores the range of $n = 1$ (linear) to $n = 10$ (peak detection). The equation used is

$$\text{Response} = [(aR)^n + (bG)^n]^{1/n} \quad (3)$$

For the first calculation, I set $a = 5$, $b = 3$ (centers in proportion to expected number of retinal cones), and $n = 2$. As Fig. 2 shows, vector summation of centers yields an acuity spectral sensitivity (solid line) which is quite similar to V_λ (points).

Difficulty of reverse engineering the luminance channel.

However, it is sobering to turn this calculation on its head. One can ask, given a certain probability summation exponent, what should the weighting of R- to G-cones be to match V_λ? I varied the ratio of a and b in Equation 3 for each value of n from 2 to 10, and fit the log of Equation 3 to the log of the Judd modified V_λ function (Wyzecki and Stiles, 1982). The results are interesting. The larger the n, the higher the crosspoint that produces the desired results. When $n = 2$, a 5R:3G cone ratio (490 nm crosspoint) produces reasonable results (Fig. 2 – solid line). However, a 560 nm crosspoint gives even better results (Fig. 2 – fine broken line). For $n = 3$, a 570 nm crosspoint is best. And

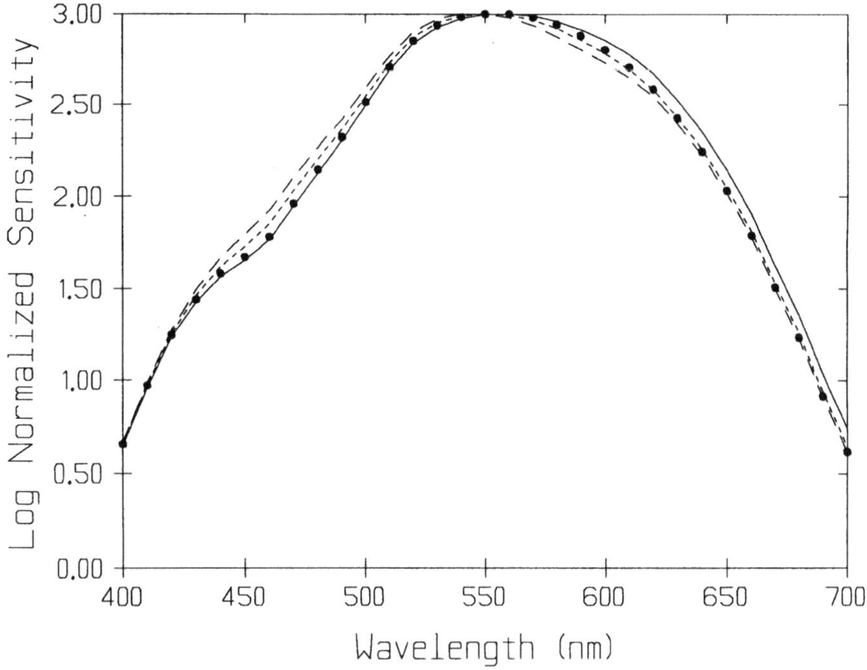

Fig. 2. Ingling and Tsou (1987) argue that acuity targets do not modulate surrounds, so only centers send signals to cortex. If centers are equal in sensitivity, are present in the proportions 5R:3G and are linearly added at the cortex, then acuity would be expected to yield V_λ. But what if centers are nonlinearly added? The solid line shows the vector sum ($n = 2$ in equation 3) of 5R and 3G. It differs very little from V_λ (points). But what if R and G centers have different weights? For vector summation ($n = 2$), the optimal fit to V_λ is achieved closer to 2R:3G. The fine broken line shows the best fitting model for cone inputs with equal sensitivity (crosspoint) near 560 nm. For exponents greater than 2, the best fitting crosspoint moves toward 580 nm (see text). The coarse broken line shows the fit for $n = 4$ and a crosspoint of 580 nm.

for $n = 4$ to 10, a 580 nm crosspoint is optimal in producing a V_λ spectral sensitivity (Fig. 2 – coarse broken line). Is the proximity of these crosspoints to the 580 nm crosspoint of the r-g channel a coincidence? Taking the result seriously would imply that the 580 crosspoint was set not to optimize the color channels, but rather to produce consistent luminance responses across the luminance channels. It also seems to imply that isolated R- and G-centers have weightings similar to those implied by the r-g channel crosspoint. I know of no evidence for this.

Discussion

In summary, nonlinear summation of receptive field centers does not contradict Ingling and Tsou's (1987) contention that the r-g parvocellular system can produce V_λ-like luminosity functions for an acuity criterion. The modelling also raises two interesting points.

First, these calculations plainly show the absurdity of trying to reverse engineer the luminance channels from estimates of spectral sensitivity. Without knowledge of the r-g channel crosspoints, a V_λ-like spectral sensitivity for acuity can be produced from a wide range of R-cone or G-cone ratios, just by varying the probability summation exponent. As Fig. 2 shows, even a 490 crosspoint is adequate, given the uncertainty in flicker estimates. Cicerone and Nerger (1989) have pointed out a similar problem for luminosity functions derived from flicker or minimum border. To emphasize the severity of this problem, note that Vos *et al.* (1990) have fit their cone spectral sensitivities to two models of luminosity.

$$V_\lambda = aR + bG + cB = R^a G^b B^c \qquad (4)$$

If two models as different as these produce similar results, then spectral sensitivity is a poor measure of linearity. In addition, Ingling and his associates have found that luminance additivity is also insensitive to nonlinearities. Many nonlinearities do not affect additivity, and detection of even the more extreme nonlinearities (such as htan compressive functions) require special analytical techniques (see Ingling and Tsou, 1985; Ingling *et al.*, 1978, 1986). Similar problems have occurred in modelling color discrimination. Color discrimination is so dependent on the shapes of the cone fundamentals and so independent of everything subsequent to cone transduction, that almost every model of color discrimination produces the correct wavelength discrimination function (Vos, 1982).

Second, it is interesting that four separate luminance criteria – critical flicker fusion, heterochromatic flicker photometry, minimum border and acuity – driving two different retinogeniculate systems (magnocellular and parvocellular) all produce the same spectral sensitivity. It may be useful for the parvo and magno systems to produce similar spectral reponses to facilitate the combination of parvo and magno signals at some later stage of processing. In a similar context, Barlow (1982) has argued that the similarity in R- and G-cone spectral sensitivity is an adaptation which results in less demodulation when R- and G-cone signals are added together. What is unclear is the cortical stage where this integration takes place.

Acknowledgements

I thank Carl Ingling for suggesting that I examine nonlinear summation of cone signals. This work was supported in part by a NRC-USAARL Research Award.

References

Barlow, H.B. (1982). What causes trichromacy? A theoretical analysis using comb-filtered spectra. Vision Res. 22: 635-643.

Cicerone, C.M. and Nerger, J.L. (1989). The relative numbers of long-wavelength sensitive to middle-wavelength sensitive cones in the human fovea centralis. Vision Res. 29: 115–128.

Cowan, W.B., Wyszecki, G. and Yaguchi, H. (1984). Probability summation among color channels. Optic News 10: 90.

Graham, N.V.S. (1989). Visual Pattern Analyzers. Oxford University Press, New York.

Guth, S.L., Massof, R.W. and Benzschawel, T. (1980). Vector model for normal and dichromatic color vision. J. Opt. Soc. Am. 70: 197–212.

Heeger, D.J. (1992). Half-squaring in the responses of cat striate cells. Visual Neurosci. 9: 427–443.

Ingling, C.R., Jr., Grigsby, S.S. and Long, G. (1992). Comparison of spectral sensitivity using heterochromatic flicker photometry and an acuity criterion. Color Res. Appl. 17: 187–196.

Ingling, C.R., Jr. and Martinez, E. (1983). The relationship between spectral sensitivity and spatial sensitivity for the r-g X-cell channel. Vision Res. 23: 1495–1500.

Ingling, C.R., Jr. and Tsou, B.H.P. (1977). Orthogonal combination of the three visual channels. Vision Res. 17: 1075–1082.

Ingling, C.R., Jr. and Tsou, B.H.P. (1985). Flicker photometry and achromatic-channel structure. J. Opt. Soc. Am. A 2: 1375–1378.

Ingling, C.R., Jr. and Tsou, B.G.P. (1987). Spectral sensitivity for flicker and acuity criteria. J. Opt. Soc. Am. A 5: 1374–1378.

Ingling, C.R., Jr., Tsou, B.H.P. and Billock, V.A. (1986). The nonlinearity of CFF as a photometric criterion. Invest. Ophthal. Vis. Sci. Suppl. 27: 93.

Ingling, C.R., Jr., Tsou, B.H.P., Gast, T.J., Burns, S.A., Emerick, J. and Risenberg, L. (1978). The achromatic channel: The nonlinearity of minimum-border and flicker matches. Vision Res. 18: 379–390.

Kranda, K. and King-Smith, P.E. (1979). Detection of coloured stimuli by independent linear systems. Vision Res. 19: 733–745.

Lynch, J.J., III, Silveira, L.C.L., Perry, V.H. and Merigan, W.H. (1992). Visual effects of damage to P ganglion cells in macaque. Visual Neurosci. 8: 575–583.

Pokorny, J., Graham, C.H. and Lanson, R.N. (1968). Effect of wavelength on foveal grating acuity. J. Opt. Soc. Am. 58: 1410–1414.

Quick, R.F. (1974). A vector-magnitude model of contrast detection. Kybernetik 16: 65–67.

Schiller, P., Logothetis, N.K. and Charles, E.R. (1990). Functions of the colour-opponent and broad-band channels of the visual system. Nature 343: 68–70.

Vimal, R.L.P., Pokorny, J., Smith, V. and Shevell, S.K. (1989). Foveal cone thresholds. Vision Res. 29: 61–78.

Vos, J.J. (1982). On the merits of model making in understanding color-vision phenomena. Color Res. Appl. 7: 69–77.

Vos, J.J., Estévez, O. and Walraven, P.L. (1990). Improved color fundamentals offer a new view on photometric additivity. Vision Res. 30: 937–943.

Walraven, P.L. (1974). A closer look at the tritanopic convergence point. Vision Res. 14: 1339–1343.

Wyzecki, G. and Stiles, W.S. (1982). Color Science (2nd ed.) Wiley, New York.

404 Melbourne Dr.
Enterprise, AL, U.S.A.
36330

32. Ganglion cell circuits in primate fovea

DAVID J. CALKINS, STANLEY J. SCHEIN[1],
YOSHIHIKO TSUKAMOTO[2] and PETER STERLING
(Philadelphia, PA and [1]Los Angeles, CA, U.S.A.; [2] Hyogo, Japan)

Abstract

In the standard view all information from the fovea is relayed via only two types of ganglion cell, P (midget) and M (parasol) thought to form respectively 90-95% and 5-10% of the ganglion cell population. We characterized all 157 ganglion cells in a small patch of macaque fovea using electron micrographs of serial sections. One hundred fifteen (73%) were midget ganglion cells and were of two types, one with 28 $\pm$ 4 bipolar synapses and the other with 47 $\pm$ 3 synapses. Forty-two (27%) were non-midget ganglion cells. Most had dendrites restricted to either sublamina *a* or sublamina *b* of the inner plexiform layer, but one quarter had dendrites in both. These cells were of two types, one with input in sublamina *b* from blue cone bipolar cells and the other with only diffuse bipolar cell input. The ganglion cells with dendrites in either sublamina *a* or sublamina *b* were of at least one type with the possibility of more. We conclude that non-midget ganglion cells are numerous and provide additional parallel arrays to brain.

Introduction

In primates more than half of the cerebral cortex is devoted to vision and about half of visual cortex is devoted to the fovea (central 5° of retina) (Van Essen and Anderson, 1990). Therefore, how foveal ganglion cells encode visual signals is of major consequence for understanding the nature of the initial input to a great part of the cerebral cortex. In the standard model for this input, information from the fovea is relayed via two types of retinal ganglion cell: P and M, so named for their sites of termination in parvocellular and magnocellular geniculate layers (reviewed by Kaplan *et al.*, 1990). The P cell is thought to have a receptive field that is both spatially and spectrally antagonistic. It is thought to correspond morphologically to the 'midget' ganglion cell and to comprise as many as 90–95% of the foveal ganglion cells (Dacey, 1993). The M cell exhibits a spectrally broad-band receptive field. It is thought to correspond morpho-

B. Drum (ed.), Colour Vision Deficiencies XII, pp. 267–274.
© 1995 *Kluwer Academic Publishers, Dordrecht.*

logically to the 'parasol' ganglion cell and to comprise 5–10% of foveal ganglion cells (Grünert *et al.*, 1993; Silveira and Perry, 1991). In the standard model, the P cell serves both high spatial acuity and color opponency, while the M cell serves high contrast, broad-band sensitivity.

Whether the midget and parasol types indeed account for all foveal ganglion cells, and what proportion they actually comprise, has not been studied directly. To do so requires that the morphology of all the ganglion cells in some defined region be determined in enough detail to classify them reliably. We prepared a small patch of fovea from the Old World primate Macaca fascicularis for electron microscopy. To study complete cells with resolution sufficient to identify synapses, we cut an unbroken series of 319 ultrathin sections (0.09 μm) and photographed each section at 2000x with a montage of 16 overlapping pictures (Tsukamoto *et al.*, 1992). Ganglion cells were reconstructed by tracing their profiles from electron micrographs of successive sections onto mylar sheets aligned on a cartoonist's jig. The tracings were then digitized, stacked by computer, and displayed with the hidden lines removed (Smith, 1987).

Results

The region of inner plexiform layer studied was 29 x 70 μm in the tangential plane. This region was served by an array of cone pedicles with inner segments at 1° eccentricity and contained the dendritic processes of 157 ganglion cells. One hundred fifteen of the ganglion cells (73%) were midget cells. Each of these had a compact dendritic tree (5–10 μm in diameter) in close association with the axon terminals of a single midget bipolar cell. Forty-two of the cells (27%) were non-midget. These were distinguished from the midget cells by their morphologies, larger dendritic trees and varying synaptic contacts. We traced in some detail the synaptic connections to determine whether cells in these two groups could be further subdivided.

Midget ganglion cells

Each M and L cone pedicle in our series was presynaptic to a flat midget bipolar cell and an invaginating midget bipolar cell (see Kolb, 1970). Furthermore, each midget bipolar cell contacted only a single cone pedicle. The flat bipolar cell terminated in sublamina *a* of the inner plexiform layer and the invaginating cell in sublamina *b*; each contacted a single midget ganglion cell (see Polyak, 1941). Every synapse from a midget bipolar terminal to a midget ganglion cell came from the same midget bipolar cell. Thus, the connections in both the outer and inner plexiform layers were exactly 1:1 without divergence or convergence of synapses (Calkins *et al.*, 1994). Such precision seems an impressive accomplishment for a biological system, but might be expected here since otherwise spatial pooling would degrade acuity.

The midget ganglion cells were paired according to their cone pedicle input,

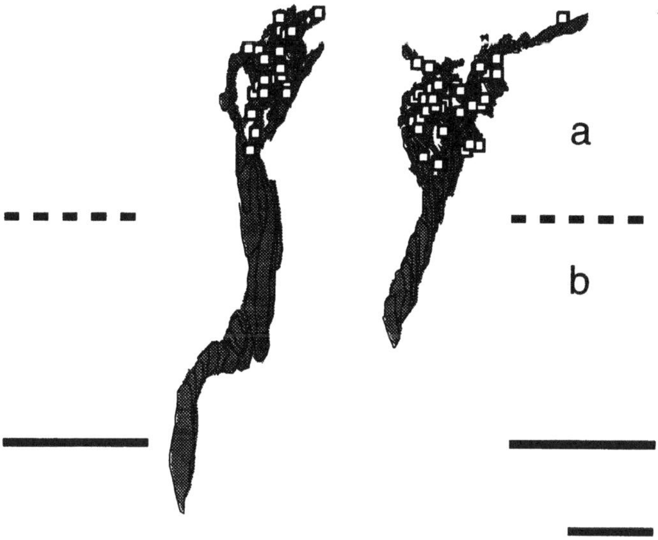

Fig. 1. Left, sublamina *a* type midget ganglion cell with 25 midget bipolar synapses (squares). Right, a second sublamina *a* type midget ganglion cell with 46 midget bipolar synapses. The two sublaminae of the inner plexiform layer are indicated. Scale bar = 5 μm.

an *a* and a *b* sublamina cell type for each M and L cone. Presumably, these correspond to OFF and ON circuits (Nelson *et al.*, 1978). Both members of a midget ganglion cell pair collected virtually identical numbers of bipolar synapses. Perhaps the greatest surprise was that the pairs of midget circuits sorted into two groups based on the number of synapses from bipolar to ganglion cell. The midget bipolar cells from some cones provied 28 ± 4 synapses to their ganglion cells (Fig. 1, left), while the midget bipolar cells from other cones provided 47 ± 3 synapses (Fig. 1, right) (Calkins *et al.*, 1994). We hypothesize that the two circuits serve M and L cones since the division into two roughly equal populations is sharp, and the only known difference between non-S cones is in their spectral sensitivities. Also, the two populations of midget circuit distributed in small, homogeneous clusters similar to those of M and L cones observed by *in situ* measurements of spectral sensitivity in Cercopithecus (Mollon and Bowmaker, 1992). The clustering pattern implied by this circuitry is such that the cones surrounding any given cone are most often mixed with regard to spectral sensitivity (Calkins *et al.*, 1994).

Non-midget ganglion cells

Eleven of the 42 non-midget ganglion cells provided dendritic branches to both sublaminae of the inner plexiform layer (Fig. 2, middle). In this sense, these cells can be called 'bistratified', though they bear a strong resemblance to Polyak's 'shrub' cell (Polyak, 1941). The synaptic connections analyzed thus far suggest

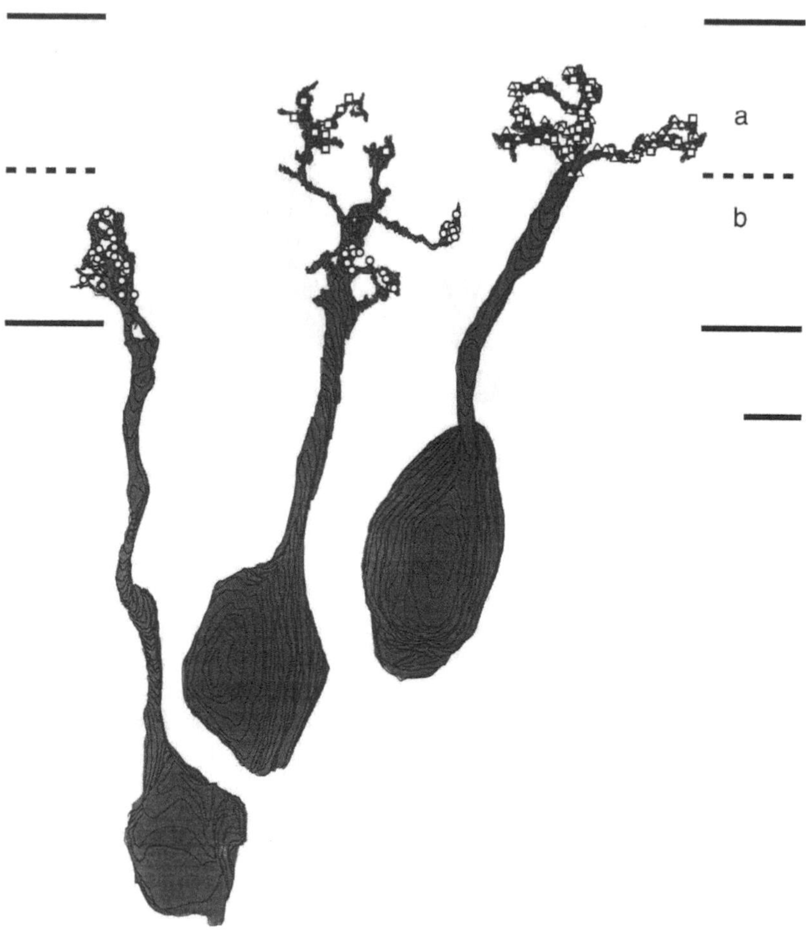

Fig. 2. Middle, bistratified ganglion cell that receives 27 synapses (circles) from blue cone bipolar cells and 12 synapses (squares) from flat diffuse bipolar cells. Right, non-midget ganglion cell stratifying in sublamina *a* that receives 120 synapses (triangles) from amacrine cell processes and 62 synapses (squares) from diffuse bipolar cell terminals. Left, shown for comprison is a sublamina *b* type midget ganglion cell with 24 midget bipolar synapses. Scale bar = 5 μm.

two types. One type collects synapses in sublamina *b* from blue cone bipolar terminals (Mariani, 1984; Kouyama and Marshak, 1992) and synapses in sublamina *a* from diffuse bipolar cell terminals (Boycott and Wässle, 1991). In one case the diffuse bipolar cell was traced back to its connections with five cone pedicles which, based on their midget cell circuitry, included both M and L cones. Thus, this type of non-midget ganglion cell seems to be wired ON(S)/OFF(M+L) (see Nelson *et al.*, 1978) and very likely corresponds to the blue-ON bistratified cell characterized by Dacey and Lee (1994). The other type of bistratified cell (not shown) collects synapses in both sublaminae from diffuse bipolar cells and not from blue cone bipolar terminals; their specific cone inputs are as of yet unknown.

Thirty-one of the non-midget ganglion cells had dendrites either in sublamina *a* or in sublamina *b* of the inner plexiform layer. We identified at least one type (Fig. 2, right) which resembled the garland cell of Polyak (1941) and whose synaptic inputs consisted of more amacrine cell synapses than diffuse bipolar cell synapses. Other cells showed clearly different patterns of connectivity that remain to be quantified. Thus, within this group there are several types, only one of which can be the parasol cell.

Discussion

Midget ganglion cells

If the two types of midget circuit truly represent the distribution of M and L cones, then nearly all cones are surrounded by a mixture of both types and only rarely by cones all of the other type. Spectrally pure antagonism from the surround (e.g. Reid and Shapley, 1992) may require spectrally selective lateral wiring. However, there is strong evidence that horizontal cells are not spectrally selective. (Wässle *et al.,* 1989), and we find that the narrow-field amacrine cell inputs to the midget ganglion cell are not spectrally selective either (unpublished observation). Thus, as suggested by Lennie *et al.,* (1991), whatever degree of spectral specificity is present in the surround of a particular midget ganglion cell may arise simply as a result of the random distribution of M and L cones in the array.

We speculate that the midget ganglion cell circuits with more synapses serve M cones and those with fewer serve L cones. Recent measurements of spectral reflectance in natural scenes show that on average M cones should collect 10% fewer photons than L cones (Osorio and Bossomaier, 1992; Nagle and Osorio, 1993). Since the midget ganglion cell array is clearly specialized by its high sampling rate for spatial acuity, the unequal spectral sensitivities of the M and L cones would create 'chromatic' noise (Williams *et al.,* 1991; Buchsbaum *et al.,* 1993). The extra synapses in the M cone midget pathway might increase the gain and thus impart to the M cells the same output as they would have if they were driven instead by L cones. Thus, the extra synapses for M cone midget circuits would tend to remove spectral information from the midget ganglion cell array. If so, one is led to reconsider the idea that these ganglion cells form the substrates of the color opponent channels.

Non-midget ganglion cells

One major finding of our study is that non-midget ganglion cells in fovea are numerous (27% of the population). In previous work, the proportion of foveal midget ganglion cells was extrapolated from more peripheral regions because the cells pack so densely in the fovea that the appropriate measurements from silver staining (Silveira and Perry, 1991) retrograde filling (Perry *et al.*, 1984) or intracellular injection (Dacey and Brace, 1992) are difficult. Our finding suggests that enough non-midget ganglion cells are available to establish multiple, parallel arrays.

Information regarding the synapses to non-midget ganglion cells is still sparse. It is uncertain, for example, whether the second type of bistratified cell with only diffuse bipolar cell inputs identified here corresponds to a chromatically antagonistic cell or to an ON/OFF luminance cell. An uneven weighting of the cone inputs to the presumed ON and OFF diffuse bipolar cells could render these cells spectrally antagonistic. If so they, like the ON(S)/OFF(M+L) bistratified cell, would be candidates for physiological 'type II' cells (Wiesel and Hubel, 1966; De Monasterio and Gouras, 1975; Derrington and Lennie, 1984). If, as suggested by Rodieck (1991), type II cells mediate B/Y and R/G color opponency, the sampling frequencies of their arrays should match psychophysical observations. Sekiguchi *et al.*, (1993) presented evidence that the neural pathways for detecting R/G isoluminant gratings and S cone isolating gratings are spatially similar. For R/G isoluminant gratings, maximal resolution was 20–27 c/degree (Sekiguchi and Williams, 1993; Sekiguchi *et al.*, 1993; but see Mullen, 1985). Using this range as an upper limit and allowing for the lower resolution of the macaque (Merigan and Katz, 1990), we would expect the color opponent channels at 1° eccentricity to resolve about 10–14 c/degree. At this order of resolution, the distribution of the ON(S)/OFF(M+L) bistratified cell and of the second type of bistratified cell in our sample is sufficiently dense to form two arrays with the requisite sampling frequencies.

Apparently multiple parallel arrays relay information from the fovea to the brain. Possibly there are enough for each array to encode only one aspect of the visual scene. Such parallelism would avoid the need to multiplex information on a single anatomical line and would allow maximum gain for a given signal component at the earliest stage. This is where the gain should be to protect against noise (Pelli, 1990, 1991).

Acknowledgements

This work was supported by R01 EY08124. We thank Pat Masarachia, Sally Shrom, and Alex Shrom for their expert technical assistance.

References

Boycott, B.B. and Wässle, H. (1991). Morphological classification of bipolar cells of the primate retina. Euro. J. Neurosci. 3: 1069–1088.

Buchsbaum, G. Calkins, D. and Sterling, P. (1993). Patchy mosaic of L and M cones enables efficient tradeoff between spatial and color vision. Ann. Meet. Opt. Soc. Amer. (Abstr.).

Calkins, D.J., Schein, S.J., Tsukamoto, Y. and Sterling, P. (1994). M and L cones in macaque fovea connect to midget ganglion cells by different numbers of excitatory synapses. Nature 371: 70–72.

Dacey, D.M. and Lee, B.B. (1994). The "blue-on" opponent pathway in private retina originates from a distinct bistsatified ganglion cell type. Nature 367: 731–735.

Dacey, D.M. (1993). The mosaic of midget ganglion cells in the human retina. J. Neurosci. 13: 5334–5355.

Dacey, D.M. and Brace, S. (1992). A coupled network for parasol but not midget ganglion cells in the primate retina. Vis. Neurosci. 9: 279–290.

De Monasterio, F.M. and Gouras, P. (1975). Functional properties of ganglion cells in the rhesus monkey retina. J. Physiol. 251: 167–197.

Derrington, A.M. and Lennie, P. (1984). Spatial and temporal sensitivities of neurones in lateral geniculate nucleus of macaque. J. Physiol. 357: 219–240.

Grünert, U., Greferath, U., Boycott, B.B. and Wässle, H. (1993). Parasol (Pα ganglion-cells of the primate fovea: immunocytochemical staining with antibodies against $GABA_A$-receptors. Vision Res. 33: 1–14.

Kaplan, E., Lee, B. and Shapley, R. (1990). New views of primate retinal function. In: Osborne, N. and Chader, J. (eds.), Progress in Retinal Research, 0th ed.: 273–336, Pergamon Press, New York.

Kolb, H. (1970). Organization of the outer plexiform layer of the primate retina: electron microscopy of Golgi-impregnated cells. Philos. Trans. R. Soc. Lond. B 258: 261–283.

Kouyama, N. and Marshak, D.W. (1992). Bipolar cells specific for blue cones in the macaque retina. J. Neurosci. 12: 1233–1252.

Lennie, P., Haake, P.W. and Williams, D.R. (1991). The design of chromatically opponent receptive fields. In: Landy, M.S. and Movshon, J.A. (eds.), Computational Models of Visual Processing: 71–82, MIT Press, Cambridge.

Mariani, A.P. (1984). Bipolar cells in monkey retina selective for the cones likely to be blue-sensitive. Nature 308: 184–186.

Meriga, W.H. and Katz, L.M. (1990). Spatial resolution across the macaque retina. Vision Res. 30: 985–991.

Mollon, J.D. and Bowmaker, J.K. (1992). The spatial arrangement of cones in primate fovea. Nature 360: 677–670.

Mullen, K.T. (1985). The contrast sensitivity of human color vision to red-green and blue-yellow chromatic gratings. J. Physiol. 359: 381–400.

Nagle, M.G. and Osorio, D. (1993). The tuning of human photopigments may minimize red-green chromatic signals in natural conditions. Proc. R. Soc. Lond. B 252: 209–213.

Nelson, R., Famiglietti, E.V. and Kolb, H. (1978). Intracellular staining reveals different levels of stratification for on-center and off-center ganglion cells in the cat retina. J. Neurophysiol. 41: 472–483.

Osorio, D. and Bossomaier, T.R.J. (1992). Human cone-pigment sensitivities and the reflectances of natural surfaces. Biol. Cybern. 67: 217–222.

Pelli, D.G. (1990). The quantum efficiency of vision. In: Blakemore, C. (ed.), Vision: Coding and Efficiency: 3–24, Cambridge University Press, Cambridge.

Pelli, D.G. (1991). Noise in the visual system may be early. In: Landy, M.S. and Movshon, J.A. (eds.), Computational Models of Visual Processing: 147–151, MIT Press, Cambridge.

Perry, V.H., Oehler, R. and Cowey, A. (1984). Retinal ganglion cells that project to the dorsal lateral geniculate nucleus in the macaque monkey. Neurosci. 12: 1101–1123.

Polyak, S.L. (1941). The Retina. University of Chicago Press, Chicago.

Reid, R.C. and Shapley, R.M. (1992). Spatial structure of cone inputs to receptive fields in primate geniculate nucleus. Nature 356: 716–718.

Rodieck, R.W. (1991). Which cells code for color? In: Valberg, A. and Lee, B.B. (eds.), From Pigments to Perception: 83–93, Plenum Press, New York.

Sekiguchi, N., Williams, D.R. and Brainard, D.H. (1993). Aberration-free measurements of isoluminant contrast sensitivity. J. Opt. Soc. Amer. A 10: 2105–2117.

Sekiguchi, N. and Williams, D.R. (1993). Efficiency for detecting isoluminant and isochromatic interference fringes. J. Opt. Soc. Amer. A 10: 2118–2133.

Silveira, L.C.L. and Perry, V.H. (1991). The topography of magnocellular projecting ganglion cells (M-ganglion cells) in the primate retina. Neurosci. 40: 217–237.

Smith, R.G. (1987). Montage: a system for three-dimensional reconstruction by personal computer. J. Neurosci. Methods 21: 55–69.

Tsukamoto, Y., Masarachi, P., Schein, S.J. and Sterling, P. (1992). Gap junctions between the pedicles of macaque foveal cones. Vision Res. 32: 1809–1815.

Van Essen, D.C. and Anderson, C.H. (1990). Information processing strategies and pathways in the primate retina and visual cortex. In: An Introduction to Neural and Electronic Networks: 43–72, Academic Press, Inc., New York.

Wässle, H., Boycott, B.B. and Röhrenbeck, J. (1989). Horizontal cells in the monkey retina: cone connections and dendritic network. Euro. J. Neurosci. 1: 421–435.

Wiesel, T.B. and Hubel, D.H. (1966). Spatial and chromatic interactions in the lateral geniculate body of the rhesus monkey. J. Neurophysiol. 29: 1115–1156.

Williams, D.R., Sekiguchi, N. Haake, P.W., Brainard, D. and Packer, O. (1991). The cost of trichromacy for spatial vision. In: Lee, B. and Valberg, A. (eds.), From Pigments to Perception: 11–22, Plenum Press, New York.

David J. Calkins
Mahoney Institute for Neurological Sciences and Dept. of Neuroscience
123 Anatomy/Chemistry Building
University of Pennsylvania, PA 19104, U.S.A.

PART FOUR

Short-wavelength-sensitive cones

33. Blue cone bipolar cells of the macaque retina

DAVID MARSHAK, DONNA STAFFORD, ROY JACOBY and NOBUO KOUYAMA
(Houston, TX, U.S.A.)

Abstract

Antisera which recognize glycine-extended precursors of the peptide cholecystokinin labeled neurons in the macaque retina with dendrites that ran obliquely and had diffuse dendritic terminals as is the case for blue cone bipolar cells described previously. They clearly formed a single population since the density and sizes of their perikarya varied monotonically with eccentricity, while the levels of their perikarya and axon terminals were constant. Short-wavelength cones were also labeled in some preparations, either with Procion Black or anti-blue opsin, and the labeled bipolar cells received input exclusively from these cones. The dendrites formed the central elements at ribbon synapses in the short-wavelength cones, and the axons of the labeled bipolar cells branched in the fifth stratum of the inner plexiform layer. Taken together, these findings indicate that the labeled bipolar cells were different from the invaginating midget bipolar cells that contact middle- and long-wavelength cones and suggest that the labeled bipolar cells depolarize in response to increments in short-wavelength stimuli. There were numerous, unlabeled bipolar cell dendrites receiving inputs from short-wavelength cones at other types of contacts, however, and these presumably belonged to bipolar cells with hyperpolarizing light responses.

Introduction

In 1984, Mariani described a new type of bipolar cell in the macaque retina using the Golgi method. Although its dendrites invaginated the cone pedicles and formed the central elements at ribbon synapses as is the case for invaginating midget bipolar cells, it differed from those cells in many respects. Rather than contacting the cone pedicle directly above its perikaryon like a typical midget bipolar cell, its dendrites typically ran obliquely for a considerable distance through the outer plexiform layer before terminating. Even when the dendrites did not show this oblique course, they had smaller and more sparse dendritic

B. Drum (ed.), Colour Vision Deficiencies XII, pp. 277–283.
© 1995 *Kluwer Academic Publishers, Dordrecht.*

terminals that were distinctive. The axons were also characteristic, and they terminated in stratum 5 of the inner plexiform layer and not stratum 4, like the axons of the invaginating midget bipolar cells. Rodieck (1988), also using the Golgi method, showed that these bipolar cells had larger axonal arbor areas than the midget bipolar cells. Unlike midget bipolar cells, the dendrites of these new bipolar cells frequently contacted more than one cone. They did not resemble diffuse bipolar cells, however, which contacted all the cones within their dendritic field (Boycott and Wässle, 1991). These bipolar cells contacted only a subset of widely-spaced cones, and, based on the spacing of those cones, Mariani (1984) proposed that these were the short-wavelength-selective or blue cones. In human retinas, Kolb *et al.,* (1992) found a very similar type of bipolar cell and a second type, which also apparently contacted blue cones selectively but terminated in the distal half of the inner plexiform layer. A number of questions remained unanswered in the studies of these blue cone bipolar cells with the Golgi method, however.

Several years later, we found that the same bipolar cells could be labeled using immunocytochemical techniques with antisera that recognized the glycine-extended precursors of the neuropeptide cholecystokinin (Marshak *et al.,* 1990). It was possible to label a large proportion, if not all, of the blue cone bipolar cells and to study their morphology, distribution and synaptic connections. The labeled bipolar cells were regularly distributed, and most cells resembled the ones described earlier. There were considerable differences in their appearance in the central and peripheral retina, however, and the first step was to determine whether there were distinct types in the different regions. We studied the diameters of the perikarya and their depths within the inner nuclear layer and found that there was no change in these parameters with eccentricity. The plot of cell density as a function of eccentrity was also smooth. Another indication that all the labeled cells were a single type was that all the axon terminals were found at the same level in the inner plexiform layer. We concluded that the morphological differences represented the variability within a single population and predicted that all the labeled bipolar cells would have the same light responses (Kouyama and Marshak, 1992).

Extending the analysis to the electron microscopic level, we found that the axons received some symmetric, non-reciprocal synapses from amacrine cells. They also made ribbon synapses, typically dyads with one process from a ganglion cell and one from an amacrine cell (Marshak *et al.,* 1990). In the outer plexiform layer, the bipolar cell dendrites formed the central elements at ribbon synapses of a subset of cones. While many of these were typical ribbon contacts, the labeled processes also made two unusual types of contacts. In one of these, there were two or more apparently identical bipolar cell dendrites as central elements at otherwise typical ribbon synapses. The opposite was true at the other unusual type, where two or more ribbons were presynaptic to a single large central element (Kouyama and Marshak, 1992). All the ribbon synapses onto labeled processes had the narrow cleft and membrane densities characteristic of sign-inverting junctions. Taken with the finding that the labeled axons terminated in the proximal half of the inner plexiform layer, this suggested that the labeled bipolar cells were the ON-type, depolarizing in response to increments in light intensity. The same pedicles also formed numerous basal junctions and triad-associated flat contacts with unlabeled dendrites that, presumably, originated from other types of bipolar cells that had OFF or hyperpolarizing responses to light. Shortly afterwards, Klug *et al.*, (1992) reconstructed these pedicles from serial electron micrographs and demonstrated that they also contacted flat midget bipolar cells.

We also did a series of experiments in which we simultaneously labeled putative blue cones using Procion Black to determine whether the presynaptic cones were the short-wavelength-sensitive type (Kouyama and Marshak, 1992). We found that the bipolar cells received input exclusively from labeled cones, and this provided further evidence that they constituted a single type. In whole mount preparations, we studied the input-output relations at the labeled synapses quantitatively. The bipolar cells were distributed randomly around blue cones in the far periphery, and the connections were spatially symmetrical. There was a tendency for the dendrites of mid-peripheral bipolar cells to be oriented both inward and outward along the radial axes of the retina, however, and in the parafovea, the labeled bipolar cells were always displaced peripherally from the cones they contacted. These results supported the hypothesis that these cones and bipolar cells were associated during the expansion of the developing retina (Kouyama and Marshak, 1993). There were 1.4 labeled bipolar cells for every labeled cone, and every labeled cone pedicle received labeled bipolar cell dendritic contacts. There were also a few labeled bipolar cells with 'blind dendrites' that did not contact any cones, but these were found in areas where the blue cones were apparently missing. The divergence at the cone to bipolar cell synapses, that is the average number of bipolar cells postsynaptic to each cone, was 1.8. More than half the cones were presynaptic to two labeled bipolar cells and approximately a third to just one bipolar cell. The convergence, that is the average number of cones presynaptic to each labeled bipolar cell, was 1.2. Almost two-thirds of the bipolar cells received input from only one cone, and most of the others received inputs from two cones. These findings suggested that the relationship of blue cones to the ON bipolar cells was not strictly one-to-one, as it was with the other types of cones and the ON midget bipolar cells. There was apparently more lateral spread of signals from the blue cones than from the red and green cones, but, since there were more bipolar cells than cones, there would not be much loss of acuity (Kouyama and Marshak, 1992). A major shortcoming of these studies, however, was that there was no direct evidence that Procion Black labeled the blue cones. In the experiments reported here, we labeled cones with an antibody to blue-cone opsin and confirmed that they contacted the labeled bipolar cells.

Methods

Tissues from three macaques (2 *Macaca mulatta* and 1 *Macaca fascicularis*) and a baboon (*Papio cynocephalus*) were used. Whole isolated retinas were fixed overnight at 4° C in 4% paraformaldehyde, 0.1% picric acid in 0.1M sodium phosphte buffer pH 7.4, either directly, in the case of the baboon retina, or after vascular perfusion fixation with 4% paraformaldehyde in the same buffer. The light microscopic immunolabeling techniques were similar to those we described previously (Kouyama and Marshak, 1992), with two exceptions. The retinas were incubated in affinity-purified antibody to blue opsin at 1:50 for 7 days at 4° C (Lerea *et al.* 1989), before treating them with antisera to G6-gly, and 0.08% nickel chloride was included in the diaminobenzidine solution in order to intensify the peroxidase reaction product. The retinas were dehydrated and embedded in epon. For studies of the ratio of blue cones to labeled bipolar cells, selected areas were re-embedded and vertical, 30 μm sections were prepared on a sliding microtome. Counts were made at regular intervals of 200 μm or more only in sections wher the labeling appeared to be uniform for both populations of cells. To analyze the convergence and divergence, areas where the labeling was optimal were selected, and serial, horizontal 55 μm sections were prepared on a sliding microtome. Otherwise the techniques were similar to those we described previously. Cone pedicles with bipolar cell dendritic specializations within 1–2 μm were considered to be contacted on the basis of our electron microscopic results (Kouyama and Marshak, 1992).

Results

Our findings with antibody to blue cone opsin confirmed that the labeled bipolar cells did, in fact, contact short-wavelength-sensitive cones. Figure 1 shows a camera lucida drawing of a patch of G6-gly-IR bipolar cells from a 55 μm horizontal section of the macaque retina with the positions of the blue opsin-IR cones indicated schematically. The drawings of the outer segments needed to be shifted a few microns in order to bring the two mosaics into alignment, but it was clear that the labeled dendrites extended toward the blue cones. There were numerous examples of dendrites from two or more labeled bipolar cells converging on single labeled cones. Figure 2 shows light micrographs from vertical sections of a double-labeled baboon retina that demonstrate that there are direct contacts. The labeling with anti-blue opsin was particularly prominent in cone outer segments, and it extended though the perikaryon and into the axon, at least partially. Although the axonal labeling was not as prominent and there was somewhat more background labeling than with Procion Black labeling, the G6-gly-IR bipolar cell dendrites could be traced directly to labeled pedicles in preparations such as these. Since we had previously shown that G6-gly-IR bipolar cells made contacts exclusively with pedicles labeled with Procion Black (Kouyama and Marshak, 1992), these

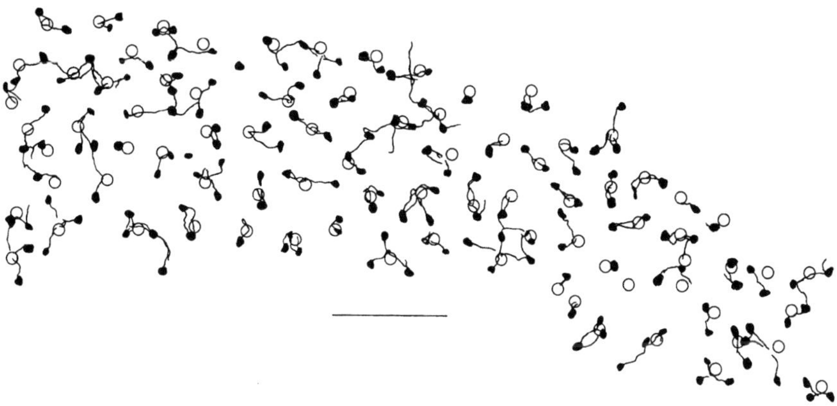

Fig. 1. Camera lucida drawing of G6-gly-immunoreactive bipolar cells from a 55 μm section of a *M. fascicularis* retina. The circles are schematic representations of outer segments labeled with anti-blue opsin. The entire mosaic of labeled cones was shifted laterally to account for the obliquity of their axons. The calibration bar is 100 μm..

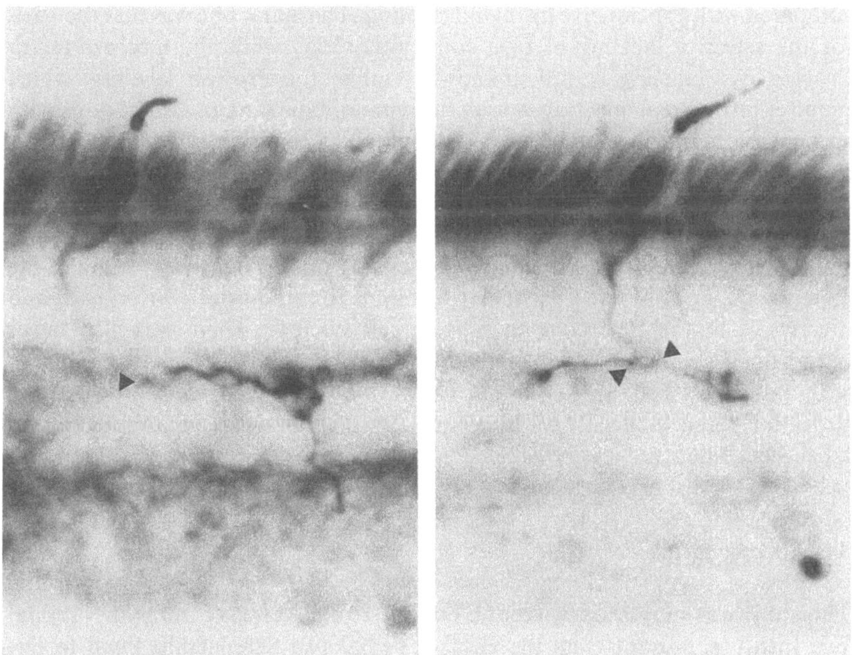

Fig. 2. Photomicrographs of 20 μm sections from a baboon retina double-labeled with anti-G6-gly and anti-blue opsin (850 X). Note that labeled bipolar cell dendrites could be traced to labeled blue cone pedicles (arrowheads).

results indicate that both techniques label the same subpopulation of cones. Thus, these experiments also provide the first direct evidence that Procion dyes label blue cones.

One macaque retina double-labeled with anti-G6-gly and anti-blue opsin was also used for quantitative studies of the input-output relations at blue cone to bipolar cell synapses. Sections from central and mid-peripheral retina like the ones illustrated in Fig. 2 were used. The ratio of labeled bipolar cells to blue cones was found to be 1.48 ± 0.09, only slightly higher than the value obtained previously from whole mounts of peripheral retina using Procion Black. In horizontal sections from mid-peripheral retina, the average convergence was quite similar, $1.15 \pm .37$. The divergence averaged $2.18 \pm .72$, slightly higher than we had previously reported. There may be more blue cones presynaptic to three or four bipolar cells, but it was uncertain whether this was attributable to the regions sampled or to individual differences.

Discussion

The immunolabeling technique for blue cones had some advantages when compared with Procion dye intravital labeling. The major one was that the basis for the selective labeling of blue cones was clear, while the mechanism for Procion dye labeling is still unknown. Unlike the Procion labeling, which requires intravitreal injection *in vivo,* the immunolabeling can also be done with specimens obtained *post mortem,* and therefore it is applicable to human retinas, as Curcio *et al.,* (1991) demonstrated. In addition, the immunolabeling can be done on individual sections, which makes it possible to label one or two other antigens in adjacent sections, while the Procion technique requires the entire retina. Nevertheless, the Procion dyes provided more extensive labeling of the cone axons, and this was particularly helpful for studies of convergence and divergence at the blue cone to bipolar cell synapse. There was also lower background staining in double-labeling experiments with Procion Black than with anti-blue opsin, at least in our hands. Now that we have shown that the Procion labeling is specific for blue cones, we recommend using this method to label them whenever it is possible.

Acknowledgements

This study was supported by grant EY-06472 to Dr. Marshak from the National Eye Institute, a grant from the Hisako Yamakawa Scholarship Fund to Dr. Kouyama and by Fight for Sight postdoctoral fellowship PD-94020 to Dr. Stafford. We are very grateful to Dr. Connie Lerea for providing the antibody to blue opsin and to Dr. DelValle and Dr. Tadataka Yamada for providing the antisera to G6-gly. We also wish to thank Dr. Robert Rodieck, Dr. Daniel Felleman and Dr. Daniel Tso for providing the tissue used in this study.

References

Boycott, B.B. and Wässle, H. (1991). Morphological classification of bipolar cells of the primate retina. Eur. J. Neurosci. 3: 1069–1088.

Curcio, C.A., Allen, K.A., Sloan, K.R., Lerea, C.L., Hurley, J.B., Klock, I.B. and Milam, A.H. (1991). Distribution and morphology of human cone photoreceptors stained with anti-blue opsin. J. Comp. Neurol. 312: 610–624.

Klug, K., Tiv, N., Tsukamoto, Y., Sterling, P. and Schein, S. (1992). Blue cones contact OFF-midget bipolar cells. Soc. Neurosci. Abst 18: 838.

Kolb, H., Lindberg, K.A. and Fisher, S.K. (1992). Neurons of the human retina: a Golgi study. J. Comp. Neurol. 318: 147–187.

Kouyama, N. and Marshak, D. (1992). Bipolar cells specific for blue cones in the macaque retina. J. Neurosci. 12: 1233–1252.

Kouyama, N. and Marshak, D. (1993). Synaptic connections and spatial relationships of blue cones and blue cone bipolar cells in the macaque retina. Jpn. J. Physiol. 43: S220.

Lerea, C., Bunt-Milam, A. and Hurley, J. (1989). Alpha transducin is present in blue, green and red sensitive cone photoreceptors in the human retina. Neuron 3: 367–376.

Mariani, A.P. (1984). Bipolar cells in monkey retina selective for the cones likely to be blue-sensitive. Nature 308: 184–186.

Marshak, D., Aldrich, L., DelValle, J. and Yamada, T. (1990). Localization of immunoreactive cholecystokinin precursor to amacrine cells and bipolar cells of the macaque monkey retina. J. Neurosci. 10: 3045–3055.

Rodieck, R.W. (1988). The primate retina. In: Steklis, H.D., Erwin, J., (eds.), Comparative Primate Biology, Vol. 4, Neurosci.: 203–278, Alan R. Liss, New York.

David W. Marshak, Ph.D.
Department of Neurobiology and Anatomy
The University of Texas Medical School
P.O. Box 20708 Houston, TX, U.S.A.

34. Short-wavelength-sensitive cones: Morphology and color-specific connections

PETER K. AHNELT and HELGA KOLB
(Vienna, Austria and Salt Lake City, UT, U.S.A.)

Abstract

The paper reviews recent work on the morphological identification of mammalian short-wavelength-sensitive (S-) cones using LM- and EM-Golgi studies and its correlation with histochemical and immunhistochemical labeling techniques. S-cones comprise less than 10% of the total cone population. The S-cone inner segments are more slender yet longer, projecting further into subretinal space, and their synaptic terminals may be distinguished by criteria such as radial position, smaller size and scarcity of telodendrial contacts to neighboring cones.

Evidence from light- and electron- microscopic reconstructions demonstrates patterns of selective connections of the S-cones to horizontal cells and to an S-cone-specific bipolar cell type as compared to middle-wavelength-sensitive (M-) and long-wavelength-sensitive (L-)cones. Thus we have evidence from cyto-morphology, topography and circuitry for a basic dichromatic design at the first synaptic level of primate and human retinas.

Introduction

Color vision of vertebrates is based on two or more subtypes of cone photoreceptors. In recent years it has become evident that the S-cone system provides a key element for tracking the pathway for color related processing in the retina. The retinas of lower vertebrates show a variety of morphological cone photoreceptor subtypes. In the retina of goldfish as well as in turtle the morphology of cones and interneurons such as horizontal cells could be related to spectral sensitivity. Correlation with data from microspectrophotometry, electrophysiology and electron microscopy has led to models of first stages of color processing in the outer retina (Stell and Lightfoot, 1975; Kamermans *et al.*, 1991; Kolb and Lipetz, 1992).

Mammalian cone subtypes had been assumed to be morphologically indistinguishable (e.g. Polyak, 1941) and thus their spectral identity could not be

B. Drum (ed.), Colour Vision Deficiencies XII, pp. 285–297.
© 1995 *Kluwer Academic Publishers, Dordrecht.*

identified at a cellular level. Marc and Sperling (1976) introduced a histochemical approach to study the spectral topography. Monochromatic stimulation and subsequent incubation for succinate dehydrogenase activity resulted in accumulation of Formazan precipitates in the ellipsoidal mitochondria of cone subtypes in goldfish. By applying this technique on isolated Baboon retinas (Marc and Sperling, 1977) they were able to demonstrate a regular mosaic of presumptive S-cones. DeMonasterio and collaborators (1983) have developed another approach in Rhesus monkey. For unclear reasons a subpopulation assumed to be the S-cones tend to accumulate Procion dyes after injection into the vitreous.

The histochemical techniques provide valuable topographical information but investigation of detailed circuitry requires specimens with minimal alterations of normal morphology. Therefore we have searched for morphological criteria to distinguish between mammalian cone types including those of primates and man.

Materials and methods

Retinas of European ground squirrels were processed according to a modified protocol for NBT-histochemistry (Marc and Sperling, 1977). Human retinas were obtained immediately after enucleation for peripheral melanoma, or post mortem from corneal transplant donors and fixed in P-buffered glutaraldehyde/formaldehyde mixtures. Golgi material was obtained following the Colonnier method. For details see references (Ahnelt, 1985; Ahnelt *et al.*, 1987, 1990).

Results

Bimorphism in non-primate mammalian cones

Assuming that the presence of morphological differences should be most likely detectable in cone-dominant retinas we have studied the receptor mosaic of sciurid rodents reported to have protanopic dichromat vision (Jacobs *et al.*, 1976).

Semithin sections of the photoreceptors in the retina of the European ground squirrel (*Citellus citellus*, L.) did indeed reveal three types of photoreceptors. Besides a minor population of rods, a regular mosaic of about 7% of cones stands out by their larger inner segments. In particular the distal ellipsoidal region, consisting of densely packed mitochondria, is more voluminous. These cones also appear darker through higher affinity to standard stains such as Stevenel's Blue, a possible analogy to their specific Procion affinity. Similar features have been described in the retina of an American ground squirrel (Long and Fisher, 1983).

We were able to support the idea that these morphological differences are indeed associated with spectral classification by using the NBT-technique (Ahnelt, 1985). After long wavelength stimulation and incubation the larger subpopulation showed intense Formazan precipitation while the minor population of larger cones remained unlabeled (and vice versa). We are thus able to distinguish individual middle-wavelength-sensitive (M-) and S-cones in this mammal. Subsequently similar morphological differences have been found in prairie dog (Ahnelt, unpublished) and in another cone-dominant species, the tree shrew (Müller and Peichl, 1990).

Human L- and S-cones

Encouraged by the findings in sciurid mammals we proceeded to look for criteria to distinguish S-cones in human and primate retinas (Ahnelt *et al.*, 1987). Studying the periphery of retinal wholemounts systematically by focusing and comparing along the cone inner/outer segments revealed two cone varieties (Fig. 1a). The inner segments of the predominant type (L) are conical,

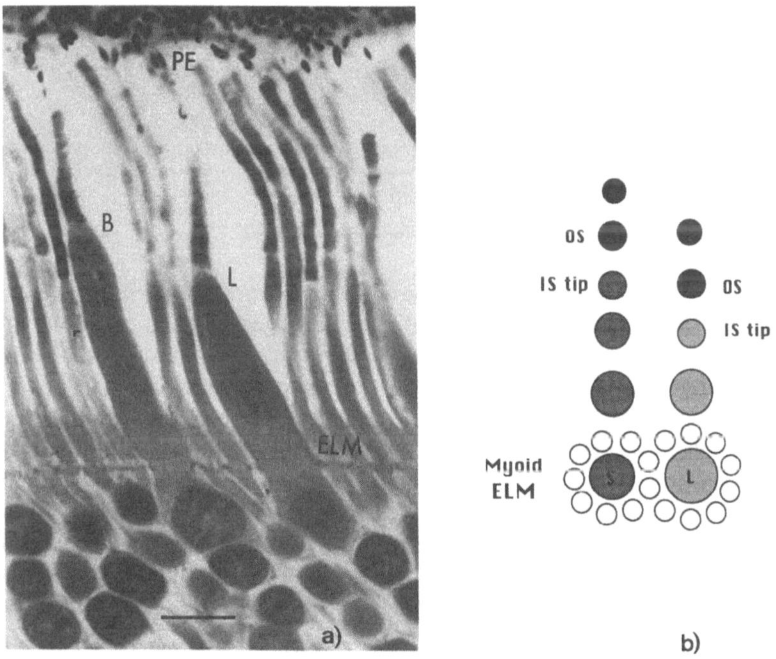

Fig. 1. Radial semithin section of human retinal photoreceptors (1μm, ca. 5 mm temporal from fovea, Stevenel's Blue staining). Short-wavelength-sensitive (B) and long-wavelength-sensitive (L) cone subtypes differ by their basal diameter at the external limiting membrane (ELM), shape and length of inner segment (IS) and position of outer segment (OS) tips. Graph on the right illustrates the appearance of consecutive tangential sections at corresponding levels r: rod, PE: pigment epithelial processes. Scale bar = 10μm.

as is familiar from standard textbook illustrations. These inner segments arise from the external limiting membrane with a large diameter similar to that of the cell body and then taper continuously towards the outer segment. A minority (ca. 8%) of cones (S) differs by having a longer, more cylindrical inner segment. Their base at the external limiting membrane is smaller and more slender. As a result of this difference S-cone cross sections appear smaller basally but larger distally when compared with their direct neighbors (Fig. 1b).

This paradox of a smaller diameter at the base, yet a longer and 'larger' appearance at the tip and the scarcity of blue cones may have been the reasons why this bimorphism has not been detected previously.

Studies using antibody labeling of S-cone outer segments have confirmed a correlation with these morphological features for rhesus monkey (Szél *et al.*, 1988) and human cones (Curcio *et al.*, 1991).

Table 1. Morphological characterization of human L- and S-cones.

Features	S-cones	L-cones
Outer segment	tips more distal	more proximal
Ellipsoid	longer, less tapering-	shorter, tapering
Myoid	tubular,	conical
	small aperture	larger aperture
Mosaic position	frequently irregular	usually regular
Affinity to basic dyes	high	normal
Affinity to Procion dyes	high	low

S- and L-cone mosaics do not complement one another

In addition to the differences in light microscopial cytomorphology peripheral S-cones frequently stand out topographically by their irregular position within the mosaic (Fig. 2). The peripheral cone mosaic is usually characterized by wide cone spacing. Several rows of rods may fill the space between adjacent cones. Both the L-cones and the members of the sparse S-cone population represent regular submosaics. However, the submosaics are frequently found to interfere. S-cones are found frequently at positions closer or even adjacent to neighboring L-cones.

While the distances between cones of the same subtype are regular, neighboring pairs of different subtypes often are found to be close or in direct contact without interspersed rods (Ahnelt *et al.*, 1987; Nork *et al.*, 1990; Curcio *et al.*, 1991).

Another complication is introduced in the fovea with its steep density gradient and a center which consists of cones only (Ahnelt *et al.*, 1987). At a certain level of the inner segments in semithin cross sections, the S-cone submosaic becomes discernible. Here their cross sections are larger (area = 180%) and more circular than those of the majority of cones. They are more frequent around the foveola (ca. 10%) while their density decreases towards its

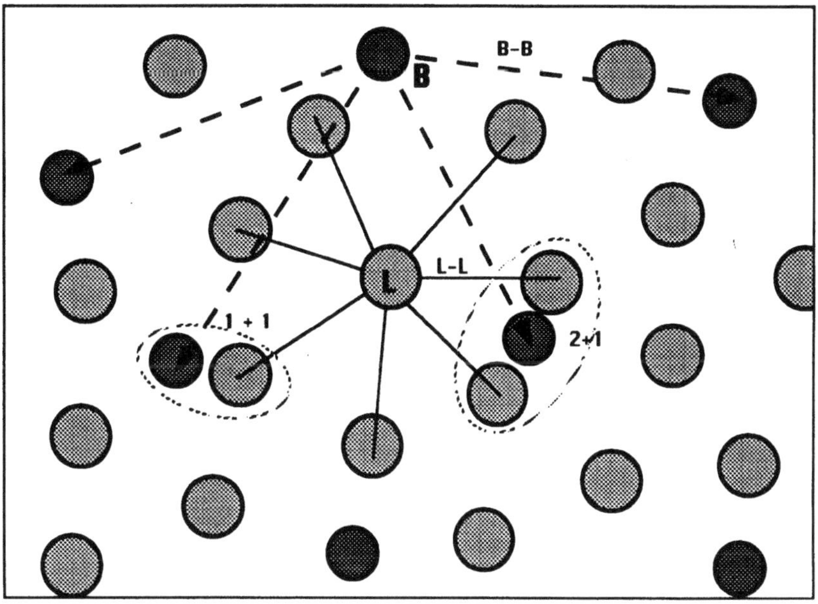

Fig. 2. Schematic drawing of humn peripheral cone mosaic organization. Both submosaics are regular but spacing for S-cones is > 200% of the L-cones. Interference between the two patterns may result in instances where S-cones lie directly near 1 or 2 L-cones (ellipses).

center. A small proportion (<3%) appears to be present even at the very center of the foveola (also for Baboon; see Marc and Sperling, 1977) but the complete absence of functional S-cones has also been reported (Curcio *et al.*, 1991). The question of existence and extension of a foveolar S-cone-free area is possibly complicated by individual and species specific variation and requires further investigation.

Foveal S-cones are also significantly associated with irregular (non-hexagonal) positions as can be shown by analysis of the foveal crystalline packing order (Pum *et al.*, 1991). We tend not to attribute this phenomenon solely to the size difference but mainly to differences during foveal ontogeny.

Cone terminals

Cones terminate, forming so-called pedicles, at the outer plexiform layer. We have studied cone and rod terminals in the near periphery of a human retina (Ahnelt *et al.*, 1990). Fig. 3 shows the outlines of photoreceptors and their connections in such an area.

S-cone pedicles of the near periphery clearly differ from those of L-cones by having a smaller size and by penetrating somewhat more deeply into the plexiform layer. S-cone pedicles have similar numbers of synaptic ribbon complexes (ca. 20) but the ribbons are shorter.

Most strikingly, S-cone pedicles are more roundish and isolated from their neighbors. On average each L-cone is connected to about 3–4 neighboring cones by telodendrial junctions (dots on Fig. 3). S-cones (B), however, are almost completely isolated from their neighbors. In only two of sixteen reconstructed pedicles were we able to identify such contacts. Due to their larger spacing, there will virtually never be direct contact between S-cones. S-cones on average had 3, and other cones had 5, rod contacts (triangles in Fig. 3) indicating no significant difference in that respect. We have not found direct rod-to-rod contacts.

Only recently have we analyzed pedicles in the far periphery. At ca. 13 mm from the fovea we found S-cones extending their horizontal range by the formation of two or three lobes. These lobes reach across 30 μm and are not telodendria since each has a set of synapses. This strategy may allow the dispersed S-cone terminals to reach equally dispersed components of the neuronal subsystem. L-cone pedicles retain their normal roundish shape at this eccentricity.

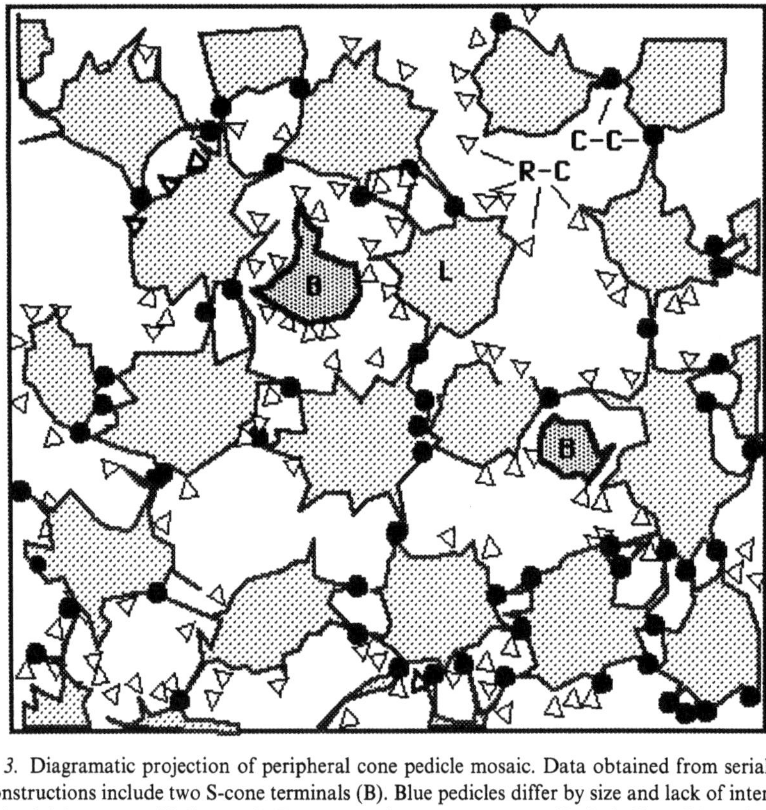

Fig. 3. Diagramatic projection of peripheral cone pedicle mosaic. Data obtained from serial EM reconstructions include two S-cone terminals (B). Blue pedicles differ by size and lack of intercone telodendrial contacts (C-C, spots). Punctate rod-to-cone contacts (R-C) indicated by triangles) occur at both cone subtypes.

Horizontal cell connections

Our recent efforts (Kolb *et al.*, 1989, Kolb and Ahnelt, 1994; Ahnelt and Kolb, 1994) have focused on the first sites of synaptic interaction in the outer plexiform layer. Lateral interaction is assumed to be mediated by horizontal cells (Fig. 4) terminating with their dendrites at the lateral position of triadic synaptic complexes. In primate retinas there are at least two subtypes of horizontal cells HI and HII (Boycott and Kolb, 1973; Kolb *et al.,* 1980). Recently a Type III has been differentiated from Type I according to significant differences in size and shape (Linberg *et al.,* 1987).

We have investigated the relations of subtypes of cones and horizontal cells at both light microscopical and EM-levels (Ahnelt and Kolb, 1994a,b). One approach based on light microscopy of Golgi-labeled cells is to project the cone inner segment mosaic on the dendritic clusters of underlying horizontal cells. This allows us to identify the likely position of corresponding pedicles. If chromatic preferences exist, L- or S-cone positions could be contacted, missed or even avoided by the dendritic clusters of the horizontal cell subtypes.

HI dendrites (Fig. 4, upper right) usually form clusters of terminals that correspond to invaginating synaptic contacts with cones. These clusters cover most of the overlying pedicle sites. Only the rare positions of S-cone pedicles are ignored by nearby dendrites. Serial reconstruction at electron microscopical resolution has confirmed these patterns. While L-cones may be loaded with invaginating contacts from an underlying HI cell, the S-cone pedicles are avoided. In 2 out of 12 pedicles situated above reconstructed HI cells a total of only 3 invaginating dendrites could be identified. Considering that every single pedicle has ca. 20 triadic synapses, this implies the existence of only sporadic connections between single HI cells and S-cones (Fig. 5, upper graph). Even taking into account the spatial overlap of several HC dendritic fields only a minor proportion (ca. 10%) of the synapses of a S-cone terminal is established with HI cells. Since this ratio has to be multiplied with the general sparsity of S-cones, the difference in the number of contacts provided by the HI plexus to S-cones and L-cones is about 1:500, or at least two orders of magnitude.

The newly differentiated HIII type (Fig. 4, bottom) does not seem to contact S-cones at all. EM reconstructions showed cases where S-cone pedicles were deformed into irregular or bi-lobed shapes by an underlying HIII dendrite without receiving any contact from it.

The HII cell (Fig. 4, upper left) clearly differs from the two previous types. It has thin dendrites that wind along the inner border of the OPL. They form delicate tips which are less easy to detect than the knobby swellings of the other types. Nevertheless, one is able to discern clustering at some locations and sometimes dendrites converge beyond the main dendritic field towards a presumptive pedicle location. In such cases, we have found S-cone inner segments at the predicted positions. A more precise analysis is possible by serial EM-sectioning. HII cell dendritic fields may contain 20–35 cones including

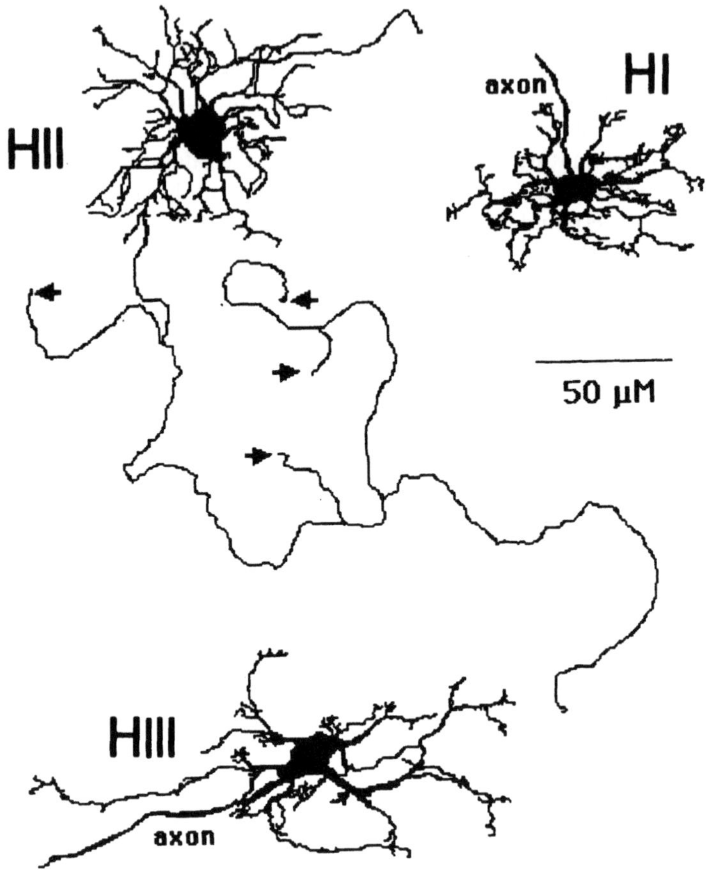

Fig. 4. Schematic illustration of horizontal cell subtypes (Camera lucida drawings from Golgi-processed human retinal wholemounts). HI and HIII have clusters of dendritic terminals indicating cone pedicle positions. The axons of HI and HIII cells (assumed to be rod-driven) are not drawn completely. Terminals of HII collaterals (arrows) have been found to contact S-cones exclusively.

1–4 S-cones. All cones receive HII dendritic contacts but the number is significantly larger to the S-cones. S-cones may receive 7–9 dendritic terminals as compared to 3–5 for L-cones at the same eccentricity.

HII horizontal cells also have an axon-like process that meanders away from or around the dendritic field. Occasional collaterals formed by this axon contact cones only. When tracing several of these axons to contacted ones we found that contacts are invariably established with S-cones. In one case such a blue pedicle could be shown to be targeted simultaneously by a labeled HII and a S-cone bipolar cell (Mariani, 1984). A telodendria-like process of the pedicle was found to run parallel to the HII collateral, even bending to briefly follow the axon itself. In our view this is an example for the strong affinity between S-cone terminals and their specific interneurons.

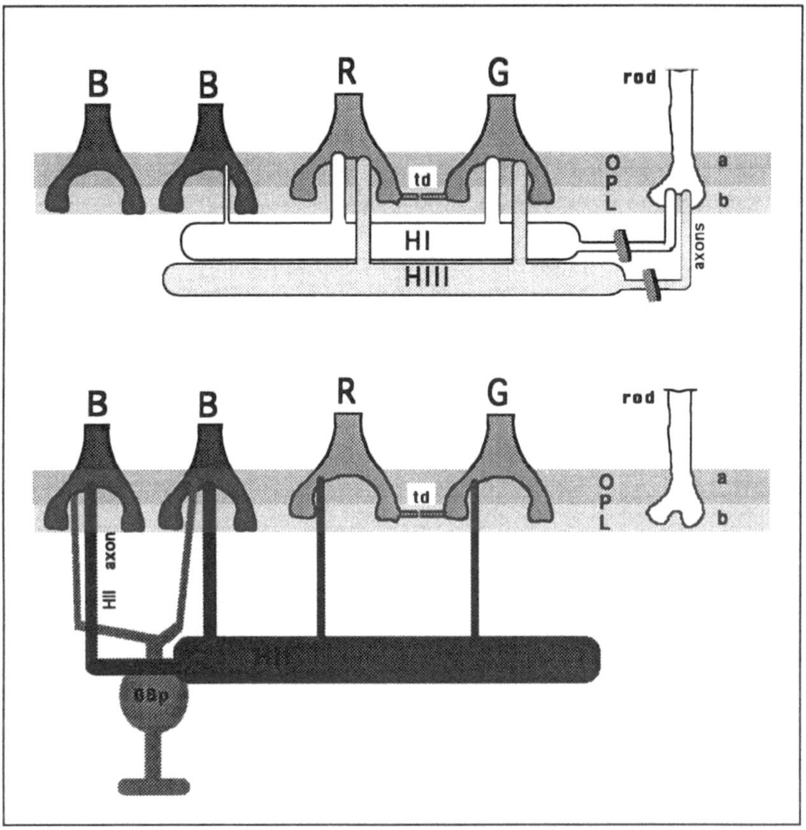

Fig. 5. Schematic radial view of horizontal cell/photoreceptor connections in human retina. S-cones (B) penetrate somewhat more deeply into the outer plexiform layer (OPL). An extensive telodendrial network (td) connects L- (R + G) cones but not S-cones. Rods are contacted by axon terminals of HI and probably also HIII cells. HI and HIII cells (upper graph) are almost exclusively associated with the pedicles of L- (R + G) cones contributing to the distal sublayer a of the OPL. The HII cell dendrites (lower graph) converge preferentially to S-cones, but also connect with L-cones. The HII axon collaterals samples S cones exclusively. The invaginating S-cone bipolar also contacts S-cones only and contributes to the proximal sublayer of he OPL.

In summary (Fig. 5, lower graph), it is the HII cell system that provides the majority of lateral triadic contacts to S-cones. The delicacy of the HII's terminals may be the reason for the dominance of smaller synaptic ribbon sizes in these S-cone pedicles.

Discussion

Some implications of S-cone morphology

The present data show that morphological differences exist between two basic subtypes of cones in a number of species of placental mammals including man. Interestingly, the S-cones of ground squirrels and tree shrews are larger than the L-cones, while primate S-cones are longer, but more slender except in the fovea. The increase in density of the foveal mosaic during development appears to affect L-cones more strongly than S-cones. Whether S-cones and the features found for human retina are present in all diurnal primates (see however Wikler and Rakic, 1990) requires systematic study. The extent of the morphological differences does appear to vary: it is less pronounced in Rhesus while it is very evident in Squirrel monkeys (Ahnelt, unpublished observations). Inner segments may be viewed as light collectors. Based on a theoretical study (Horowitz, 1981) an altered shape as described here for the (peripheral) S-cones may have specific functional consequences:

- collection of directed light will be less efficient, so that directional sensitivity may be less than in L-cones,
- the absorption peak will shift to shorter wavelengths and
- the absorption curve will broaden

It should be possible to test this prediction experimentally.

The finding of a bimorphism is in good agreement with the recent evidence for an ancient genetic separation of the S-cone pigment as revealed by genetic sequencing. The morphological differences may merely be a result of slow random drift, but they may also be a result of selective pressure due to spectrally different energies and optics. Trichromacy as manifested in red and green cone pigments arose only recently by minor mutations in primates and may not have yet resulted in qualitative morphological differences.

Mosaic and spatial vision

The mosaic organization may have consequences for visual functions such as resolution, color or computation of movement (Marc, 1983). The two major subpopulations of the mammalian mosaic are spaced regularly but they do not cooperate to create an optimized spatial sampling matrix. Instead a general strategy of vertebrate color vision has evolved: a sparse S-cone mosaic leaves most sampling positions to the L-cones (and rods) while providing chromatic hue information at low spatial frequency (Williams *et al.,* 1983). The apparent independence of the two cone submosaics points to different spatial and/or temporal patterns of differentiation, possibly from different precursor cells.

Previous studies based on general topographic correlations (Boycott *et al.*, 1987; Wässle *et al.*, 1989) were not able to reveal the minute yet distinct patterns originating from the two major subtypes of primate cones. We attribute our results mainly to the new possibility of identifying individual cones at both the light- and electron microscopic level.

As summarized in the diagrams of Fig. 5 there is a discrete subsystem related to the primate S-cones which cannot be compared directly to patterns found in lower-vertebrate retinas. It manifests itself in the existence of qualitatively specific subtypes of interneurons (S-cone bipolar cells, HII horizontal cells,...? and results in a bi-layering of the outer plexiform layer. The proximal b-layer contains the S-cone terminals, HII processes and related bipolar dendrites. The L-system and the majority of rod terminals occupies what we have proposed to call the distal a-sublayer.

The only intracellular recordings so far available are those from primate HI cells (Dacheux and Raviola, 1990), and their response to chromatic stimuli showed a single maximum at long-wavelengths as one would expect for a luminosity system (Fig. 5, upper graph).

On the role of the HII cell we may only speculate at present. One viewpoint is that the axon serves to pool the dispersed S-cone signals. Since the spacing of S-cones does not allow for direct communication via telodendria, the HII axons may provide such a network.

Blue illumination may propagate a pooled S-cone response via the S-cone on-bipolar. The radial transmission of the surround response could be mediated by the HII central connections with L-cones or by the other horizontal cell system and by L-cone off-bipolars. This could require a bi-stratification of the associated ganglion cell.

S-cone pedicles are isolated almost entirely from their neighbors (Fig. 3). Some averaging of potential changes in neighboring cones via lateral telodendrial connections as may be the case for the L-system, may not be available for S-cones. We need to know more about the other elements involved in color processing, the direction of information flow and its modulation (e.g. at gap junctions) before we can evaluate the definite role of this framework for different aspects of color vision. Certainly some differential cone weighting or perhaps even antagonism is already provided by hardwiring at the first synaptic level in the outer plexiform layer.

Acknowledgements

Parts of the study on above results have been supported by Austrian Science Foundation grants 4520 and 6412 and the Dreher Foundation to P.A. and by N.E.I. grant EY03323 to H.K.

References

Ahnelt, P.K. (1985. Characterization of the color-related receptor mosaic in the ground squirrel retina. Vision Res. 25: 1557–1567.

Ahnelt, P.K., Keri, C. and Kolb (1990). Identification of pedicles of putative blue-sensitive cones in the human retina. J. Comp. Neurol. 293: 39–53.

Ahnelt, P.K., Kolb, H. and Pflug, P. (1987). Identification of a subtype of cone photoreceptor, likely to be blue sensitive, in the human retina. J. Comp. Neurol. 255: 18–34.

Ahnelt, P.K. and Kolb, H. (1994a). Horizontal cells and cone photoreceptors in primate retina: a Golgi-light microscope study of spectral connectivity. J. Comp. Neurol.: 387–405.

Ahnelt, P.K. and Kolb, H. (1994b). Horizontal cells and cone photoreceptos in primate retina: a Golgi-electron microscope study of spectral connectivity. J. Comp. Neurol.: 406–427.

Boycott, B.B., Hopkins, J.M. and Sperling, H.G. (1987). Cone connections of the horizontal cells of the rhesus monkey's retina. Proc. R. Soc. Lond. B 229: 345–379.

Boycott, B.B. and Kolb, H. (1973). The horizontal cells of the rhesus monkey retina. J. Comp. Neurol. 148: 115–140.

Curcio, C., Allen, K.A., Sloan, K.R., Lerea, C.L., Hurley, J.B., Klock, I.B. and Milam, A.H. (1991). Distribution and morphology of human cone photoreceptors stained with anti-blue opsin. J. Comp. Neurol. 312: 610–624.

Dacheux, R.F. and Raviola, E. (1990). Physiology of HI horizontal cells in the primate retina. Proc. R. Soc. Lond. B 239: 203–221.

DeMonasterio, F.M., Schein, S.J. and McCrane, E.P. (1981). Staining of blue-sensitive cones of the Macaque retina by a fluorescent dye. Science 213: 1278–1281.

Horowitz, B.R. (1981). Theoretical considerations of the retinal receptor as a waveguide. In: Enoch, J. and Tobey, F.L., Jr. (eds.), Vertebrate Photoreceptor Optics: 219–300, Springer, Berlin.

Jacobs, G.H., Tootell, R.B.H., Fisher, S.K. and Anderson, D.H. (1976). Spectral sensitivity and colour vision in the ground dwelling sciurids: results from Golden mantled ground squirrel and comparison for five species. Anim. Behav. 26: 409–421.

Kamermans, M., Van Dijk, B.W. and Spekreijse, H. (1991). Color opponency in cone-driven horizontal cells in carp retina. J. Gen. Physiol. 97: 819–843.

Kolb, H. and Lipetz, L.E. (1992). The anatomical basis for colour vision in the vertebrate retina. In: Cronly-Dillon, J. (ed.), Vision and Visual Dysfunction VI, Gouras, P. (ed.), The Perception of Colour: 128–145, Macmillan Press, London.

Kolb, H., Ahnelt, P.K., Fisher, S.K., Linberg, K.A. and Keri, C. (1989). Chromatic connectivity of the three horizontal cell types in the human retina. Invest. Ophthalmol. Vis. Sci. (Suppl.) 30: 348.

Kolb, H., Mariani, A. and Gallego, A. (1980). A second type of horizontal cell in the monkey retina. J. Comp. Neurol. 189: 31–44.

Linberg, K.A., Fisher, S.K. and Kolb, H. (1987). Are there three types of horizontal cells in the human retina? Invest. Ophthal. Vis. Sci. (Suppl.) 28: 262.

Long, K.O. and Fisher, S.K. (1983). The distribution of photoreceptors and ganglion cells in the California ground squirrel, *Spermophilus beecheyi.* J. Comp. Neurol. 221: 329–340.

Marc, R. (1983). Chromatic organization of the retina. In: LaVail, M. and Hollyfield, J. (eds.), Cell Biology of the Eye: 435–473, Academic Press, New York.

Marc, R. and Sperling, H.G. (1976). Color receptor identities of goldfish cones. Science 191: 487–489.

Marc, R. and Sperling, H.G. (1977). The chromatic organization of primate cones. Science 196: 454–456.

Mariani, A.P. (1984). Bipolar cells in monkey retina selective for cones likely to be blue-sensitive. Nature 308: 184–186.

Müller, B. and Peichl, L. (1989). Topography of cones and rods in the tree shrew. J. Comp. Neurol. 282: 581–594.

Nork, M.T., McCormick, S.A., Chao, G.-M. and Odom, V.J. (1990). Distribution of carbonic anhydrase among human photoreceptors. Invest. Ophthalmol. Vis. Sci. 31: 1451–1458.

Polyak, S.L. (1941). The Retina. University of Chicago, Chicago.
Pum, D., Ahnelt, P.K. and Grasl, M. (1990). Iso-orientation areas in the foveal cone mosaic. J. Vis. Neurosci. 5, 511–523.
Stell, W.K., Lightfoot, D.O., Wheeler, T.G. and Leeper, H.F. (1975). Goldfish retina: Functional polarization of cone horizontal cell dendrites and synapses. Science 190: 989–990.
Szél, A., Diamantstein, T. and Röhlich, P. (1988). Identification of the blue sensitive cones in the mammalian retina by anti-visual pigment antibody. J. Comp. Neurol. 273: 593–603.
Wässle, H., Boycott, B.B. and Röhrenbeck, J. (1989). Horizontal cells in the monkey retina: cone connections and dendritic network. Eur. J. Neurosci. 1: 421–435.
Williams, D.R., Collier, R.J. and Thompson, B.J. (1983). Spatial resolution of the short wavelength mechanism. In: Mollon, J.D. and Sharpe, L.T. (eds.), Colour Vision: 487–503, Academic Press, London.

Peter K. Ahnelt
Dept. General & Comparative Physiology
Medical School, University of Vienna
Schwarzspanierstr. 17, A-1090-Vienna, Austria

35. Assessment of S-cone sensitivity

JOEL POKORNY and VIVIANNE C. SMITH
(Chicago, Il., U.S.A.)

Abstract

Several recent analyses of the Short-wavelength-sensitive (S-) cone threshold vs. retinal illuminance (TVR) function suggest that it should be possible to differentiate a number of potential types of insult with psychophysical procedures. An assumption frequently made in modelling sites of disease action is that S-cone sensitivity is independent of the activity of L- and M-cones. We review increment threshold S-cone isolation techniques that have been used in clinical measurements. These experimental conditions produce S-cone sensitivity estimates which include sensitivity regulation of the S-cone system by the L- and M-cone types. TVR functions measured with a discrimination paradigm along a tritan line yield a different pattern of sensitivity regulation by L- and M-cones. For some purposes paradigms incorporating equiluminant stimuli along tritan lines may be of utility in estimating S-cone sensitivity.

Introduction

Since the pioneering work of Marion Marré beginning in the early 1970's, spectral sensitivity functions obtained under conditions of selective chromatic adaptation have been used extensively in clinical populations both to aid in diagnosis and to understand the functional consequences of disease processes. Marion and Ernst Marré (Marré and Marré, 1978) noted that all disease states which produce sensitivity losses under selective chromatic adaptation conditions involve a reduction in the sensitivity of the S-cone mechanism. This reduction can be selective in only affecting S-cone sensitivity (as in diabetes, Greenstein *et al.*, 1992); relatively selective (as in central serous choroidopathy, Alpern, *et al.*, 1983; Marré, 1972) or concomitant with reduced sensitivity for the L- and M-cone mechanisms (as in juvenile macular degeneration, Marré and Marré, 1978).

One of the great strengths of chromatic adaptation methodologies is the apparent robustness of findings with widely varying stimulus conditions.

B. Drum (ed.), Colour Vision Deficiencies XII, pp. 299–308.
© 1995 *Kluwer Academic Publishers, Dordrecht.*

Table 1. Experimental adaptation conditions employed in some clinical studies of S-cone sensitivity.

Reference	Approximate dominant wavelength	Approximate retinal illuminance (TD)
Marré, 1972	582	1,500
Went and deVries-deMol, 1976	593	12,500[a]
King-Smith *et al.*,, 1980	586	10,000
Adams *et al.*, 1987	586	2,200[a]
Lutze *et al.*, 1987	599	50,000
Zisman and Adams, 1982	589	31,600
Ohta *et al.*, 1982	593	14,000
Greenstein *et al.*,	600	900
		9,000
		15,800
Kalloniatis and Harwerth, 1989	586	10,000

[a] Estimated from published luminance levels and conversion factors given by LeGrand (1968).

Researchers have used a variety of experimental situations all directed to measuring the sensitivity of the S-cone system. Table 1 shows a survey of some adaptation conditions described in the literature. To create this table, we have limited the selection to articles published in the Proceedings of the IRGCVD. This list is not meant to be comprehensive, its purpose is to indicate the diversity of conditions that have been employed. Estimates of dominant wavelengths of the background fields are approximate; the majority of the studies employed glass or gelatin cut-on filters whose published characteristics are typical but not necessarily representative of individual filters.

The intent of this communication is 1) to explore a model directed at localizing the locus of insult in the S-cone pathway, 2) to describe what aspects of S-cone function are measured by various experimental techniques and 3) to ask which of these procedures are most suitable for clinical measurement of S-cone sensitivity.

Models of sensitivity loss

Several recent reports have sharpened the analysis of selective chromatic adaptation techniques. Early in the chain of events leading to vision, factors operate to preserve and maintain discrimination over a large dynamic range (Boynton, 1979; Hood and Finkelstein, 1986). These factors include multiplicative adaptation which scales the input signals to extend the dynamic range of vision, and subtractive adaptation which removes much of the adapting signal. The retinal signal itself may be subject to saturation, a property revealed for cones by pulse-on-pulse paradigms. By multiplicative adaptation we include factors such as the pupil and a physiological mechanism that changes its transfer properties from linear at low illumination levels (ΔI = a constant) to Weber behavior ($\Delta I/I$ = a constant) at high illumination levels.

Processes active prior to or following the retinal mechanisms regulating sensitivity

Measurement of the threshold-vs-retinal illuminance (TVR) function has the potential to differentiate between potential loci of disease insult. Two patterns of sensitivity change with luminance have been attributed to eye disease (Greenstein, *et al.*, 1987; Hood and Greenstein, 1988, 1990; Yeh, *et al.*, 1989): a reduction in absolute sensitivity (shift of the TVR function along the 45° diagonal) and/or a reduction of increment sensitivity (upward vertical shift in the TVR function). Absolute sensitivity loss would occur for any insult occurring prior to the physiological site at which the TVR function is determined. This is typically envisioned as a loss in quantum catch efficiency (Fig. 1, top). An

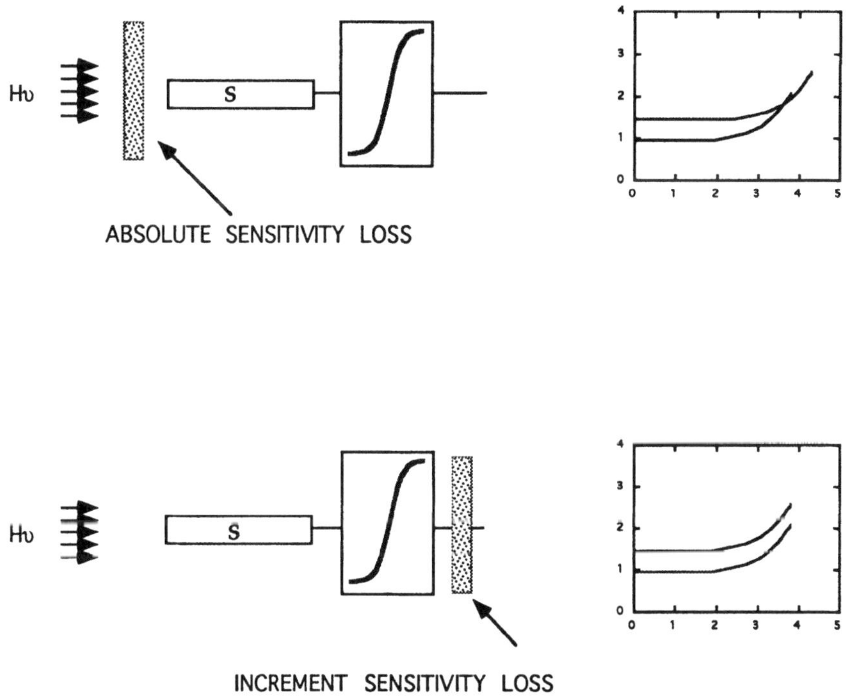

Fig. 1. Top: Characterization of absolute sensitivity loss. A loss occurring prior to the locus where the TVR function is generated affects sensitivity to both the background and the increment test. Thus at background levels which do not raise threshold, this mechanism produces a loss of sensitivity. If the background level is sufficiently high so that thresholds are determined by contrast (Weber behavior, $\Delta I/I$=C), the absolute sensitivity loss mechanism affects background and test equally and the contrast is not changed.
Bottom: Characterization of increment sensitivity loss. A loss occurring subsequent to the locus where the TVR function is generated affects sensitivity to the increment test at all background levels.

example is the sensitivity reduction that accompanies cone receptor disorientation though a multitude of receptoral insults may produce a similar TVR picture (Hood and Greenstein, 1990). In these cases, absolute thresholds would be elevated but thresholds would appear normal at high adaptation levels where test threshold is proportional to field level (Weber region). Increment sensitivity loss may occur when the physiological insult is at a locus more central than where the TVR function is determined. An example would be the loss in sensitivity accompanying optic nerve disease. Here test thresholds would be elevated at both absolute threshold and high field levels since the site of sensitivity loss is subsequent to the locus of generation of the TVR function (Fig. 1, bottom).

Processes active within the retinal mechanisms regulating sensitivity

A defect may occur within the multiplicative gain mechanism itself. In this case absolute thresholds would remain normal, but increment thresholds would accelerate from the Weber line, resembling those measured with pulsed adaptation fields. Similarly, a defect in the subtractive gain mechanism itself would result in normal absolute thresholds with increment threshold rising with the unit Weber slope at lower steady adaptation levels compared with normal. We are not aware of literature concerning insults to processes affecting the retinal mechanisms regulating sensitivity and we will not consider them further here.

Current measurement techniques for assessing S-cone sensitivity

The measurements of sensitivity from the diverse experimental conditions described in Table 1 each represent single points on TVR functions describing the relation between S-cone threshold and long-wavelength background light level. How do the experimental conditions shown in Table 1 map onto TVR data in the literature? Figure 2 shows a theoretical template which summarizes threshold data for the $\pi 1$ and $\pi 3$ mechanisms collected by Stiles (1946, 1953) and Pugh and Mollon (1979) for field chromaticities of 580–610 nm and test wavelengths of 420–490 nm. In this and the following figure, the quantities of light for both field and test expressed in units of S-cone trolands (Boynton and Kambe, 1980; Smith *et al.*, 1993b). This metric, defined by the efficiency of different lights in exciting S-cones, allows an economical format to view data obtained with diverse long-wavelength field and short-wavelength test chromaticities. The Pugh and Mollon (1979) interpretation of the π mechanisms included the ideas that the $\pi 1$ branch was mediated by a post-receptoral opponent channel influenced by activity of L- and M-cones and the $\pi 3$ branch represented adaptation within the S-cone pathway before the opponent site.

It is not possible to plot in a sensible way the experimentally determined delta S-cone trolands from the various studies cited in Table 1 because of

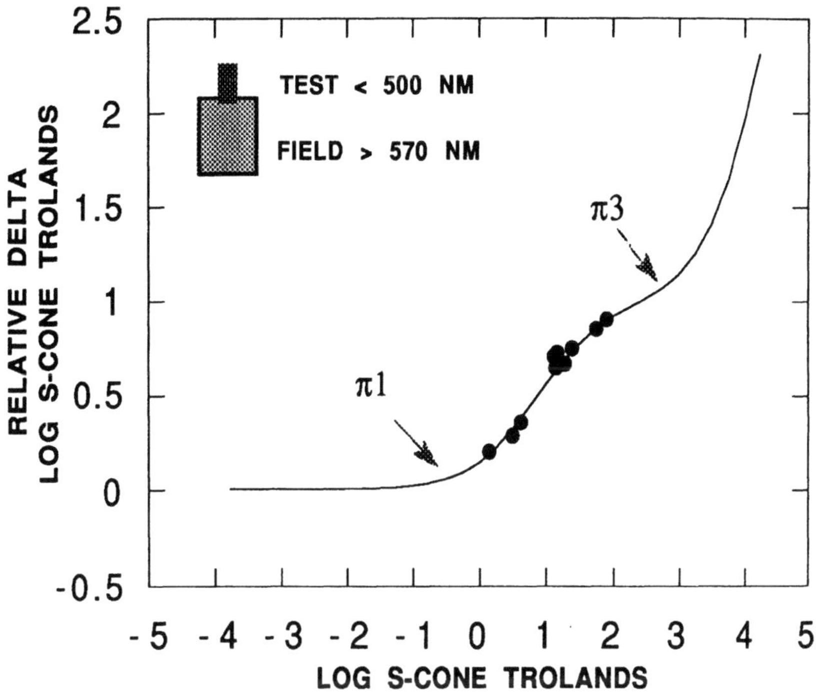

Fig. 2. Template for increment threshold vs, retinal illuminance function for a short wavelength (460–490 nm) increment on a long wavelength (580–610 nm) background. Both field and test levels are expressed in units of S-cone trolands. The superimposed filled circles represent the thresholds expected from a typical observer for the background levels employed in the experiments cited in Table 1.

experimental differences in field size and test duration. Further, many of the studies expressed thresholds in relative units. To circumvent these problems we plot as circles the sensitivity of the template observer to the various backgrounds used by the experimenters in Table 1. The plot of Fig. 2 can be thought of as representing the behavior of a typical color-normal observer under a standard set of experimental variables.

The purpose of expressing test sensitivity data in the format of Fig. 2 is to show that for the typical clinical conditions of S-cone sensitivity measurement, the measurements fall in the $\pi 1$ region of the TVR function. The thresholds thus represent a region where S-cone sensitivity is regulated not by S-cone activation but principally by L- and M-cone activation (Pugh and Mollon, 1979). Thus clinical changes in sensitivity of L- and M-cones may affect estimates of S-cone sensitivity defined by field sensitivity techniques. This complicates the analysis of threshold loss models described above.

TVR functions measured with discrimination paradigms

There is an alternative technique for obtaining TVR functions which yields further insight into the nature of the mechanisms governing S-cone mediated thresholds. If a discrimination threshold is measured along a tritan line in an equiluminant plane, the result may be expressed as a difference in S-cone stimulation relative to the S-cone activation produced by the starting chromaticity. Yeh *et al.* (1993) show that changes in S-cone activation achieved by variation of chromaticity (yellow to white to violet along a tritan line) and by variation of luminance (low- to high-luminance background) do not result in equivalent changes in sensitivity. Figure 3 shows a family of templates for TVR functions derived from discrimination data (Yeh *et al.*, 1993; Smith *et al.*, 1993a). Note that the pattern of template shifts with luminance appear the same as that shown in Fig. 1 for an absolute sensitivity loss. Here the explanation of the reduction in S-cone sensitivity is not associated with a change in quantal sensitivity of the cone type; but rather as due to a regulation of S-cone sensitivity by the L- and M-cone types early in the visual system. As Yeh *et al.* (1993) pointed out, the most parsimonious model for this behavior involves interaction

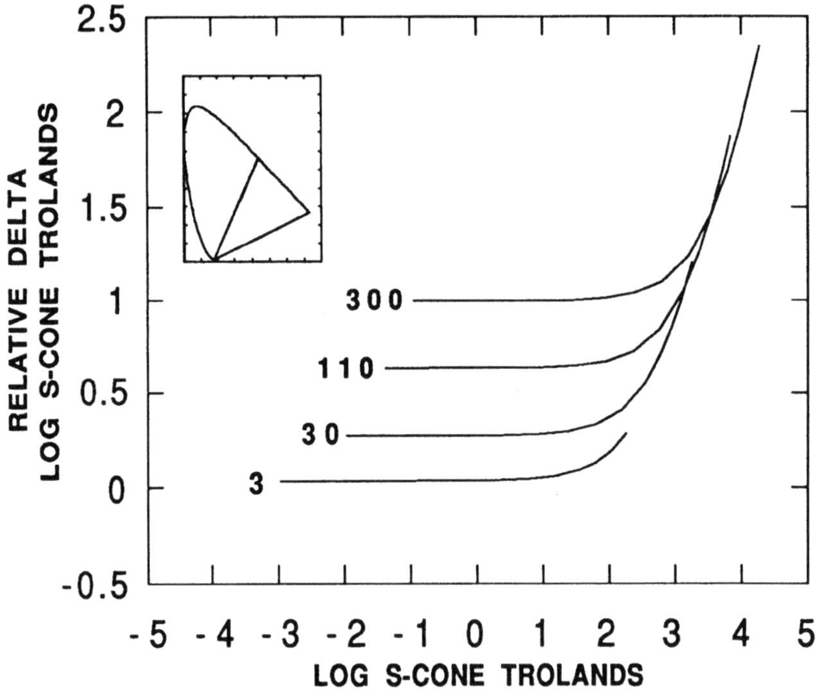

Fig. 3. Templates for discrimination data obtained along a tritan line at different retinal illuminance levels. The metric for the starting chromaticity and the difference in S-cone excitation between the starting and the just discriminable chromaticities are in S-cone trolands.

between the S- and the L- and M-cone types prior to the physiological site at which the TVR function is determined.

The templates of Figs. 2 and 3 representing increment threshold and discrimination data are plotted in the same coordinate system, ΔS-cone trolands vs. S-cone trolands. The two types of data traverse the stimulus space in different ways. If the two were compatible, the Stiles $\pi 1$ function would follow the path defined by points near the left most origins of each of the family of discrimination functions (the origins are defined in this plot for a wavelength of 570 nm). This is clearly not the case. At intermediate S-cone troland levels the $\pi 1$ function sensitivity is substantially greater than that expected from the functions derived from discrimination data. The dynamic range of the $\pi 1/\pi 3$ and discrimination functions is virtually the same. The functions converge at low S-cone retinal illuminances where S-cone sensitivity is not affected by L- and M-cone excitation and at high S-cone troland levels where sensitivity regulation is determined by S-cone activation levels for both the $\pi 3$ (direct S-cone activation by the long-wavelength background) and the discrimination (high levels of S-cone activation on tritan lines) functions. We believe that the disparate sensitivities at intermediate S-cone troland levels originate from the different spatial configurations for the two paradigms. Figures 2 and 3 are plotted with the same aspect ratio so that the interested reader may photocopy and compare them.

Strategies for measurement of S-cone sensitivity

As outlined above, the techniques frequently used to assess clinically S-cone sensitivity involve S-cone isolation procedures which include S-cone sensitivity regulation by the L- and M-cone types. As a consequence, S-cone sensitivity estimates may be biased by the consequences of disease processes affecting the L- and M-cone types. For example, photoreceptor disorientation, a process which yields an absolute sensitivity loss for all cone types would result in an underestimation of S-cone sensitivity loss. The reduction in S-cone sensitivity due to decreased S-cone quantal catch would in part be compensated for by a decrease in the efficacy of L- and M-cone quantal catch in the background.

In assessing S-cone sensitivity, it is necessary to decide whether information is needed about unadapted sensitivity (sensitivity in the linear region of the TVR function), sensitivity in the Weber region and/or sensitivity under conditions where the S-cone system begins to go into saturation. Discrimination measurements on tritan lines (fig. 3) allow estimates of unadapted sensitivity with low luminance levels and Weber sensitivity with higher light levels. With this technique it is possible to move well into the Weber region at moderate luminance levels, levels that can be accomplished on CRTs.

There are two potential issues in the use of discrimination techniques to assess S-cone sensitivity. The first concerns the possibility of rod participation in the low luminance discriminations. Rod intrusion was not evident in our data for normal observers under the foveal viewing conditions but needs to be evaluated in clinical populations. The second potential issue involves the use of a representative observer (e.g. the Judd observer) to set tritan lines. The alternative is to establish the tritan line of each individual observer. Some possible problems of using a representative tritan line include 1) differences in sensation luminance among observers and 2) variation in the tritan confusion lines caused by movement of the tritan copunctal point. Smith and Pokorny (in press) review possible sources of variability and consider their effect on chromatic discrimination. For example, variation of the tritan copunctal point could be caused by lens and macular pigment absorption and variation in sensation luminance could be caused by variation in the L- and M-cone populations or by polymorphism in the L- and M-cone photopigments. For normal variation in such factors, calculations indicate that there should be no problem in the use of a standard observer metric for the collection of chromatic discrimination data. However, these factors may become significant for observers who have substantially impaired S-cone discrimination. In these cases, highly abnormal S-cone discrimination thresholds would be measured. If quantitative information about discrimination loss is desired, the individual's tritan line could be established and the S-cone discrimination loss could be reevaluated.

In situations where it is desirable to operate in the Weber region of the S-cone TVR function, placing stimuli on a tritan line is an effective isolation strategy. An example where this technique could be useful is in the measurement of S-cone spatial contrast sensitivity (Swanson *et al.*, 1994). A final paradigm which warrants further evaluation is the pulse-on-pulse paradigm. It is unknown if S-cone sensitivity in the Weber region measured with this technique is comparable to that measured with steady backgrounds.

Acknowledgements

Supported in part by NEI grants EY00901 and EY07390. We thank Donald Hood, Kenneth Knoblauch, Arthur Shapiro and William Swanson for comments on a draft of the manuscript.

References

Adams, A.J., Huie, K.E., Schefrin, B.E., Bresnick, G.H. and Zisman, F. (1987). A simple clinical test of blue cone sensitivity in early eye disease. In: Verriest, G. (ed.), Colour Vision Deficiencies VIII, Doc. Ophthalmol. Proc. Ser. 46: 237–242, Martinus Nijhoff/Dr W. Junk, Dordrecht.

Alpern, M., Kitahara, K. and Krantz, D.H. (1983). Classical tritanopia. J. Physiol. (Lond.) 335: 655–681.

Boynton, R.M. (1979). Human Color Vision. Holt, Reinhart and Winston, New York.

Boynton, R.M., and Kambe, N. (1980). Chromatic difference steps of moderate size measured along theoretically critical axes. Color. Res. Appl. 5: 13–23.

Greenstein, V.C., Hood, D.C. and Carr, R.E. (1989). A comparison of S cone pathway sensitivity loss in patients with diabetes and retinitis pigmentosa. In: Drum, B. and Verriest, G. (eds.), Colour Vision Deficiencies IX, Doc. Ophthalmol. Proc. Ser. 52: 233–241, Kluwer Acad. Publ., Dordrecht.

Greenstein, V.C., Shapiro, A., Zaidi, Q., and Hood, D.C. (1992). Psychophysical evidence for post-receptoral sensitivity loss in diabetics. Invest. Ophthalmol. Vis. Sci. 33: 2781–2790.

Greenstein, V.C., Starter, B., Carr, R. and Hood, D. (1989). Diabetes and S (blue) cone pathway sensitivity. In: Noninvasive Assessment of the Visual System. Optical Society of America Technical Digest Series 7: 7–10.

Hood, D.C. and Finkelstein, M.A. (1986). Sensitivity to Light. In: Boff, K.R., Kaufman, L. and Thomas, J.P. (eds.), Handbook of Perception and Human Performance, Vol. I, Sensory Processes and Perception: 5–1 and 5–66, John Wiley and Sons, New York.

Hood, D.C. and Greenstein, V.C. (1988). Increment threshold (tvi) data and the site of disease action In: Noninvasive Assessment of the Visual System. Optical Society of America Technical Digest Series 3: 2–5.

Hood, D.C. and Greenstein, V. (1990). Models of the normal and abnormal rod system. Vision Research 30: 51–68.

Kalloniatis, M. and Harwerth, R.S. (1989). Differential adaptation of cone mechanisms explains the preferential loss of short-wavelength cone sensitivity in retinal disease. In: Drum, B. and Verriest, G. (eds.), Colour Vision Deficiencies IX, Doc. Ophthalmol. Proc. Ser. 52: 353–364, Kluwer Acad. Publ., Dordrecht.

King-Smith, P.E., Rosten, J.G., Alvarez, S.L. and Bhargava, S. K. (1980). Human vision without tonic ganglion cells? In: Verriest, G. (ed.), Colour Vision Deficiencies V: 99–105, Adam Hilger Ltd., London.

LeGrand, Y. (1968). Light, Colour and Vision, 2nd ed. Chapman and Hall, London.

Lutze, M., Pokorny, J. and Smith, V.C. (1987). Improved clinical technique for Wald-Marré functions. In: Verriest, G. (ed.), Colour Vision Deficiencies VIII, Doc. Ophthalmol. Proc. Ser. 46: 259–265, Martinus Nijhoff/Dr W. Junk, Dordrecht.

Marré, M. (1972). Clinical examination of the three color vision mechanisms in acquired color vision defects. In: Verriest, G. (ed.), Acquired Colour Vision Deficiencies, Mod. Probl. Ophthalmol. 11: 224–227, S. Karger, New York.

Marré, M. & Marré, E. (1978). Different types of acquired colour vision deficiencies on the base of CVM patterns in dependence upon the fixation mode of the diseased eye. In: Verriest, G. (ed.), Colour Vision Deficiencies IV, Mod. Probl. Ophthalmol. 19: 248–252, S. Karger, New York.

Ohta, Y., Myamoto, T., Seki, A., & Kajigaya, Y. (1982). On spectral sensitivity of the retinal receptors in central serous chorioretinopathy. In: Colour Vision Deficiencies VI, Doc. Ophthalmol. Proc. Ser. 33: 389–398, Dr W. Junk, The Hague.

Pugh, E.N.J. and Mollon, J.D. (1979). A theory of the π-1 and π-3 color mechanisms of Stiles. Vision Res. 19: 293–312.

Smith, V.C., Pokorny, J. and Couropmitree, N. (1993a). S-cone discriminations as a function of luminance and background. Invest. Ophthalmol. Vis. Sci. Suppl. 34: 750.

Smith, V.C., Pokorny, J. and Yeh, T. (1993b). Chromatic Discrimination and Pigment Tests. J. Opt. Soc. Am. A 10: 1773–1784.

Smith, V.C. and Pokorny, J.: Chromatic discrimination axes, CRT phosphor spectra and individual variation in color vision. J. Opt. Soc. Am. A (in press).

Stiles, W.S. (1946). Separation of blue and green mechanisms of foveal vision by measurements of increment thresholds. Proc. Roy. Soc. Lond. B 133: 418–434.

Stiles, W.S. (1953). Further studies of visual mechanisms by the two-colour threshold method. In: Coloquio Sobre Problemas Opticas de la Vision: 65–103, Union Int. Phys. Pure. Appl., Madrid.

Swanson, W.H., Fellman, R.H., Lynn, J.R. and Starita, R.J. (1994). S-cone contrast sensitivity in glaucoma as a function of mean luminance. In: Drum, B. (ed.), Colour Vision Deficiencies XII, Doc. Ophthalmol. Proc. Ser. 57: 63–71, this volume.

Went, L.N. and deVries-deMol, E.C. (1976). Genetics of colour vision. In: Verriest, G. (ed.), Colour Vision Deficiencies III, Mod. Probl. Ophthalmol. 17: 96–107, Adam Hilger Ltd., London.

Yeh, T., Smith, V.C. and Pokorny, J. (1989). The effect of background luminance on cone sensitivity functions. Invest. Ophthalmol. Vis. Sci. 30: 2077–2086.

Yeh, T., Pokorny, J. and Smith, V.C. (1993). Chromatic Discrimination with variation in chromaticity and luminance: Data and theory. Vision Res. 33: 1835–1845.

Zisman, F. and Adams, A.J. (1982). Spectral sensitivity of cone mechanisms in juvenile diabetics. In: Verriest, G. (ed.), Colour Deficiencies VI, Doc. Ophthalmol. Proc. Ser. 33: 127–131, Dr W. Junk, The Hague.

Joel Pokorny
Visual Sciences Center
University of Chicago
939 East 57th Street
Chicago, IL
U.S.A.

36. The range of simultaneous scotopic contrast colors

STEVEN L. BUCK and JEFFREY L. BRANDT
(Seattle, WA, U.S.A.)

Abstract

We sought to determine the range of hues produced by simultaneous scotopic color contrast (hue induced by a mesopic field into a simultaneously presented, purely scotopic test field) in order to provide clues about how rod signals are processed in the pathways that mediate color vison.

Observers viewed either (1) a 1°-square scotopic test field contiguous with a 1°-square mesopic inducing field or (2) a 2°-diameter scotopic test field surrounded by an annular mesopic inducing field (8° outer diameter). Observers adjusted the wavelength of a 1°-diameter foveal comparison spot, visible only to the contralateral eye, to match the hue induced into the test field.

The two stimulus arrangements gave comparable results, although hue judgments were easier to make with the larger, concentric stimuli. The scotopic test field, colorless when presented alone, was described as shades of violet, blue, or green, and was matched dichoptically with 450 to 540 nm lights, in the presence of the various inducing fields. Simultaneous scotopic contrast hues did not include yellow, orange, or red. Inducing field wavelengths <500 nm induced the greener colors, while longer wavelengths induced blues and violets. Thus, inducing field wavelengths associated with greater S-cone activation produced contrast colors associated with lesser S-cone activation, and vice versa.

Introduction

Although the sensations associated with purely scotopic, rod-mediated vision are normally hueless, some exceptions have been reported. Willmer (1950) reported that a red appearing, long-wavelength patch induced a blue appearance in an adjacent purely rod-detected patch. This simultaneous scotopic color contrast was confirmed and explored further by Stabell and Stabell (1973), who reported that they were not able to detect a scotopic contrast hue in the presence of shorter-wavelength inducing stimuli because of scotopic brightness contrast.

B. Drum (ed.), Colour Vision Deficiencies XII, pp. 309–316.
© 1995 *Kluwer Academic Publishers, Dordrecht.*

In an attempt to overcome this, Stabell and Stabell (1978) used a preadapting bleach that covered the 1°-square inducing field but not the contiguous 1°-square test field. They described their results as showing that all spectral hues could be obtained by means of scotopic color contrast, although their data do not actually show scotopic contrast hues in most of the blue-green to green range. They concluded that the scotopic contrast hues were similar to, but not exactly the same as normal photopic complementary colors.

Scotopic contrast hues are theoretically significant because they provide clues about how rod signals are processed in the pathways that mediate color vision. The range of these hues and their relation to photopic contrast hues can help determine whether rod signals are processed in all of the parallel color pathways (r/g and b/y opponent processes, for example) or in only a subset (b/y, for example). The range of scotopic colors may also help to identify the locus of interaction or insertion of rod signals into the color pathways (for example, at the receptor level).

Because of the importance of these issues we sought to reexamine the range of simultaneous scotopic contrast hues using larger, concentric stimuli that might produce stronger contrast hues without using the preadapting bleach, which might induce photopic color perceptions and successive contrast effects that could confound the scotopic contrast colors.

Methods and results

Bipartite fields

We first attempted to replicate the results of Stabell and Stabell (1978) using their bipartite-field stimulus configuration. We used a computer automated Maxwellian-view optical system for all conditions (see Buck, 1985, for details). The test and inducing fields were contiguous 1° squares, with the test field located above the inducing field. Both test and inducing stimuli were presented for 0.5 sec every 2.5 sec. The inducing field was set to be about 0.75 log unit above absolute cone threshold for each wavelength tested from 400–700 nm. The 500-nm test field was set to be 0.3 log unit below the cone plateau measured during dark adaptation. All observers were color-normal males (SB, JB, MA) or females (DG) ranging in age from 23 to 43. Observers were dark adapted for at least 30 minutes prior to testing, unless otherwise indicated.

Observers adjusted the wavelength of a 1°-diameter foveal comparison spot, visible only to the contralateral eye, to match the hue induced into the test field as closely as possible. Observers could freely control both the brightness and the wavelength of the comparison spot, which was formed by a B&L grating monochromator. The comparison spot was continuously illuminated but observers could only see stimuli with one eye at a time: an eye movement was necessary to switch between test/inducing stimuli and the comparison spot. Subjects were instructed to disregard any simultaneous or successive contrast effects that were not temporally and spatially associated with the presentation of the scotopic center.

Stabell and Stabell (1978) argued that a 7°-diameter semicircular preadapting bleach was needed to see simultaneous scotopic colors using short- or middle-wavelength inducing fields. The afterimage of the preadapting field was positioned so that its straight edge was collinear with the shared border between the test and inducing fields. The preadapting bleach was intended to ensure that only cones detected the inducing field.

In contrast, we saw simultaneous scotopic contrast colors with inducing fields of any wavelength without using the preadapting bleach. For us, the preadapting bleach made the scotopic contrast colors difficult or impossible to see. For example, we had great difficulty keeping the area of the bleach superimposed over the 1° inducing field but not over the contiguous 1° scotopic test field, which would then disappear. Unable to collect a complete set of data using the bleach, we abandoned its use. Thus, in contrast to Stabell and Stabell, our inducing fields were mesopic and stimulated both rods and cones.

Figure 1 compares the results of Stabell and Stabell (1978) with those for our observers JB and DG using the bipartite stimuli. Each data point represents the mean of two observations. JB made similar hue judgments with both 'central fixation' fixation (2° eccentricity) and 'eccentric' fixation (6° eccentricity, that used by Stabell and Stabell). DG reported reliable scotopic contrast colors only in the central condition. Despite some differences between our two observers, their results generally agree with those of Stabell and Stabell for inducing fields >490 nm. The scotopic contrast hues were described as shades of violet or blue and were matched to foveally-presented wavelengths from about 450–495 nm.

However, for inducing fields <490 nm, our results were quite different from those of Stabell and Stabell (1978), who reported yellow, orange, and red scotopic contrast hues, with increasing redness as inducing field wavelength increased from 400–470 nm. Our observers never saw these scotopic contrast hues and instead saw blue-green or green.

Concentric stimuli

In an attempt to produce stronger scotopic contrast colors, to make the judgment task easier, and to extend the generality of our findings, we changed the stimulus arrangement visible to the test eye. The test field was a 2°-diameter disc that was surrounded by an annular inducing field having outer and inner diameters of 8° and 2°, respectively. Our observers found that this stimulus arrangement was easier to use than the bipartite field and yielded nearly identical results, as shown in Fig. 2. Each data point represents the mean of 4–8 observations. Although not shown here, neither variations of stimulus timing and fixation nor the presence or absence of a second scotopic comparison spot centered 10° to the left of the concentric stimuli had systematic effects on the results, so the data are collapsed across those conditions.

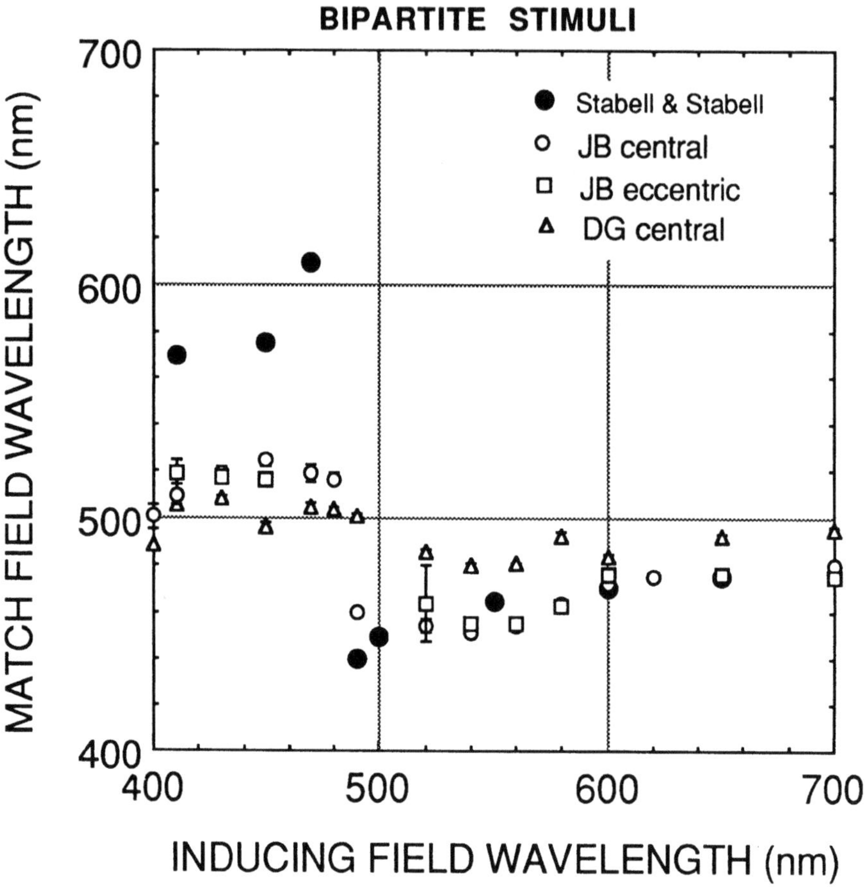

Fig. 1. Wavelengths matched to simultaneous scotopic contrast hues obtained using 1° by 2° bipartite field, for inducing wavelengths shown. Observers JB and DG made matches in the green/blue/violet range, regardless of whether fixation was 2° eccentric (central) or 6° eccentric (eccentric). Results from Stabell and Stabell (1978) agreed with ours for middle- and long-wavelength inducing fields but not for short-wavelength inducing fields. For all Figs., error bars are ±1 s.e.m. and are smaller than symbol size if not shown, except that no error bars are available for the data replotted from Stabell and Stabell (1978).

Photopic contrast hues

The relative consistency of judgments of scotopic contrast hues in the conditions described above gives little sense of the complexity of our observers perceptions of these stimulus conditions. Thus, we instructed observers to classify as scotopic contrast colors only those hues that were temporally and spatially specific to the test spot.

One color-contrast percept that failed these criteria is shown in Fig. 3 for observer SB (filled squares). Spectral matches to these hues increased from

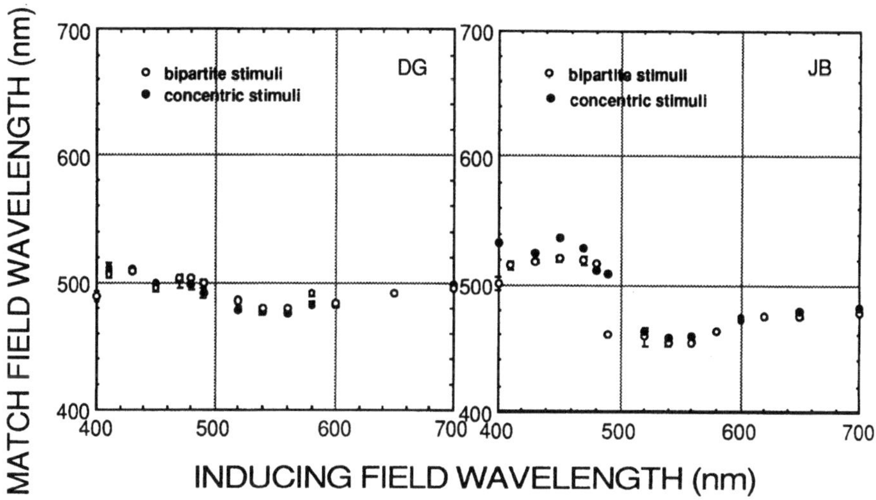

Fig. 2. Comparison of wavelengths matched to simultaneous scotopic contrast hues obtained using 1° by 2° bipartite stimuli or concentric stimuli composed of a 2°-diameter test spot surrounded by a contiguous annulus with 8° outer diameter.

about 535 nm to 600 nm as annulus wavelength increased from 400 nm to 500 nm. True scotopic contrast colors typically appeared as soon as the stimuli were presented, were confined to the exact location of the test spot, and required the presence of the illuminated (scotopic) test spot. In contrast, this 'photopic-like' percept appeared more gradually, was sometimes seen both inside the annulus and along its outer border, and was visible in the absence of the scotopic test both when the observer was either dark adapted or flash bleached. Only two (SB and MA) of four observers reported these percepts which appeared to involve a successive photopic contrast effect produced by small eye movements.

This possibility is supported by the measurements of simultaneous photopic color contrast for observer DG shown by (open circles) in Fig. 3. For these measurements, light levels for both annuli and test spots were increased substantially, the test spot was composed of tungsten light that was color-balanced to a perceptually neutral white by means of Wratten color-correction filters, and the observer was fully room light adapted.

The photopic contrast data make three points. First, for all wavelengths of inducing field, our observers' scotopic contrast hues fall outside the range of the corresponding photopic contrast hues (compare Figs. 2 and 3). Second, for short-wavelength inducing fields, both the 'photopic-like' percepts seen by SB and the data of Stabell and Stabell (1978) fall in a range similar to the photopic contrast data. Finally, for longer-wavelength inducing fields, our scotopic contrast hues agree fairly well with those reported by Stabell and Stabell (1973, 1978).

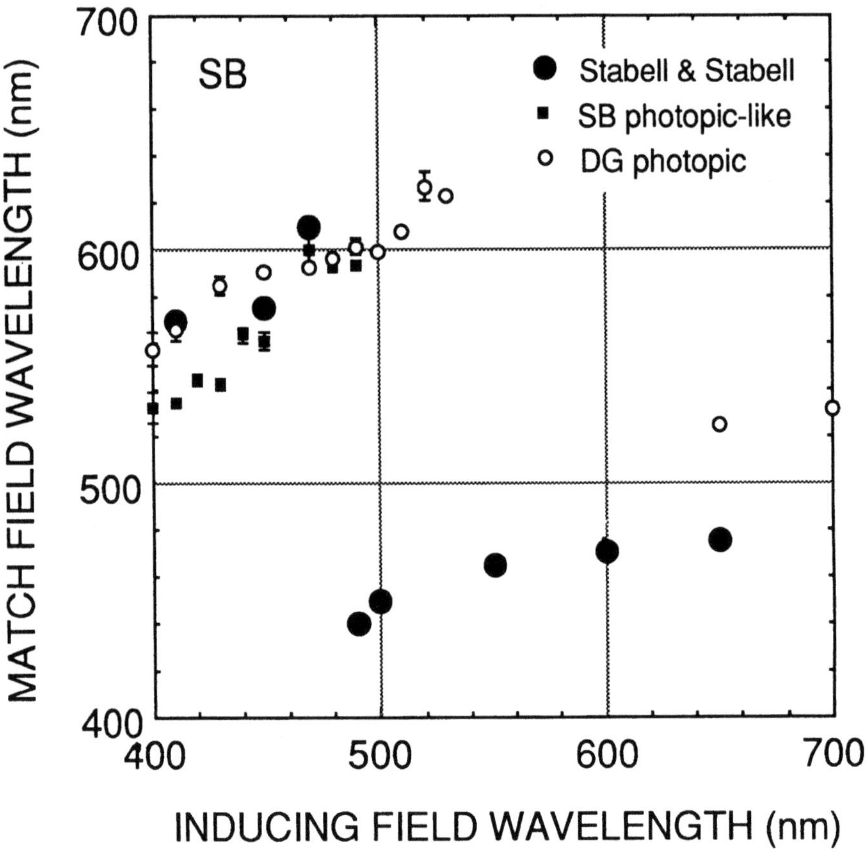

Fig. 3. Comparison of simultaneous scotopic contrast hues reported by Stabell and Stabell (1978) with simultaneous photopic contrast hues measured for observer DG and a 'photopic-like' percept seen by observer SB under scotopic contrast testing conditions, possibly caused by successive photopic contrast.

Discussion

We find a consistent range of simultaneous scotopic contrast colors (shades of violet, blue, and green) using different stimulus arrangements. Our results agree with all previous reports of blue scotopic contrast hues produced by long-wavelength inducing fields (Willmer, 1949; Stabell and Stabell, 1973, 1978). However, our results contradict the only previous report of simultaneous scotopic contrast hues produced by short-wavelength inducing fields (Stabell and Stabell, 1978).

What may account for this discrepancy? For short-wavelength inducing fields, the Stabells' stimulus arrangement, including preadaptation, may have produced (1) simultaneous photopic contrast or (2) successive contrasts, either scotopic or photopic, that masked the simultaneous scotopic contrast. This second possibility is strengthened by the close correspondence between the hues that the Stabells report for both simultaneous and successive scotopic contrast and by their use of preadapting bleaches for both procedures (Stabell and Stabell, 1972, 1978). However, it is not clear why we, but not the Stabells, were able to see scotopic contrast colors without the preadapting bleach.

We can reject one hypothesis concerning the discrepancy between our results and those of Stabell and Stabell (1978) for short-wavelength inducing fields. Even though our 490-nm test spot was kept below the cone plateau for detection, there is the possibility that cones might have become more sensitive in the more dark-adapted state of our experimental conditions than during the cone plateau. If this had happened, then the color sensations produced by cone detection of the 490-nm test could have mixed with or masked the scotopic contrast colors. We cannot determine the true state of cone sensitivity when rods are more sensitive than cones, as was the case in our experimental procedures. However, we did determine that the wavelength of the test spot had little or no effect on the hue perceived under our scotopic contrast conditions. For observers SB and MA, matches to the scotopic contrast hues varied by less than 5 nm when the 490-nm test was replaced by 450-nm or 560-nm tests, for both short-wavelength (400, 410, 420 nm) and long-wavelength annuli (700 nm). Since the three test wavelengths produced very different hues at photopic levels, we can reject the hypothesis that our scotopic contrast colors were influenced by photopic detection of the test.

Simultaneous scotopic contrast hues are different from corresponding photopic contrast hues and must reflect altered or incomplete operation of the chromatic pathways. We cannot yet explain scotopic colors at any single level of the visual system but some aspects are apparent. Inducing field wavelengths associated with greater S-cone activation produced contrast colors associated with lesser S-cone activation, and vice versa. However, because of the actual hues observed, neither variation of a tritan (S-cone) channel nor a b/y opponent channel seems sufficient to explain the results. Violet and greenish yellow, but not blue, could be explained by tritan variation. Blue, but not violet or green, could be explained by b/y variation. Similarly, short-wavelength redness, but not long-wavelength redness, induces greenish scotopic contrast colors; thus, variation of a r/g opponent channel is also an insufficient explanation.

Acknowledgements

This work was supported by a grant (EYO3221) from the National Eye Institute.

References

Buck, S. (1985). Cone-rod interaction over time and space. Vision Res. 25: 907–916.
Stabell, B. and Stabell, U. (1972). Chromatic rod vision-II. Duration of pre-stimulation varied, Scand. J. Psychol. 13: 136–140.
Stabell, U. and Stabell, B. (1973). Chromatic rod vision-VIII. Simultaneous contrast. Scand. J. Psychol. 14: 16–19.
Stabell, U. and Stabell, B. (1978). Scotopic hues of simultaneous contrast. Vision Res. 18: 1491–1496.
Willmer, E.N. (1950). Low threshold rods and the perception of blue. J. Physiol., Lond. 111: 17P.

Department of Psychology, NI-25
University of Washington
Seattle, WA 98195, U.S.A.

37. The necessary intensity of the white background when measuring the response of the S cone system

HISATO GUNJI, KENJI KITAHARA, JUN NOJI and AKIKO KUBO

(Tokyo, Japan)

Abstract

In order to investigate the necessary intensity of the white background when measuring the response of the S cone system, we measured the spectral sensitivities on various intensities of the white background at 15° from the fovea on two normal observers. The sensitivity curves on 10 and 31.4 photopic troland backgrounds for both observers have only one peak. However, from the sensitivity curve on 100 photopic troland background for both observers, the peak of the S cone response appears around the short wavelength range, then it becomes much clearer on the curve with a 1000 photopic troland background. From these results, it was suggested that the necessary intensity of the white background in order to measure the S cone system is at least 100 photopic trolands.

Introduction

The vulnerability of the S cone system in various eye diseases has been pointed out. As a procedure of detecting each cone response, test sensitivity measurements on a high intensity white background with a 200 msec test light exposure is most commonly used. The spectral sensitivity curve at the center of the retina in this method shows a three peak pattern, that is, the sensitivity peak at the short wavelength indicates the S cone system, while the other two peaks at the longer wavelength have been attributed to linear subtractive interaction between the M and L cones (Sperling *et al.*, 1971; Kranda *et al.*, 1979; Thornton *et al.*, 1983). Using this method we can study the characteristics of each type of cone system damage clinically. However, it is quite difficult to apply this method directly for clinical examination. It would be more convenient to use color perimetry to measure the S cone response for early diagnosis. Unfortunately, although yellow backgrounds are usually used to measure S cone responses, clinical perimeters normally do not provide enough intensity to separate S cone mechanisms from other mechanisms.

B. Drum (ed.), Colour Vision Deficiencies XII, pp. 317–320.
© 1995 *Kluwer Academic Publishers, Dordrecht.*

In this paper, we investigated the intensity of the white background necessary to measure the response of the S cone system. We measured the spectral sensitivities on various intensities of the background at 15 degrees from the fovea in order to reveal any interaction of rod and S cone responses and considered the changes of the patterns as a function of background intensity.

Methods

In this experiment, we used a 2-channel Maxwellian view optical system with a 150 Watt xenon arc for the light source. A 1° diameter circular test light that was filtered with narrow-band (10 nm half-band width) interference filters (400–700 nm) was superimposed in the center of an 8° circular background field. The observer's fixation was controlled by a small red light from an accessory system. The biting board was adjusted at eccentric position in order to allow the test flash and the background light to pass directly through the center of the pupil. The test flash duration was set at 200 msec. Using two normal subjcts aged 28 and 35 years, we measured the thresholds for each background condition 5 times, calculated the averages and then determined the mean sensitivity for each condition.

Results

The spectral sensitivities were measured on various intensities of the background at 15 degrees from the fovea. The test sensitivities (log photons^{-1} sec deg^{2}) were plotted as a function of wavenumber (cm^{-1}). The sensitivity curves for both observers with a 10 photopic troland background are shown at the top of Fig. 1. These are followed by curves for 31.4, 100 and 1000 photopic troland backgrounds respectively. Each result has been shifted 1 log unit toward the bottom for a clearer presentation. The top two curves for both observers have only 1 peak, however, these curves correspond with the luminous efficiency function for scotopic vision at short wavelengths and for photopic vision at middle to long wavelengths, which indicates these curves show the response of the rod function at short wavelengths. From the third curves for both observers, the peak of the S cone response appears short wavelengths, then it becomes much clearer on the 4th curve which represents the sensitivity curve on a 1000 photopic troland background. And simultaneously at middle and long wavelengths the peak which indicates the opponent channel from the M and L cones appears.

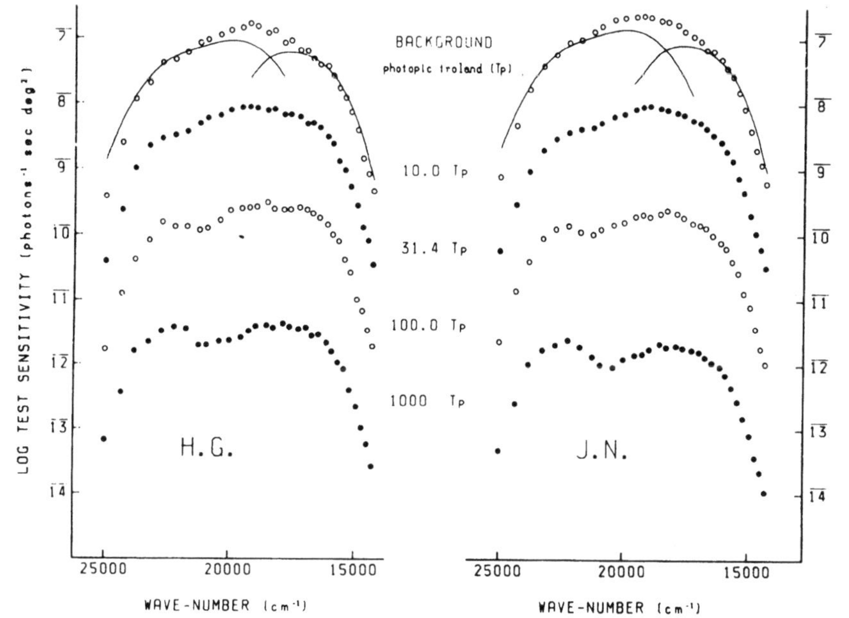

Fig. 1. The test sensitivity (log relative test sensitivity) for normal observers H.G. (on the left) and J.N. (on the right). Each result has been shifted 1 log unit toward the bottom for a clear presentation. The solid lines through the topmost data are V_λ and V'_λ.

Discussion

Kitahara *et al.*, (1987) studied the spectral sensitivities on a white background of illuminance 1000 photopic trolands up to 15° from the fovea. They found that the two peaks in the long wavelength region gradually decreased with increasing retinal eccentricity. However, below 500 nm the prominent peak attributed to the S cones was found even at the 15° temporal retina.

In this experiment, we investigated the intensity of the white background necessary to measure the S cone system. We found that the sensitivity curves on 10 and 31.4 photopic troland backgrounds have only one peak for both observers. However, from the sensitivity curve on a 100 photopic troland background for both observers, the peak of the S cone response appears around the short wavelength range, then it becomes much clearer on the curve with a 1000 photopic troland background. Therefore, we concluded that the necessary intensity of the white background when measuring the response of the S cone system is at least 100 photopic trolands.

References

Kitahara, K., Kandatsu, A., Tamaki and Matsuzaki, H. (1987). Spectral sensitivities on a white background as a function of retinal eccentricity. Doc. Ophthalmol. Proc. Ser.49: 651–655.

Kranda, K. and King-Smith, P.E. (1979). Detection of coloured stimuli by independent liner systems. Vision Res. 9: 733–745.

Sperling, H.G. and Harwerth, R.S. (1971). Red-green cone interactions in the increment-threshold spectral sensitivity of primates. Science, 172: 180–184.

Thornon, J.E. and Pugh, E.N. (1983). Red/green color opponency at detection threshold. Science 219: 191–193.

Hisato Gunji
Department of Ophthalmology
The Jikei University School of Medicine
3-25-8 Nishi-Shinbashi, Minato-ku
Tokyo 105, Japan

38. S-cone light adaptation: Effects of moderately intense adapting flashes

FRANKLIN NAARENDORP, PAUL KORTICK and GRAIG SPENCE
(Boston, MA, U.S.A.)

Abstract

The course of light adaptation of parafoveal S-cones to a moderately intense steady red field to which a pedestal had been added was studied by psychophysical means. Upon exposure to the red field, the observer adjusted the intensity of a 430 nm probe (flashed with the pedestal) to threshold visibility continuously until equilibrium was reached. With a short wavelength pedestal, S-cone sensitivity decreased initially. About 10 seconds after onset of the adapting stimuli, sensitivity began to increase and reached a maximum about 40–50 seconds later. Then, in the second minute of adaptation, S-cone sensitivity again decreased and stablilized eventually (below maximum sensitivity) in the fourth minute of adaptation. The transient increase in sensitivity during the early stage of adaptation became more pronounced with increased intensity of the short wavelength pedestal. With a long wavelength pedestal present, S-cone sensitivity decreased considerably at onset of the steady field and equilibrium was reached gradually without any sign of a transient increase in sensitivity. The results are discussed in terms of a shift in the balance of short and long wavelength signals reaching a chromatically opponent site.

Introduction

The sensitivity of the S-cone pathway is believed to be regulated at two successive, presumably, retinal sites (for review, see Mollon, 1982). At the first site, the fate of S-cone signals is determined by the rate of photon absorption in S-cones themselves. At the second site, beyond the receptors, sensitivity is regulated by antagonistic interaction between signals from S-cones and those from L- and M-cones. There is substantial evidence that S-cone signals isolated on long wavelength backgrounds are processed in an opponent channel (e.g. Polden and Mollon, 1980). Psychophysical experiments suggest that in the steady state, the opponent channel attains maximum sensitivity at an intermediate value of its response range, when subjectively chromatic cancellation

B. Drum (ed.), Colour Vision Deficiencies XII, pp. 321–326.
© 1995 *Kluwer Academic Publishers, Dordrecht.*

occurs in the field. In a state of transition, such as during the course of adaptation to a field that appears achromatic, S-cone sensitivity does exceed the level of sensitivity observed in the steady state. It has been suggested that this transient increase in sensitivity originates in a bleach signal from L- and M-cones delivered to the opponent site, during the first minute after onset of the field (Mollon and Polden, 1980).

The present study examines S-cone light adaptation to a steady red field that bleaches less than 0.3% of cone visual pigments at equilibrium. A 480 nm adapting flash (or pedestal) was added to the red field and the course of S-cone light adaptation was tracked with a 430 nm test flash. S-cone sensitivity was greater in the first minute of light adaptation than in the fourth minute (at equilibrium). The results suggest that pigment bleaching in L- and M-cones is not a precondition for observing the transient increase of S-cone sensitivity during light adaptation.

Methods

A four-channel Maxwellian system described previously by Battersby *et al.* (1964) was used. The test flash was 430 nm in wavelength, 200 ms in duration and 7.5° in diameter. In all experiments, the observer adjusted the intensity of the test flash to threshold. The pedestal (adapting flash) was either 480 nm or 550 nm in wavelength, 500 ms in duration and 10° in diameter. The steady adapting field was red in color (Kodak Wratten 29) and 21° in diameter. The Wratten 29 filter had a cut-on wavelength of about 600 nm and peaked at 632 nm. Bandwidths at half height for the 429.6 nm ($\approx$430 nm) and 550.4 nm ($\approx$550 nm) intereference filters were 9.4, 11.3 and 10.1 nm, respectively. A fourth channel provided a fixation target located near the edge of the steady red field. The stimuli were delivered, in concentric arrangement, to the temporal retina (10°) of the left eye. The pupil was dilated with 2 drops of neo-synephrine (10%). The inset of Fig. 1 shows the spatial arrangement of the stimuli. The onset of the pedestal always preceded test flash onset by 100 ms. Test and pedestal were presented once every 3 seconds against the steady red field. The stimulus parameters presented in this report were chosen in the context of a study that sought to isolate rod as well as S-cone responses. For the data shown here, the 480 nm (or 550 nm) beam entered the pupil off-axis (2.5 mm nasally) but was always intense enough to stimulate cones.

Light adaptation experiments began after a short period of dark adaptation. While maintaining fixation, the observer was exposed abruptly to the adapting field, the pedestal and test flash, all at the same time. The test flash and pedestal recurred on the red field once every 3 seconds. The observer adjusted the intensity of the 430 nm flash to threshold continuously, for about 4 minutes. A computer recorded at various times, during the course of adaptation, threshold intensities for the test flash. Two emmetropic, color normal, experienced males were run. Results for one observer only are shown here.

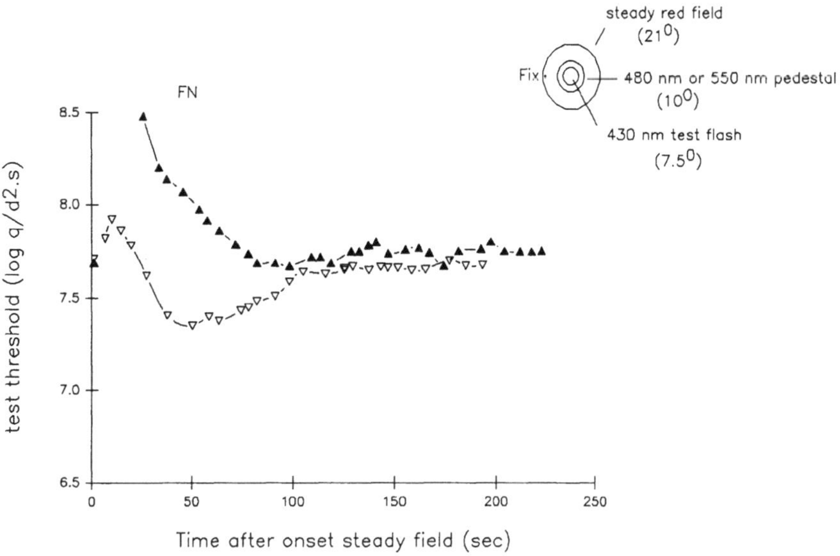

Fig. 1. Time course of adaptation to a steady red field of 8.9 log quanta $deg^{-2}.s^{-1}$ on which a pedestal was flashed. A 430 nm test flash with a duration of 200 ms was used. Pedestal duration was 500 ms. Pedestal onset preceded test flash onset by 100 ms. The two stimuli recurred once every three seconds. Observer FN. Open inverted triangles: 480 nm pedestal of 7.61 log quanta $deg^{-2}.s^{-1}$. Filled triangles: 550 nm pedestal of 9.46 log quanta $deg^{-2}.s^{-1}$. Inset: spatial configuration of stimuli. All measurements were at a retinal eccentricity of 10°.

Results

Figure 1 shows the temporal variation of test flash threshold after onset of a red field (8.9 log quanta $deg^{-2}.s^{-1}$) to which either a 480 nm pedestal (open inverted triangles) or a 550 nm pedestal (filled triangles) had been added. The intensities of both pedestals were selected so that they had equal effects on test flash threshold at equilibrium. Intensities were 7.6 and 9.46 log quanta $deg^{-2}.s^{-1}$ for the 480 nm and 550 nm pedestal, respectively. The field sensitivity of Stiles' pi-1 mechanism at the fovea (Wyszecki and Stiles, 1982) appeared to account reasonably well for the difference in intensity between the pedestals. The field sensitivity of the mechanism mediating test flash threshold suggested that the 430 nm test flash itself was detected by pi-1 which, in its short wavelength portion, reflects S-cone spectral sensitivity (Pugh, 1976).

With the 480 nm pedestal (7.61 log quanta $deg^{-2}.s^{-1}$) present, S-cone threshold increased first. This trend was reversed after about 5–10 seconds: threshold began to fall quickly and reached a minimum, about 60 seconds after adapting field onset. Thereafter, threshold began to rise again, gradually, and stabilized (after 180–240 seconds) about 0.3 log units above its lowest level attained in the course of adaptation. Similar results were obtained using a 440 nm, 460 or 500 nm pedestal. With the 550 nm pedestal (9.46 log quanta deg^{-2}.s^{-1}) present, the initial loss of sensitivity (= 1/threshold) was substantial and equilibrium was reached gradually without any sign of a transient increase in sensitivity during the course of adaptation. The difference in the results obtained with the 480 nm and 550 nm pedestals suggests that the lowering of threshold may be related to a shift in the balance of short and long wavelength signals reaching a chromatically opponent site.

Figure 2 shows, to the left, the course of light adaptation of S-cones to the steady red field (8.9 quanta deg^{-2}.s^{-1}) in the presence of 480 nm pedestals of various intensity. The abscissa (to the right) gives the range of pedestal intensities used. The data shown, to the left, represent a single set of measurements and are typically obtained in one session. The curves (except the upper one) show a small but clear decrease in sensitivity in the very beginning of the light adaptation process (this effect is usually present at the highest pedestal intensity as well but was missed by the observer in this particular data set). About 10 seconds after adapting field onset, the trend toward decrease of S-cone sensitivity reversed: sensitivity began to increase. This is apparent in all curves. This phenomenon, i.e. the increase in sensitivity, became more pronounced and peaked faster with increased pedestal intensity. An effect of 0.6 log units was observed with the highest pedestal intensity. In the second minute of adaptation threshold began to

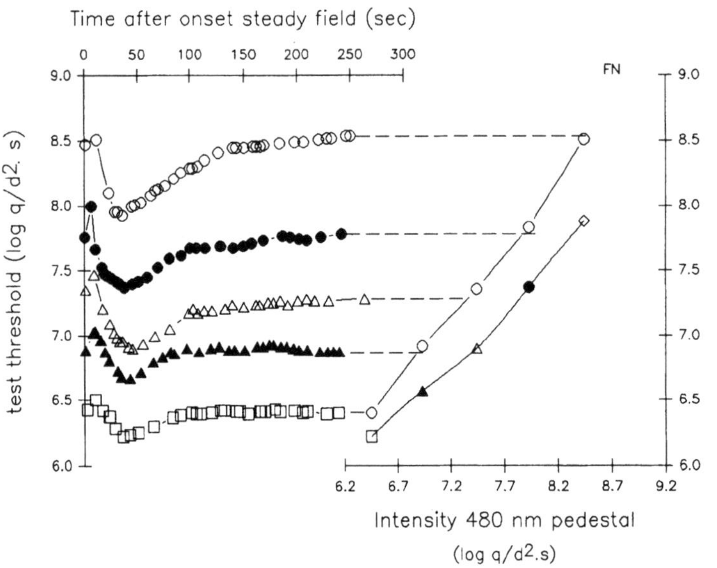

Fig. 2. Left: Time course of adaptation to a steady red field of 8.9 log quanta deg^{-2}.s^{-1} on which a 480 nm pedestal of variable intensity was flashed. A 430 nm, 200 ms test flash was used. Pedestal duration was 500 ms. Pedestal onset preceded test flash onset by 100 ms. The two stimuli recurred once every three seconds. Observer FN. Right: Increment threshold functions. Open circles: Threshold intensity as a function of pedestal intensity. A red field of 8.9 log quanta deg^{-2}.s^{-1} was constantly present; each measurement was made after 5 minutes of preadaptation. Lower curve: Minimum threshold obtained during the course of adaptation (to the left) as a function of pedestal intensity. All measurements were made at a retinal eccentricity of 10°.

rise again. The horizontal lines in Fig. 2 indicate roughly equilibrium values of S-cone thresholds. These values correspond closely to the increment thresholds (Fig. 2, open circles, to the right) measured on 480 nm pedestals of variable intensity against a steady red field of 8.9 log quanta $deg^{-2}.s^{-1}$. The increment threshold data were obtained in separate sessions in which the observer pre-adapted for 5 minutes to the red field. The slope of the curve described by the open circles was almost 1.4 for data points obtained with pedestal intensities >7.5 log quanta $deg^{-2}.s^{-1}$, which suggests that S-cones were being driven into 'saturation' by the pedestal while no substantial bleaching could have occurred in these cones. Prior to being driven into 'saturation', S-cones exhibited Weber-Fechner behavior, as demonstrated by the lower curve (right panel) which relates the lowest S-cone thresholds observed during light adaptation (left panel) to the intensity of the pedestal. The slope of the lower curve was slightly smaller than 1.

Discussion

The present results show that S-cone sensitivity first decreased (slightly), then rapidly increased, and again gradually decreased during the course of adaptation to a moderately intense red field to which a short wavelength pedestal had been added. The temporal variation of S-cone sensitivity became more pronounced with increased pedestal intensity (Fig. 2). The transient increase of S-cone sensitivity is not observed when a 550 nm pedestal is used which suggests that this phenomenon is specifically associated with a combination of adaptive signals from S-, L- and M-cones at some neural site (see Introduction). The present results resemble strikingly those reported by Mollon and Polden (1980) but were obtained under different experimental conditions. Most importantly, they used intense steady fields (their yellow field bleached approximately 85% of the visual pigments in L- and M-cones; the blue field had a radiance of about 10 log quanta $deg^{-2}.s^{-1}$ but did not bleach S-cones substantially). They presumed that signals from L- and M-cones delivered to the opponent site were changing more rapidly than those from S-cones, due to changes in the proportion of unbleached pigment in the first minute after field onset (Mollon and Polden, 1980). In the present case, the mechanism responsible for changes in the balance of signals at the opponent site is not likely to be a bleach signal. The retinal illuminance of the steady field used was 2.1 log td and is estimated to bleach about 0.3% of cone visual pigments at equilibrium (Hollins and Alpern, 1973).

In our experiments, the striking variation in sensitivity occurred only when the blue field was flashed. Possibly, bursts of S-cone signals produced by the short wavelength pedestal affected a post-receptoral saturation site which may have the identity of a chromatically opponent site (see Stromeyer *et al.*, 1979). Based on the antagonistic nature of the second site, we would expect S-cone threshold to be lowest in an early stage of adaptation (to a red field with added blue pedestal) because the repeatedly flashed pedestal would drive the opponent site out of the polarized state. On the other hand, adapting flashes tend to saturate the S-cone pathway. But 'saturation' will be abolished in the initial stages of adaptation due to the strength of opponent signals from L- and M-cones at field onset. In our experiments, adaptation to the red field was continuous. According to the Pugh and Mollon model (1979), following the onset of a long wavelength field a restoring force, with a long time constant, will reduce the polarization of the opponent site. The effect of the 'slow' acting restoring force and the burst of adaptive signals from S-cones (responding to the pedestal) begins to reverse the balance of signals from the three cone types at the opponent site. The balance shifts through a point of neutrality and sensitivity is transiently maximal. The remaining minute of adaptation to the red field is then dominated by the development of second site saturation. In brief, bursts of adaptive signals from S-cones may fail to 'saturate' the S-cone pathway initially due to strong signls from other cones at the opponent site. The strength of L- and M-cone signals diminishes (restoring force). A fragile balance of signals is temporarily achieved before post-receptoral saturation begins to dominate.

Acknowledgements

We thank Drs. Adam Reeves and Rhea Eskew for their comments on an earlier version of this manuscript.

References

Battersby, W.S., Oesterreich, R.E. and Sturr, J.F. (1965). Neural limitation of visual excitability. VII. Nonhomonymous retrochiasmial interaction. Am. J. Physiol. 206: 1181–1188.

Hollins, M. and Alpern, M. (1973). Dark adaptation and visual pigment regeneration in human cones. J. Gen. Physiol. 62: 430–447.

Mollon, J.D. (1982). Color vision. Ann. Rev. Psychol. 33: 41–85.

Mollon, J.D. and Polden, P.G. (1980). A curiosity of light adaptation. Nature. 286: 59–62.

Polden, P.G. and Mollon, J.D. (1980). Reversed effect of adapting stimuli on visual sensitivity. Proc. R. Soc. Lond. B 210: 235–272.

Pugh, E.N.(1976). The nature of the pi-1 colour mechanism of W.S. Stiles. J. Physiol. 267: 713–747.

Pugh, E.N. and Mollon, J.D. (1979). A theory of the pi-1 and pi-3 color mechanisms of Stiles. Vision Res. 19: 293–312.

Stromeyer, C.F., Kronauer, R.E. and Madsen, J.C. (1979). Response saturation of short-wavelength cone pathways controlled by color-opponent mechanisms. Vision Res. 19: 1025–1040.

Wyszecki, G. and Stiles, W.S. (1982). Color Science: Concepts and Methods, Quantitative Data and Formulae, 2nd ed., Wiley, New York.

Frank Naarendorp
Department of Psychology
125 Nightingale Hall
Northeastern University
Boston, MA 02115, U.S.A.

PART FIVE

Rod-cone interaction

39. Suppressive rod-cone interactions: Underlying mechanisms and practical application

T.E. FRUMKES, G. LANGE, F. NAARENDORP and THOR EYSTEINSSON
(New York, NY, and Boston, MA, U.S.A.; Reykjavik, Iceland)

Abstract

Dark adapted rods exert tonic suppressive influences upon cone-mediated vision, i.e. suppressive rod-cone interactions (SRCI). We have intensively studied two forms: (1) sensitivity to rapid flicker progressively decreases during rod-dark adaptation, and increases if rods are adapted to increasing intensity backgrounds. This 'flicker-SRCI' is mediated in distal retina and involves a prominent role for horizontal cells; (2) sensitivity to high spatial frequency gratings similarly decreases during rod-dark adaptation, and increases if rods are adapted to increasing intensity backgrounds. However, this 'grating-SRCI' is as prominent with interocular adapting fields as with one-eyed stimulation, and psychophysical data are inconsistent with an electrical coupling model accounting for flicker-SRCI. Contrast sensitivity is well-known to increase as the size of the visual display increases. For higher spatial frequencies, we show this influence of display size is likely to reflect a removal of SRCI rather than an effect of display size *per se*. Although counterintuitive, our data strongly suggests that night-time sensitivity to fine spatial and temporal detail should be improved under everyday conditions by maintaining rods in a slightly light-adapted state.

Introduction

Vision is mediated by two different types of photoreceptors, the rods and cones. Statement of the 'duplicity theory', is often accompanied by three incorrect, corollary conclusions which maintain that (1) rods and cones act independently, (2) 'night-time vision' is synonymous with 'rod vision' and (3) dark-adaptation is a passive recovery process. Two hundred literature references in the past 25 years dealing with 'rod-cone interaction' suggest that the first conclusion is incorrect. The second conclusion may be true in a rural environment on a dark night with no artificial illumination, but the limit for rod-mediated visual acuity is about 6/30 or 20/100 (Hess, 1990), below the limit defining legal blindness in

B. Drum (ed.), Colour Vision Deficiencies XII, pp. 329–334.
© 1995 *Kluwer Academic Publishers, Dordrecht.*

many countries and American states; hence, rod-monochromats have substandard night-time *and* daytime vision. The practical implication is that night vision of greatest consequence in modern society is mediated by cones.

Rods still play an important role in night vision in modern urban society. This is because the third corollary is also wrong and dark adaptation is more than a passive recovery process, and cone vision is subject to different tonic suppressive influences from dark-adapted rods, i.e. suppressive rod-cone interactions (SRCIs). Alexander and Fishman (1984), Coletta and Adams (1984), and Goldberg *et al.*, (1983) showed that sensitivity to cone-mediated flicker decreases throughout the time-course of rod-dark adaptation, or increases as a dim, rod-stimulating background increased in luminance. This involves a slight change in sensitivity for slow flicker, but increases in normal observers to about a 1 $\log_{10}$ unit effect if flicker frequency $\geq$ 15 Hz. Neurophysiological studies (e.g. Eysteinsson and Frumkes, 1989; Frumkes and Eysteinsoon, 1987; Frumkes and Wu, 1990; Pflug *et al.*, 1990), show that 'flicker-SRCI' involves distal retinal mechanisms in which horizontal cells play a particularly important role. It also principally involves a response to the offset and not to the onset of illumination (Frumkes and Wu, 1990; Frumkes, *et al.*, 1992). In an analogous fashion, cone-mediated spatial sensitivity decreases throughout the time-course of rod-dark adaptation, and increases as a dim, rod-stimulating background increased in luminance (Naarendorp *et al.*, 1988). This 'grating-SRCI' involves a slight sensitivity change for low spatial frequency gratings, but increases in normal observers to about a 1 $\log_{10}$ unit effect if grating frequencies $\geq$ 15 cycles per degree (cpd).

Methods

Recent results indicate that grating- and flicker-SRCI reflect distinct retinal mechanisms (Lange and Frumkes, 1992, 1993). For the data illustrated in Fig. 1, we examined contrast sensitivity to a test display with an average luminance of 10 cd/m^2 and modulated sinusoidally in both temporal and spatial domains; we used a total of 42 stimulus combinations involving spatial frequencies between 1–16 cycles per degree (cpd), and 0.2–16 Hz. The overall test display was shaped as a 1° sided square and presented 4° parafoveally to one eye. The observer adjusted display contrast so that the flickering grating could just be detected. Data were obtained when both eyes were otherwise dark adapted, or in the additional presence of a large, 0.03 cd/m^2 background field; since stimuli were viewed through 2.5 mm artificial pupils to rule out any role of pupillary dilation, this luminance is decidedly below a liberal estimate of cone-dectection threshold. The conditions used to collect the data in Fig. 2 were identical, save that the dimensions of the both test and adapting stimuli were slightly different as specified below, and the test grating was not flickered sinusoidally but flashed 'on' for 3 second duration viewing trials.

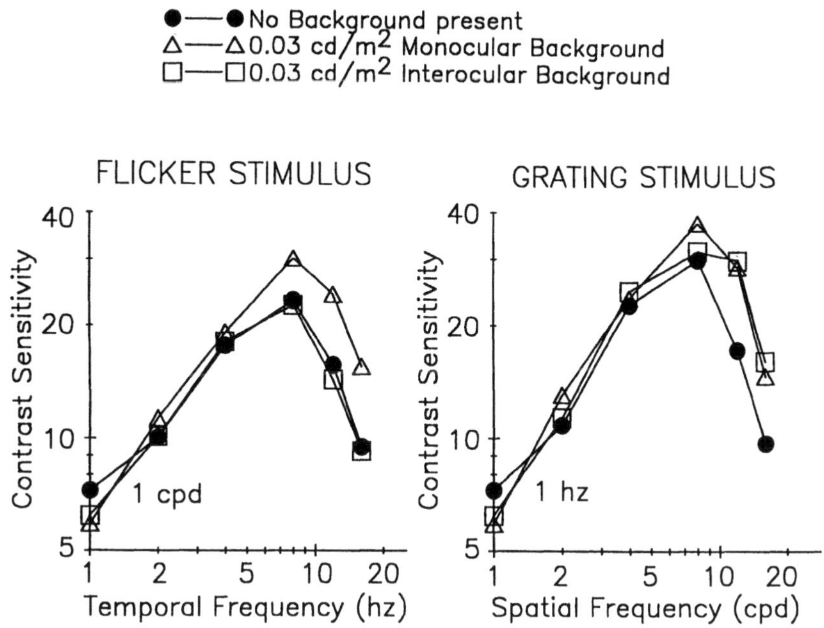

Fig. 1. Sensitivity to 1° square displays modulated concomitantly in both spatial and temporal domains. See text for further explanation.

Results and discussion

The left coordinates in Fig. 1 show contrast sensitivity to a 1 cycle per degree (cpd) spatial frequency display as a function of the temporal frequency in Hz; on the right, contrast sensitivity to a 1 Hz temporal frequency is plotted as a function of the spatial frequency of the display in cpd. The closed symbols plot the sensitivity for the display when no adapting field was present and replicate well-known findings (e.g. Kelly, 1977); the open triangles indicate that an adapting field was additionally presented to the same eye as the test display. This rod-background increased sensitivity of displays combining low spatial and high temporal frequencies (the 8–16 Hz data on the left coordinates) or high spatial and low temporal frequencies (the 8–16 cpd data on the right coordinates). However, with other temporal/spatial frequency combinations including many not illustrated here (e.g. any spatial frequency combined with 4 Hz stimulation, any temporal frequency combined with 4 cpd stimulation), this background exerted little influence upon display visibility (Lange and Frumkes, 1992). These results suggest that dim rod-stimulating backgrounds enhance visibility to either gratings or to flicker, but not to both.

We used two approaches to tease apart grating- and flicker-SRCI. In early experiments we parametrically varied the size, shape, and retinal position of both test and adapting stimuli. This revealed some characteristic differences

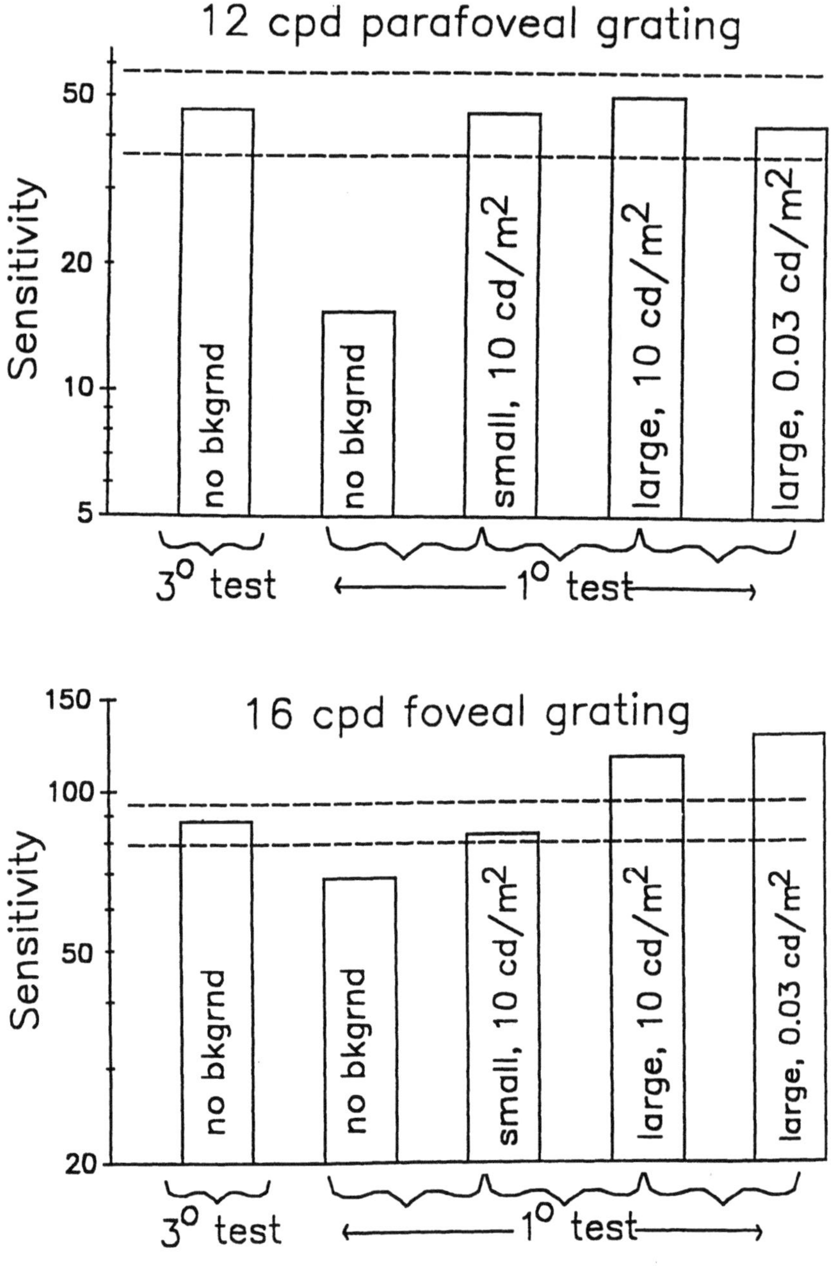

Fig. 2. Sensitivity to 1° and 3° gratings modulated sinusidally in the spatial domain, as a function of their overall dimensions, retinal position, and the adapted state of the surrounding retinal area. See text for further explanation.

between flicker and grating effects which we plan to publish soon. However, more dramatic results were obtained by comparing the influence of monocular and interocular backgrounds. The open squares in Fig. 1 illustrate the influence of a rod-stimulating background field presented to the eye contralateral to that viewing the test display. This interocular background has no effect upon sensitivity to a low spatial frequency display, regardless of temporal frequency (left coordinates), but produced essentially the same enhancement in sensitivity to high spatial frequency displays flickered slowly as the monocular background (right coordinates). Although we currently lack a model for grating-SRCI, it has different spatial limitations than flicker-SRCI, and obviously, is unlikely to involve exclusively a retinal-horizontal cell mechanism.

There are other SRCI influences. Lie (1963) showed that the specific threshold (minimal intensity necessary for correct hue-detection) for a stimulus increases throughout the rod-recovery stage of dark-adaptation. The 'Lie effect' is strongly influenced by interocular light-adaptation (Prestude *et al.,* 1978) although we are unaware of purely interocular demonstrations comparable to that for grating SRCI described above. Moreover, we (Denny *et al.,* 1991) have some informal obervations suggesting a comparable influence of dark-adapted rods upon the cone-mediated contribution to brightness which is much easier to show with interocular than monocular adaptation.

SRCI has important practical implications. Figure 2 compares the visibility of non-flickering, sinusoidally modulated gratings of 10 cd/m^2 average luminance and shaped as either 1° or 3° sided squares. The upper coordinates show sensitivity for a 12 cpd grating presented 4° parafoveally. The left most bar shows sensitivity to a 3° grating when no background field was present: the horizontal dashed lines show 95% confidence intervals for this sensitivity value. The second bar to the left shows sensitivity to a 1° grating, also with no adapting field present. Sensitivity to the smaller grating is less as is commonly reported in the literature (e.g. Kelly, 1977). The right three bars also indicate sensitivity to the 1° test grating but when surrounded by annular adapting fields presented to the same eye as the test stimulus. For the third bar from the left, the outer dimensions of this annulus was a 3° sided square so that the total luminance flux presented by the test-plus-adapting-field-complex was the same as provided by the 3° grating; this complex produces a sensitivity which approximates that of the 3° grating. Essentially the same sensitivity is achieved when the outer dimensions of this annulus were a 15° sided square, regardless of whether its luminance is 10 cd/m^2 (fourth bar from left) or when it was reduced to a sub-cone luminance level of 0.03 cd/m^2 (right most bar). These results indicate that the increase in contrast sensitivity produced by increasing display size is attributable to a removal of SRCI rather than to display size *per se.* The bottom set of coordinates in Fig. 2 show results of a comparable experiment with foveally centered gratings of 16 cpd, and show even more striking results: the layout of the lower coordinates is identical to that described above for parafovea. In the fovea, the difference in sensitivity between the 1° and 3° gratings in the dark-adapted eye is still significant, and presentation of the 3°

sided annulus increases sensitivity to the 1° grating up to a level approaching that of the 3° grating. However, a rod-stimulating 15° sided annulus causes the sensitivity to the 1° grating to greatly exceed the sensitivity to the 3° grating in the dark-adapted eye.

Since light adapting rods improves cone-mediated vision, sensitivity involving the detection of spatial and temporal detail could be optimalized under night-time driving situations by light-adapting rods. Although counterintuitive, removal of SRCI probably explains why 'glare sources' sometimes improve dark-adapted visual acuity (e.g. van den Berg, 1986).

References

Alexander, K.R. and Fishman, G.A. (1984). Rod-cone interaction in flicker perimetry. Brit. J. Ophthalmol. 68: 303–309.

Coletta, N.J. and Adams, A.J. (1984). Rod-cone interactions in flicker detection. Vision Res. 24: 1333–1340.

Denny, N., Frumkes, T.E. Barris, M.C. and Eysteinsson, T. (1991). Tonic interocular suppression and binocular summation in human vision. J. Physiol. 437: 449–460.

Eysteinsson, T. and Frumkes, T.E. (1989). Physiology and pharmacological analysis of suppressive rod-cone interaction in Necturus retina. J. Neurophysiol. 61: 866–873.

Frumkes, T.E. and Eysteinsson, T. (1987). Suppressive rod-cone interaction in distal vertebrate retina: Intracellular records from *Xenopus* and *Necturus*. J. Neurophysiol. 57: 1361–1382.

Frumkes, T.E., Lange, G., Denny, N. and Beczkowska, I. (1992). Influence of rod-adaptation on cone-responses to light offset in humans: I. Results in normal observers. Visual Neurosci. 8: 83–89.

Frumkes, T.E. and Wu, S.M. (1990). Independent influences of rod adaptation on cone-mediated responses to light onset and offset in distal retinal neurons. J. Neurophysiol. 64: 1043–1054.

Goldberg, S.H., Frumkes, T.E. and Nygaard, R.W. (1983). Inhibitory influence of unstimulated rods in human retina: evidence provided by examining cone flicker. Science 221: 180–182.

Hess, R.F. (1990). Post receptoral sensitivity of the achromat. In: Hess, R.F., Sharpe, L.T. and Nordby, K. (eds.), Night Vision: 390–414, Cambridge, New York.

Kelly, D.H. (1977). Visual contrast sensitivity. Optica Acta 24: 107–129.

Lange, G. and Frumkes, T.E. (1992). Influence of rod-light adaptation upon the visibility of displays modulated concurrently in the temporal and spatial domains. Invest. Ophthalmol. Vis. Sci. Suppl. 33: 1261.

Lange, G. and Frumkes, T.E. (1993). Separate spatial and temporal forms of suppressive rod-cone interaction (SRCI). Invest. Ophthalmol. Vis. Sci. Suppl. 34: 818.

Lie, I. (1963). Dark adaptation and the photochromatic interval. Doc. Ophthalmol. 17: 411–510.

Naarendorp, F., Denny, N. and Frumkes, T.E. (1988). Rod light and dark adaptation influence cone-mediated spatial acuity. Vision Res. 28: 67–74.

Pflug, R., Nelson, R. and Ahnelt, P.K. (1990). Background-induced flicker-enhancement in cat retinal horizontal cells. I: Temporal and Spectral Properties. J. Neurophysiol. 64: 313–325.

Prestude, A.M., Watkins, L. and Watkins, J. (1978). Interocular light adaptation effect on the 'Lie' specific threshold effect. Vision Res. 18: 855–857.

Van den Berg, T.J.T.P. (1986). Importance of pathological interocular light scatter for visual disability. Doc. Ophthalmol. 61: 327–333.

Professor Thomas E. Frumkes
Dept. of Psychology
Queens College of CUNY
Flushing, NY 11367, U.S.A.

40. Rod and cone signal processing in mesopic heterochromatic photometry

F. VIÉNOT and A. CHIRON
(Paris, France)

Abstract

Heterochromatic photometric measurements have been compared as assessed by two colour normal observers using heterochromatic flicker photometry (HFP) and direct comparison brightness (DCB) matching at 10 mesopic levels ($3*10^{-3}$ to 10^2 phot td) and five short wavelengths (413, 432, 445, 465 and 480 nm) against 506 nm.

The results show that allowing for a physical temporal phase shift facilitates HFP matches and increases the precision of settings. Modelling rod and cone signal processing during one physical cycle leads to a complex vectorial representation where the physical phase shifts can be explained but the amplitudes of the settings cannot.

The difference in settings as a result of the two matching methods reflects antagonistic processes rather than a simple additional S-cone contribution.

Introduction

Visual photometry requires an individual observer to assess a match between a test stimulus and a reference stimulus of different chromatic content.

At photopic illuminance levels, it is known that the visual match depends upon the experimental technique (CIE, 1978; Lennie *et al.*, 1993). It is generally accepted that when the technique relies on high spatial or temporal frequencies, as in heterochromatic flicker photometry (HFP), the result of the visual match is additive, like luminance, and can be modelled as the arithmetic sum of the long wavelength sensitive cone (L-cone) signals and the medium-wavelength sensitive cone (M-cone) signals. When the technique relies on low spatial or temporal frequencies, as in direct comparison brightness matching (DCB), the result of the visual match is non-additive, differs from the HFP match, and presumably reflects the activity of three mechanisms: an achromatic one and two chromatically antagonistic ones. The difference between the results yielded by the two techniques may reflect the additional contribution of two

B. Drum (ed.), Colour Vision Deficiencies XII, pp. 335–342.
© 1995 *Kluwer Academic Publishers, Dordrecht.*

chromatically antagonistic mechanisms to brightness, as compared with luminance (see e.g. Yaguchi *et al.,* 1993).

At mesopic illuminance, the experimental situation becomes even more complex, since four types of photoreceptors are active: L-cones, M-cones, S-cones and rods.

In a previous study, we have shown that the brightness match undergoes smooth changes over the full mesopic range but that the flicker results exhibit a step-like transition in the mid-mesopic range (Viénot and Chiron, 1992). The possibility that cone signals and rod signals could be out of phase and either positively or negatively interacting, depending upon the flicker frequency, was discussed.

In this paper, we intend, on the one hand, to examine the possibility of reducing the discontinuity of the HFP assessments in the mid-mesopic range by facilitating rod-cone cooperation, and on the other hand, to sort out the various contributions of photoreceptors in both the HFP and DCB methods.

Methods

The procedure is designed firstly to provide comparison between HFP and DCB matches, secondly to compensate for the phase relationship between receptor signals in the matches, and thirdly to address the contributions of specific photoreceptors to the matches.

Two colour normal observers (the authors) participated in the experiment. Each observer performed all visual matches between the test and reference stimuli, at 10 mesopic illuminance levels, using HFP and DCB matching. Ten illuminance levels were tested with a fixed reference stimulus (from 100 to 0.003 photopic trolands (phot. td), decreasing by 0.05 log unit steps). At each level, 10 adjustments were made using HFP and 10 using DCB, with 3 sessions at each test wavelength. The diameter of the field was 10° viewed with central vision.

Before any heterochromatic flicker photometry adjustment, the subject determined the temporal frequency and the phase shift that yielded the most satisfactory flicker percept by an iterative method. In practice, too high a frequency would extend the range of phase shift associated with the flicker null, and too low a frequency would make the flicker impossible to null. The observer was invited to choose the frequency that made the phase shift adjustment as precise as possible. Once the frequency was chosen, and once the two minutes preadaptation period was over, the subject made 10 matches, adjusting the phase shift at each trial.

Phase shift relative to strict reference-test alternation (exact counterphase) is referred to in this paper as 'physical phase shift'. The physical phase shift is transformed to 'test advance' in milliseconds using the following equation:

test advance [ms] = 1000 * (physical phase shift [deg] / 360) / Frequency [Hz]

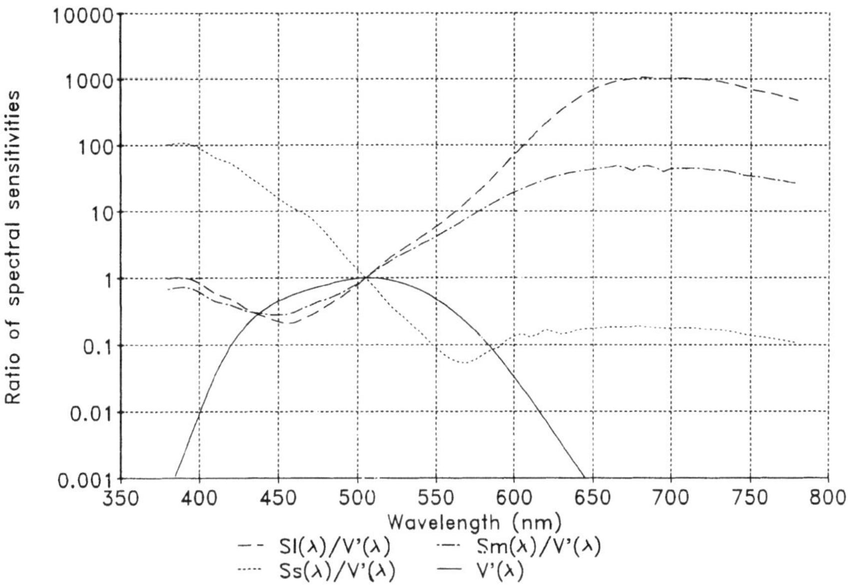

Fig. 1. Ratio of the cone action spectra $S_S(\lambda)$, $S_M(\lambda)$, $S_L(\lambda)$ (after Vos, 1988) to the scotopic spectral luminancy efficiency $V'(\lambda)$, normalized at 506 nm. The scotopic spectral luminous efficiency $V'(\lambda)$ is also shown.

The reference wavelength is 506 nm, where the relative scotopic luminous efficiency, $V'(\lambda)$, is maximum. five test wavelengths were selected at less than 506 nm (413, 432, 445, 465, 480 nm). In this spectral range, the ratio of the S-cone action spectrum to $V'(\lambda)$ increases as wavelength decreases, but the reverse occurs for the ratios of the L-cone and M-cone action spectra to $V'(\lambda)$ (with a minimum around 445 nm) (Fig. 1).

Results

Allowing for a physical phase shift enables the observer to assess a precise and unique HFP match over full mesopic range and reduces the discontinuity in the mid-mesopic region (Fig. 2).

The test advance which is derived or calculated from the physical phase shift is independent of the temporal frequency chosen by the observer. In this experiment when the test wavelength is shorter than 506 nm, the test advance is always positive and undergoes the same variation for the two observers. It reaches its maximum value between 0.3 and 1 phot. td for each wavelength. However, this maximum value depends significantly upon the wavelength. It attains a value of 6 ms for a 432 nm test (Fig. 3).

The results yielded by the two matching techniques differ over the full mesopic domain for the two observers. In these experiments where test

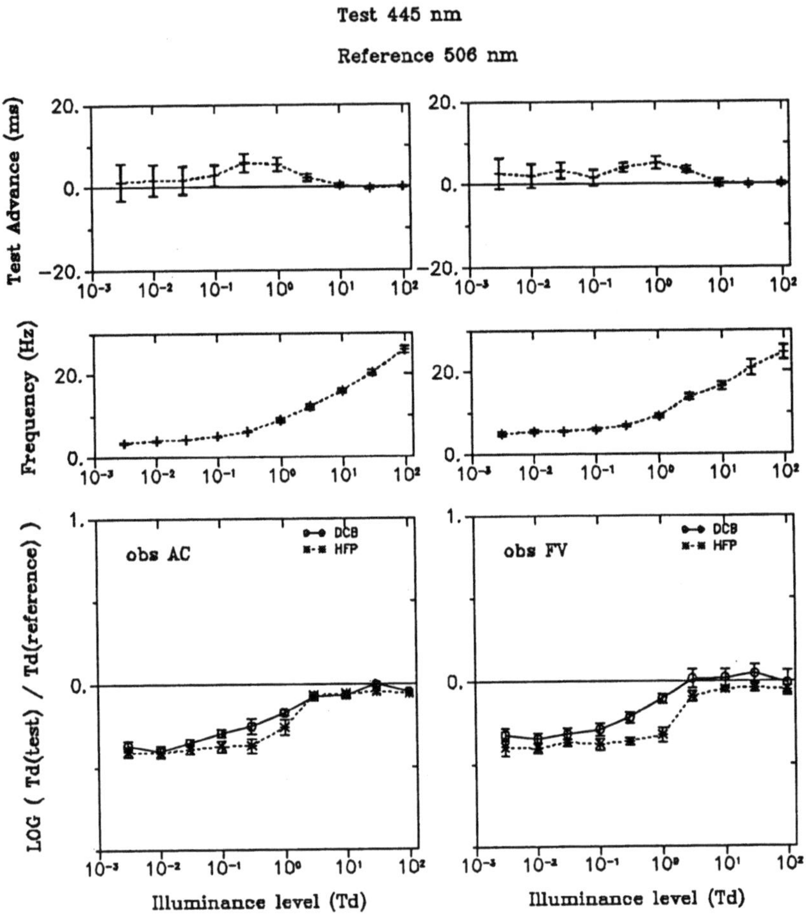

Fig. 2. Results from two observers. *Top*: Test advance relative to exact counterphase alternation with the reference, as a function of retinal illuminance. *Middle:* Frequencies chosen by the observer for heterochromatic flicker photometry as a function of retinal illuminance. *Bottom:* Relative spectral sensitivity at 445 nm compared with 506 nm. Abscissae: retinal illuminance produced by the 506 nm reference in photopic trolands. Ordinates: log ratios of retinal illuminances of test lights and reference lights. Stars: heterochromatic flicker photometry. Circles: direct comparison brightness matching. Error bars: standard deviation of 30 matches recorded during 3 sessions.

wavelength is shorter than 506 nm, extra light is always required using DCB matching, compared with HFP. At 3 phot. td, the difference in log values increases as the wavelength increases. It reaches 0.5 log unit for observer FV, a value which is twice as large as that for observer AC (Fig. 4).

The maximum value of the difference shifts as the test wavelength changes. It occurs between 0.1 and 0.3 phot. td for a 413 nm test and between 0.3 and 3 phot. td for a 480 nm test.

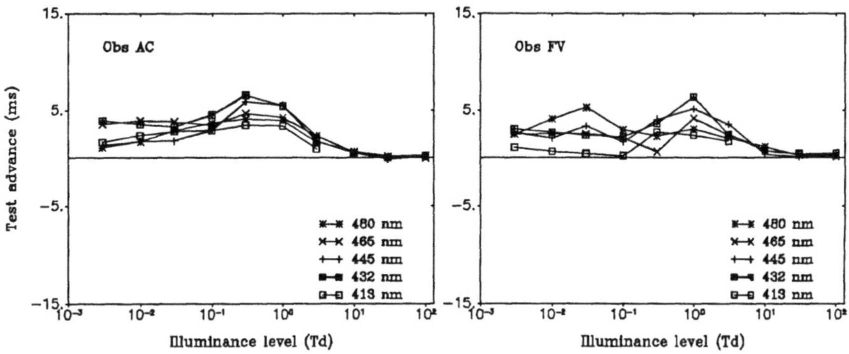

Fig. 3. Test advance in milliseconds relative to exact counterphase alternation with the reference, as a function of retinal illuminance. Results from two observers performing HFP matches at 413, 432, 445, 465 and 480 nm against the 506 nm reference. Abscissae: retinal illuminance produced by the 506 nm reference in photopic trolands.

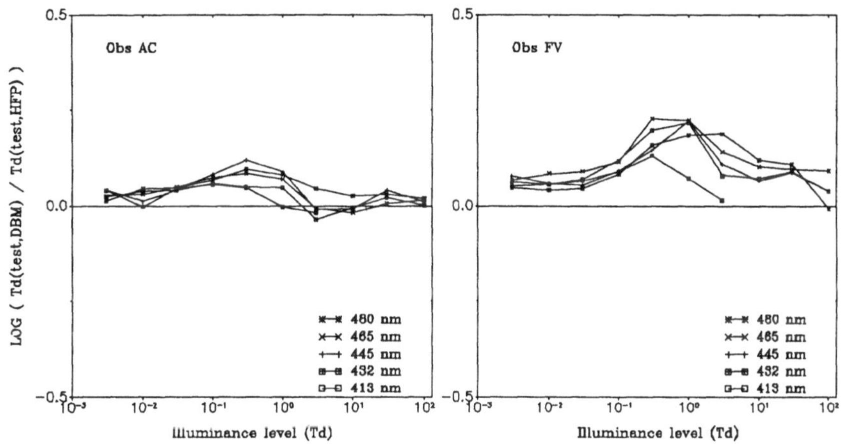

Fig. 4. Difference in log values between settings using HFP and using DCB matching. Results from two observers performing HFP and DCB matches at 413, 432, 445, 465 and 480 nm against the 506 nm reference. Abscissae: retinal illuminance produced by the 506 nm reference in photopic trolands.

Discussion

HFP matches

We show that introducing a physical phase shift reduces the discontinuity of HFP matches. The two observers achieve precise and repeatable matches. Moreover, the observers have noticed that introducing a physical phase shift makes the flickering appearance of the field look homogeneous.

In a previous experiment, we have shown that in the mid-mesopic range, the exact antiphase of the test and reference stimuli led the observer to determine either a photopic-like match or a scotopic-like match, depending upon the temporal frequency he chose (Viénot and Chiron, 1992). Any intermediate match would have been highly variable. It was suggested that counterphase rod and cone signals might be responsible. Assuming a rod-cone latency difference of about 75-80 ms (MacLeod, 1972; Sharpe *et al.,* 1989), rod signals and cone signals should be in counterphase around 6.25 Hz and should null when they are equal; they should be in phase around 12.5 Hz and should add. No intermediate frequency would have led to a straightforward combination of rod and cone signals.

In the present experiment, the effect of introducing a physical phase shift is to change the temporal distribution of cone and rod signals originating from test or reference stimulus. We have tried to fit a simple vectorial model to our data: one cone signal only is assumed to represent the total cone population response; rod-cone latency difference is assumed to be 80 ms; experimental physical phase shift and experimental flicker frequency were used. An example is given at 432 nm but similar conclusions can be drawn for all the test wavelengths and for the two observers. The four vectors represent the rod and cone signals originating from the test and the reference stimuli (Fig. 5). The angle between the rod signal vectors is equal to the angle between the cone signal vectors and is determined by the physical phase shift. The angle between the rod and the cone signal vectors originating from the same stimulus is determined by the flicker frequency and by the rod-cone latency difference. From 3 to 0.3 phot. td, it can be noted that the cone receptor signal vectors turn round relative to the rod receptor signal vectors, thus following a critical change in flicker frequency (from 9.4 Hz to 6.9 Hz). At the intermediate frequency (7.3 Hz) which is set at a 1 phot. td illuminance level, the vectors offer a characteristic pattern of relative phases like an exploding figure. There (under the restriction that the vector amplitudes are of the same order of magnitude), the angular configuration is stable only if strict phase relationships are maintained. This explains the high precision of the phase setting at 1 phot. td for every wavelength and for the two observers.

A further step in modelling requires reporting the signal amplitude as the vector length. For our demonstration, we consider inequalities between amplitudes only. On the basis of the photopic match for which cone signals only are acting, and the scotopic match for which rod signals only are acting, we find that, at mid-mesopic illuminance levels, cone signals associated with the test are smaller than cone signals associated with the reference, and rod signals associated with the test are larger than rod signals associated with the reference. In this case, the equilibrium of the vectorial configuration cannot be reached. Therefore, it appears impossible to explain the choice of phase shifts on the basis of a simple vectorial combination of rod and cone signals. Recent experiments (Sharpe *et al.*, to be published) show that a 80 ms rod-cone latency difference is the maximum delay that is observed between rod and cone signals. A possible improvement in vectorial modelling might arise from introducing a luminance-dependent rod-cone latency difference.

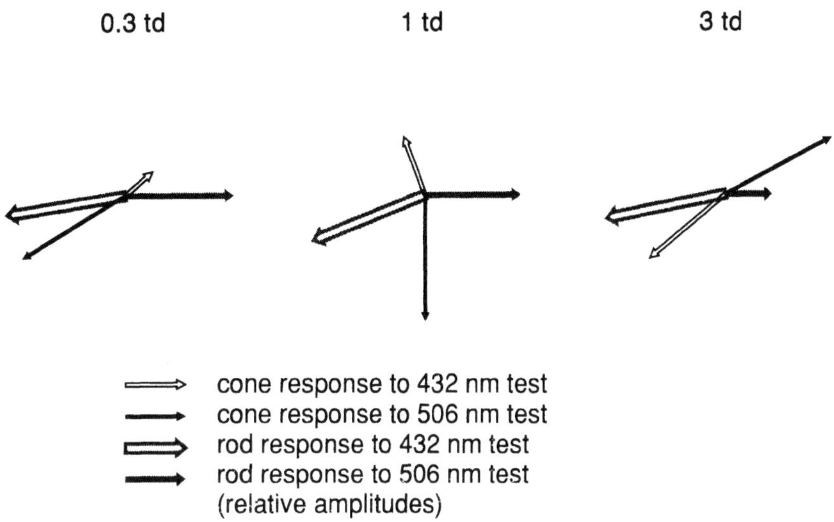

Fig. 5. Example of a vectorial model fitted to the data obtained using HFP. The four vectors represent rod and cone signals originating from the test or the reference stimulus. One cone signal only is supposed to represent the total cone population response. Rod-cone latency difference is assumed to be 80 ms. Experimental physical phase shifts (2.5 ms at 3 phot. td, 6.3 ms at 1 phot. td, 3.6 ms at 0.3 phot. td) and experimental flicker frequencies (9.4 Hz at 3 td, 7.3 Hz at 1 phot. td, 6.9 Hz at 0.3 phot. td) are used for modelling. Test wavelength 432 nm. The modelling is for the data of observer FV.

Comparison of DCB and HFP matches

The results obtained with the two matching methods significantly differ throughout the mesopic range. We restrict ourselves to asking which extra mechanisms contribute to mesopic DCB matching. The answer is to be found in the difference between results yielded by DCB matching or by HFP: for test wavelength shorter than 506 nm, more test light is always required with DCB matching than with HFP. The maximum amount of extra light increases significantly as the test wavelength increases (0.2 log unit at 413 nm and 0.4 log unit at 480 nm). Such a variation should originate from a mechanism the influence of which undergoes the same variation. The simplest assumption is that S-cones are excluded from HFP matching but not from DCB matching. However, the ratio of S-cone action spectrum to $V'(\lambda)$ decreases as wavelength increases. Consequently, extra S-cone contribution alone cannot be responsible for the differences between the two matching methods and such a simple explanation is implausible.

Furthermore, it may be noted that, for the two observers, whereas the retinal illuminance corresponding to the maximum difference increases monotonically as test wavelength increases, the curves showing the difference steepen in the mid-mesopic range for 432, 445 and 465 nm. Such a qualitative observation leads us to suppose that two extra mechanisms at least are operating under DCB matching as compared with HFP matching. One of these may be post-receptoral and have a slow response. A possible candidate is an antagonistic cone mechanism, still operating at mid-mesopic illuminance levels, the maximum spectral sensitivity of which is located between 432 and 465 nm.

Conclusion

It has been shown that it is necesssary to introduce a physical phase shift between the test and the reference to ensure precise heterochromatic flicker photometry over the full mesopic range.

Some aspects of the complex problems associated with the combination of receptor signals using either heterochromatic flicker photometry or direct comparison brightness matching have been indicated. The S-cone contribution has not been identified, but the influence of one or more antagonistic cone mechanisms using DCB matching is implied.

Acknowledgements

We are grateful to L.T. Sharpe for suggestions.

References

Commission Internationale de l'Eclairage (1978). Light as a true visual quantity: principles of measurement, CIE Publication no. 41.

MacLeod, D.I.A. (1972). Rods cancel cones in flicker. Nature 235: 173–174.

Lennie, P., Pokorny, J. and Smith, V.C. (1993). Luminance. J. Opt. Soc. Am. A 10: 1283–1293.

Sharpe, L.T., Stockman, A. and MacLeod, D.I.A. (1989). Rod flicker perception: Scotopic duality, phase lags and destructive interference. Vision Res. 29: 1539–1559.

Viénot, F. and Chiron, A. (1992). Brightness matching and flicker photometric data obtained over the full mesopic range. Vision Res. 32: 533–540.

Vos, J.J. (1978). Colorimetric and photometric properties of a 2° fundamental observer. Color Res. Appl. 3: 125–128.

Yaguchi, H., Kawada, A., Shioiri, S. and Miyake, Y. (1993). Individual differences of the contribution of chromatic channels to brightness. J. Opt. Soc. Am. A 10: 1373–1379.

Laboratoire de Physique Appliquée
CNRS UPR 257
Muséum National d'Histoire Naturelle
43 rue Cuvier, 75005 Paris, France.

41. Rod-mediated cone sensitization in a population of off-centre ganglion cells in cat retina in the presence of large adapting backgrounds

ELKE GUENTHER and EBERHART ZRENNER
(Tübingen, Germany)

Abstract

Response properties of off-centre retinal ganglion cells (RGCs) were investigated in the cat by means of extracellular recordings. The stimulus parameters were selected so as to demonstrate interactions between the rod and the cone systems. Response vs. log stimulus intensity functions were determined for the receptive field centres while both the test stimulus irradiance and the background illumination were varied over a range of up to 7 log units. In most off-centre cells, an increase in background illuminance produced a shift of the R-log I functions to higher irradiances of test stimuli indicating a normal desensitization of the centre response in the presence of background lights. In 20% of the off-centre cells, however, a strikingly different adaptation behaviour was observed. The receptive field centre became *more* sensitive in the presence of dim adapting backgrounds than in the dark adapted state. Accordingly, the threshold versus intensity function did not increase monotonically, but rather showed a 'dip' at dim backgrounds. The sensitization was observed only with a cone-stimulating test light in presence of a large, rod-adapting background. The sensitization disappeared when the background diameter was reduced to less than threefold the diameter of the receptive field. Sensitization could not be observed under any other chromatic conditions nor in on-centre cells.

Introduction

Most animals have a rod system for dim ambient illuminations, and a cone system for higher illuminations. The two different types of vertebrate photoreceptors were believed by early anatomists to act independently of each other. More recent results, however, have uncovered some extended and complex interactions between rod- and cone-signals in the visual system (e.g. Steinberg, 1969; Enroth-Cugell *et al.*, 1977; Goldberg *et al.*, 1983; Stromeyer and Hill, 1983; Frumkes and Eysteinsson, 1987).

A summatory interaction between rod and cone-mediated signals could be

B. Drum (ed.), Colour Vision Deficiencies XII, pp. 343–350.
© 1995 *Kluwer Academic Publishers, Dordrecht.*

shown at almost all levels of the visual system of higher vertebrates, from the photoreceptor layer (Nelson, 1977) through higher visual centres in the brain (Virsu *et al.*, 1987). In physiological studies, Steinberg (1969) found a linear combination of rod and cone signals in cat horizontal cells. In cat ganglion cells Enroth-Cugell *et al.* (1977) reported a nearly linear combination although they could not rule out the possibility of some nonlinear interactions. A non-linear form of rod-cone interaction has been described in recent years. Dark adapted, photically unstimulated rods exert an inhibitory influence upon cone-pathways that can be removed by selective rod-light adaption (Alexander and Fishman, 1984; Goldberg *et al.*, 1983). Frumkes and Eysteinsson (1987) have referred to this phenomenon as suppressive rod-cone interaction (SRCI). SRCI is described in amphibians, cats and humans (Arden and Frumkes, 1986: Frumkes and Eysteinsson, 1987; Pflug and Nelson, 1986). The data suggest that SRCI reflect rod modulation of horizontal cell influences upon cone pathways. SRCI could best be demonstrated with flickering test stimuli beyond 20 Hz.

We describe a form of rod-cone interaction that has some similarities with SRCI but is present only in a subpopulation of off-centre ganglion cells under distinct conditions of chromatic adaptation. In these off-centre cells a large, dim short-wavelength background induces a prominent centre response sensitization when the test stimulus was designed to stimulate the cone system. This sensitization could never be observed under any other chromatic conditions nor in on-centre cells. The sensitization is only obtained with adaptation backgrounds that considerably exceed the field centre diameter. A reduction of the background diameter to about three-fold the receptive field centre size results in a loss of sensitization.

We, therefore, conclude that there are large spatial integration areas which are involved in the sensitization effect in this population of off-centre cells. Only the surrounding rods can suppress L-cone signals in the receptive field centre: The suppressive influence on the L-cone system is abolished after rod light-adaptation.

Methods

Extracellular recordings were made from retinal ganglion cells of cats using a microelectrode inserted through a scleral guide tube. After initial anaesthesia, the animals were artificially respirated and pentobarbital-sodium was given i.v. in a continuous flow of 0.4 ml/kg*h. Additionally, alcuroniumchloride was administered at a rate of 0.09 mg/kg*h. Recordings were made directly in the area centralis or nearby (max. angle off-axis; + 5°).

For light stimulation a dual beam optical bench with a 900 W Xenon arc lamp was used. Test spot size was adjusted to match the receptive field (RF) centre diameter of each individual ganglion cell. The centre of the background was coincident with the centre of the test spot and subtended a visual angle of 25° or was reduced to the size of the test stimulus. The light stimuli were projected in Maxwellian view.

Variation of the chromaticity of the test beam (I) was achieved by narrowband interference filters (half bandwith 12 nm). For efficient stimulation of cat rods, the test stimulus had a peak energy at 501 nm. L-cones were stimulated with a wavelength of 575 nm. This is a trade-off condition as the L-cone threshold is higher than the rod threshold over a wide spectral range. A blue-green broadband filter with a transmission maximum at 456 nm (BG28) and a steep orange cut-off filter (OG580) were used in the adaptation beam (μ). Light intensity could be varied by means of neutral density filters. The test spot was presented as a square-wave pulse of 500 ms duration (D) with an interstimulus interval (ISI) of 1.8 s. Response versus Intensity functions (R-log I) to a sequence of test flashes were recorded first without a background present and then at different levels of background illuminance. The spectral sensitivity of the receptive field centres was determined for each cell, both in the dark adapted state and in the presence of different background illuminations. Spectral sensitivity and tvi-functions were derived from R-log I functions by using a 40 Hz centre-response criterion.

For data analysis, the amplified action potentials were fed into a window discriminator and a frequency analyzer. The analog signals obtained from the frequency analyzer were digitized and averaged by a Nicolet Averaging System (1072, Fabri-Tec.), and were used to provide post-stimulus time histograms (PSTHs).

Results

Most off-centre ganglion cells showed a course of adaptation as indicated by the R-log I functions in Fig. 1. A test stimulus wavelength of 575 nm was chosen as it is efficient for stimulating the L-cone system in the RF centre at high irradiances. However, at low irradiances and in the dark adapted state the spectral sensitivity of the receptive field centre is mediated by the rod system. This is shown in Fig. 1b (filled circles) where a Dartnall nomogram (Dartnall, 1957) with a rod absorption maximum (501 nm; continuous line through filled circles) fits the data best. The wavelength of the superimposed background was selected to adapt the rod-system in the entire visual field illuminated. The response threshold for the 575 nm test stimulus is lowest without any background (filled circles). For increasing test stimulus irradiances, the R-log I function rises monotonically over an intensity range of nearly 2.5 log units until a plateau is reached. An increase in background illumination light-adapted the rods and decreased their sensitivity to test stimuli in the receptive field centre. This resulted in a rightward shift of the R-log I functions along the abscissa to higher irradiances. For a bright blue-green adaptation light of 0.5 log $lm*m^{-2}$ (triangles), the peak spectral sensitivity shifts to longer wavelengths. A Dartnall nomogram with a peak at 555 nm (solid line through base up triangles) fits these data best, indicating the L-cone system as the generator of the centre response. In most cells, the transition from rod- to cone-mediated vision occurred at background illuminances between 0.5 to 1.0 log $lm*m^{-2}$

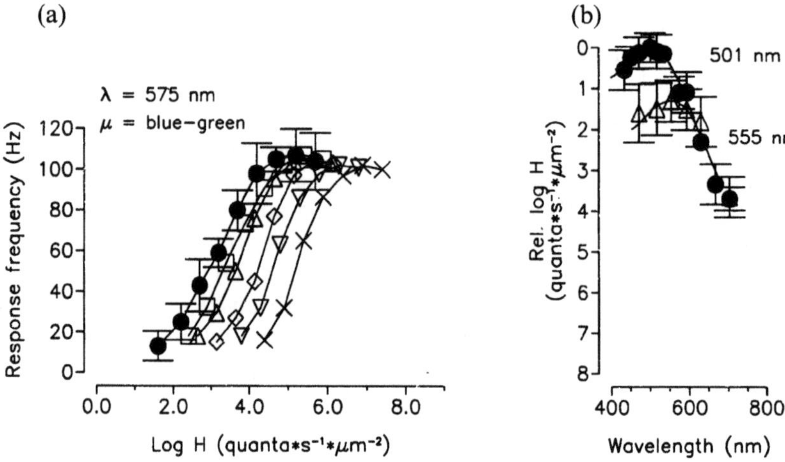

Fig. 1. (a) Mean R-log I functions (n = 13) of the receptive field centres of off-centre cells obtained with a 575 nm test stimulus. The test stimulus was presented in the dark (filled circles) and against different blue-green (BG 28) backgrounds (log E: □ – 0,5 Δ 0.5, ◇ 1.5, ∇ 2.5, × 3.5 log lm*m^{-2}). The background-induced light adaptation of the rod system results in a transition of the R-log I functions to the right along the irradiance axis. In presence of a background illumination of 0.5 log lm*m^{-2} the slopes of functions become steeper. This is paralleled by a transition on the spectral sensitivity from rod-mediated vision (Fig. 1b, filled circles) to L-cone-mediated vision (Fig. 1b, triangles). The wavelengths to the right of curves in Fig. 1b refer to the peak wavelength of the Dartnall nomogram.

Centre sensitization

About 20% of off-centre ganglion cells within the area centralis showed an entirely different course of adaptation (Fig. 2). While the position of the R-log I function in the dark adapted state (filled circles) remained unchanged, superposition of dim blue-green backgrounds (field size: 25°) resulted in a transition of the R-log I functions to the *left* (arrow). This sensitization of the centre response is most pronounced for a background illuminance of −0.5 log lm*m^{-2} (squared). The test irradiance necessary to elicit a 40 Hz criterion response is about 30 times *lower* than in the dark adapted state. At higher illuminances, the response sensitivity returns to the pre-adapted level and thereafter desensitizes in the usual manner. Notive that in the presence of backgrounds that induce the sensitization, the centre response is already L-cone-mediated (Fig. 2b, squares).

The tvi function for off-centre cells with centre sensitization shows a 'dip' at dim adaptation light in contrast to tvi functions for off-cells without centre sensitization where it rises monotonically (Fig. 3a).

Influence of background diameter

In a further experiment we investigated the effect of background diameter on the centre reponse sensitization. For this purpose the adapting background for an

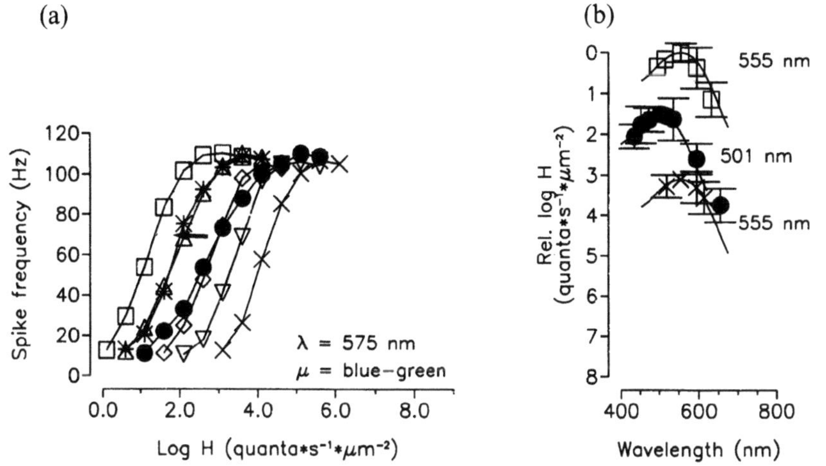

Fig. 2. (a) Background-induced sensitization of the receptive field centre; mean R-log I functions of 15 off-centre ganglion cells. Test stimulus wavelength was 575 nm, background was blue-green (● no background, −1.5, □ −0.5, Δ 0.5, ◇ 1.5, ∇ 2.5, 3.5 log lm*m^{-2}). The centre response is most sensitive in the presence of dim backgrounds in the range of − 1.5 to − 0.5 log lm*m^{-2} and *not* in the pre-adapted state (filled circles). This results in a shift of the coresponding R-log I functions to the *left* (stars and squares). Only at higher background illuminances (> 0.5 log lm*m^{-2}, triangles to crosses) does the R-log I functions shift to the right, indicating a desensitization of the centre response as one would expect in the regular course of adaptation. (b) Mean spectral sensitivity curves (n = 10) in the dark-adapted state (filled circles) and at background illuminances of −0.5 (squares) and 3.5 (crosses) log lm*m^{-2}. Even at background illuminances where the centre sensitization is most pronounced (squares), the spectral sensitivity is L-cone mediated.

off-centre cell showing a centre sensitization was reduced in diameter from 25° down to a size matching the receptive field centre (Fig. 3b; open circles). At about three times the centre size the sensitization disappears and the tvi function rises monotonically as in other off-centre cells. Thus, the data indicate large spatial integration areas being involved in the sensitization of the centre response in a subpopulation of retinal ganglion cells.

Discussion

The centre-sensitization observed in a subpopulation (20%) of off-centre ganglion cells reveals non-linear interactions between the rod and cone pathways at dim levels of background illuminance. The effect is only observed with backgrounds that mainly affect rods and with test stimuli that already are detected by the L-cone system under the given chromatic adaptation. It has not been observed in on-centre cells in the area centralis nor in the retinal periphery in any other cell type (Guenther and Zrenner, 1993). We therefore conclude that the effect is restricted to the off-pathway in the area centralis and that it involves a tonic suppression of cone-driven signals by rods, which only can be removed by rod light adaptation.

(a)

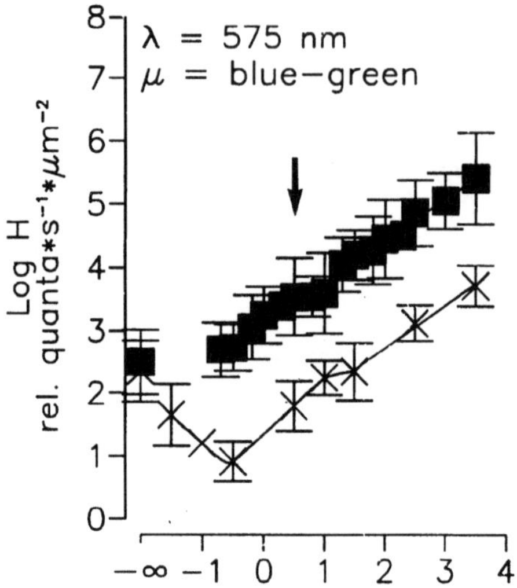

(b)

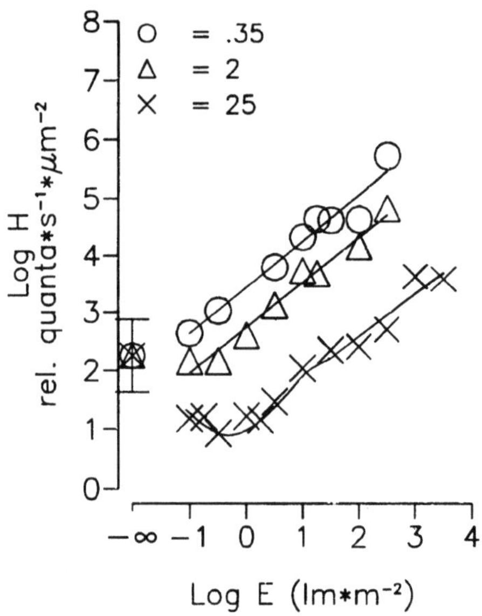

Fig. 3. (a) The mean tvi function of off-centre cells (n = 15) without centre sensitization (squares) rises monotonically; linear regression lines are fitted to the rod-mediated part (slope: 0.75) and the cone-mediated part (slope: 0.66) separated by the arrow. The tvi function of off-centre cells (n = 19) showing centre sensitizarion (crosses, $SD_{max} = \pm 0.49$) has a 'dip' at background illuminations in the range of $-$ 1,5 to $-$ 0.5 log lm*m^{-2}. The centre response is 1 to 1.5 log units more sensitive than in the no-background state. For higher background illumination, sensitivity first reaches dark-adapted levels and then decreases with a slope of 0.69.

(b) Effect of background size upon centre sensitization. The tvi functions are plotted for one cell

In recent years, a growing body of evidence suggests that rod adaptation can influence cone-mediated signals in vertebrate retina. In amphibians, Frumkes and Eysteinsson (1987, 1988) have shown that rod light adaptation enhances cone-mediated responses to rapid flicker. The authors referred to this phenomenon as suppressive rod-cone interaction (SRCI). The effect was greatest when flicker frequencies higher than 20 Hz were used (Goldberg, *et al.*, 1983). Intracellular recordings from retinae of amphibians (Frumkes and Eysteinsson, 1987; Frumkes and Wu, 1990) and cats (Pflug and Nelson, 1987) indicate that SRCI takes place in the distal retina and is mediated by horizontal cells.

Recent measurements of SRCI show that when using 1-2 Hz sawtooth-flicker with either rapid-on or –off waveforms, rod light adaptation predominantly influences cone-responses at flicker offset rather than at flicker onset (Lange *et al.*, 1992; Frumkes *et al.*, 1992). Both Lange *et al.* (1992) and Frumkes *et al.* (1992) argue that it would be advantageous for an off-system, or at least part of it, to be suppressed under dark adapted conditions when it can provide no meaningful signal; this possibility is also supported by recent anatomical and physiological data in mammals (see Daw *et al.*, 1990). Although there are some similarities we do not think that the SRCI is related to the form of rod-cone interaction described in this paper. Differences in the temporal parameters of stimulation in the two effects argue against the presence of a similar mechanism. In addition, an involvement of horizontal cells would point to the involvement of a more general mechanism of rod-cone interaction than the one obviously present only in a small population of off-centre cells.

A sensitization of the field centre of off-centre cells in the presence of dim, rod-adapting backgrounds was obvious only when the background diameter considerably exceeded the test stimulus diameter. This suggests that large spatial integration areas are necessary for the effect to occur.

A background-induced sensitization of test responses has also been reported by Crawford (1940), Westheimer (1967) and others, who show that the response thresholds in the presence of a steady, small field background is higher than the response threshold in the presence of a large field background. Their term 'sensitization', however, refers to the lowering of thresholds when the size of the background field is increased. It is certainly not related to our term 'centre sensitization' since it is not restricted as in our case to only a small cell population. The steeper slope of the tvi functions in the photopic region as measured in our experiments, however, may be explained on the basis of the finding that small backgrounds can overload the cone system and produce saturation at low light levels (Buss *et al.*, 1982).

---

showing centre sensitization in the presence of a large field of blue-green background (25°, crosses). The slope of the function in the linear part is 0.74. Reducing the background to a diameter of 2° (triangles) results in a loss of sensitization in the receptive field centre and the tvi function rises with a slope of 0.77. When the background matches the diameter of the test stimulus, the slope of the function is even steeper (0.81). This result suggests that a lateral mechanism with a large spatial integration mediates the centre sensitization.

References

Alexander, K.R. and Fishman, G.A. (1984). Rod-cone interaction in flicker perimetry. Br. J. Ophthal. 68: 303–309.

Buss, C.M., Hayhoe, M.M. and Stromeyer, C.F. (1992). Lateral interactions in the control of visual sensitivity. Vision Res. 22: 693–709.

Crawford, B.H. (1940). The effect of field size and pattern on the change of visual sensitivity with time. Proc. Royal Soc. B 129: 94–106.

Datnall, H.J.A. (1953). The interpretation of spectral sensitivity curves. British Med. Bull. 9: 24–30.

Daw, N.W., Jensen, R.J. and Brunken, W.J. (1990). Rod pathways in mammalian retinae. TINS 13: 110–115.

Enroth-Cugell, C., Hertz, B.G. and Lennie, P. (1977). Convergence of rod and cone signals in the cat's retina. J. Physiol. 269: 297–318.

Frumkes, T.E. and Eysteinsson, T. (1987). Suppressive rod-cone interaction in the distal vertebrate retina: intracelluar records from Xenopus and Necturus. J. Neurophysiol. 57: 1361–1383.

Frumkes, T.E. and Eysteinsson, T. (1988). The cellular basis for suppressive rod-cone interaction. Visual Neurosci. 1: 263–273.

Frumkes, T.E. and Wu S.M. (1990). Independent influences of rod-adaptation on cone-mediated responses to light onset and offset in distal retinal neurons. J. Neurophysiol. 64: 1043–1054.

Frumkes, T.E., Lange, G. Denny, N. and Beczkowska, I. (1992). Influence of rod adaptation upon cone responses to light offset in humans: I. Results in normal observers. Visual Neurosci. 8: 83–89.

Goldberg, S.H., Frumkes, T.E. and Nygaard, R.W. (1983). Inhibitory influence of unstimulated rods in the human retina: evidence provided by examining cone flicker. Science 221L 180–182.

Guenther, E. and Zrenner, E. (1993). Differences in adaptation between on- and off-centre ganglion cells and rod-mediated cone sensitization in cat retina. J. Physiol. 464: 613–627.

Lange, G., Frumkes, T.E., Denny, N. and Beczkowska, I. (1992). Influence of rod adaptation upon cone-responses to light offset in humans: I. Results in normal observers. Visual Neurosci. 8: 91–95.

Nelson, R. (1977). Cat cones have rod input: a comparison of the response properties of cones and horizontal cell bodies in the retina of the cat. J. comp. Neurol. 172: 109–135.

Pflug, R. and Nelson, R. (1987). Enhancement of red cone flicker by rod selective backgrounds in cat horizontal cells. Neurosci. Abstr. 16: 402.

Pflug, R. and Nelson, R. (1987). Background enhancement of cone signals in cat horizontal cells. Invest. Ophthalmol. Vis. Sci. Suppl. 28: 240.

Steinberg, R.H. (1969). Rod-cone interaction in S-potentials from the cat retina. Vision Res. 9: 1331–1344.

Stromeyer, C.F. and Hill, T.L. (1983). The cone threshold: spatial interactions of rod and cone adapting signals. Vision Res. 23: 713–722.

Virsu, V., Lee, B.B. and Creuzfeld, O.D. (1987). Mesopic spectral responses and the Purkinje shift of macaque lateral geniculate nucleus cells. Vision Res. 27: 191–200.

Westheimer, G. (1967). Spatial interaction in human cone vision. J. Physiol. 190: 139–154.

Westheimer, G. (1970). Rod-cone independance for sensitizing interaction in the human retina. J. Physiol. 206: 109–116.

Elke Guenther
Department of Experimental Ophthalmology
Röntgenweg 11
D-72076 Tübingen, Germany

PART SIX

Electrophysiology of colour vision

42. Retinal and brain responses of S-cones

P. GOURAS, C.J. MACKAY, M. ROY, M. SAEKI, and S. YAMAMOTO
(New York, NY, U.S.A.)

Abstract

S-cone mediated responses can be identified in the human ERG and VER by using ganzfeld spectral flashes in the presence of moderately strong ganzfeld white (tungsten) adapting fields. The S-cone mediated responses are produced by short wave flashes (404 to 482 nm). The evidence that these responses are due to S-cones comes from action spectra and their unique presence in an S-cone achromat who has only S-cone and no L- and M-cone function. These short wavelength flashes produce both S- as well as L- and M-cone responses. The S-cone responses are distinguishable by their greater delay. At the ERG the S-cone b-wave is about 15 msec later than the L-M cone b-wave. In the VER the S-cone response is about 33 msec later than the L-M cone responses. The greater delay in the S-cone response at the brain compared to the retina may be due to antagonistic interactions between cone mechanisms at this point and/or a slower conduction time.

Inrtroduction

This paper reviews research we have been doing to identify short-wavelength-sensitive (S-) cone mediated visually evoked responses in human subjects (Gouras and MacKay, 1990; Gouras *et al.*, 1993a; Gouras *et al.*, 1993b). The ganzfeld S-cone electroretinogram (ERG) can be exposed by strongly light-adapting the retina with white or yellow light (Gouras and MacKay, 1990; Gouras *et al.*, 1993a). White light is interesting because it also light-adapts S-cones. Nevertheless when a thite (tungsten) adapting light exceeds about 5000 photopic trolands, the S-cone ERG begins to emerge out of an ERG that is otherwise dominated by L and M cones. The S-cone ERG appears as a corneal positive b-wave riding on a quicker b-wave produced by the L and M cones. In addition there is a relatively large but later negative wave that follows the b-wave much like the slow negativity that follows the b-wave of rods.

Our proof that this ERG is due to S-cones is based on its action spectrum and

B. Drum (ed.), Colour Vision Deficiencies XII, pp. 353–364.
© 1995 *Kluwer Academic Publishers, Dordrecht.*

on the fact that an S-cone achromat who has no L or M cone function generates only this response (Gouras and MacKay, 1990; Gouras *et al.*, 1993). The finding that the S-cone ERG is slower than the L-M cone response is unexpected because recordings from primate cone outer segments have revealed that S-cones do not respond more slowly than the other cones (Baylor, 1986). In this paper we examine an even greater delay in the S-cone response of the brain.

Methods

The methods used in this research have been published elsewhere (Gouras and MacKay, 1990; Gouras *et al.*, 1993a; Gouras *et al.*, 1993b). They involve the use of ganzfeld flashes on relatively strong (17,000-20,000 photopic trolands) ganzfeld adapting fields. Kodak Wratten filters are used to modify the test flash; the strength of the background is controlled by the voltage across a 12 volt, 80-W halogen lamp. For all of the major conclusions of this paper the voltage was set to produce a retinal illumination of 15,000-20,000 photopic trolands. In Fig. 2 and 3 the influence of reducing the adapting field on the ERG is illustrated. This produces both a reduction in illumination and a change in color temperature. Evidence obtained with the use of spectral filters (yellow) in the adapting beam indicate that retinal illumination rather than color temperature is the major factor changing these ERGs. Young human, normal subjects with widely dilated pupils, diameter 8 mm, are used. The visual evoked response (VER) of the brain is recorded with a gold cup electrode on the scalp about 2 cm above the inion and a gold clip reference electrode on an ear lobe. Responses are detected with a Nicolet IIA system giving us a sensitivity of at least 0.1 microvolt.

Results

Figure 1 illustrates how we currently measure the S-cone ERG. This S-cone response is best detected with short wavelength stimuli near 450 nm. It is measured as a b-wave riding on top of a quicker b-wave of the L and M cones. There are sometimes oscillations in the L-M cone response, one of which occurs almost contemporaneously with the S-cone b-wave. This osciallation, however, has the action spectra of L and M cones and does not grow significantly with increase in flash intensity (Gouras *et al.*, 1993a).

Figure 2 shows the cone ERG of an S-cone achromat to the same spectral flashes. Only an S-cone response is apparent and stimuli longer than about 482 nm elicit no ERG even at much greater energies. There is no augmentation in the amplitude of this S-cone ERG implying that the L-M cones do not strongly suppress the S-cone response under these testing conditions.

Figure 3 illustrates how white light adaptation exposes this response in the normal ERG. At relatively dim backgrounds, the cone ERG is relatively large but the S-cone b-wave remains small or becomes smaller. As the background

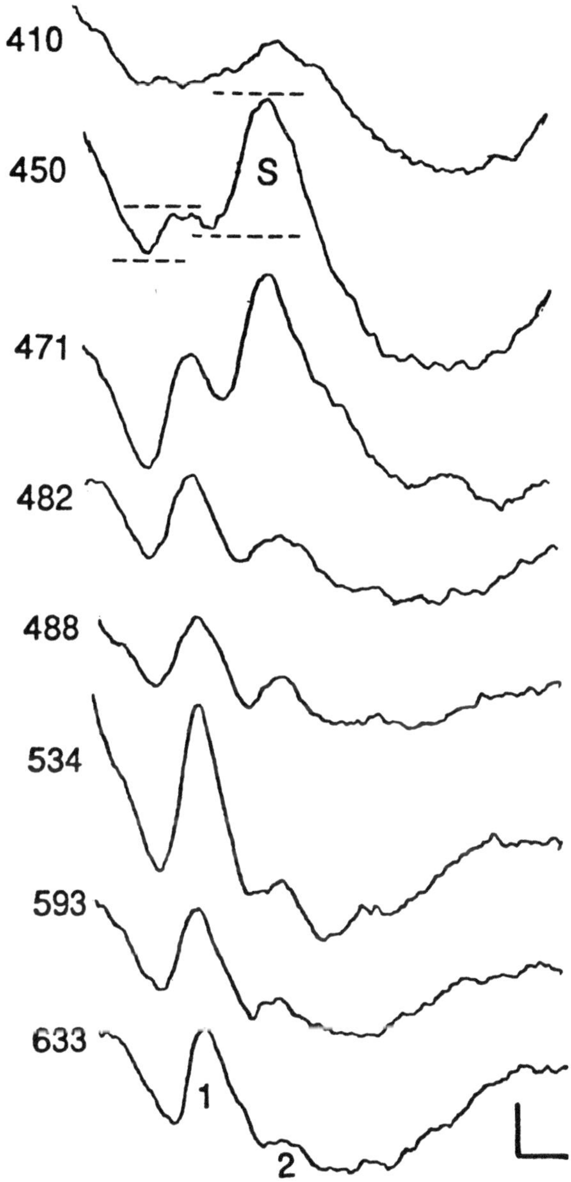

Fig. 1. Flash ERGs to spectral stimuli (wavelength in nanometers on left) in the presence of a 17,000 photopic troland background. The hatched lines show the measured b-wave of the L-M cones (1st) and the S-cones (2nd). The S-cone response is only detectable at 410, 450 and 471 nm. The numbers 1 and 2 below show the initial and second oscillation of the L-M cone b-wave. Each trace is 100 msec. The calibration is 1 microvolt vertically and 11 msec horizontally.

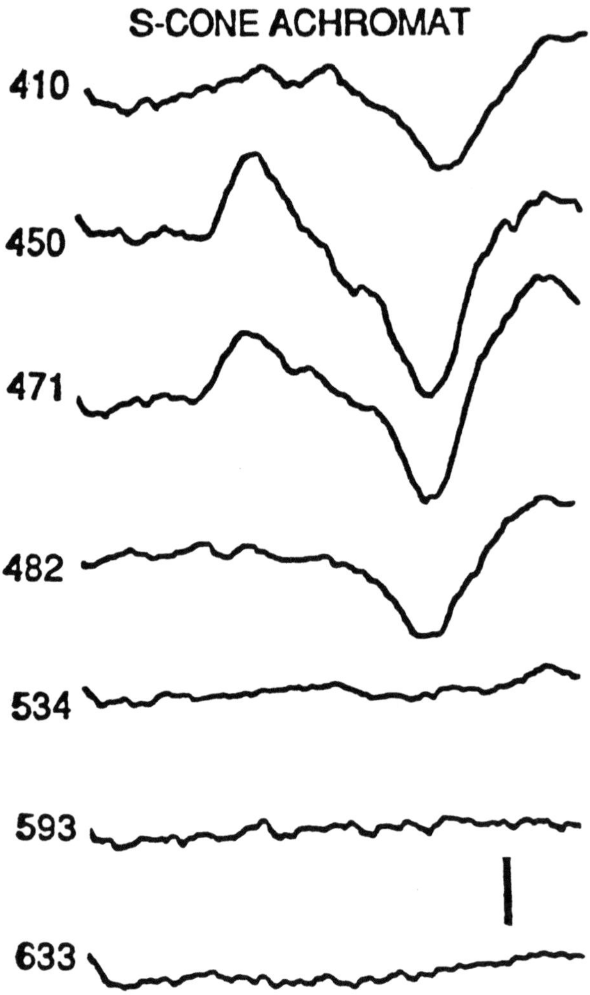

Fig. 2. Flash ERGs of an S-cone achromat under the same conditions of Fig. 1. No responses are detectable to flashes of wavelengths longer than 482 nm. Each trace is 0.1 sec.

light is increased, this S-cone b-wave becomes more conspicuous as the L-M cone response diminishes.

Figure 4 illustrates the relationship between the amlitude of the S-cone and L-M cone b-waves and the intensity of the adapting light. There appears to be an optimum background of about 20,000 trolands where the S-cone b-wave can be separated from the response of the L and M cones.

Figure 5 illustrates the S-cone ERG obtained by subtracting the L-M cone ERG obtained by a long wavelength flash, which was been matched by using neutral density filters to produce approximately an equal L-M cone b-wave to that produced by the 450 nm flash (above). The subtracted wave shows a small

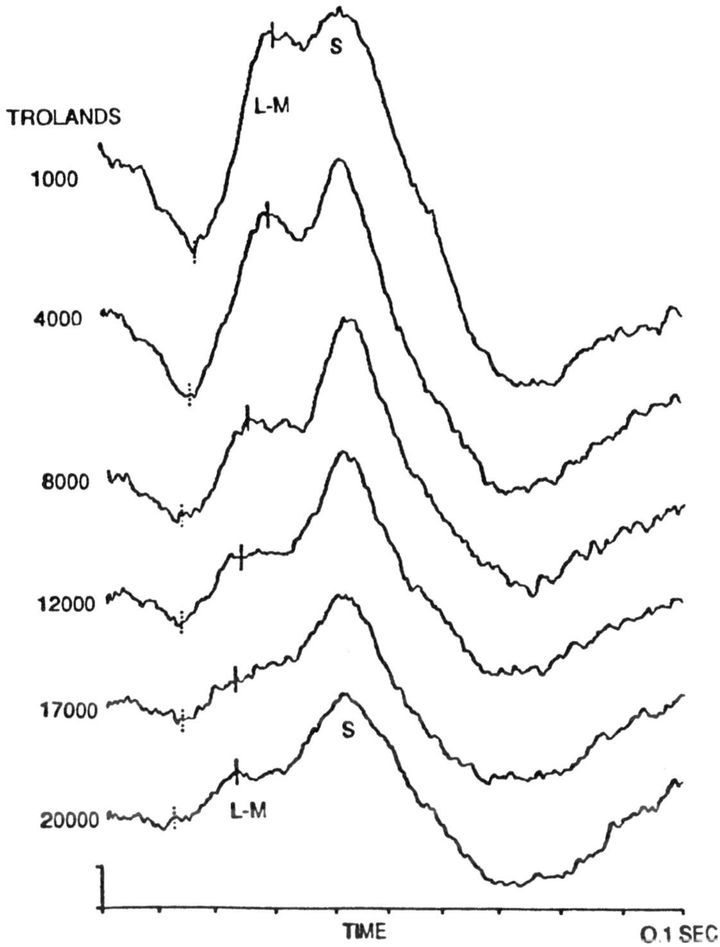

Fig. 3. Flash ERGs of a normal subject to a 450 nm flash in the presence of different amounts of background light, shown as photopic trolands on the left.

a-wave a slow b-wave followed by a slow negative wave. If we assume superposition of these ERG responses, then this is a close facsimile of the isolated S-cone ERG.

By matching the L-M cone ERG to be approximately identical at seven different spectral flashes we can facilitate detecting the S-cone VER. Figure 6 illustrates the use of such stimuli to produce VERs. All stimuli which affect only L and M cones produce relatively similar, not identical, VERs which have two obvious inion positive waves; these stimuli fail to produce a larger later positive wave which is conspicuous with short wave flashes. Figure 7 shows that increasing or decreasing the flash intensity around the match point for equal L-M cone ERGs does not produce this late wave with flashes that do not affect S-cones. With flashes that do affect S-cones, the late wave is graded with light intensity.

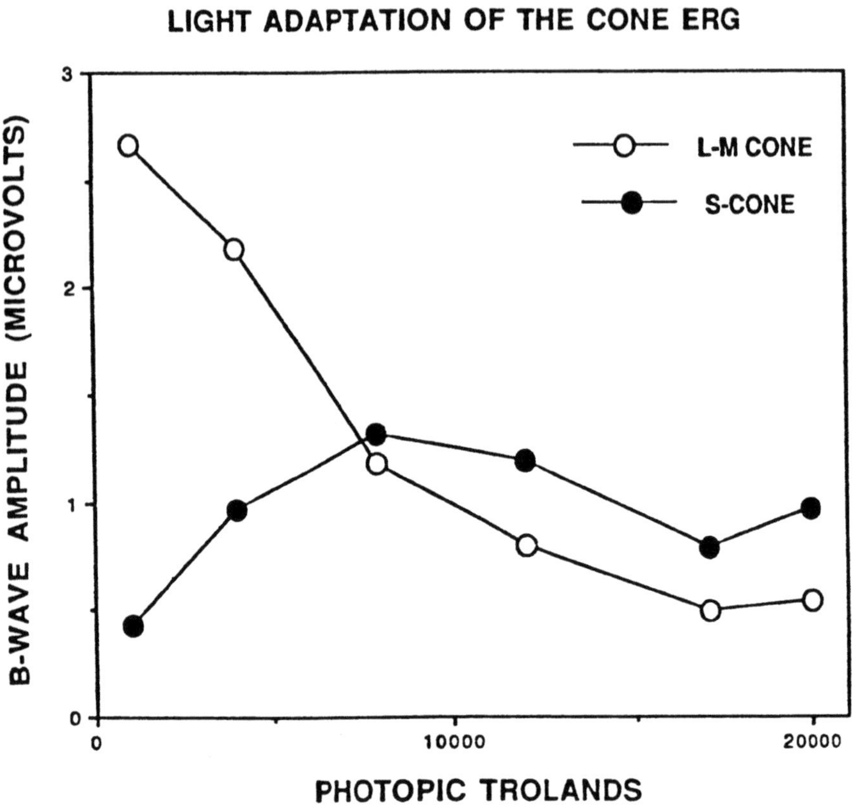

Fig. 4. The relationship between cone b-wave amplitude as measured in Fig. 1 for the L-M and S-cone response and the strength of the background light (abscissa) in photopic trolands.

Figure 8 shows a significant short wavelength contribution to the VER using stimuli matched for L-M cones. The ordinate in Figure 8 compares the maximum trough to peak response in the VER at a particular wavelength of stimulation to the maximum trough to peak response to the 633 nm flash. With all flashes that produce no S-cone ERG, this ratio is essentially unity. With flashes that do generate S-cone ERGs, this ratio becomes larger than 1. It is maximum with 450 nm flashes which are most effective for eliciting S-cone ERGs.

Figure 9 is a remarkable example of how the activation of the S-cone system can alter the flash VER. In this subject flashes which are absorbed by S-cones generate a prolonged oscillation in the VER that follows the appearance of component 3. White light, which also influences S-cones, does not generate this oscillation, however. We believe that it does not because it had to be dimmed too much by neutral filtering in order to produce equal L-M cones ERGs. This idea is supported by the fact that such dim white flashes do not generate an S-cone ERG.

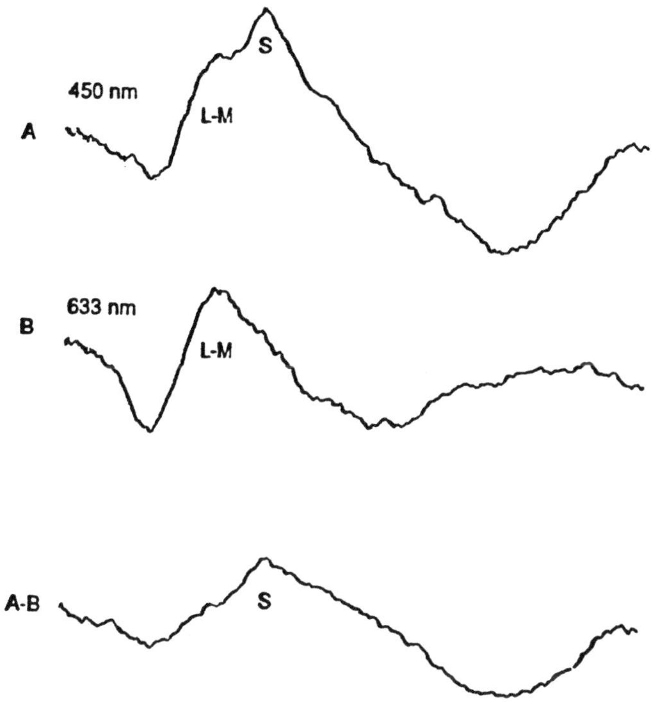

Fig. 5. The ERG by subtracting an L-M cone ERG to a long wavelength (633 nm) flash from a mixed L-M and S-cone response to a short wavelength (450 nm) flash when the amplitude of both L-M cone b-waves are approximately equal. Each trace is 0.1 sec.

Discussion

The results indicate that S-cone responses of the brain can be identified by arranging the stimulus parameters appropriately. These S-cone responses are relatively large but later than those generated by L and M cones. This delay is intriguing because it is even greater than that observed at the ERG level. The S-cone ERG b-wave is about 15 msec later than that of the L and M cones (Gouras and MacKay, 1990; Gouras *et al.*, 1993). In the VER, the S-cone response is about 25 msec later than the most comparable L and M cone mediated response, component 2. Although these values may only indirectly reflect the underlying neuronal events, they must depend on a unique difference in the way S-cone signals are conducted and/or processed in visual cortex.

This delay could represent a slower conduction time of the S-cone subserving neurons in the visual pathway. But current evidence suggests that the retinal axons are faster than many of those subserving L and M cones (Malpeli and Schiller, 1978; DeMonasterio, 1979). S-cone signals are thought to be transmitted from the retina to the brain by a relatively small subset of ganglion cells, perhaps about 10% of all optic nerve fibers, in which the responses of S-

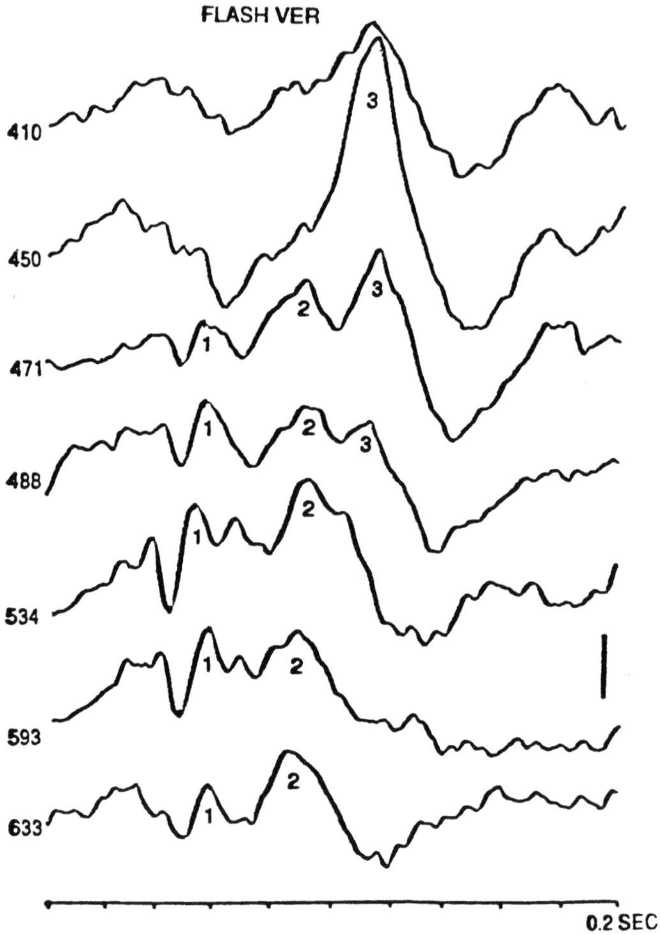

Fig. 6. Flash VERs to stimuli matched to produce L-M cone ERGs of approximately equal amplitude. Component 3 is uniquely produced by S-cones. The wavelength of the flash in nanometers is shown on the left. Each trace is 0.2 sec.

cones are opposed by L and/or M cones. In some of these neurons S-cones excite and L and/or M cones inhibit; in a smaller fraction the converse occurs (Kaplan *et al.*, 1990; Gouras, 1992). It has been suggested that the former are on- and the latter are off-channels (Gouras, 1992). If this hypothesis is correct there must be L and M cone signals being transmitted by these same S-cone channels and consequently arriving at visual cortex with the same delay as the S-cone signal. To a flash stimulus, however, the on-channels would be favored in the sense that it would set off an excitatory S-cone signals along the on-channels and an inhibitory signal from L-M cone stimulation. The off-channel would respond conversely but would be weakly provoked by a flash. On the other hand a decrement of light would provoke excitation by L-M cones in the on-channel and

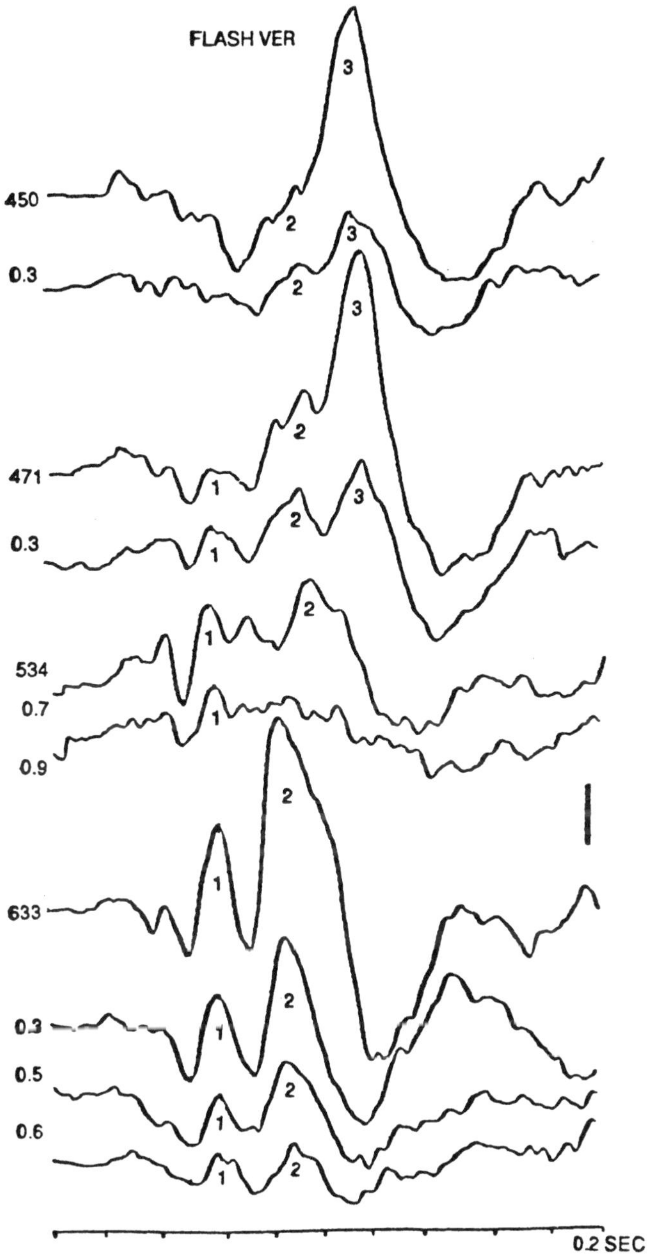

Fig. 7. Flash VERs to four different spectral stimuli. The energy of each spectral stimulus is varied slightly around the match point for producing equal L-M cone b-waves (neutral density filter values shown). Despite these changes in energy only the short wavelength flashes produce component 3. Each trace is 0.2 sec.

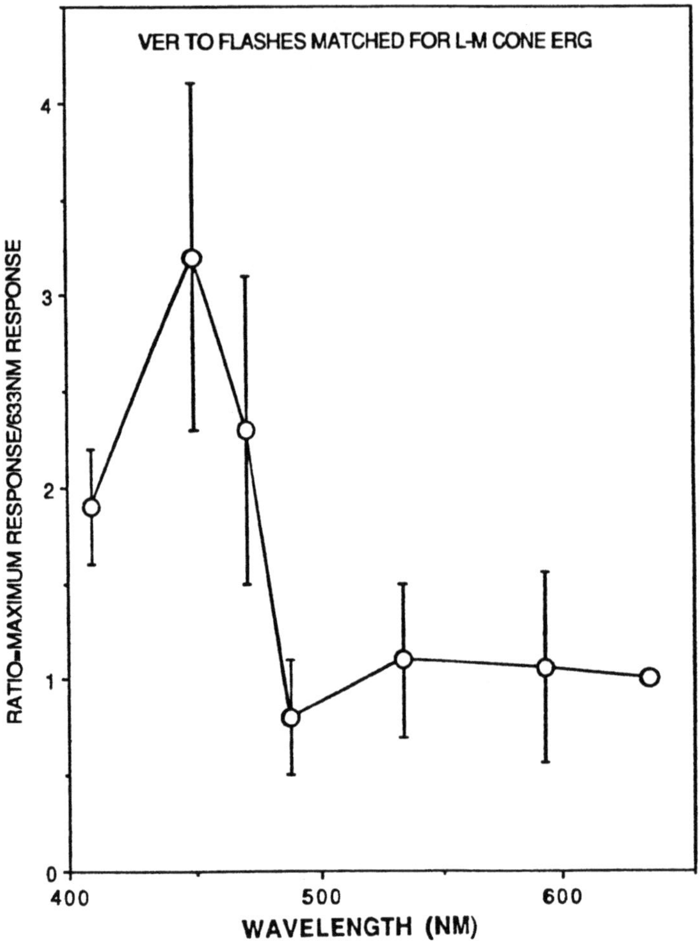

Fig. 8. The ratio of the peak to peak VER response to 7 different spectral stimuli (abscissa) to the peak to peak response to 633 nm stimulation (ordinate) with stimuli approximately matched to produce equivalent L-M cone ERGs for 7 different subjects. The vertical lines indicate the ± standard deviations.

if there were the same conduction delay in this channel, it would generate a later L-M cone VER comparable to that of component 3 in our experiments. This prediction is a viable test of the conduction velocity hypothesis.

Another explanation for the greater delay of the S-cone VER would be one that depends on antagonistic interactions between L and M versus S-cone systems which leads to a delay in appearance of the S-cone response. There is a suggestion that such interaction is occurring between L and M cone systems. For example, the positive peak of component 2 of the VER to the green (534 nm) flash is later than that to the red (633 nm) flash (Figs. 6 and 7). Because both of these response are due only to L and M and not to S-cones, this difference in waveform

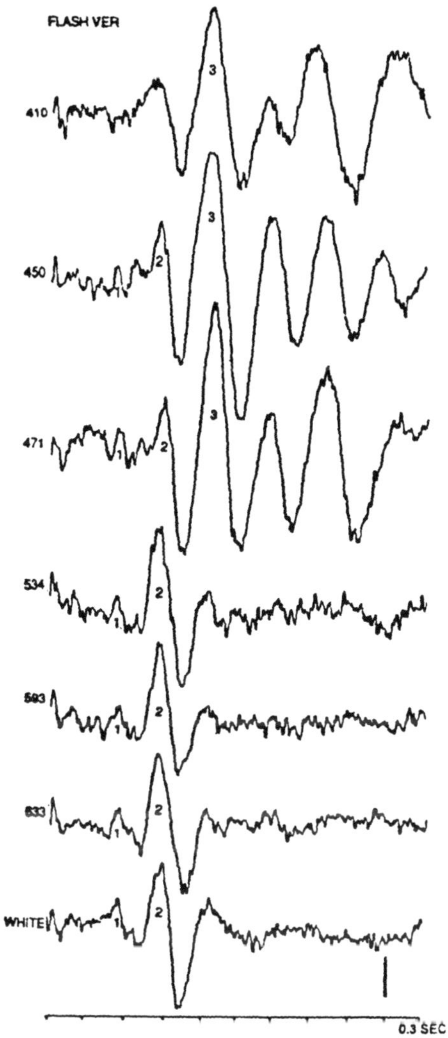

Fig. 9. Cone VERs to 6 spectrally different stimuli all approximately equivalent for producing L-M cone ERGs. A VER to a white light stimulus also equated for producing an equivalent L-M cone ERG is included (below). Only the stimuli which affect the S-cone mechanism generate component 3 and a unique oscillation. The white light stimulus must be dimmed by 1.6 neutral density filtering which makes it extremely weak for generating an S-cone response. Each trace is 0.3 seconds in duration.

must reflect either a difference in conduction time or interactions between L and M cones. At the ERG level the response to the green flash is extremely simular to that from the red flash, on neutral white adapting fields, implying there is no significant delay of the L and M signals at the inner and outer layers of the retina. At the ganglion cell level there are opponent interactions between L and M cone signals in many ganglion cells which could alter the waveform of the VER.

Therefore there is the possibility that the VER delay of the S-cone response is due to cone interactions at the retinal ganglion cell, geniculate or cortical cell level. One way to test this hypothesis is to examine the S-cone VER of an S-cone achromat. Since such subjects have no L or M cone function, the VER should be as delayed as normal subjects if the delay is due to conduction. If it is due to L-M cone interactions, this delay should disappear.

A ganzfeld flash is presumably a weak stimulus for eliciting responses from neurons in visual cortex, which are thought to be uniquely responsive to borders, slits or bars oriented in space. Nevertheless these flashes generate relatively large VER responses. A large number of cortical cells must be responding to these flashes. It could be argued that these responses reflect only presynaptic geniculate input fibers but the fact that our subjects perceive the presence and color of the flashes indicates that they must activate cortical cells. Perhaps what is a weak response to a ganzfeld flash by many cortical neurons, which respond more strongly but more individually to oriented stimuli, give rise to these VERs as well as to these percepts.

The fact that there is a fast and a slower component to the L and M cone VER could be due to a faster magno (phasic) and a slower parvo (tonic) system. Perhaps some conditions in which there is a selective loss of one or the other of these subsystems (King-Smith *et al.*, 1980) could be used to test this hypothesis.

References

Baylor, D.A. (1986). Photoreceptor signals and vision. Invest. Ophthalmol. Vis. Sci. 28: 34–49.

De Monasterio, F.M. (1979). Asymmetry of on- and off-pathways of blue-sensitive cones of the retina of macaques. Brain Research 166: 39–48.

Gouras, P. and MacKay, C.J. (1990). Electroretinographic responses of the short wavelength sensitive cones. Invest. Ophthalmol. Vis. Sci. 31: 1203–1209.

Gouras, P., MacKay, C.J. and Yamamoto, S. (1993a). The human S-cone electroretinogram and its variation among subjects with and without L and M cone function. Invest. Ophthalmol. Vis. Sci. 34: 2437–2442.

Gouras, P., MacKay, C.J., Roy, M. and Yamamoto, S. (1993b). Brain responses of short-wavelength cones. Clin. Vis. Sci. 8: 519–527.

Gouras, P. (1992). Retinal circuitry and its relevance to diagnostic psychophysics and electrophysiology. Current Opinions in Ophthalmol. 3: 803–812.

Kaplan, E., Lee, B.B. and Shapley, R.M. (1990). New views of primate retinal function. In: Osborne, N. and Chader J. (eds.), Progress in Retinal Research: 273–336, Pergamon Press, Oxford.

King Smith *et al.* (1980). Human Vision without tonic ganglion cells? In: Verriest, G., (ed.), *Colour Vision Deficiencies V*: 99–105, Adam Hilger Ltd., Bristol.

Malpeli, J.G. and Schiller, P.H. (1978). Lack of blue OFF-center cells in the visual system of the monkey. Brain Research 141: 385–389.

Peter Gouras, M.D.
Columbia University
Department of Ophthalmology
630 W. 168 St.
New York, NY 10032, U.S.A.

43. Evaluation of macular function by red-flicker electroretinogram in optic media opacities

M. FIORETTO, G.P. FAVA, C. BURTOLO, C. ORIONE and M. ZINGIRIAN

(Genoa, Italy)

Abstract

The evaluation of macular function in patients with opacities of the optic media is very difficult.

We have studied the reliability of the red flicker electroretinogram (FLIERG) in 173 patients with visual acuity less than or equal to 0.1 due to corneal opacities (19 cases), cataract (117 cases) and uveitis (37 cases).

Results were compared to those obtained in 82 age-matched normal subjects and to those obtained in the same 173 patients after surgery or when optic media were clear in uveitis patients. FLIERGS were recorded after 15 minutes of dark adaptation; full field red flash stimuli were 475 LUX in intensity at the frequency of 31 Hz. Amplitudes measured from peak to peak in normal subjects averaged 85 ± 11.5 μV.

Results suggest that the red FLIERG is effective in the diagnosis of maculopathies when optic media are opaque.

Introduction

Evaluation of macular function in patients with optic media opacities is rather difficult, even with electrophysiologic methods (Biersdorf and Diller, 1969; Birch and Fish, 1988; Foerster and LI, 1986; Porciatti *et al.*, 1989; Seiple *et al.*, 1986). The impossibility of using the chequerboard for low vision, does not allow for a detailed electrophysiologic analysis of the visual pathways. Flash visual evoked potentials (VEP), due to their considerable inter- and intra-individual variability, do not provide precise information. The VEP tracing is a graphic representation of the changes in potential on the scalp induced by visual stimuli, and is modified by any disease affecting the visual pathways, and can thus not be considered to be an index solely for macular function. Considering the importance of assessing central retinal function for the prognostic value of surgery, we have evaluated the reliability of flicker

B. Drum (ed.), Colour Vision Deficiencies XII, pp. 365–368.
© 1995 *Kluwer Academic Publishers, Dordrecht.*

electroretinogram (FLIERG) in a group of patients with cataract, in a group with corneal opacities and in a group with uveitis.

Materials and methods

One hundred and seventy-three patients aged 65 to 84 years (mean 74.8 ± 5.1), with optic media opacities reducing the best corrected visual acuity to 0.1, were examined. Out of these, 117 suffered advanced senile cataract, mostly bilateral, 19 presented corneal opacities due to previous keratopathy, and 37 suffered from posterior uveitis. Diabetic subjects and those who had a history of tapetoretinal degeneration or thrombosis of the central vein of the retina were excluded from the study.

The FLIERG results were compared with a group of 82 healthy subjects aged 58 to 82 years (mean 68.6 ± 8.4).

The FLIERG examination was carried out with the following method:

- dark adaptation for 15 minutes
- pupils dilated with 1% tropicamide and 10% phenylephrine
- full-field stimulation (AMPLAID SD 15)
- red light flash: intensity 475 LUX – frequency: 31 Hz
- dermal Ag/AgCl electrodes on the inferior orbital margin – reference: linked mastoids – ground: Fpz (10-20 International System)
- band pass 0.5 – 500 Hz
- sampling time: 100 msec.

The reliability of the method was compared with the examination of the macular area carried out with a binocular ophthalmoscope or with a Volk lens at the biomicroscope when the clarity of the optic media was recovered.

Results

Flicker ERG amplitude was measured from peak to peak, and was found to be 85 ± 11.5 μV in normal subjects.

In patients with cataract the amplitude was slightly lower (76 ± 11.2 μV), and this may be due to the filter effect of the cataract, but it may be more convenient, instead of considering the global value, to observe that out of 117 patients, in 96 cases the result was similar to that of the control group (81 ± 13.4 μV), whereas in the remaining 21 cases, in 12 patients the amplitude was significantly reduced (41 ± 13.4 μV, $p < 0.0001$) and in 9 patients the trace was not recognizable from the noise.

Thirty days after extracapsular cataract extraction surgery with introduction of IOL in posterior chamber the examination was repeated in all patients.

In the 96 cases with almost normal FLIERG tracings, the second record showed amplitudes equaling 84 ± 14.6 μV, which were thus comparable to those of the control group.

In the 12 cases with a FLIERG of decreased amplitude the results obtained subsequently were similar to the previous ones in 9 cases (44 ± 8.6 μV), whereas in 3 cases the tracings were not detectable. The tracing which was not detectable in the 9 patients remained undetectable even after surgery.

The ophthalmoscopic examination carried out after cataract surgery showed that out of the 96 cases with normal FLIERG, 91 cases did not present maculopathies, whereas the other 5 cases presented a moderate senile maculopathy which in any case allowed for a visual acuity of 0.4–0.6.

In the 12 cases in whom FLIERG was significantly decreased the ophthalmoscopic examination has shown the presence of a dry macular degeneration, and in the 3 cases in which the tracing was not detectable there was a wet macular degeneration.

In the 9 cases whose tracing was not detectable both before and after surgery the ophthalmoscopic examination has shown the presence of a macular degeneration with neo-vascularization which reduced visual acuity to less than 0.1.

In the 19 patients with keratopathy FLIERG amplitude was slightly reduced with respect to the normal subjects (81 ± 11.3 μV, T = 1.42, n.s.), and the values normalized in all patients in a 6-month follow up after surgery (87 ± 9.1 μV, T = 0,43, n.s.). Funduscopy did not reveal maculopaties. Visual acuity was greater than or equal to 0.7.

In the patients with uveitis (37 cases) FLIERG amplitude values were slightly reduced with respect to the normal subjects (75 ± 11.1 μV, T = 3.75 $p < 0.001$) in 22 cases. In 8 cases there was a significant reduction (45 ± 8.7, T = 1.25 $p < 0.001$). In the remaining 7 cases the tracing was not detectable.

The patients were re-examined with transparent optic media, with the following results: the first 22 cases had tracings similar to those of the normal group (83 ± 13.7 μV). Ophthalmoscopy did not reveal the presence of maculopathy and visual acuity was greater than or equal to 0.9. Out of the 8 cases with significantly reduced amplitude, 3 normalized (80 ± 6.5 μV) and 5 remained significantly reduced (52 ± 8.6 μV, T = 10.15 $p < 0.001$). The 3 normalized ones were ophthalmoscopically normal and visual acuity was greater than or equal to 0.8. The 5 cases with reduced amplitude presented moderate wet maculopathies and visual acuity was between 0.4 and 0.8. Out of the 7 cases in which the tracing was not detectable, in 4 it was possible to obtain tracings with reduced amplitudes (38 ± 6.6 μV, T = 13.60 $p < 0.001$) whilst in the remaining 3 the tracing remained indistinguishable from the background noise.

The ophthalmoscopic examination revealed 3 severe wet macular degenerations, 2 pseudo-macular holes and 2 dry macular degenerations with atrophic areas. In all cases visual acuity was less than 0.3.

Discussion

The results obtained with the FLIERG method carried out with optic media opacities such as to make an exploration of the macular area impossible, compared with an accurate examination of the fundus after transparency has been restored, allow us to point out the excellent diagnostic and prognostic reliability of the FLIERG ($x = 0.44$, n.s.).

A good standardization of the recording method underlies the clinical use of the examination. We would like to point out that in our experience lower stimulation frequencies (21 Hz), tested only in 6 patients whose 31 Hz stimulation frequency tracing was not detectable, provide poorly reliable results, since the tracing was recordable in 4 cases, albeit of low amplitude. It is likely that these results are provided by portions of the paramacular retina in which the frequency of 21 Hz is below the critical fusion frequency. A better standardization of the method and a larger normative study tested at different frequencies could probably permit a better prognostic evaluation of macular function in subjects with optic media opacities, and thus enabling us to distinguish also the amplitude of the macular area affected by the degenerative disease.

References

Biersdorf, W.R. and Diller, D.A. (1969). Focal electroretinogram in macular degeneration. Am. J. Ophthalmol. 68: 296–303.

Birch, D.G. and Fish, G.E. (1988). Focal cone electroretinograms: Aging and macular disease. Doc. Ophthalmol. 69: 211–220.

Foerster, M.H. and Li, X.X. (1986). Evaluation of the central retina and optic nerve function in media opacities. Doc. Ophthalmol. 63: 101–106.

Porciatti, V., Falsini, B., Fadda, A. and Bolzani, R. (1989). Steady-state analysis of the focal ERG to pattern and flicker: relationship between ERG components and retinal pathology. Clin. Vis. Sci. 4: 323–332.

Seiple, W.H., Siegel, I.M., Carr, R.E. and Mayron, C. (1986). Evaluating macular function using the focal ERG. Invest. Ophthalmol. Vis. Sci. 27: 1123–1130.

Mauro Fioretto M.D.
Clinica Oculistica dell'Università
Ospedale S. Martino – Pad. IX
16132 Genova, Italy

44. The cerebral generator of the colour evoked component N87 of the visual evoked potential: Localization by application of the regional source technique

H. PLENDL, D. PRÖCKL*, S. SCHULZE*, M. MAYER, K. BÖTZEL* and W. PAULUS**

*(Bad Aibling, *München and **Göttingen, Germany)*

Abstract

We have recorded human visual evoked potentials (VEPs) evoked by homogeneous red and yellow stimuli of 4° diameter presented centrally and in the left lower quadrant of the visual field. The surface data picked up from 32 electrode positions showed an early colour related component N87 for red but not for yellow.

We used a regional source location technique on these data in order to find the cortical generator of N87. It was possible to demonstrate in all seven subjects that for stimulation of the left lower quadrant N87 is generated in the contralateral visual area V1. For central stimulation the same result was achieved in six of the seven subjects.

These data support the hypothesis that N87 reflects the incoming activity to the primary visual cortex.

Introduction

A colour dependent component of the human visual evoked potential (VEP) occuring at 87 msec (N87) has been described by Paulus *et al.* (1984). This component can be evoked by homogeneous red and blue stimuli (Paulus *et al.*, 1988), whereas the less saturated green and yellow stimuli were less effective. N87 for equiluminant stimuli is qualitatively similar, but additional luminance steps enhance the amplitude of N87 for red and decrease it for green (Paulus *et al.*, 1984).

Mixed colour stimuli of low spatial frequency do not produce an N87 (Plendl and Paulus, 1989), whereas fine gratings or checker-boards evoke a comparable component, which was named N1 by Plant (Plant *et al.*, 1983).

N87 can be recorded with maximal amplitude around electrode position Oz. Timing and location of N87 were compatible with a generation in V1 respectively area 17 and less likely in area V4 which is assumed to be located at the base of the occipital cortex (Paulus *et al.*, 1984).

B. Drum (ed.), Colour Vision Deficiencies XII, pp. 369–374.
© 1995 *Kluwer Academic Publishers, Dordrecht.*

In order to answer the question if N87 really stems from area V1 we performed new experiments using the same stimulus equipment as in 1984.

Methods

The visual stimulus consisted of a circular array of red and green light emitting diodes (LEDs) behind a translucent screen. The stimulated visual area had a diameter of 4° and appeared to the subject as a homogeneous green, red or yellow spot, depending on the relative intensities of the green and red LEDs. The room in which the experiments were performed was illuminated by an electrical bulb.

Subjects had to fixate a black spot of 0.5 mm diameter in the centre of the stimulus field in the first experimental condition. In the second experiment the fixation spot was mounted at the crossing of the horizontal upper tangent and the right vertical tangent of the stimulus area. So in this condition the stimulated field was located in the left lower quadrant of the visual field.

In the interstimulus interval the screen appeared yellow (wavelength 567 nm, luminance 21.5 cd/m^2). Then it changed for 400 msec to a bright red (wavelength 595 nm, luminance 26 cd/m^2) or a bright yellow (wavelength 567 nm, luminance 32.5 cd/m^2).

These two stimuli were presented 300 times.

The EEG was recorded from 32 electrode positions (see Fig. 1), all electrodes were referenced to an electrode located in the midline at a position between Fz and Fpz. The raw data were digitized with a sampling rate of 500/sec and a resolution of 12 bits and stored continuously on hard disc.
The data were filtered (high pass frequency 0.5 Hz, low pass frequency 35 Hz) and averaged. The resulting VEPs were displayed as common average reference data.

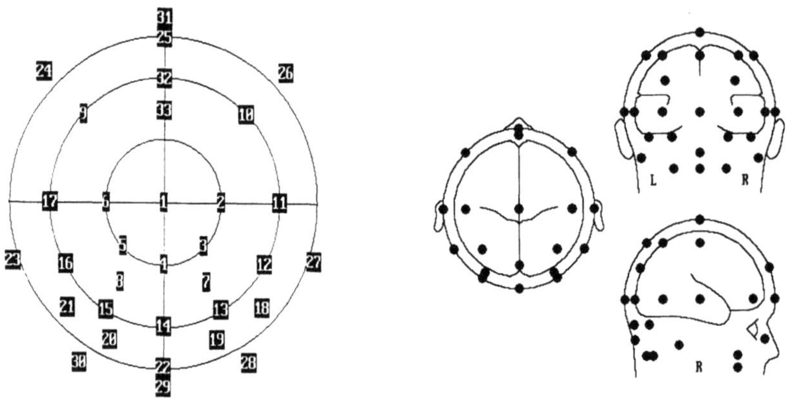

Fig. 1. At the left side the electrode positions are shown in top view. All 32 electrodes were referenced to electrode 33. At the right side the electrode locations with respect to the head are demonstrated.

A dipole analysis was performed by using this data set. We applied the method developed by Scherg, which is distributed as software package BESA (Scherg and von Cramon, 1986).

In principle the method works as follows. If we assume a spherical head model it is possible to calculate from a given dipole of a certain strength the surface potentials which are the result of this generator. These data are compared with the measured data. In an iterative way the dipole is moved and rotated until the difference between the measured data and the predicted by the model are minimized.

There are many pitfalls of this method (Snyder, 1991). The most important is that totally wrong solutions will be obtained if the number of generators is underestimated. If for instance N87 were the result of the overlapping fields generated by two sources it would be misleading to analyse the data by only one dipole.

Therefore we performed before dipole modelling a factor analysis of the data in order to estimate the number of generators. Furthermore we tried out solutions with up to five dipoles. These procedures demonstrated that modelling N87 with only one regional source explained the surface data with a residual variance of about 90%. A regional source is a set of three orthogonal dipoles with one origin which is able to localize precisely the location of a generator but not its orientation (for details see Scherg and Picton, 1991).

We recorded seven subjects. Dipole analysis was performed for all seven subjects in the time range of onset to peak of N87.

Results

In Fig. 2 the surface potentials for the sum of all subjects (grand average data) are shown.

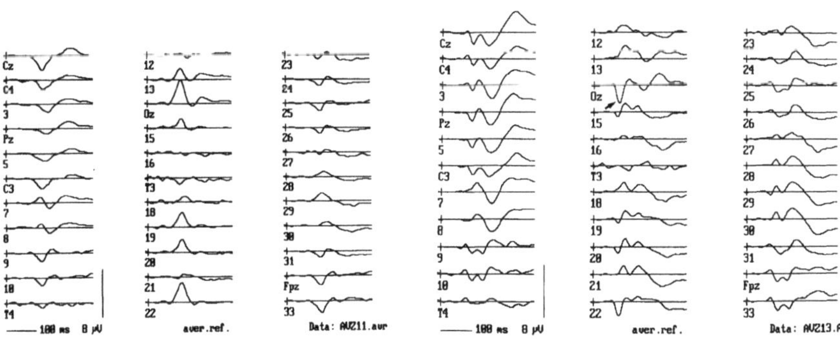

Fig. 2. At the left side the surface potentials for the yellow stimulus and central fixation are shown. Negative potentials are plotted downwards. All data are calculated as common average reference potentials. On the right side the data for the red stimulus are presented. N87 is seen only in the potentials evoked by red. It is pronounced at electrode position Oz (arrow).

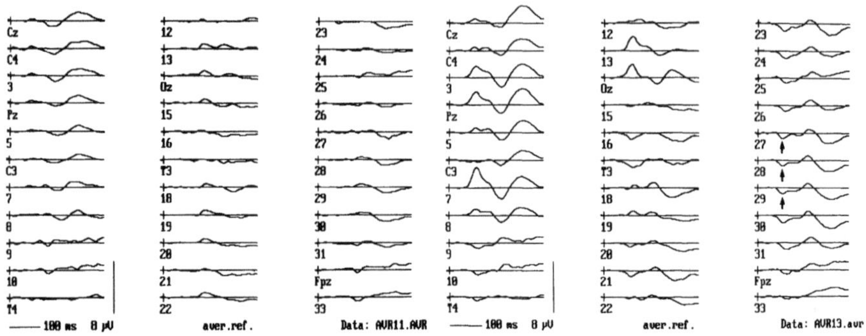

Fig. 3. The figure presents the data for stimulation in the left lower quadrant of the visual field. The data are presented in the same way as in Fig. 2. Again no clear cut N87 is seen in the data evoked by the yellow stimulus. For stimulation with red a N87 is seen at basal electrodes (arrows).

In the first experimental condition with central fixation, a marked N87 for stimulation with red is found at electrode position Oz. This component is missing for stimulation with yellow.

In condition 2 (see Fig. 3), stimulation in the left lower quadrant of the visual field, N87 can be recorded at more basal electrode sites. Again yellow evokes no N87.

In Fig. 4 the locations of the generator of N87 for all subjects are plotted. For central fixation only six subjects could be analysed because in one subject the appearance of alpha-rhythm prevented a stable solution. All six locations can be assumed to lie in the region of the striate cortex, even when taking into account the variability of striate cortex (Brindley, 1972).

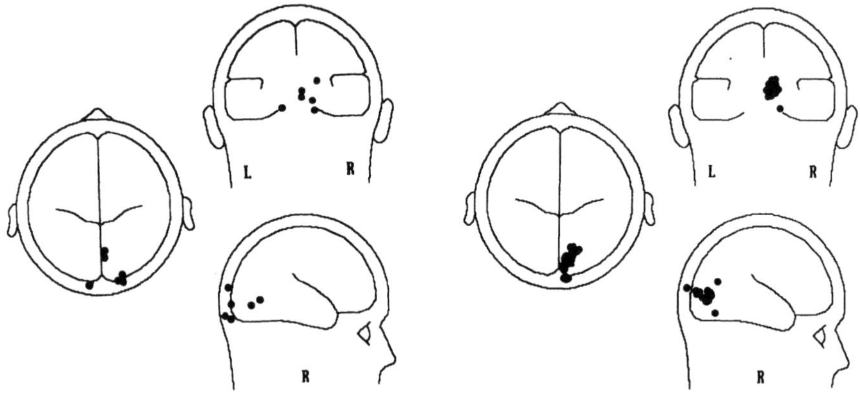

Fig. 4. At the left side the locations of the generator of N87 evoked by central stimulation are presented. The locations cluster around the occipital pole and are located in the region of area V1. At the right side the source-locations of N87 for stimulation in the left lower quadrant are shown. This time the locations lie in the region of the striate cortex on the right side, contralateral to the stimulus.

In addition, the remaining variability between subjects might reflect interhemispheric interactions of the electric generators which could not be resolved by one regional source.

For our second experiment the sources lie closer together in the region of the right striate cortex, extending from the area adjacent to the calcarine fissure to the occipital pole.

Discussion

The present study confirmed the relation to colour of the early component of the human VEP, namely N87. The previously described features of N87 could be confirmed, namely the restriction to occipital electrode sites and the high amplitude obtained by using red stimuli.

For the first time N87 was studied using dipole analysis. Our assumption that N87 is generated in area V1 could be supported by our data which fitted well in the region of V1 but not in the region of V4.

From experiments using Mondrian-like stimuli we know that activity in the human VEP in the time range of about 80 to 100 msec stems from striate cortex, whereas later components reflect activity of extrastriate areas (Plendl *et al.*, 1993).

Therefore colour specific components in the human VEP arising from extrastriate areas, e.g. area V4, should occur in a time range later than 100 msec. Indeed Murray and coworkers recorded colour dependent components using isoluminant gratings in a time range of about 120–140 msec (Murray *et al.*, 1987).

In contrast to our findings, Ossenblok and Spekreijse (1991) localized the generator underlying the first component of the checkerboard-onset VEP in area V2. Using a current source density analysis Kraut investigated flash-VEPs in macaques. He concluded that the human component N70 of the flash-VEP is an excitatory response in lamina 4C of the primary visual cortex (Kraut *et al.*, 1985). These findings together with the results of the present study give support to the hypothesis of Paulus *et al.* (1984) that N87 reflects the activity of the incoming geniculate fibres on the dendrites of cortical units in area V1. Because this hypothesis cannot be derived from scalp recordings alone, intracranial recordings of colour evoked potentials in humans are planned during the presurgical evaluation of epileptic foci.

Acknowledgements

We would like to thank Dr. M. Scherg for supplying us with the dipole source analysis program BESA.
Financially supported by the Bundesministerium für Forschung und Technologie (BMFT).

References

Brindley, G. (1972). The variability of the human striate cortex. J. Neurophysiol. 225: 1-3.
Kraut, M.A., Arezzo, J.C. and Vaughan, H.G. (1985). Intracortical generators of the flash VEP in monkeys. Electroenceph. Clin. Neurophysiol. 62: 300–312.
Murray, I.J., Parry, N.R.A., Carden, D. and Kulikowski, J.J. (1987). Human visual evoked potentials to chromatic and achromatic gratings. Clin. Vis. Sci. 3: 231–244.
Ossenblok, P. and Spekreijse, H. (1991). The extrastriate generators of the EP to checkerboard onset. A source localization approach. Electroenceph. Clin. Neurophysiol. 80: 181–193.
Paulus, W.M., Hömberg, V., Cunningham, K., Halliday, A.M. and Rohde, N. (1984). Colour and brightness components of foveal visual evoked potentials in man. Electroenceph. Clin. Neurophysiol. 58: 107–119.
Paulus, W.M., Plendl, H. and Krafczyk, S. (1988). Spatial dissociation of early and late colour evoked components. Electroenceph. clin. Neurophysiol. 71: 81–88.
Plant, G.T., Zimmern, R.L. and Durden, K. (1983). Transient visually evoked potentials to the pattern reversal and onset of sinusoidal gratings. Electroenceph. Clin. Neurophysiol. 56: 147–158.
Plendl, H. and Paulus, W. (1989). Interaction between colour- and pattern evoked potentials. In: Drum B. and Verriest, G. (eds.), Colour Vision Deficiencies IX, Doc. Ophthalmol. Proc. Ser. 52: 155–160, Kluwer Acad. Publ. Dordrecht.
Plendl, H., Paulus, W., Roberts, I.G., Bötzel, K. Towell, A., Pitman, J.R., Scherg, M. and Halliday, A.M. (1993). The time course and location of cerebral evoked activity associated with the processing of colour stimuli in man. Neurosci. Lett. 150: 9–12.
Scherg, M. and von Cramon, D. (1986). Evoked dipole source potentials of the human auditory cortex. Electroenceph. Clin. Neurophysiol. 65: 344–360.
Scherg, M. and Picton, T.W. (1991). Separation and identification of event related potential components by brain electric source analysis. EEG suppl. 42: 24–37.
Snyder, A. (1991). Dipole source localization in the study of EP generators: a critique. Electroenceph. Clin. Neurophysiol. 80: 321–325.

Dr. Helmut Plendl
Neurologische Klinik Bad Aibling
Kolbermoorer Str. 72
83043 Bad Aibling, Germany

45. Diagnosis of protan and deutan color vision deficiencies with pattern-ERG and VEP

JÜRGEN GERLING, THOMAS MEIGEN and MICHAEL BACH

(Freiburg, Germany)

Abstract

We compared electrophysiological responses [pattern-ERG (PERG) and VEP] and psychophysical measures to color stimuli to separate different forms of anomalous color vision. PERG and VEP were simultaneously recorded from 7 normals and 11 volunteers with congenital color vision deficiencies. Stimuli were color checkerboards with 0.5° check size, phase reversing at 34 revs/s. The luminance of the red and green parts were varied in opposite directions from 0 to 30 cd/m^2 while hue of individual squares and space-averaged luminance were held constant. This allowed for one equal luminance condition where flicker appeared fused. In the 7 normals, the subjective equiluminance was reached at a luminance ratio red/(red+green) = 0.53 to 0.56. At the same point, the PERG amplitude was moderately, and the VEP amplitude sharply reduced. In 11 color anomalous subjects both the PERG and VEP were sharply reduced at equiluminance. These dips were shifted compared to normals and the dip position corresponded to the predicted values obtained by calculations from the Smith-Pokorny transformation. As we found a close correlation of the VEP-dip position and the anomalous quotient, these electrophysiological measures may allow objective assessment of color vision deficiencies.

Introduction

Congenital red/green color vision deficiencies are x-linked genetic abnormalities. The physiological defect is localized in the photopigment of the cones. So far these deficiencies have been measured psychophysically, e.g. using the anomaloscope with Rayleigh matches. Advantages in molecular genetics have led to the cloning and sequencing of the human X-chromosomal cone pigments (Nathans *et al.*, 1986). This ultimately leads to correlation of the genotype and the clinically defined color vision defect. According to recent work (Merbs and Nathans, 1992) it is possible to find the typical absorption spectra of different

B. Drum (ed.), Colour Vision Deficiencies XII, pp. 375–380.
© 1995 *Kluwer Academic Publishers, Dordrecht.*

genotypes even in the red/green hybrid situation of pigments produced in cultured cells.

Besides psychophysical and genetical examination it might be useful to obtain electrophysiological data on color vision deficiencies as objective correlates. Previously, the photopic b-wave and the early receptor potential of the ERG were analyzed in congenital color vision deficiencies (Dodt *et al.*, 1958; Copenhaver and Gunkel,1959; Yokoyama *et al.*, 1973; Kawasaki, 1987) for differentiation. In this study we focus on responses of the retinal ganglion cells (PERG) and the visual cortex (VEP) for objective diagnosis of dyschromatopsia.

Methods

7 visual normals and 11 color anomalous subjects were examined. Acuity was ≥ 1.0, age ranged from 23 to 67 years. In all subjects the diagnosis was based on the Rayleigh match obtained with Nagel's anomaloscope.

Stimuli were presented on a visual color display unit (NEC Multisync II) controlled by a color graphics board (miro 510, resolution 6 bits per gun). Calibration and gamma correction were performed using a photodiode with near V_λ-sensitivity (BPW 27), which in turn was calibrated with a photometer (Minolta). The stimulus screen subtended 26° x 34° at a distance of 57 cm. The stimulus patterns were presented within a circle of 9.5° radius surrounded by a yellow field of 15 cd/m^2. Checkerboards with a mean luminance of 15 cd/m^2 and a check size of 0.5° were counterphased at 34 reversals/sec (i.e. 17 Hz). The luminance of the red (x = 0.611, y = 0.346) and green (x = 0.303, y = 0.591) checks varied in opposite directions from 0 to 30 cd/m^2. A stimulus sequence began with dark red and bright green checks, passed equiluminance and ended with bright red and dark green checks in 15 steps.

PERG responses were recorded with DTL electrodes (Dawson *et al.*, 1979) above the margin of the lower lid, referenced to the outer canthus. Simultaneously the VEP was recorded from Oz vs Fz. A stepwise-sweep computer program allowed us to present the stimuli in an interleaved block design. A cycle of 15 stimuli was repeated 10 times, resulting in a total of 150 sweeps for each condition. Subjects indicated equiluminance, where the flicker appeared absent. Evoked potential amplitudes were measured in the frequency domain as the magnitude at the reversal frequency, using Fourier analysis.

Results

Figure 1a shows the grand average result for 7 color normals. Here, the amplitudes (Fourier component at 34 Hz) are plotted vs the luminance ratio

$$t = \frac{\text{red}}{\text{red+green}} \cdot$$

Amplitudes are normalized relative to the mean response at the two pure luminance conditions [$t = 0$ and $t = 1$] because of different absolute amplitudes of PERG and VEP. This 'equiluminance tuning curve' shows a dip for the VEP at the subjective equiluminance point. The PERG is lower around equiluminance as compared to pure luminance contrast, but the dip is less pronounced. Thus, the relative attenuation at equiluminance is more pronounced in the VEP as compared to the PERG.

Figure 1b (top) shows the normalized equiluminance tuning curve for a protanope. Protanopes report flicker fusion for a condition where normal observers perceive the red fields much brighter than the green ones. VEP and PERG show an amplitude dip at the equiluminance point. At this point both amplitudes drop to noise level. Comparable results (t = 0.70 to 0.71) were obtained for another 3 protanopes (not shown here). In protanomaly equiluminance occurred at $t = 0.65$ together with a VEP- and PERG-amplitude reduction (Fig. 1b, bottom).

Deutan subjects reported flicker fusion for a brighter green and a darker red. Accordingly, their dip in the equilumince tuning curves is shifted to the left. In the 3 deuteranopes the VEP and the PERG showed a broad trough. Figure 1c

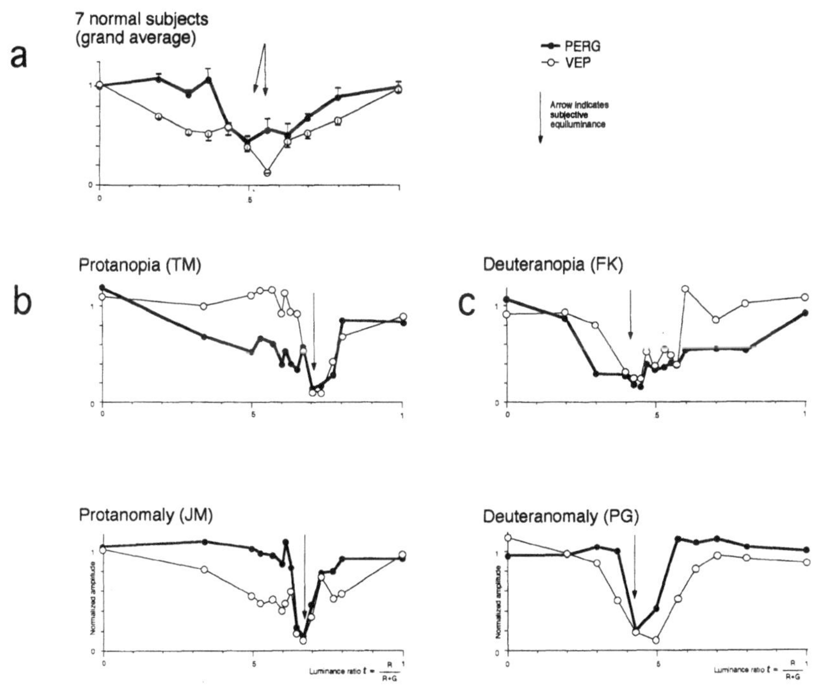

Fig. 1. Equiluminance tuning curves for (a) 7 color normals (grand average), (b) two protan subjects and (c) two deutan subjects. At subjective equiluminance (arrow), a dip in the VEP (thin line) and the PERG (thick line) is observed. For protans, dip and subjective equiluminance is shifted to brighter red compared to normal subjects. For deutans, dip and subjective equiluminance is shifted to brighter green compared to normals.

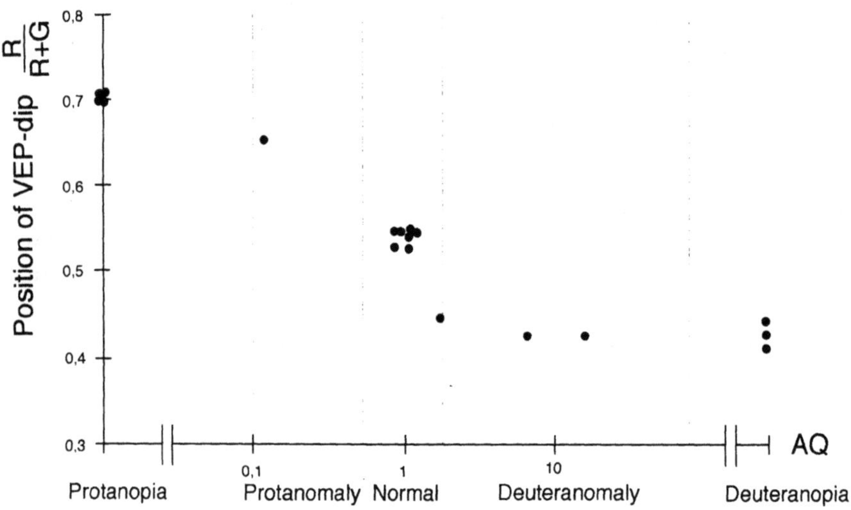

Fig. 2. Individual values of the VEP-dip position plotted vs the anomalous quotient (AQ) as measured by Nagel's anomaloscope. The protanopes are depicted on the extreme left and the deuteranopes on the extreme right, as the AQ proper is not defined in dichromates. A close correlation of the electrophysiological and the psychophysical results is seen.

(top) shows a broad amplitude trough with a minimum at t = 0.41. The two deuteranopic subjects not shown here had a minimum at 0.43 and 0.45 respectively. In deuteranomaly (Fig. 1 c, bottom), the VEP-and PERG-dip occurs at the luminance ratio t = 0.43.

In Figure 2 the psychophysical anomalous quotient is plotted vs the position of the VEP-dip.

Discussion

At equiluminant color flicker fusion the VEP shows a dip in color normal subjects. This correlate of flicker fusion at the level of the visual cortex is in accordance with a report by Regan (1970). The PERG, however, is only moderately reduced at equiluminant heterochromatic flicker fusion and reflects residual activity of the retinal ganglion cells in flicker fusion (Lee *et al.*, 1989; Bach and Gerling, 1992).

In congenital color vision deficiencies, both PERG and VEP show a deep amplitude dip at equiluminance. Compared to normals this dip is shifted (Fig. 1). In our 11 subjects with color vision deficiencies we found a close correlation between the position of the dip in the equiluminance tuning curve and the psychophysical anomaloscope result (Fig. 2).

For dichromats the dip position can be theoretically predicted as follows: Dip position occurs at the luminance ratio

$$t = \frac{\text{red}}{\text{red+green}}$$

where both colors activate the remaining cone type by the same degree (silent substitution, Estevez and Spekreijse, 1982). Cone activations are calculated from the Smith-Pokorny transformation (Wyszecki and Stiles, 1982) of the tristimulus values X, Y, Z for the red and green fields. In our experimental setup, the tristimulus values of our stimuli were:

$$red: \begin{pmatrix} X_R \\ Y_R \\ Z_R \end{pmatrix} = t\bullet \begin{pmatrix} 52.8 \\ 30.0 \\ 3.73 \end{pmatrix}, \quad green \begin{pmatrix} X_G \\ Y_G \\ Z_G \end{pmatrix} = (1-t)\bullet \begin{pmatrix} 15.3 \\ 30.0 \\ 5.39 \end{pmatrix} \quad with\ t \in [0,1].$$

Smith-Pokorny transformation leads to the respective cone activations L, M, S:

$$red: \begin{pmatrix} L_R \\ M_R \\ S_R \end{pmatrix} = t\bullet \begin{pmatrix} 24.4 \\ 5.62 \\ 0.03 \end{pmatrix}, \quad green \begin{pmatrix} L_G \\ M_G \\ S_G \end{pmatrix} = (1-t)\bullet \begin{pmatrix} 18.5 \\ 11.5 \\ 0.04 \end{pmatrix}$$

If we neglect the S-cone contribution, the equiluminance points for protanopes and deuteranopes are:

$$\text{Deuteranopes: } L_R \stackrel{!}{=} L_G \Leftrightarrow t_D = 0.43$$
$$\text{Protanopes: } M_R \stackrel{!}{=} M_G \Leftrightarrow t_P = 0.67.$$

In normal subjects both L and M must be considered

$$\text{Normals: } L_R + M_R \stackrel{!}{=} L_G + M_G \Leftrightarrow t_N = 0.50.$$

These calculated values for normals and dichromats correspond closely to the observed dip positions. In the case of anomalous trichromats we found a PERG- and VEP-reduction that was close to the dip position found in dichromats. This difference is probably caused by the small shift in spectral sensitivity of the human hybrid cone pigments (DeMarco *et al.*, 1992).

After the experiments we calibrated the Minolta photometer with a spectroradiophotometer and found deviations from the V_λ-curve. When corrections for these are introduced in the above calculations, the following dip predictions result: $t_D = 0.45$, $t_P = 0.69$, $t_N = 0.52$, providing an even closer fit with the experimentally obtained values.

As we found a close correlation between the amplitude dip in the equiluminance tuning curves and the psychophysical anomaloscope result (Fig. 2), this method might be used for objective determination of congenital color vision deficiencies. Discrimination between dichromats and anomalous trichromats, however, might be questionable since flicker photometric measures of spectral sensitivity show an overlap between dichromats and anomalous trichromats (Verriest, 1971).

References

Bach, M.and Gerling, J. (1992). Retinal and cortical activity in human subjects during color flicker fusion. Vision Res. 32: 1219–1223.

Copenhaver, R.M. and Gunkel, R.D. (1959). The spectral sensitivity of color defective subjects determined by electroretinography. Arch Ophthalmol. 62: 55–68.

Dawson, W.W., Trick, G.L. and Litzkow, C.A. (1979). Improved electrode for electroretinography. Invest Ophthalmol. Vis Sci. 18: 988–991.

DeMarco, P., Pokorny, J. and Smith, V.C. (1992). Full-spectrum cone sensitivity functions for X-chromosome-linked anomalous trichromats. J. Opt. Soc. Am. A 9: 1465–1476.

Dodt, E., Copenhaver, R.M. and Gunkel, R.D. (1958), Photopischer Dominator und Farbkomponenten im menschlichen Elektroretinogramm. Arch. Ges. Physiol. 267: 497–507.

Estévez, O. and Spekreijse, H. (1982). The 'silent substitution' method in visual research. Vision Res. 22: 681–691.

Kawasaki, K. (1987). Electrodiagnosis of red-green colour deficiency. Jpn. J. Ophthalmol. 31: 50–60.

Lee, B.B., Martin, P.R. and Valberg, A. (1989). Nonlinear summation of M- and L-cone inputs to the phasic retinal ganglion cells of the macaque. J. Neurosc. 9: 1433–1442.

Merbs, S.L. and Nathans, J. (1992). Absorption spectra of the hybrid pigments responsible for anomalous color vision. Science 258: 464–466.

Nathans, J. Thomas, D. and Hogness, D.S. (1986). Molecular genetics in human color vision: the genes encoding blue, green and red pigments. Science 232: 193–202.

Regan, D. (1974). Evoked potential indications of colour blindness. Vision Res. 14: 89–95.

Verriest, G. (1971). Les Courbes spectrales photopiques d'efficacité lumineuse dans les déficiencies congénitales de la vision des colours. Vision Res. 11: 1407–1434.

Wyszecki, G. and Stiles, W.S. (1982). Color science: Concepts and Methods, Quantitative Data and Formulae. John Wiley and Sons, New York.

Yokoyama, M., Yoshida, T. and Ui, K. (1973). Spectral responses in the human electroretinogram and their clinical significance. Jpn. J. Ophthalmol. 17: 113–124.

Dr. J. Gerling
Universitäts-Augenklinik
Killianstr. 5
D-79106 Freiburg
Germany

46. Spectral sensitivity by flash, flicker and pattern ERG

HARRY. G. SPERLING and JASLEEN MISHRA
(Houston, TX, U.S.A.)

Abstract

Rhesus b-wave spectral sensitivity was measured for 12°, 50 msec flashes against very large, intense neutral backgrounds. On humans, pattern ERG was measured with 1 cpd checkerboards alternating between spectral and neutral at frequent intervals in the spectrum. Flicker ERG spectral sensitivity was obtained by the same method with the checkerboard reticles removed. The flicker ERG spectral sensitivities and monkey b-wave spectral sensitivities have the same features, showing a small amount of red-green opponency and little or no blue sensitivity. The pattern ERG spectral sensitivities are radically different showing as much red-green opponency and blue sensitivity as in final common path psychophysical data. It is argued that they must represent ganglion cell stage response, since the b-wave represents a bipolar stage response.

Introduction

In a continuing series of studies, we are trying to trace the anatomical origins of color opponency by measuring spectral sensitivity at different afferent neural stages. Of particular emphasis in these studies is the presence and depth of the 580 nm notch – denoting the beginning and degree of red-green opponency and the presence and size of the peak in the blue near 450 nm – denoting the beginning and magnitude of amplification of blue response in the visual pathways.

In fig. 1, from previous work by Sperling and Mills (1987), on rhesus, we compare spectral sensitivity at the first afferent neural stage, the photoreceptor, derived from the a-wave of the focal flash ERG, with the spectral sensitivity at the second neural stage – the bipolar level – obtained from the b-wave of the focal flash ERG. These are compared with the spectral sensitivity of the final common path measured by the increment-threshold spectral sensitivity (of 3 of the 4 animals).

At the receptor stage (A in Fig. 1) the function is fit by the arithmetic sum of

B. Drum (ed.), Colour Vision Deficiencies XII, pp. 381–390.
© 1995 *Kluwer Academic Publishers, Dordrecht.*

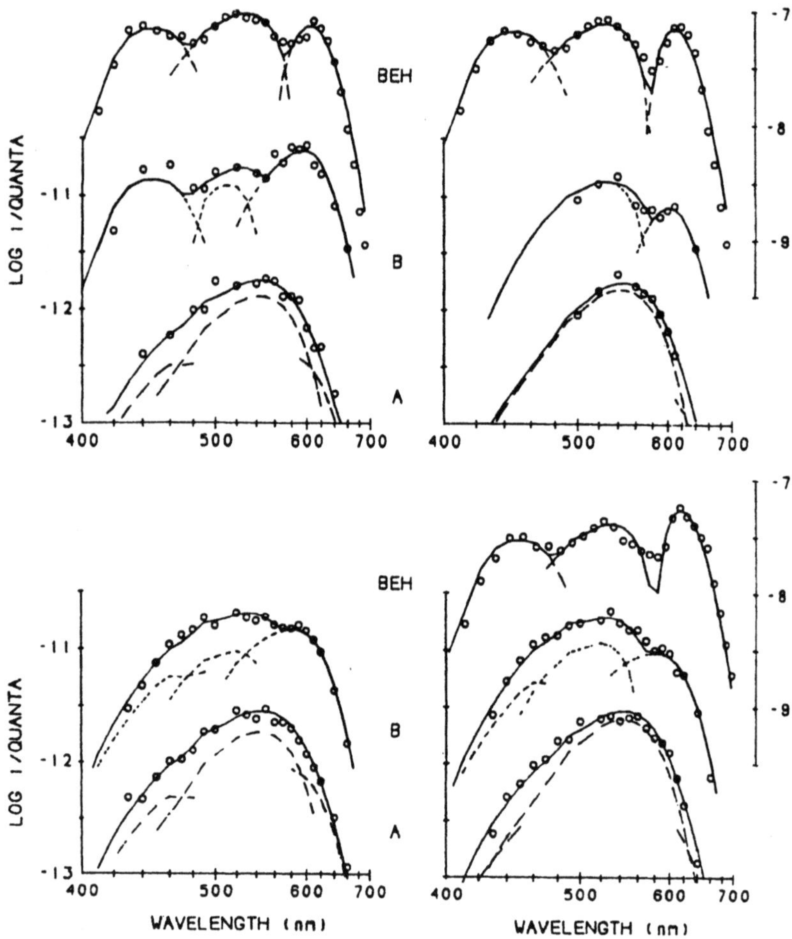

Fig. 1. Comparison of a-wave (A) and b-wave (B) flash ERG and behavioral increment-threshold (BEH) spectral sensitivity functions for four rhesus monkeys. From Sperling and Mills (1987).

red-plus-green cone response functions plus some intrusion of rods from outside the focal area (which was 12° centered on the fovea). There is no 580 nm notch, hence no red-green opponency. At the bipolar neural stage (B, Fig. 1), there is the beginning of a 580 nm notch. The data are better fit by the scalar sum of R-G and G-R, with some intrusion of rods, than by R+G.

The presence of red-green opponency in the b-wave function was confirmed (Fig. 2) when we injected bicuculline into the vitreous. After 4 hours with 64 mM bicuculline, the 580 nm notch disappeared and the functions superimposed on the a-wave function. Since bicuculline blocks GABA mediated inhibitory interactions, these results (Mills and Sperling, 1991), establish that R/G opponency at the bipolar level is based on a GABA synapse. It, therefore, involves feedback on photoreceptors or on bipolar cells via a chemical synapse.

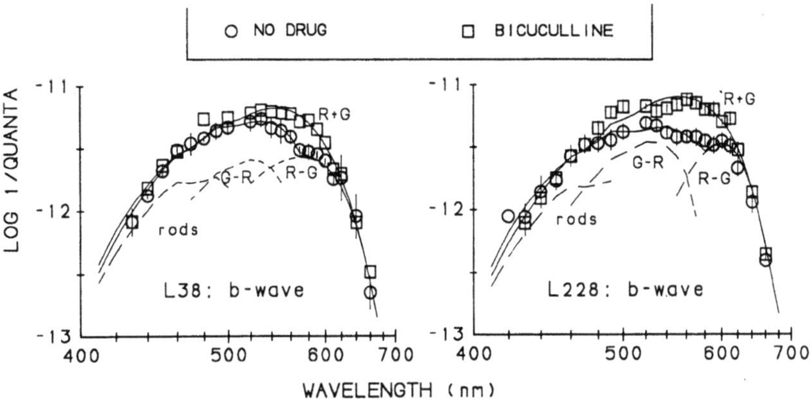

Fig. 2. Flash ERG spectral sensitivity functions for two rhesus monkeus, before and after injection of bicuculline (from Mills and Sperling, 1990).

This would necessarily involve either horizontal or amacrine interneurons in the feedback circuit.

The increment-threshold data (BEH, Fig. 1) show a much deeper 580 nm notch than at the bipolar level. This curve also shows a blue peak, over which there is nearly the same sensitivity as over the red and green peaks. Since fewer than one-tenth of the cones are short-wave sensitive cones, there is a substantial neural amplification of short-wave sensitive cone response by the time we reach these final common path data.

In this work we seek to obtain spectral sensitivity data on the third afferent neural stage – the ganglion cell level – to see if red-green opponency is further contributed to, perhaps completed, at this final retinal level, or perhaps contributed to at higher, brain levels. We seek, also, to see how much of the amplification of blue response is accomplished at this third retinal stage, or at higher levels.

The pattern ERG (PERG) is an ERG wave form that has been shown to be largely generated by the ganglion cells, because it disappears when the ganglion cells degenerate either in diseases which cause ganglion cell loss or in retrograde degeneration following severance of the optic nerve (Fiorentini *et al.*, 1981; Maffei and Fiorentini, 1981). Also, distribution currents at different retinal depths in response to alternating patterns are localized to proximal retina, the region of ganglion cells, while those generated by homogeneous flashes are localized to distal retina, the region of photoreceptors and bipolar cells (Baker *et al.*, 1988).

Spectral sensitivity functions using PERG criteria have previously been measured by Niepel and Dodt (1987) and Korth and Rix (1993). Unlike our work, they segrate 'on' and 'off' responses. Niepel and Dodt found that the off component showed red-green opponency. Korth and Rix found a large blue response but measured too infrequently in the spectrum to see the red-green notch.

Methods

Our technique uses a corneal thread electrode placed under the lower eyelid, and we record the potential difference between the cornea and the upper cheek bone, with a ground electrode placed on the contralateral ear. Our pattern is focused on the retina by a triple-path Maxwellian view optical system which produces intense spectral patterns. The work reported here used a 17° diameter checkerboard target with 1 cpd luminance alternating neutral and spectral checks, centered on the fovea. The checks alternated at 8 Hz on a 50–50 duty cycle.

In our method of deriving spectral sensitivity values, the neutral checks were held constant in intensity at 12,000 td as a standard; the spectral cheks were varied in intensity from dimmer then the neutral to brighter in a number of small steps. In the green through red part of the spectrum, as shown in Fig. 3, the voltage response of the pattern ERG amplitude went through a minimum (measured peak to peak in the sine-wave like response wave-form). The spectral sensitivity functions were derived as the reciprocal number of quanta at these minima for each wavelength.

In the violet and blue regions, the calculations were complicated by the fact that there were two minima. We showed in an auxiliary experiment, with a series of yellow adaptations, that one minimum was contributed by middle-

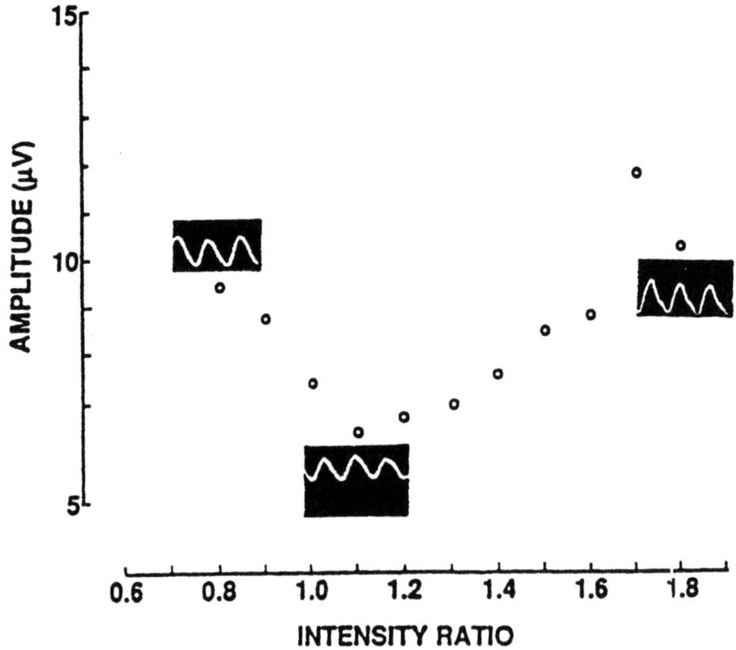

Fig. 3. Pattern ERG voltage response to an alternating (580 nm) spectral-white 1 cpd checkerboard when the white is constant at 12.000 Tds and the spectral varies in density steps from 0.6 to 1.8 times that of the white.

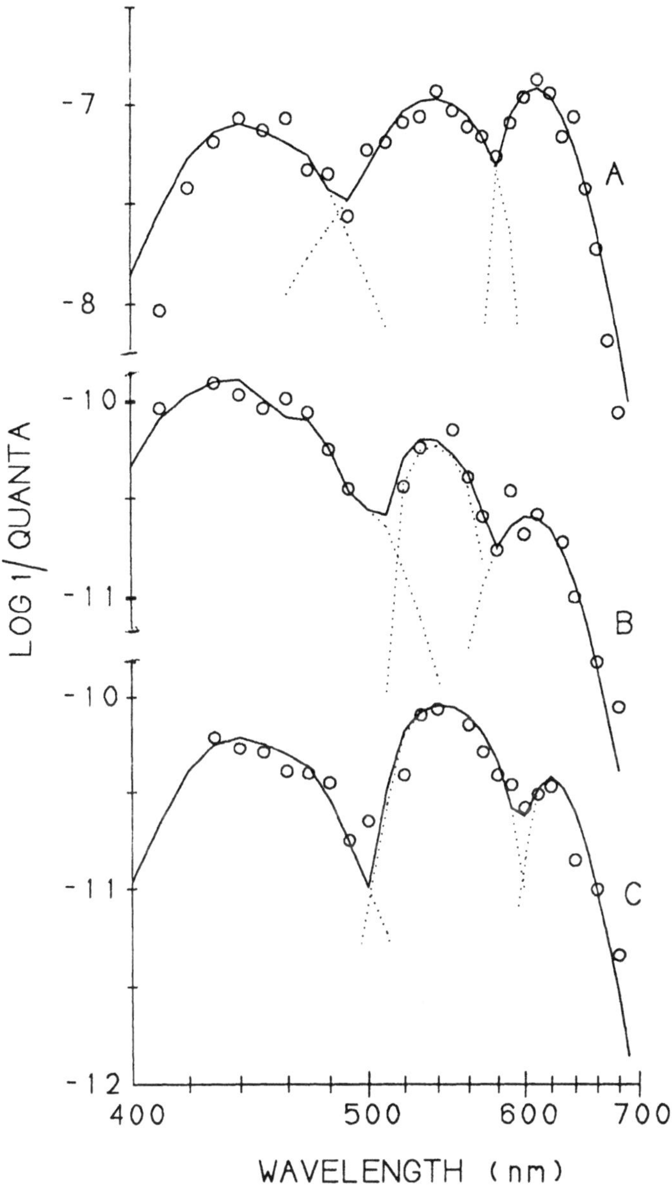

Fig. 4. Pattern ERG spectral sensitivity for two human subjects (B and C) and increment-threshold psychophysical spectral sensitivity (A) measured on the same eye as in C.

wavelength sensitive cones, the other by short-wavelength sensitive cones. For all wavelengths below 500 nm, the reciprocal number of quanta at the minima contributed by the short wavelength sensitive cones were calculated and plotted in the spectral sensitivity functions of Fig. 4.

The intention of the null point measurement, as illustrated in Fig. 3, was to equate or balance the luminance channel on the hypothesis that if the pattern stimulus drives both luminance and chromatic channels, then the voltage response would not go to zero and the residual response would be that of the chromatic channel. This hypothesis is confirmed since the signal does not go to zero and the residual voltages, converted to spectral sensitivities agree with the main features of the color opponency dominated behavioral increment-threshold spectral sensitivity function.

Results

The spectral sensitivity data are shown in Fig. 4 on a $\log_{10}$ scale as a function of wavelength. The bottom two functions (B and C) are PERG spectral sensitivities obtained on two normal human observers. They are compared with final common path increment-threshold psychophysical data (A), obtained on the same eye as measured in (C). Both PERG sets (B and C) show a 580 nm notch of comparable depth with the psychophysical data (in A) – albeit with different sized peaks. The PERG data of B and C also show comparable, if not greater, sensitivity over the peak in the blue relative to the peaks in the green and red with the psychophysical data of A.

Fig. 5 allows direct comparisons with earlier stages of color processing. Here, the small circles are the same data as shown in Fig. 4 B and C. The large data points are spectral sensitivities obtained on these two subjects with the same technique (holding the neutral stimuli constant and obtaining minima for each wavelength), but with the checkerboards removed. Neitz and Jacobs (1984) call this ERG flicker photometric spectral sensitivity. These (large symbol) data are on the same true scale as the PERG (small symbol) data. The small filled symbols are the mean monkey b-wave data from Fig. 1 (adjusted on the ordinate for fit the PERG data in the red). Clearly, both flicker photometric and b-wave data show little or no blue amplification. Although they show red-green opponency as compared with the scalar sum of middle and long-wave cone response (the top curves in Fig. 5), it is not great enough to yield a clear 580 nm notch and is much smaller than in the PERG data. In the PERG data, there is apparently fully completed blue amplification and the full red-green notch, as compared with the psychophysical data of Fig. 4. Since ERGs occur in the retina, these features must be contributed by neural processing beyond the bipolar (b-wave) level, at the third order ganglion cell level.

The discrepancy between the b-wave and flicker photometric data in the 500 nm region (filled points vs large circles, Fig. 5) can be accounted for by substantial rod intrusion from outside the focal stimulus area in the flash ERG b-wave data. As shown in Fig. 2, some amount of rod sensitivity must be added to model the ERG b-wave spectral sensitivity. In the case of the flicker photometric data (large circles, Fig. 5) extra-focal rods are much less sensitive to spectral and neutral stimuli alternated at a high rate, especially since the

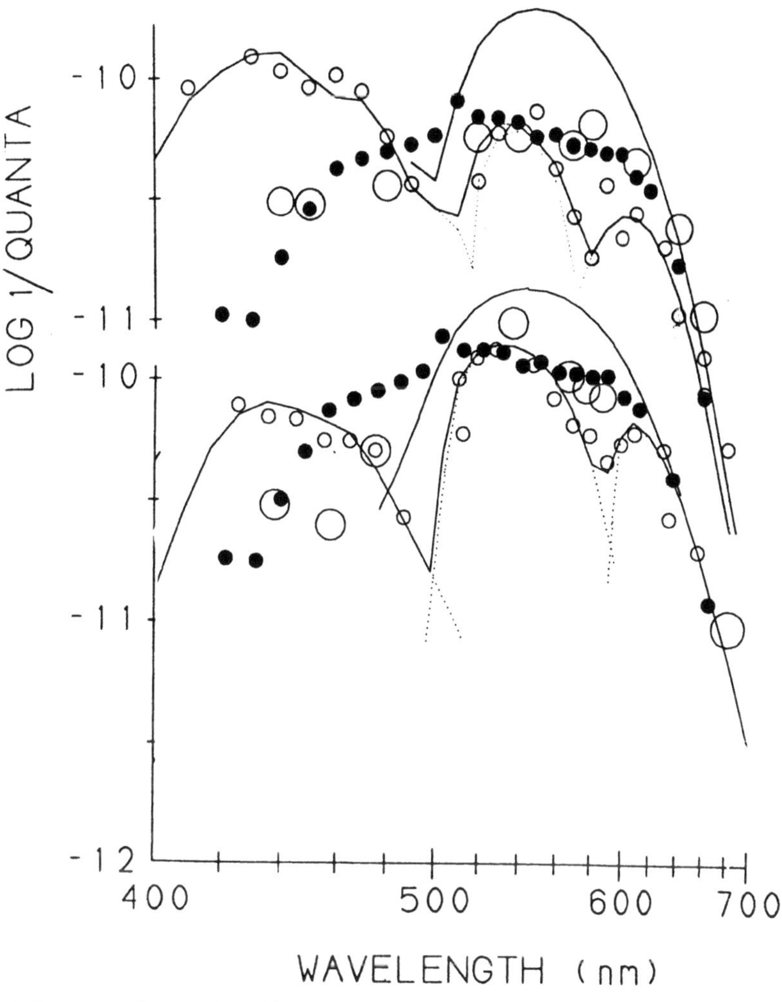

Fig. 5. Pattern ERG spectral sensitivity (the same as in Fig. 4) shown as the small circles. The same measurements on the same scale with the same size stimulus area for the same eyes with the checkerboard patterns removed shown as the large circles. These are compared with the mean flash ERG b-wave spectral sensitivity (as in Fig. 1) shown as the filled circles.

spectral sensitivity values were taken at the point of equal photopic luminance. The flicker photometric spectral sensitivities should be contaminated by rods only to the extent of the difference between photopic and scotopic luminances, and that minimized by the fact that rod temporal response is low at 8 Hz.

Discussion and summary

The flash ERG b-wave spectral sensitivity, Fig. 1 and 5, and the threshold obtained with our minimum voltage criterion if the patterns are removed yield functions with some of the same characteristics, no blue peak and little or no yellow notch. The b-wave thresholds were interpolated at the median response voltage in an intensity series of 12° foveated flashes at each wavelength. It is not surprising that the two results agree if the intensity series succeeded, as intended, in bracketing the linear range of the luminance response and the null measurement (voltage minimum) with the checkerboard removed was of the luminance channel.

Throughout, we refer to three levels of neural processing, which we equate to the three afferent neural levels of the retina: photoreceptor, bipolar and ganglion cell. The flash ERG waveform has clear features, the a-wave and b-wave, which have been associated with photoreceptor and bipolar responses (Tomita and Yanigada, 1981; *et al.*, 1985; Heynen and van Norren, 1985). The pattern ERG, as cited above, disappears with loss of ganglion cells with no accompanying change in the b-wave and has been shown by current source density changes, with progression of a moving electrode, to be generated in the proximal retina, the region of the ganglion cells. The studies reported here of changes in spectral sensitivity using a-wave, b-wave and pattern ERG criteria offer clear-cut differences, which we therefore feel justified in ascribing to the three levels. Just as the ERG components result from current loops involving other classes of cells, the spectral sensitivity function components result from interactions with other cells. The 580 nm notch results from interaction between the L and M cone pathways, via interneurons. These must be horizontal and/or amacrine cells. Present evidence favors amacrine cells, since it has been shown that horizontal cells, which feed back onto photoreceptors, contact every cone within their dendritic fields (Boycott *et al.*, 1986).

Why the pattern ERG is dominated by color opponent ganglion cell signals and the ERG flicker sensitivity is not, but reflects luminance is the big question posed by these results. It seems to us that the answer must lie in the presence or absence of local spectral alternation across sharply focused boundaries. Our experiments isolate this fact with respect to spectral stimuli in a highly controlled way. Others have pointed out that local luminance alternation of non-spectral stimuli is a spatial ERG response contributed at the ganglion cell level (Maffei and Fiorentini, 1981). We have implicated spectral opponency with this pattern or spatial dimension contributed at the ganglion cell level. A reasonable extension of this statement is that the spectral pattern ERG results reported here are the result of the response of ganglion cells which are both spectrally and spatially opponent.

The amplification of blue response could result from an anatomical mechanism such as greater convergence of S cones pathways. The fact that amplification is not seen at the b-wave level argues that convergence of S cone bipolars on ganglion cells would be the locus, if this were the mechanism.

Alternative to anatomical convergence as the mechanism of amplification would be possible physiological mechanisms. One such mechanism would result from the imbalance in the number of blue on- and off-center ganglion cells (deMonastario and Gouras, 1975). If there were fewer off-center ganglion cells than on-center ganglion cells and these tended to cancel each other's responses, there would be a relative increase in blue on response at the ganglion cell level as compared with red and green responses, which are more closely balanced between on and off centers.

In summary, we have shown that at the photoreceptor level neural processing of spectral information involves the simple summation of red and green responses. At the bipolar level, we see some inhibitory interaction between red and green responses but much less than seen in the final common path. At the ganglion cell level, we see nearly the same degree of red-green inhibitory interaction as in the final common path and in addition we see amplification of blue response to at least equal that seen in the final common path.

It is important to underscore that the differences between spectral sensitivity ascribed to ganglion cell level processing, bipolar and photoreceptor level processing are from well controlled experiments in which the major effects were reversed by withdrawing or reversing the hypothesized major variables. The difference between photoreceptor and bipolar spectral sensitivity was reversed in the same preparation by blocking the GABA synapses. The difference between the ganglion cell level spectral sensitivity and bipolar level spectral sensitivity was reversed in the same preparation by removing the pattern and finding that the resulting spectral sensitivity most closely resembled the bipolar level function with respect to the features of interest (depth of the red-green notch and sensitivity in the blue region).

References

Baker, C.L., Hess, R.F., Olson, R.T. and Zrenner, E. (1988). Current source density analysis of linear and non-linear components of the primate electroretinogram. J. Physiol. (Lond) 407: 155–176.

Dick, E., Miller, R.F. and Bloomfield, S. (1985). Extracellular K+ activity changes related to electroretinogram components. II. Rabbit (E-type) retinal. J. Gen. Physiol. 85: 911–931.

Dick, E., Miller, R.F. and Dacheux, R.F. (1979), Neuronal origin of b- and d-waves in the I-type ERG. Invest. Ophthalmol. Vis. Sci. Suppl. 18: 34.

Fiorentini, A., Maffei L., Pirchio, M., Spinelli, D. and Porcellati, V. (1981). The ERG in response to alternating gratings in patients with diseases of the peripheral visual pathway. Invest. Ophthalmol. Vis. Sci. 21: 490–493.

Heynen, H. and van Norren, D. (1985). Origin of the electroretinogram in the intact macaque eye, II: Current source density analysis. Vision Res. 25: 709–717.

Korth, M., Nguyen, N.X., Rix, R., Sembritzki, O. (1993). Spatial and chromatic interactions in the human pattern electroretinogram. Vision Res. 33: 275–287.

Maffei, L. and Fiorentini, A. (1981). Electroretinographic responses to alternating gratings before and after section of the optic nerve. Science 211: 953.

Mills, S.L. and Sperling, H.G. (1990). Red/green opponency in the rhesus macaque ERG spectral sensitivity is reduced by bicuculline. Visual Neuroscience 5: 217–221.

de Monastario, F.M. and Gouras, P. (1975). Functional properties of ganglion cells of the rhesus monkey. J. Physiol. (Lond.) 251: 167–196.

Neitz, J. and Jacobs, G.H. (1984). Electroretinogram measurements of cone spectral sensitivity in dichromatic monkeys. J. Opt. Soc. Am. A 1: 1175–1180.

Niepel, G. and Dodt, E. (1987). Spektrale Empfindlichkeit in menschenlichen Elektroretinogramm Darbeitung von musterreizen. Fortschr. Ophthalmol. 84: 635– 640.

Sperling, H.G. and Mills, S.L. (1987). ERG and behavioral analysis of spectral sensitivity in normal and blue-blind rhesus monkeys. In: Verriest, G. (ed.), Colour Vision Deficiencies VIII, Doc. Ophthalmol. Proc. Ser. 46: 365–374, Nijhoff/Junk Publishers, Dordrecht.

Tomita, T. and Yanagida, T. (1981). Origins of the ERG waves. Vision Res. 21: 1703–1707.

Harry G. Sperling
University of Texas Health Sciences Center at Houston
Sensory Sciences Center
6420 Lamar Fleming Avenue
Houston, TX 77030 U.S.A.

47. Psychophysical and occipital responses to aberration-free blue/yellow and red/green gratings

D.J. McKEEFRY and J.J. KULIKOWSKI
(Manchester, U.K.)

Abstract

The blue/yellow colour opponent system was selectively stimulated by a blue/yellow isoluminant coarse grating with hues corresponding to a tritanopic confusion line; chromatic aberration was eliminated using a system which focused different wavelengths on one plane (Mullen, 1985), thus avoiding luminance intrusions. Under such conditions the spatio-temporal low-pass characteristics appeared to be reasonably well matched for the blue/yellow and red/green colour opponent systems.

The Visual Evoked Potentials (VEPs) elicited by the onset of blue/yellow and red/green gratings were similar (negative), but the off-waves differed. The off-wave (though smaller) was similar (negative) to the onset blue/yellow VEP, as though originating from one source. In the red/green VEP the off-wave (positive) resembles the achromatic offset VEPs. This suggests that red/green, but not blue/yellow isoluminant gratings can also stimulate the achromatic (magnocellular) system.

Introduction

Colour vision in primates is subserved by the responses of the three cone types sensitive to short- (S cones), middle- (M cones) and relatively long- (L cones) wave spectral light. Chromatic contrast is provided by the antagonistic interactions of cone output at some post-receptoral stage. Achromatic or luminance contrast sensitivity is subserved by a summation of L and M cone outputs, the existence of an S-cone input into the luminance system is contentious (see e.g. Eisner and MacLeod, 1980; Lee and Stromeyer, 1989, for opposing views). These interactions produce two colour opponent systems (red/green and blue/yellow) and a luminance system, the existence of which have been verified psychophysically (Jameson and Hurvich, 1955) and electrophysiologically (e.g. Derrington *et al.*, 1984; Crook *et al.*, 1987).

The colour processing stream contains neurones which are characterised by

B. Drum (ed.), Colour Vision Deficiencies XII, pp. 391–398.
© 1995 *Kluwer Academic Publishers, Dordrecht.*

responses which are sustained (tonic) and colour opponent; they project to the parvocellular layers of the lateral geniculate nucleus (LGN), as opposed to the larger transient (phasic) cells that project to the magnocellular layers of the LGN and have broadband characteristics (e.g. Wiesel and Hubel, 1966; DeMonasterio and Gouras, 1975; Perry *et al.*, 1984; Shapley and Perry, 1986; Merigan, 1989; Schiller *et al.*, 1990; Kaplan *et al.*, 1990).

It is within the framework of the opponent model of colour vision that many of the psychophysical investigations into the spatio-temporal properties of colour vision have taken place. Various studies have revealed the colour system to be low-pass in both the spatial and temporal domains (e.g. Mullen, 1985; Kelly, 1974, 1983, 1989).

Classically the S-cone mechanism, in comparison to the L and M cone mechanisms, has been considered to possess low spatial and temporal resolution limits and low sensitivity (Brindley *et al.*, 1966; Green, 1968; Williams and Collier, 1983). These properties presumably reflect the relative sparsity of the S-cones in the retina (Marc and Sperling, 1977). However, analysis of the electrophysiological (Schnapf *et al.*, 1990) and psychophysical (Stockman and MacLeod, 1992) responses from the S-cones has revealed little difference between the temporal properties of the L, M and S cones.

Previous studies which have examined the properties of the red/green and blue/yellow colour opponent systems have used stimuli which have been prone to either the effects of chromatic aberration or have modulated cones along non-selective colour axes. These two factors are considered to be especially important for the examination of the blue/yellow system, as chromatic aberration is increased for colour combinations that involve shorter wavelengths. Also, selective activation of the blue system, requires stimulation along a tritanopic confusion line in order to preclude contributions from L and M cones (tritanopic stimulation equalises the quantum catches in these two cone types).

The psychophysical and electrophysiological responses of the blue/yellow and red/green colour opponent systems were investigated utilising aberration-free isoluminant, chromatic, sinusoidal gratings the constituent hues of which modulated cones along appropriate axes in colour space. Such stimuli, when of low contrast, have been shown to isolate selectively the colour opponent activity of the parvocellular pathway (see for review Kulikowski, 1991). When the gratings are of low spatial frequency and presented in on/off or pattern reversal modes, it is easy to dichotomize both psychophysical and electrophysiological experiments. Comparison of onset with pattern reversal may reveal the existence of transient and sustained mechanisms, which mean that the contrast thresholds are determined by a transient change in contrast or by standing contrast respectively (Kulikowski and Tolhurst, 1973; Kulikowski, 1978).

Methods

Stimulus generation

Monochromatic gratings were produced on two Joyce CRT displays and the images of these gratings were combined at the observer's eye by the optical system in antiphase, to produce a chromatic grating of mean retinal illuminance 60 photopic trolands. Isoluminance was established for each observer using heterochromatic flicker photometry.

One of the displays could be placed at a variable distance from the eye in order to compensate for the dioptric interval of focus between the constituent colours of the stimulus and the other display was at a fixed distance from the observer. This differential placement of the displays from the eye allowed compensation for axial chromatic aberration. Adjustment of the time-base and raster size of this display also allowed compensation for differences in spatial frequency, caused by transverse chromatic aberration.

Monochromatic interference filters were interposed between the screens and the observer in order to produce gratings of appropriate hues. For the blue/yellow stimulus, filters were used which modulated along the cardinal axis of Krauskopf *et al.*, (1982), which is a tritanopic confusion line. For the red/green stimulus, the filters modulated along a line parallel to the spectral locus of hues rather than the red/green cardinal axis.

The observers (both colour vision normals), JJK a 58-year-old corrected myope and DMcK a 26-year-old corrected myope, viewed the stimuli with head position stabilized by means of a dental bite bar. Small fixation points in the centres of each display also had to be accurately aligned during the experiments. These procedures were employed to minimize the introduction of luminance contrast into the stimulus by any misalignment of the gratings.

Data collection

Psychophysical thresholds were set by method of adjustment, each data point representing the mean of four settings. Chromatic contrast was taken to be equivalent to achromatic contrast, defined as:

$$C = (L_{min} - L_{max})/(L_{min} + L_{max}).$$

VEP recordings were made from the occipital electrode (O_z) referenced to mid-frontal (MF), using a Medelec averager ER94. Corner frequencies of the amplifiers were set at 0.3 – 30 Hz.

Results

Psychophysics

Figures 1 and 2 illustrate spatial and temporal contrast sensitivity functions, respectively, for blue/yellow, red/green and achromatic gratings. For chromatic stimuli, the functions appear to be low-pass in both space and time compared to the band-pass nature of the achromatic functions. Spatial and temporal sensitivities for the blue/yellow and red/green stimuli do not appear to be very different.

A comparison of the on/off and reversal sensitivities in the spatial domain allows the assessment of the contribution to contrast thresholds by transient mechanisms (Kulikowski and Tolhurst, 1973). For achromatic gratings of spatial frequency up to 2.5 c/degree, reversal sensitivity is twice that of on/off sensitivity. This fact indicates that transient mechanisms, which are sensitive to change in contrast, subserve detection. For red/green stimuli reversal sensitivity is observed to be slightly higher than on/off sensitivity but by less than a factor of two. For blue/yellow stimuli reversal and on/off sensitivities are seen to be virtually identical. Thus it would appear that colour sensitivity is governed largely by sustained mechanisms (i.e. by mechanisms that are sensitive to standing levels of contrast). However, there may be a contribution from transient mechanisms to the detection of red/green stimuli.

In the temporal domain the absence of a localised peak in the chromatic function at around 5 Hz, as exhibited by the achromatic temporal function, indicates that there is little contribution to detection from the first transient filter which is optimally sensitive around this temporal frequency (Kulikowski and Tolhurst, 1973). Colour temporal modulation sensitivity therefore appears to be subserved by a single low-pass, presumably sustained temporal channel.

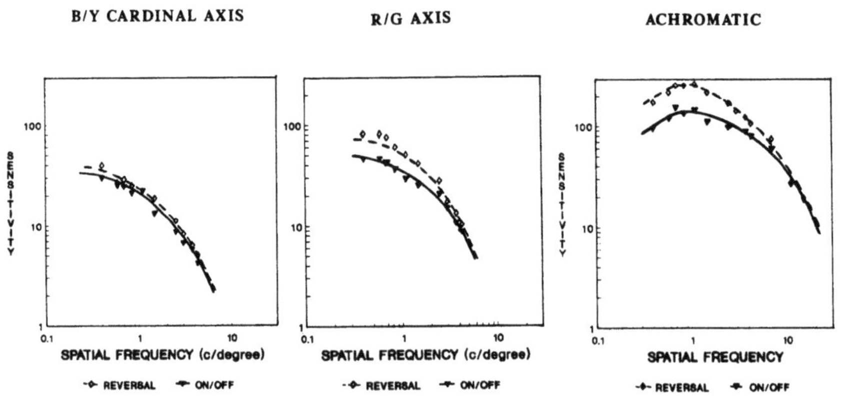

Fig. 1. Spatial contrast sensitivity functions for blue/yellow, red/green and achromatic gratings. The stimuli had a mean retinal illuminance of 60 Td (photopic) and were presented in on/off (solid lines) and reversal (dotted lines) at a temporal rate of 1 Hz (subject DMcK).

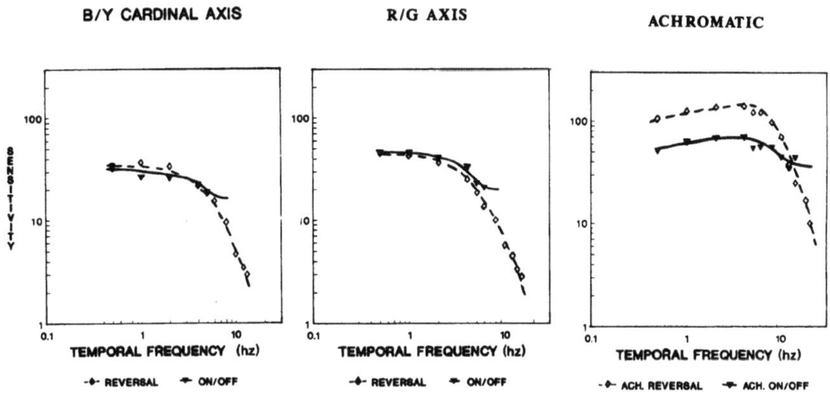

Fig. 2. Temporal modulation contrast sensitivity functions for blue/yellow, red/green and achromatic gratings. The stimuli had a mean retinal illuminance of 60 Td and were presented in on/off (solid lines) and reversal (dotted lines) and had a spatial frequency of 0.85 c/degree (subject JJK).

Visual evoked potentials

The onsets of either red/green chromatic or achromatic gratings elicit VEPs which have different waveforms (e.g. Carden *et al.*, 1985; Murray *et al.*, 1987). Similar differences are illustrated in Fig. 3. The achromatic VEPs are similar for the grating onset, offset and reversal, i.e. depend solely on contrast *change*, presumably reflecting the transient nature of the activity of the broadband neurones of the magnocellular pathway. Conversely, the sustained nature of the neurones that contribute to red/green and blue/yellow chromatic VEPs, is revealed by the marked difference between chromatic onset and chromatic reversal. These neurones reflect the activity of the colour opponent neurones of the parvocellular system (c.f. Kulikowski, 1991).

However, longer onset and offset durations reveal some differences between both the chromatic VEPs. Red/green VEP (Fig. 4a) exhibits a positive offset response, similar to the achromatic offset and reversal response, even when chromatic aberration is corrected (Murray *et al.*, 1987). Our new finding is that the blue/yellow VEP has a negative offset wave similar to onset, but of smaller amplitude. The presence of this negative offset response gives the blue/yellow VEP an appearance that closely resembles the slow-field potentials that are typically recorded from small, localised cortical areas (Kulikowski and Vidyasagar, 1987). Figures 4b and c illustrate that this negative offset response is most apparent for isoluminant gratings and is positive or absent when chromatic aberration is not eliminated (see also Kulikowski and Walsh, 1993).

The positive offset response shows a small transient contribution to red/green VEPs (not seen in the onset VEP only because it is swamped by a large negativity). This indicates that the red/green VEP does not reflect the activity of as highly a selective population of neurones, with the same degree of cortical localisation, as the neurones which subserve the blue/yellow response.

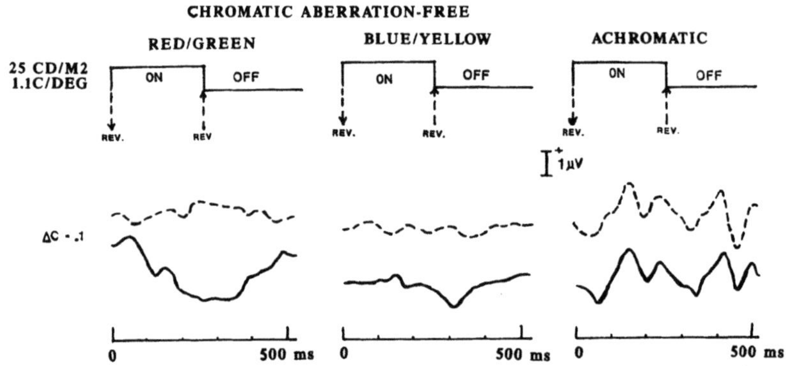

Fig. 3. VEPs elicited by red/green (R/G), blue/yellow (B/Y) and achromatic gratings (2.5 c/degree). The solid lines represent on/off responses and the dashed lines reversal responses (for the same contrast change).

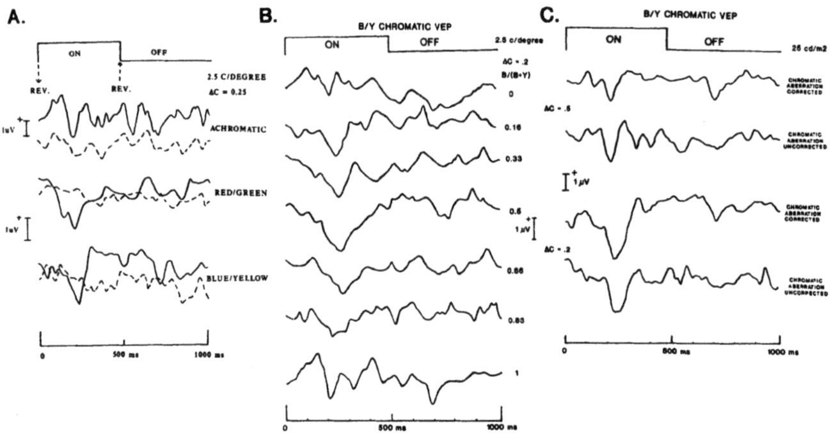

Fig. 4. (a) Achromatic, red/green and blue/yellow onset (solid lines) and reversal (dashed lines) VEPs. (b) Blue/yellow onset VEPs elicited by stimuli of different luminance contrast content, as defined by the B/(B+Y) ratio. Isoluminance occurs at B/(B+Y) = 0.5. (c) Blue/yellow onset VEPs elicited by stimuli with and without correction for chromatic aberration, note the absence of the negative offset wave in the latter case.

Discussion

The psychophysical and electrophysiological data presented here demonstrate that the colour opponent mechanisms are sustained in nature with low-pass spatio-temporal characteristics. However, red/green isoluminant borders may also stimulate transient mechanisms, as revealed psychophysically and by the positive offset wave in the red/green chromatic VEP. In the case of isoluminant stimuli, free from achromatic contrast induced by chromatic aberrations and by misalignment of the gratings due to head movements, such contributions may arise from the activity of magnocellular neurones which receive strong inputs from L and M cones. Single-unit studies have confirmed that magnocellular cells are responsive to isoluminant red/green borders (Schiller and Colby, 1983; Lee *et al.*, 1988; Hubel and Livingstone, 1990).

Thus the aberration-free system of generating blue/yellow isoluminant gratings is critical to obtain a stimulus with a greater degree of selectivity than that of flat screen stimuli, or comparable red/green gratings. Consistent with this hypothesis is the appearance of blue/yellow borders (along a tritanopic confusion line) which are not as sharp as red/green. The presence of a negative onset and smaller offset VEP suggests that the same localized group of neurones generate the responses. There appears to be no contribution from any transient (magnocellular) responses to stimulus offset, which would distort this negativity. In view of this evidence it appears likely that there is no conspicuous input from the S cones into the achromatic system.

Acknowledgements

We would like to express our thanks to John Simpson of the Visual Sciences Lab., UMIST for his technical expertise and assistance with the apparatus. Purchase of the equipment was made possible by the Visual Sciences Fund.

References

Brindley, G.S., DuCroz, J.J., and Rushton, W.A.H. (1966). The flicker fusion frequency of the blue sensitive mechanism of colour vision. J. Physiol. 183: 497–500.

Carden, D., Kulikowski, J.J., Murray, I.J. and Parry, N.R.A. (1985). Human occipital potentials evoked by the onset of equiluminant chromatic gratings. J. Physiol. 369: 44P.

Crook, J.M., Lee, B.B, Tigwell, D.A. and Valberg, A. (1987). Thresholds to chromatic spots of cells in the macaque geniculate nucleus as compared to detection sensitivity in man. J. Physiol. 392: 193–211.

DeMonasterio, F.M and Gouras, P. (1975). Functional properties of ganglion cells of the rhesus monkey retina. J. Physiol. 251: 167–195.

Derrington, A.M., Krauskopf, J. and Lennie, P. (1984). Chromatic mechanisms in the lateral geniculate nucleus of the macaque. J. Physiol. 357: 241–265.

Eisner, A. and MacLeod, D.I.A. (1980). Blue sensitive cones do not contribute to luminance. J. Opt. Soc. Am. 70: 121–123.

Green, D.G. (1968). The contrast sensitivity of the colour mechanisms of the human eye. J. Physiol. 196: 415–429.

Hubel, D.H. and Livingstone, M.S. (1990). Colour and contrast sensitivity in the lateral geniculate body and the primary visual cortex of the macaque monkey. J. Neurosci. 10: 2223–2237.

Jameson, D. and Hurvich, L.M. (1955). Some quantitative aspects of an opponent colours theory. I: Chromatic responses and spectral saturation. J. Opt. Soc. Am. 45: 546–552.

Kaplan, E., Lee, B.B. and Shapley, R.M. (1990). New views on primate retinal function. In: Osborne and Chader (eds.), Progress in Retinal Research, Vol. 9. Pergamon Press, Oxford.

Kelly, D.H. (1974). Spatio-temporal frequency characteristics of colour vision mechanism. J. Opt. Soc. Am. 64: 983–990.

Kelly, D.H. (1983). Spatiotemporal variation of chromatic and achromatic contrast thresholds. J. Opt. Soc. Am. 73: 742–750.

Kelly, D.H. (1989). Spatial and temporal interactions in colour vision. J. Imaging Tech. 15: 82–89.

Krauskopf, J., Williams, D.R. and Heely, D.M. (1982). The cardinal directions of colour space. Vision Res. 22: 1123–1131.

Kulikowski, J.J. (1978). Pattern and movement detection in man and rabbit: Separation and comparison of evoked potentials. Vision Res. 18: 183–189.

Kulikowski, J.J. (1991). On the nature of VEP's, unit responses and psychophysics. In: Valberg, A. and Lee, B.B. (eds.), From Pigments to Perception: Advances in Understanding visual processes: 197–209. Plenum Press, New York and London.

Kulikowski, J.J. and Tolhurst, D.J. (1973). Psychophysical evidence for sustained and transient detectors in human vision. J. Physiol. 232: 149–162.

Kulikowski, J.J. and Vidyasagar, T.R. (1987). Neuronal responses and field potentials evoked by gratings in the macaque striate cortex. J. Physiol. 392: 45P.

Lee, B.B., Martin, P.R. and Valberg, A. (1988). The physiological basis of heterochromatic flicker photometry demonstrated in the ganglion cells of the macaque retina. J. Physiol. 404: 323–347.

Lee, J. and Stromeyer, C.F. (1989). Contribution of human short wavelength cones to luminance detection. J. Physiol. 413: 563–593.

Marc, R.E. and Sperling, H.G. (1977). Chromatic organisation of primate cones. Science 196: 454–456.

Merigan, W.H. (1989). Assessing the role of parallel pathways in primates. In: Kulikowski, J.J., Dickinson, C.M. and Murray, I.J. (eds.), Seeing Contour and Colour. Pergamon Press, Oxford.

Mullen, K.T. (1985). The contrast sensitivity of human colour vision to red-green and blue-yellow chromatic gratings. J. Physiol. 359: 381–400.

Murray, I.J., Parry, N.R.A., Carden, D. and Kulikowski, J.J. (1987). Human visual evoked potentials to chromatic and achromatic gratings. Clin. Vis. Sci. 1: 231–244.

Perry, V.H., Oehler, R. and Cowey, A. (1984). Retinal ganglion cells that project to the dorsal lateral geniculate nucleus in the macaque monkey. Neuroscience, 12: 1101.

Schiller, P.H. and Colby, C.L. (1983). The responses of single cells in the LGN of the rhesus monkey to colour and luminance contrast. Vision Res. 23: 1631–1641.

Schiller, P.H., Logothetis, N.K. and Charles, E.R. (1990). Functions of the colour opponent and broad-band channels of the visual system. Nature 343: 68–70.

Schnapf, J.L., Nunn, B.J., Meister, M. and Baylor, D.A. (1990). Visual transduction in cones of the monkey. J. Physiol. 427: 681–713.

Shapley, R. and Perry, V.H. (1986). Cat and monkey retinal ganglion cells and their visual functional roles. Trends in Neuroscience 9: 229–235.

Stockman, A. and MacLeod, D.I.A. (1992). Faster than the eye can see: the blue cones respond to chromatic flicker. Perception 21 (suppl 2): 10.

Wiesel, T. and Hubel, D.H. (1966). Spatial and chromatic interactions in the lateral geniculate body of the rhesus monkey. J. Neurophysiol. 29: 1115–1156.

Williams, D.R. and Collier, R. (1983). Consequences of spatial sampling by human photoreceptor mosaic. Science 221: 385-387.

D.J. McKeefry
Department of Anatomy
University College, London
Gower Street
London WC1E 6BT

48. Receptive field dimensions of macaque retinal ganglion cells

JAN KREMERS, BARRY B. LEE and TSAIYAO YEH
(Göttingen, Germany)

Abstract

Receptive field dimensions of parafoveal ganglion cells of the parvocellular (PC-cells) and magnocellular pathways (MC-cells) were measured using bipartite field stimulation. Selective M- or L-cone modulation (silent substitution) was used to differentiate the centers and the surrounds of PC-cells, thereby enabling separate measurements of their dimensions. For PC-cells, our data showed no signs of mixed cone input to the surrounds. Assuming Gaussian profiles, we found that radii of the center and surround Gaussians varied between 0.5 and 23 min of arc. Surround-to-center ratios varied between 1 and 10 (median 2.8), with large ratios being associated with very small centers. For MC-cells, data with luminance modulation suggested center radii as small as 1 min of arc. MC-cell surrounds were much larger, so that they were difficult to precisely measure under our conditions. We estimate them to have at least five times the center diameter.

Introduction

Receptive field dimensions of retinal ganglion cells influence visual information transmission to the brain. Their dimensions, as well as their number, constrain the spatial resolution of the visual system. For instance, if a cell summates linearly across its receptive field, then it will respond to gratings up to a spatial period about half the center diameter (Peichl and Wässle, 1979). On the other hand, the spatial signal also depends on the responsivity within the receptive field, so that the response of MC-pathway cells to small displacements enables performance in the hyperacuity range (Lee *et al.*, 1993). Lastly, responsivity to chromatic stimuli will be determined by spatial distribution and strengths of cone inputs to ganglion cell's centers and surrounds.

The aim of this study was to measure receptive field dimensions of centers and surrounds of PC- and MC-cells. We were particularly interested in exploring the possibility of mixed cone surrounds in PC-pathway cells (Lennie

B. Drum (ed.), Colour Vision Deficiencies XII, pp. 399–406.
© 1995 *Kluwer Academic Publishers, Dordrecht.*

et al., 1991). We used as a stimulus a bipartite field (Kelly, 1975). By scanning across the receptive field, we could explore the spatial distribution of L- and M-cone inputs with selective L- and M-cone modulation, and test linearity of spatial integration and of cone summation.

Methods

Animal preparation

Details of animal preparation and cell classification are described elsewhere (Smith *et al.*, 1992). Briefly, after an initial intramuscular injection of ketamine, anesthesia was maintained with isoflurane in a 70%/30% N_2/O_2 mixture, 1–2% during surgery and 0.5–1.0% during recording. Local anesthetic was applied at the points of surgical intervention. The EEG and EKG were monitored continuously as a control for anesthetic depth. Muscular relaxation was maintained by intravenous infusion of gallamine triethiodide (5 mg/kg/hr) together with ca. 3 ml/hr dextrose Ringer. The end-tidal PCO_2 was kept near 4% by adjusting the rate and depth of ventilation, and body temperature was maintained near 37.5°C. At the end of the experiment, the animal was killed with an overdose of barbiturate.

Visual stimulation

Stimuli were presented through a two-channel Maxwellian-view system with a red and a green light emitting diode (LED) as light sources for each channel (Swanson *et al.*, 1987; Lee *et al.* 1990). Dominant wavelengths of the LEDs were 554 and 638 nm. Time-averaged chromaticity was 595 nm and luminance was 2000 td. Best focus was achieved by optimizing the response to a drifting grating of 6 or 12 cycles per degree. A half-mirrored cube combined the light of the two channels, so that one half of the bipartite field was formed by the first channel and the other half by the second channel, which was modulated in counterphase. A stepping motor positioned the edge across the receptive field. The edge was first positioned to give a minimal response, in which it was assumed to bisect the receptive field. The stepping motor returned the edge to this position after data collection, enabling us to check on spatial stability. We report here on responses to luminance modulation for MC-cells and selective L- and M-cone modulation (calculated from the Smith and Pokorny (1975) cone fundamentals) for PC-cells. Temporal frequencies were 2.44, 4.88 and 19.5 Hz.

Results and Discussion

In Figure 1 the response of a red-on center PC-cell to two cycles of 2.44 Hz selective L- and M-cone modulation is displayed for three different edge

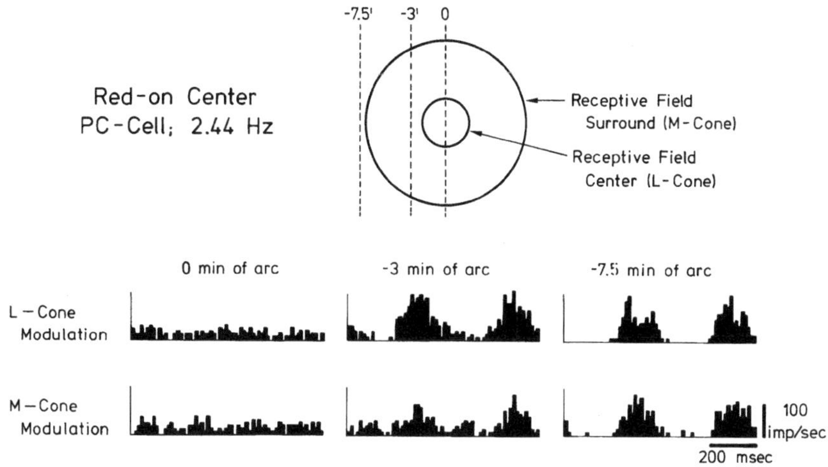

Fig. 1. Responses of a red-on PC-cell to two cycles of 2.44 Hz 53% contrast L-cone modulation and 53% contrast M-cone modulation, each at three different locations of the edge of the bipartite field. The cell does not respond to either L- or M-cone modulation when the edge is centered on the receptive field. Displacement of the edge to the border of the center results in an increase of the center (L-) cone.

locations. When centered on the receptive field, the cell responds neither to L-cone modulation nor to M-cone modulation with no sign of higher harmonic response components, indicating linear spatial summation. Displacement of the edge by 7.5 min of arc yielded responses which were identical to full field stimulation. If the edge was displaced by 3 min of arc, the response to modulation of the L-cone was similar to that with full field modulation. The response to M-cone modulation, however, was smaller than with full field modulation, indicating that the spatial extent of M-cone input to the receptive field was larger than the spatial extent of L-cone input.

Amplitude of the first harmonic components of the cell response to 2.44 Hz L- and M-cone modulation is plotted in Fig. 2 as a function of location of the edge for the red-on center PC-cell. Assuming that the spatial distribution of the cone inputs has a Gaussian profile, the relation between response and edge location is described by the absolute value of the difference between two identical Gaussian integrals. Least-square fits with this function are shown as the solid lines in Fig. 2. Fits are satisfactory. It has recently been argued that PC-cells might have mixed cone input to the surround (Lennie *et al.*, 1992). It this were the case, we would expect that the L-cone input would have a difference-of-Gaussians profile. This would be apparent in the response as a funcion of edge location in Fig. 2, but this is not the case. Further, at the higher frequencies tested (19.5 Hz), a significant interaction of edge location with response phase would be expected, due to a latency difference between center and surround cone inputs (Gouras and Zrenner, 1979; Smith *et al.*, 1992). In none of the PC-cells studied were either of these effects observed. Thus our data confirm

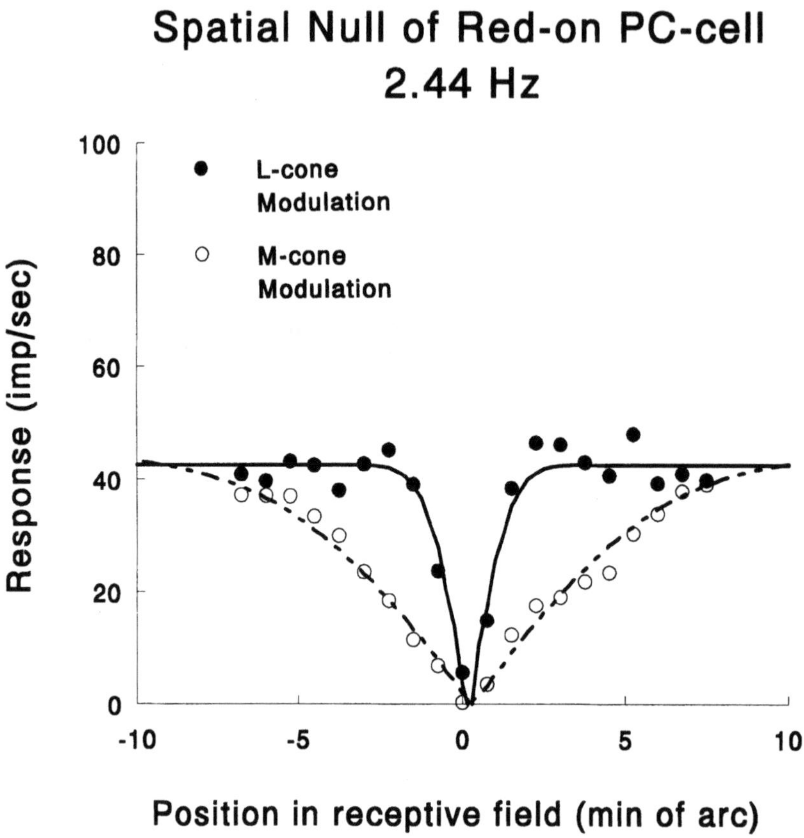

Fig. 2. Response amplitude of a red-on cell to 53% contrast L- and M-cone modulation at different edge locations. The minimal response is at the same location for L- and M-cone modulation, so center and surround are concentric. The curves are least-square fits with a function which is expected to describe the data when assuming Gaussian receptive field profiles. The center Gaussian had a radius of 1 min of arc. The surround Gaussian had a radius of 4.6 min of arc.

earlier findings (Reid and Shapley, 1992; Smith *et al.*, 1992) that PC-cells do not show mixing of cone inputs to the surround.

Data in Figs 1 and 2 were typical of all PC-cells studied. If PC-cell' centers and surrounds can be selectively stimulated by the L- and M-cone modulation, the standard deviations of the Gaussians used in the fits for L- and M-cone responses are a measure of center and surround dimensions. In 21 PC-cells with only L- and M-cone input, we found center radii between 0.5 and 23 min of arc. Surround radii were between 2 and 25 min of arc. Fig. 3 shows the surround radius as a function of the center radius. The 45° line indicates equal center and surround radii expected of type II cells (Wiesel and Hubel, 1966). Some cells fall close to this. The dashed line indicates surrounds twice as large as centers. Cells with centers smaller than about 5 min of arc had surrounds at least twice the center size, whereas for cells with larger centers the surround and center sizes

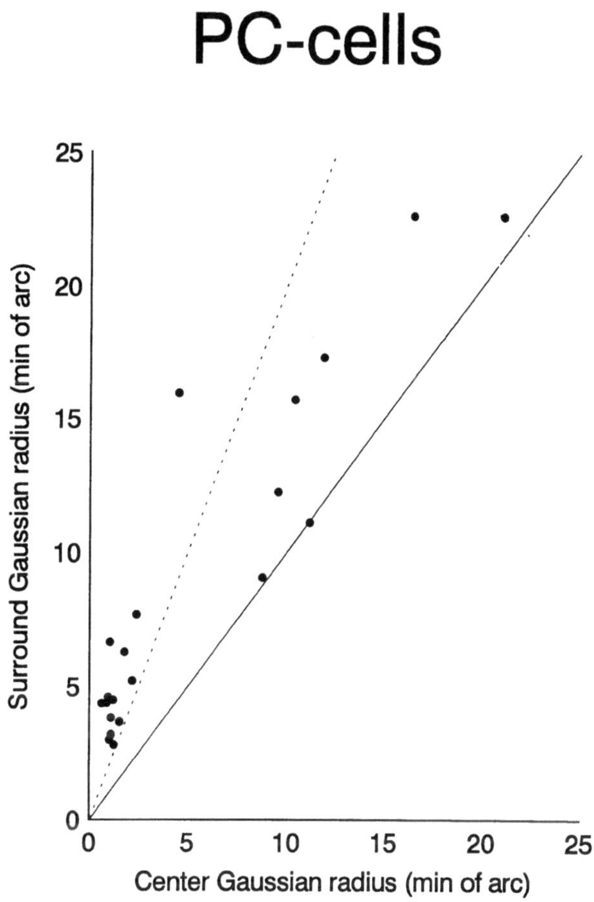

Fig. 3. Surround vs. center radius of Gaussians of different PC-cells from curve fits similar to those shown in Fig. 2. For some cells the mean of two measurements at different frequencies were used. The line under 45° indicates equal center and surround sizes. The dashed line indicates surrounds twice as large as centers.

were more similar. We interpret these data as suggesting that the ratio of center to surround sizes decreases with increasing center size.

It is not possible to separate center and surround responses of MC-cells with cone isolating stimuli. Also, to stimuli with a chromatic component, MC-cells responded with substantial higher harmonic distortions. We thus restrict our analysis to luminance modulation. Even with luminance modulation, at the first harmonic null there was often a small frequency-doubled response, indicating nonlinearity of spatial summation. Since both center and surround respond to luminance modulation, a difference-of-Gaussians model is expected to describe the data. However, receptive field surrounds of MC-cells appeared much larger than the ca. 0.4 degree range usually explored, since in all MC-cells the response to full field stimulation was smaller than the response to bipartite field

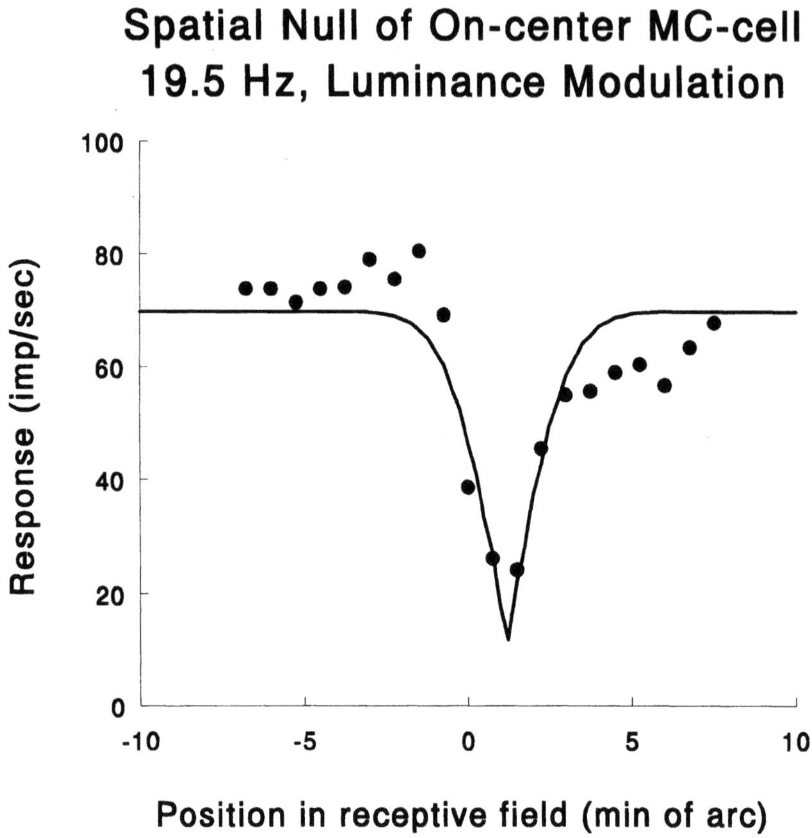

Fig. 4. Responses of an on-center MC-cell to 19.5 Hz 50% contrast luminance modulation of the bipartite field with the edge at different locations on the receptive field. The same function as was used for PC-cells was fitted to the data. Radius of the Gaussian was 1.4 min of arc.

stimulation at the extremes of the range tested. Therefore surround effects can be neglected, but data do reveal center dimensions. We measured MC-cells with luminance modulation at 2.44 and 19.4 Hz, each at 25% and 50% contrast. Fig. 4 shows the data of an on-center MC-cell for 19.5 Hz 50% contrast luminance modulation. The fitted Gaussian had a radius of 1.4 min of arc. In measurements on 8 MC-cells we found center Gaussian radii between 1 and 15 min of arc, comparable to PC-cell center size.

Our data indicate that PC-cell center radii lie between ca. 0.5 and 25 min of arc in the parafovea and MC-cell radii between 1 and 15 min of arc. Part of this variability is eccentricity related for PC and MC-cells. These radii are similar, and consistent with other measurements in the literature (Crook *et al.*, 1988). Surrounds of PC-cells did not show any sign of mixed cone input. Surround to center radius varied from 10:1 to 1:1, with a median of 2.8:1. There was no indication of a clear division into Type I and Type II cells (as proposed by Rodieck (1991)), which appear to represent extremes of a continuum. This is consistent with earlier results of Reid and Shapley (1992). Surrounds of MC-cells could not be measured reliably, but the data indicate that their surrounds were much larger than the centers, with Gaussian radii larger than 30 min of arc.

Acknowledgements

Jan Kremers was supported by DFG Grant II B7-Le 524/6-1.

References

Crook, M.J., Lange-Malecki, B., Lee, B.B. and Valberg, A. (1988). Visual resolution of macaque retinal ganglion cells. J. Physiol. 396: 205–224.

Derrington, A.M. and Lennie, P. (1984). Spatial and temporal contrast sensitivities of neurones in lateral geniculate nucleus of macaque. J. Physiol. 357: 219–240.

Gouras, P. and Zrenner, E. (1979). Enhancement of luminance flicker by colour opponent mechanisms. Science 205: 587–589.

Kelly, D.H. (1975). Luminous and chromatic flickering patterns have opposite effects. Science 188: 371–372.

Lee, B.B., Martin P.R. and Valberg, A. (1989). Nonlinear summation of M- and L-cone inputs to phasic retinal ganglion cells of the macaque. J. Neuroscience 9: 1433–2442.

Lee, B.B., Pokorny, J., Smith, V.C., Martin, P.R. and Valberg, A. (1990). Luminance and chromatic modulation sensitivity of macaque ganglion cells and human observers. J. Opt. Soc. Am. A 7: 2223–2236.

Lee, B.B., Wehrhahn, C. Westheimer, G. and Kremers, J. (1993). Macaque ganglion cell responses to stimuli that elicit hyperacuity in man: detection of small displacements. J. Neuroscience 13: 1001–1009.

Lennie, P., Haake, P.W. and Williams, D.R. (1991). The design of chromatically opponent receptive fields. In: Landy, M.S. and Movshon, J.A. (eds.), Computational Models of Visual processing: 71–82, MIT Press, Cambridge, MA.

Peichl, L. and Wässle, H. (1979). Size, scatter and coverage of ganglion cell receptive fields centres in the cat retina. J. Physiol. 291: 117–141.

Reid, C. and Shapley, R.S. (1992). Spatial structure of cone inputs to receptive fields in primate lateral geniculate nucleus. Nature, 356: 716–718.

Rodieck, R.W. (1991). Which cells code for color? In: Valberg, A. and Lee, B.B. (eds.), From pigments to perception. Advances in Understanding Visual Processes: 83–93, Plenum Press, New York and London.

Smith, V.C. and Pokorny, J. (1975). Spectral sensitivity of the foveal cone photopigments between 400 and 500 nm. Vision Res. 15: 161–171.

Smith, V.C., Lee, B.B., Pokorny, J., Martin, P.R. and Valberg, A. (1992). Responses of macaque ganglion cells to the relative phase of heterochromatically modulated lights. J. Physiol. 458: 191–221.

Swanson, W.H., Ueno, T., Smith, V.C. and Pokorny, J. (1987). Temporal modulation sensitivity and pulse-detection thresholds for chromatic and luminance pertubations. J. Opt. Soc. am. A 4: 1992–2005.

Wiesel, T.N. and Hubel, D. (1966). Spatial and chromatic interactions in the lateral geniculate body of the rhesus monkey. J. Neurophysiol. 29: 1115–1156.

Dr. Barry Lee
Department of Neurobiology
Max-Planck-Institute for Biophysical Chemistry
37 Göttingen, Germany

49. Objective assessment of short wavelength sensitive (SWS) mechanisms with the spatio-chromatic VEP: X-linked achromatopsia and transient tritanopia

M.A. CROGNALE, E. SWITKES,[1] J. RABIN, M.E. SCHNECK, G. HÆGERSTRÖM-PORTNOY and A.J. ADAMS

(Berkeley and [1] Santa Cruz, CA, U.S.A.)

Abstract

As an extension of recent investigations of the VEP produced by the SWS mechanism, we have studied two cases of visual sensitivity changes involving the SWS mechanisms: transient tritanopia, an induced loss in the SWS mechanism, and X-linked achromatopsia (blue cone monochromacy). Utilizing stimuli with spatiochromatic variations which selectively activate specific chromatic and achromatic channels, we have measured the VEP and characterized the SWS response in these two conditions.

Introduction

A large body of psychophysical, electrophysiological, anatomical and genetic evidence suggests that the SWS mechanism has properties which differ from those of the longer wavelength mechanisms (e.g. Brindley *et al.*, 1966; Marc and Sperling, 1977; Mollon, 1982). The SWS mechanism has also been shown to be selectively damaged in many ocular diseases, and psychophysical tests of SWS mechanism losses have proven to be clinically useful.

The VEP has been demonstrated to be an objective and reliable indicator of suprathreshold visual function (see Regan, 1989 for a comprehensive review). Isolation of chromatic or achromatic mechanisms is obtained in a state of neutral adaption by employing gratings which vary along selected axes in a cone activation space (Crognale *et al.*, 1992, 1993a-c; Rabin *et al.*, in press). A robust chromatic VEP is obtained by the use of low spatial frequencies and an onset temporal presentation (also see Murray *et al.*, 1987; Kulikowski *et al.*, 1989; and Berninger *et al.*, 1989). This technique has been shown to be useful in investigations of acquired and congenital color anomalies (Crognale *et al.*, 1993b). In these earlier studies we have collected normative latency and amplitude data and have shown that, in particular, latency provides a useful metric in comparing normal and anomalous visual function. Here we describe two diverse applications of the technique. One, a study of an extremely rare

B. Drum (ed.), Colour Vision Deficiencies XII, pp. 407–413.
© 1995 *Kluwer Academic Publishers, Dordrecht.*

congenital defect, and the other, a phenomenon involving interactions between photoreceptor types.

Methods

Monocular VEPs were recorded using standard techniques. Stimuli were sinusoidal gratings modulated along various directions of a color space (MBDKL) described by Krauskopf *et al.*, (1982) and based upon the cone activation space of MacLeod and Boynton (1978). Our utilization of this space and methods are detailed elsewhere (Rabin *et al.* in press; Crognale *et al.*, 1993b). The directions of isoluminant modulation used in the current study are: the LM axis (at ø = 0° in the MBDKL space), which modulates the L and M cones antagonistically but does not modulate the S cones, and the S axis (at ø = 90° in the MBDKL space) which modulates only the S cones. Modulations along intermediate directions within the isoluminant plane, modulating all 3 cone types in varying amounts, were also used. The contribution of rods to the spatio-chromatic VEP will be discussed below. Cone activations and luminosity were calculated from the Smith and Pokorny (1975) fundamentals and the CIE (1951) photopic luminosity function. Rod contrasts were calculated using V_λ/V_λ' ratios measured for our monitor phosphors using a Pritchard spectral radiometer (model 1980B). VEPs were recorded in response to the onset (100 msec on, 400 msec off) of 1 cpd vertical sine wave gratings in a 18° circular patch.

Results and discussion

We have previously characterized onset VEPs in normal observers for a number of spatial and chromatic parameters (Crognale *et al.*, 1992; Crognale *et al.*, 1993b, Rabin, *et al.*, in press). The waveforms of the spatio-chromatic VEP are in agreement with those reported by others using similar stimuli (e.g. Murray *et al.*, 1987; Kulikowski *et al.*, 1989; Berninger *et al.*, 1989). Onset isoluminant chromatic VEPs show mainly a large negative component, unlike VEPs to achromatic reversals.

Transient tritanopia

Transient tritanopia is characterized by a loss in SWS mechanism sensitivity following the offset of a long wavelength adaptation light (Mollon and Polden, 1975, 1977). It is believed that this effect is mediated by signals from the longer wavelength cones at a secondary adaptation site, at or beyond the opponent site, since losses can be observed even when the adaptation light has no effect upon the cones themselves (Pugh and Mollon, 1979). In addition, it has recently been reported that transient tritanopia can also be produced by rod interaction with

SWS cone mechanisms in x-linked achromatopsia (Hægerström-Portnoy and Verdon, 1991). Transient tritanopia is typically studied using a short wavelength test stimulus with auxiliary long wavelength adaptation to be more confident that the test is being detected by the cones (e.g. decrementing the adapting light rather than terminanting it). Nonetheless, without measuring test spectral sensitivity it is often unclear which mechanism is detecting the stimulus after the offset of the long wavelength adapting light. We have shown that transient tritanopia can be observed using the spatio-chromatic VEP in conjunction with a cone activation color space (Rabin *et al.*, in press). The advantage afforded by this technique is that one can isolate SWS cone responses without auxiliary adaptation and still be assured that detection is by the SWS cones.

Transient tritanopia in the VEP is illustrated in Fig. 1a. These waveforms were obtained by recording the responses to the onset of S-axis gratings before, during and after (within 5 seconds of the offset) of a long wavelength adapting light (Kodak wratten 23a filter; 200 cd/m^2). The responses obtained during and before the long wavelength adaptation are similar suggesting that the adapting light had little effect on the cones and that the S cones were reasonably isolated. However, the VEP responses obtained following extinction of the adapting light are greatly reduced in amplitude.

Fig. 1b indicates VEP amplitude reduction in transient tritanopia as a function of direction within the isoluminant plane. The isoluminant plane was determined individually for each subject by reversing the pattern at 16 hz and adjusting the relative luminance of the grating bars to obtain minimum flicker. The pattern mean luminance was 37 cd/m^2. The adapting luminance was 166 cd/m^2. The color space for each subject was scaled such that the contrast of the stimuli along the LM and S axes were each 10 $\times$ threshold. Stimuli in other directions were combinations of these gratings. To construct this plot, VEPs obtained from isoluminant grating onsets were measured during and after superimposition of a long wavelength adapting light. The amplitudes of the responses obtained after extinction of the adapting light was subtracted from the amplitude of the response obtained during the long wavelength adaption. This was repeated for color directions at 15° increments within the isoluminant plane.

In the polar plot of Fig. 1b, the axis which modulates only the S cones is vertical and the axis which isolates the L and M cones is horizontal. The inner gray circle indicates zero differences in amplitude during and after the adaptation. The outer gray circle indicates an amplitude decrease at offset of 5 μV. Complete data sets for 2 subject (MC and MS) are shown along with LM and S axis data for a third subject (SF).

Although differing in magnitude, the data set for each subject shows a similar pattern: the amplitudes during adaptation are smaller than those after adaptation for modulations along the LM axis, while the amplitudes during adaptation are greater than those after extinction of the long wavelength adapting light for modulations along the S axis. It can be concluded that in

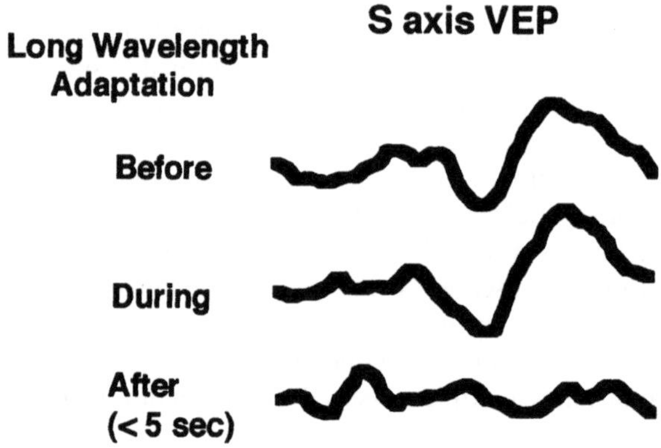

(a)

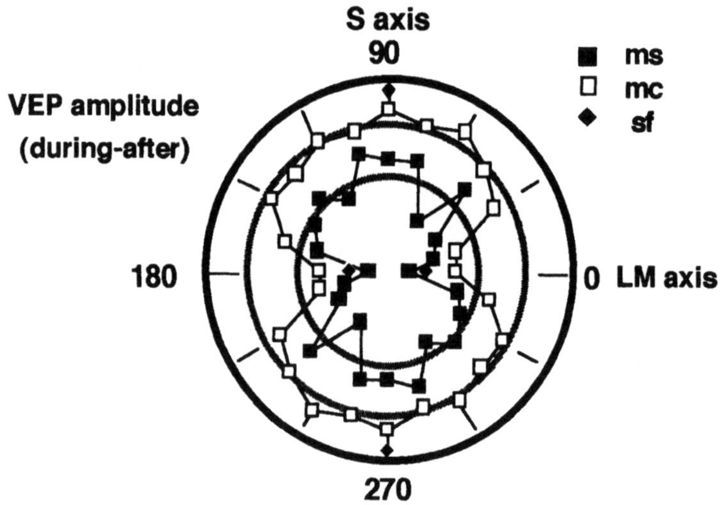

(b)

Fig. 1. (a) Transient tritanopia in the VEP. Waveforms obtained for S axis modulation before, during, and immediately following the extinction of a long wavelength adapting light (Figure modified from Rabin *et al.*, in press).
(b) Plot of the difference in VEP amplitude obtained during and after long wavelength adaptation as a function of stimulus direction within the isoluminant plane (determined individually) for 3 subjects (MS, MC and SF). The inner gray ring corresponds to zero difference and the outer gray ring + 5 μV.

addition to detection of congenital cone deficits, the spatio-chromatic VEP is a reliable indicator of post-receptoral chromatic interactions such as transient tritanopia in normal trichromats.

X-linked achromatopsia

We have previously reported how the spatio-chromatic VEP is useful for characterization of a variety of common color anomalies (Crognale *et al.*, 1993b). In particular, VEP waveforms were obtained from a congenital tritanope for modulations along 5 axes (S, LM, L, M and luminance) in a cone activation color space. The responses were robust, with normal amplitudes and latencies along all of the axes except the S axis. For modulations along the S axis no response could be obtained. Similarly, when responses latencies are plotted relative to normals as a function of direction within the isoluminant plane, latencies increase toward the S axis. Although possessing normal rod function, the tritanope gives no response to a stimulus varying along the S axis, which also strongly modulates rods. The results suggest that the VEP technique isolates the SWS mechanism and that under these conditions rods contribute little to the isoluminant VEP.

Here we present results of another form of congenital defect: X-linked achromatopsia (commonly called 'blue cone monochromatism'). The stimuli were modulated along the S* (silent rod, and L), LM (silent S) and rod (silent S) axes. The mean luminance was 37 cd/m^2. Receptor contrasts for the different axes were: S* axis- l = 0, m = .18, s = .781, rod = 0; LM axis- s = 0, l = .094, m = .18, rod = .24; Rod axis- l = .937, m = .857, s = 0, rod = .60. The LM axis was the same as used in our previous reports in trichromats and dichromats and was chosen to silence the S cones. The gratings along this axis appear to be very high contrast to a trichromat but were not visible to 2 of the 3 achromats. The third achromat reported that he could see that 'there seems to be someting, there' but he could not discern a pattern or correctly identify the orientation of the grating. When the luminance of the pattern was lowered by 1.5 log units, the subject noted that the pattern became more visible suggesting that rods were responsible for the residual detection. Figure 2 shows waveforms obtained from one of 3 monochromats tested. The results from the two other subjects were similar. As can be seen, the LM axis VEP waveforms differ little from the noise, despite producing significant rod contrast. These results indicate that if these achromats prossess residual L or M cones, they do not contribute significantly to detection or to the pattern onset VEP. The results also suggest that as in trichromats and dichromats, rods contribute little to the achromat's VEP.

Based on the previous result, the 'rod' axis utilized here was chosen to silence the S cone and produce maximum modulation of the rods without concern for L or M cone modulation. This pattern appeared to the trichromat as a high contrast grating that was almost achromatic, with the dark stripes being slightly bluish and the light stripes having a tinge of yellow. This pattern was described as 'slightly visible' by the achromats. All 3 subjects reported that the pattern was much more apparent at lower light levels. The VEP waveforms produced by modulations along this axis (fig. 2) show small peaks at long latencies. The magnitude of the response was variable and was smaller for the other two subjects. This result supports the previous conclusion that the L M cones do not

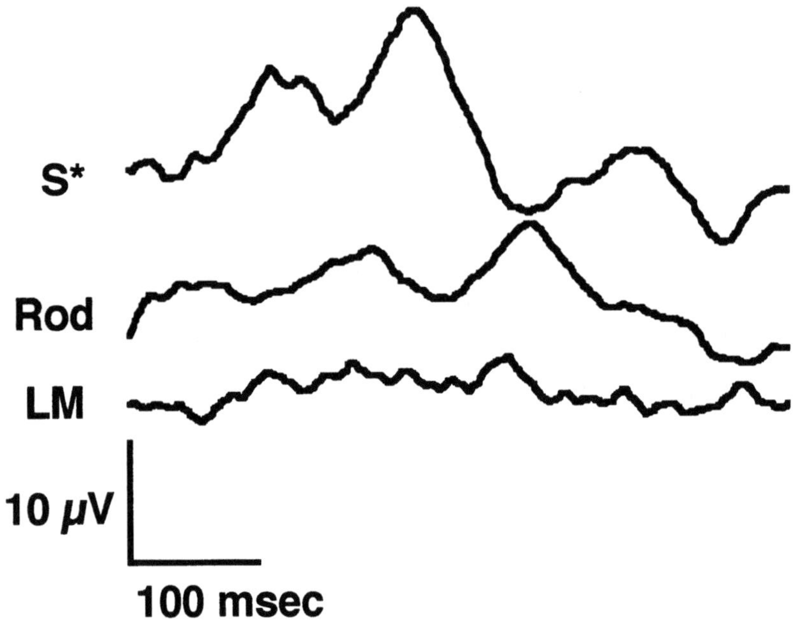

Fig. 2. VEP waveforms obtained from modulations along S* (silent rod, and L), LM (silent S) and rod (silent S) axes in an x-linked achromat (SH).

contribute significantly to the achromat's VEP and also that rods produce little onset response.

The S* axis used here silenced the rods and the L cones while maximally stimulating the S cones and, to a much lesser degree, the M cones. This pattern had the least apparent contrast of the three gratings to the trichromat and was somewhat similar in appearance to an isoluminant S axis. As expected, the achromats reported that this pattern was very visible. The S* axis in the achromat produces VEP waveforms that are robust and show prominent *positive* peaks. The waveforms are complex and do not show the characteristic large negative wave indicative of isoluminant VEP responses in trichromats and dichromats (e.g. see 'before' in Fig. 1a). This result indicates that achromats differ from trichromats and dichromats in the type of waveform produced by S cone isolation and provides further evidence for major differences in processing of achromatic vision. The failure to observe a large negative component in the onset VEP in X-linked achromats suggests that the negative component of the onset VEP in may be a product of intact chromatic opponency.

Acknowledgements

This research was supported by grants EY02271, EY07043, EY06461, EY00014 and EY07345.

References

Berninger, T.A., Arden, G.B., Hogg, C.R. and Frumkes, T. (1989). Separable evoked retinal and cortical potentials from each major visual pathway: preliminary results. Brit. J. Ophthalmol. 73: 502–511.

Brindley, G.S., Du Croz, J.J. and Rushton, W.A.H. (1966). The flicker fusion frequency of the blue sensitive mechanism of colour vision. J. Physiol. 183: 497–500.

Crognale M.A., Rabin J., Adams A.J. and Switkes E. (1992). VEP correlates of spatio-chromatic processing. Invest. Ophthalmol. Vis. Sci. (Suppl.) 33: 701.

Crognale M.A., Adams A.J. and Switkes, E. (1993a). Temporal tuning of color vision mechanisms determined with the spatio-chromatic VEP. Invest. Ophthalmol. Vis. Sci. (Suppl.) 34: 744.

Crognale M.A., Switkes, E., Rabin, J., Schneck, M.A. Hægerström-Portnoy G. and Adams A.J. (1993b). Application of the spatio-chromatic visual evoked potential to detection of congenital and acquired color deficiencies. J. Opt. Soc. Am. A. 10: 1818–1825.

Crognale, M.A., Rabin, J., Switkes, E. and Adams, A.J., (1993c). Selective loss of S-pathway sensitivity in central serous choroidopathy revealed by spatio-chromatic visual evoked cortical potential (VECP). In: Drum, B. (ed.), Colour Vision Deficiencies XI, Doc. Ophthalmol. Proc. Ser. 56: 229–239, Kluwer Acad. Publ., Dordrecht.

Hægerström-Portnoy, G. and Verdon W. (1991). Transient tritanopia in a blue-cone monochromat. Clin. Vis. Sci. 6, (3), 229–240.

Krauskopf, J. Williams, J.D.R., and Heeley, D.W. (1982). Cardinal directions of color space. Vision Res. 22: 1123–1131.

Kulikowski, J.J., Murray, I.J., and Parry, N.R.A. (1989). Electrophysiological correlates of chromatic-opponent and achromatic stimulation in man. In: Drum, B., and Verriest, G. (eds.), Colour Vision Deficiencies IX, Doc. Ophthalmol. Proc. Ser. 52: 145–153, Kluwer Acad. Publ., Dordrecht.

Marc, R.E. and Sperling, H.G. (1977). Chromatic organization of primate cones. Science, 196: 454–456.

MacLeod D.I.A. and Boyton, R.M. (1978). Chromaticity diagram showing cone excitation by stimuli of equal luminance. J. Opt. Soc. Am. 69: 1183–1186.

Mollon, J.D. (1982). A taxonomy of tritanopias. In: Verriest, G. (ed.), Colour Vision Deficiencies VI, Doc. Ophthalmol. Proc. Ser. 33: 87–101, Dr. W. Junk, The Hague.

Mollon J.D. and Polden P.G. (1975). Colour illusion and evidence for interaction between cone mechanisms. Nature (Lond.) 258: 421–422.

Mollon J.D. and Polden P.G, (1977). An anomaly in the response of the eye to light of short wavelengths. Phil. Trans. R. Soc. Lond. B. 278: 207–240.

Mollon, J.D. and Polden P.G. (1979). Post-receptoral adaptation. Vision Res. 19: 435–440.

Murray, I.J., Parry, N.R.A., Carden, D. and J.J. Kulikowski, J.J. (1987). Human visual evoked potentials to chromatic and achromatic gratings. Clin. Vision Sci. 1: 231–244.

Rabin, J., Switkes, E., Crognale M.A., Schneck, M.E and Adams A.J. Visual evoked potentials in three-dimensional color space: Correlates of spatio-chromatic processing (Vision Res., in press).

Regan D. (1989). Human Brain Electrophysiology. Elsevier, New York.

Smith, V.C. and Pokorny, J. (1975). Spectral sensitivity of the foveal cone photopigments between 400–500 nm., Vision Res. 15: 161–171.

Dr. Michael Crognale
School of Optometry
University of California
Berkeley, CA. 94720, U.S.A.

50. Analysis of EEG to colored lights by the maximum entropy method

MUNEHIRA AKITA and SHIMIZU MIYAHARA
(Kyoto and Osaka, Japan)

Abstract

In a color normal male subject, EEGs were recorded while the subject viewed red, green or achromatic lights (14 cd/m^2) presented on a CRT, and also in a dark room with the subject being instructed to keep his eyes open or closed. Data analysis was performed by the maximum entropy method (MEM) through which event-related spectral power density (PSD) was obtained with its color topographic brain mappings. Based on the EEG mapping it was shown that cortical activation patterns, in general, exhibited fewer differences corresponding to differences of colored lights stimulation, but distinct changes were found between eye closed and eye open conditions in the dark room.

Introduction

In 1967 Burg proposed a new method for the estimation of power spectra for time series random data based on an original idea of making information entropy maximum. He developed an effective algorithm to compute and obtain a high-resolution maximum entropy power spectrum.

Often in applications of Fourier transforms, and particularly with spectrum analyses, brain waves to be investigated consist of a finite collection of events to be regarded as pulses solely encoding the relative position of the events. In particular, the fast Fourier transform (FFT) algorithm devised by Cooley and Tukey (1965) employs an integer representation of brain waves.

However, the principal intent of FFT algorithms is to increase the speed of computation to obtain Fourier components of finite discrete data since the sine and cosine terms generally consume considerably less computer time. In essence this is not a method of spectral analysis.

The inevitable fault of the FFT procedure is that a finite observed wave is hypothesized to repeat infinitely, so many false extremes flank a main peak in spite of adopting a window function. But in MEM this intrinsic fault is solved so that MEM can have a very high resolution power (Tanaka *et al.*, 1993).

B. Drum (ed.), Colour Vision Deficiencies XII, pp. 415–419.
© 1995 *Kluwer Academic Publishers, Dordrecht.*

The spectral analysis of MEM is nonlinear: the more the number of data points and degrees of auto-regression increase, the more the heights of the peaks approximate the true values. Therefore, a mapping of brain potentials is possible (Akita and Miyahara, 1993). Space constraints prevent the details from being given here.

Method

Subjects

One male subject (aged 42 years) who was the co-author and had normal color vision and visual acuity participated in this experiment.

Stimulus

Each of two colored lights, red (x = .5985, y = .3531) and green (x = .3152, y = .5854) was presented on a CRT display for 60 seconds duration alternatively with an achromatic light (x = .313, y = .328) of 60-sec duration. The sequence was repeated five times. Stimuli were stored in a video memory (Shibasoku VM02B1) which was controlled by a computer (NEC PC9801VM). The subject viewed the display window binocularly. The stimuli – red, green and achromatic lights – were presented with an effective luminance of 14 cd/m^2. The subject sat in a light-proof shielded experimental booth.

Procedure

After 10 min dark adaptation, the subject was asked to open and close his eyes for alternate 60-sec time periods with the alternations being repeated five times. This was followed by the presentation of the colored and achromatic lights. EEG responses were recorded by 12 electrodes, (that is, Fp1, Fp2, F7, F8, C3, C4, T5, T6, 01, 02, Fz, and Pz) using a stretch cap made of an elastic spandex-type fabric that contained electrodes in the International 10/20 system pattern. Resistance was measured and accepted if below 15k ohm. The electrical activity from each lead was fed into separate Nihon Koden amplifiers (Models 7314). Activity for each lead was displayed on-line using a Canopus Instrument A/D board and acquisition software. The EEG underwent 0.53–30 Hz band pass filtering and was digitized at 100 samples per second with 12 bit accuracy for each of the 12 channels. Raw data were stored for later analysis.

The digitized EEG data were analyzed by both MEM-PSD with software developed by the authors and a discrete FFT using a Hanning window of 10-sec width and 50% overlap. The results were displayed as a color topographic pattern of cortical activities on CRT. The power was expressed as square microvolts.

Results

Figure 1 shows the result of EEG color topographic mappings obtained first when the MEM analysis was applied. These mappings indicate that there are distinct changes in the power and frequency pattern of the EEG across 25 regions of the brain.

First, power foci in alpha waves (8–10 Hz and 10–13 Hz) are clearly exhibited in both the eye closed and eye open conditions in occipital and frontal areas respectively.

Second, there is a great deal of change for both conditions in beta wave activity (18–20 Hz, 20–25 Hz, and 25–30 Hz). The changes include an increase in power with a distinct focusing predominance in the right frontal region in eye open condition. The eye-closed condition resulted in a decrease in power to a less distinctive more diffuse pattern. The condition in which a light stimulus was presented resulted in an intermediate power strength.

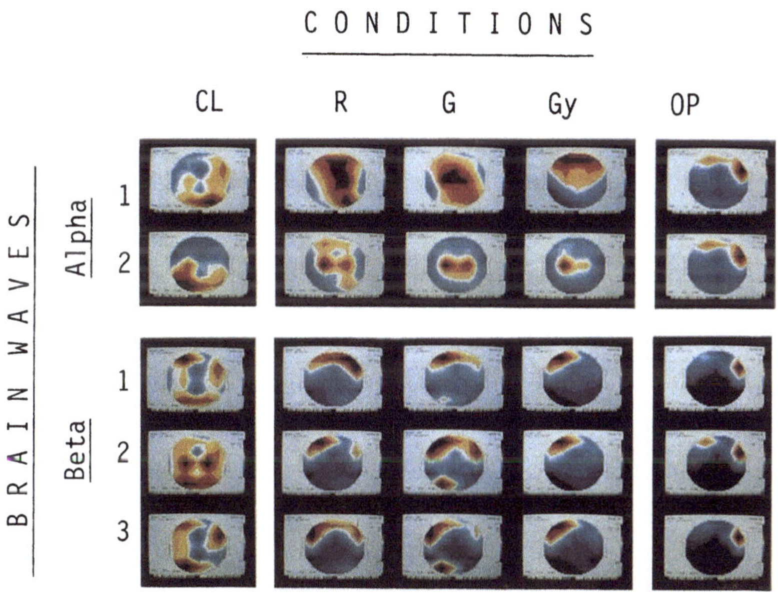

Fig. 1. EEG mapping patterns obtained by MEM. The condition codes CL, R, G, Gy and OP represent closed-eye, red light, green light, grey light and open-eye experimental conditions, respectively. Alpha 1 and 2 and Beta 1, 2 and 3 correspond to 8-10, 10-13, 18-20, 20-25 and 25-30 Hz frequency bands, respectively. The upper part of each pattern is the frontal region of the brain. A color code scale at the right of each pattern indicates the power of cortical activity at each EEG frequency band and experimental condition. Each mapping pattern is produced by the MEM power spectral sensitivity calculated on the basis of electrical potentials measured by 12 electrodes on the scalp. The color code consists of 15 grades in which dark brown represents the top, white the middle and dark blue the bottom (O) of the power scale. The highest values of each pattern are given in Table I.

Third, little differences among the light stimulus conditions are evident in the color topographic mapping. There is a general trend toward an increase of power in both alpha waves in frontal and parietal, and beta waves in (left) frontal independent of the color or the light stimulus.

Discussion

The topographic mapping patterns of EEG in power differ according to the method of analysis used on the same data. As the result of mapping based on the FFT analysis inevitably includes many false power spectra flanking a main peak (in spite of our using a Hanning window of 10 sec width and 50% overlap), a resolution of power spectrum of FFT is inferior to that of MEM. As can be seen in Fig. 2 there is a power focus as well as a broad diffusing pattern in the left frontal region of the FFT map which is absent in the MEM map. This is likely due to the false power spectra. If this is true, a power spectrum pattern based on the MEM is the better method of revealing genuine cortical activity. To document this, further careful and through examinations will be required.

Table 1. The highest value of power corresponding to Figure 1.

BRAIN WAVES		Conditions				
		CL	R	G	Gy	OP
Alpha	1	11.87	9.79	12.84	16.54	23.82
	2	41.87	16.50	28.78	39.02	43.95
Beta	1	3.21	12.46	10.20	12.79	66.06
	2	8.49	39.39	19.37	24.91	111.0
	3	2.83	18.80	15.90	16.24	176.0

FFT MEM

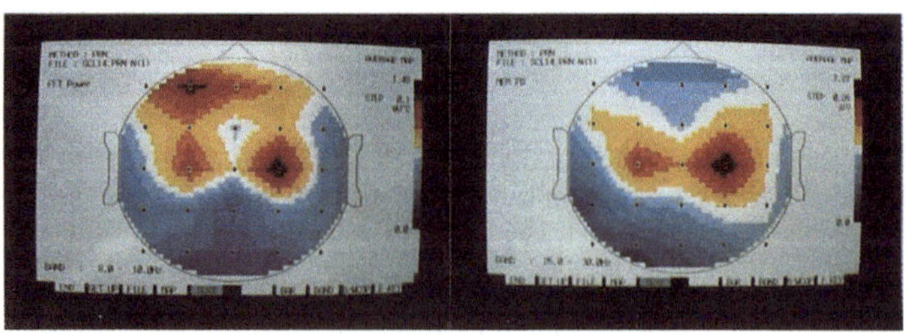

Fig. 2. Comparison of EEG color topography obtained by FFT (Fast Fourier Transform) and by MEM (Maximum Entropy Method). The highest power value is indicated as brown in the color code scale, appearing at the right margin of each pattern.

It is interesting to note that the distinct strong focus is specific to the right frontal region in the open eye conditions as shown in Fig. 1 and the highest value of power is also specific to the same region across brain waves as shown in Table 1. These results may be associated with keeping one's eyes open in darkness which seems to require more effort for the subject.

This may suggest accordingly some relationship between cortical activity in the right frontal region and the intentional effort of the subject.

As far as color perception is concerned, red, green and achromatic light stimulation could not be discriminated in general on the basis of the topographic analysis of EEG spectral power computed by the MEM. The analysis, however, is effective in discriminating cortical activity as a whole in terms of amplitudes, components, and regions as both spatial and temporal changes of power spectra of the EEG appear more precisely characterized by the MEM rather than the FFT algorithm.

Acknowledgements

The authors thank Joel Pokorny for editorial assistance in preparation of the manuscript. The work supported in part by the Japanese Ministry of Education, Science, and Culture. Grant No. 03451017 to M. Akita and S. Miyahara.

References

Akita, M. and Miyahara, S. (1993). Psychophysical and psychophysiological study on temporal and spatial factors in human color perception (Grant No. 03451017). Kyoto: Kyoto Institute of Technology, Psycology Lab., March (in Japanese).

Burg, J.P. (1968). A new analysis technique for time series data. Presented paper at Advanced Study Institute on Signal Processing. NATO, Enschende, the Netherlands.

Cooley, J.W. and Tukey, J.W. (1965). An algorithm for the machine calculation of complex Fourier series. Mathematics of Computation. 19: 297–301.

Tanaka, Y., Ohtomo, N., and Terachi, S. (1993). Nonlinear time series analysis, 6. On the MEM spectrum for cyclic functions. Bulletin of the Faculty of Engineering, Hokkaido University, Sapporo, No. 163, 47–52 (in Japanese).

Psychology Lab.
Faculty of Engineering and Design
Kyoto Institute of Technology
Gosho Kaido-cho, Matsugasaki
Sakyo-ku, Kyoto 606
Japan

PART SEVEN

Colour vision testing

51. The use of CRT displays in research on colour vision

J.D. MOLLON and M.R. BAKER
(Cambridge, U.K.)

Abstract

An introduction is given to the use of colour graphics monitors in visual research. The choice, setting-up, and calibration of graphics boards and displays are reviewed, and a survey is given of the non-additivities, non-uniformities and temporal instabilities that complicate the experimental use of such systems.

Introduction

A computer-controlled display offers the visual scientist a wonderful freedom to manipulate stimuli in time, space, and colour space, to present decrements as well as increments, and to hold constant the excitation of two classes of cone – even perhaps, of two putative post-receptoral channels – while modulating the third (Rodieck, 1983). It also offers a wonderful variety of snares and puzzles to entrap the innocent. The purpose of the present account is to supply some background on cathode ray tubes (CRTs) and graphics boards, to outline the setting-up and calibration of such a display system, to list the several problems that may then exercise the user, and to gather together the wisdom already available in a scattered literature. We write from the viewpoint of the consumer rather than the engineer, and we hope that our own experience may serve to guide newcomers in their choice of equipment. But ultimately, buying a graphics display is like buying a domestic appliance: one needs to own it for twelve months before one knows how to buy it.

Monitors and graphics boards

The cathode ray tube is a living fossil. It is approaching the centennial of its invention[1] and is probably in an evolutionary cul-de-sac; but the demand for

[1] Ferdinand Braun (1897) describes a recognizable ancestor of the modern CRT – a flask-shaped vacuum tube with cathode at one end and phosphorescent screen at the other, an anode, and electromagnetic coils for deflection in x and y dimensions.

B. Drum (ed.), Colour Vision Deficiencies XII, pp. 423–444.
© 1995 *Kluwer Academic Publishers, Dordrecht.*

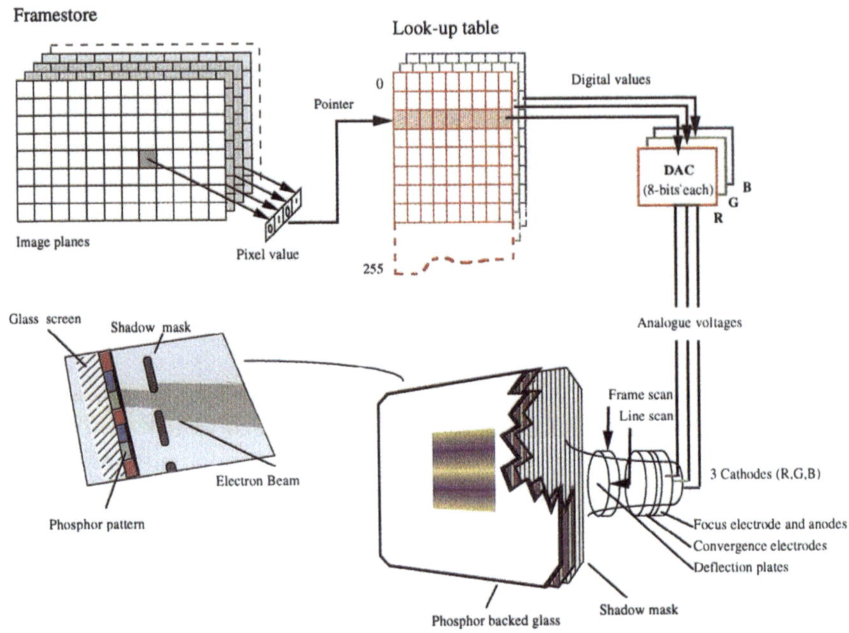

Fig. 1. Schema of a graphics card (above) and CRT monitor (below). In the framestore, the value stored for a particular pixel points to one entry in the look-up table. The look-up table depicted here has 256 entries (requiring 8 planes of framestore memory) and for each entry the gun voltage can be specified with a resolution of 8 bits. The CRT shown is based on the Sony Triniton arrangement, in that the shadow mask or 'aperture grill' consists of vertical 'wires' and the phosphor pattern consists of vertical stripes.

high-resolution computer displays has led to continued adaptations, and flat-screen displays have still not gained their long-heralded dominance. A schematic of a CRT colour raster display is shown in the lower part of Fig. 1. Within a vacuum tube, a stream of electrons is emitted from the heated cathodes of each of three guns and focussed through a shadow mask to excite phosphor dots or stripes on the faceplate of the screen. The shadow mask lies a few millimetres behind the faceplate, and the perforations in it ensure that the 'red', 'green' and 'blue' guns excite (more or less exclusively) the appropriate subset of phosphor elements. The electron beams are deflected electromagnetically from left to right, and, more slowly, from the top to the bottom of the screen, so that the modulation of the electron flux in time draws out a detailed two-dimensional image.

There exist several arrangements of guns, perforations, and phosphor elements (Wheeler and Clark, 1992). In the older 'delta gun' design, the three guns are arranged in a triangle around the neck of the vacuum tube and the perforations are circular. In the Trinitron design, introduced by Sony, the electron beams are drawn from three horizontally-aligned cathodes and the corresponding perforations are nearly continuous, vertical apertures.[2] Many other modern monitors have three horizontally-aligned guns and a shadow mask perforated with vertical slots. Clearly the purity of the colours obtained will be impaired if the shadow mask becomes misaligned relative to the triads of phosphor elements or becomes magnetized or becomes distorted through heating.

Any distortion of the temporal modulation of the electron beam – introduced by the stimulus generator, the cables, the electronics of the monitor, or nearby equipment – will necessarily be translated into a spatial distortion; and this is something that the new user should constantly keep in mind. The rate at which the entire screen is successively redrawn is the 'frame rate' and has a typical value of 70 Hz for modern display systems. An obvious point – but one sometimes forgotten when multiple targets are presented on a CRT and reaction times or eye-movement latencies are recorded – is that stimuli higher on the screen are presented several milliseconds earlier than lower ones.

Bits per pixel and bits per gun

Typically, the visual scientist drives the colour monitor either from a 'frame store' that allows picture elements (pixels) to be addressed individually, or from a specialized waveform generator that is dedicated to producing repetitive stimuli with high precision. In both cases, a stream of numerical values is transformed by three digital-to-analogue converters (DACs) into three corresponding voltage signals that specify the required outputs of the red, green and blue guns of the monitor. Digital-to-analogue converters take a finite time to settle at a new voltage level, and a longer time is required the greater the resolution with which the voltage is specified. Waveform generators may typically drive DACs with 14-bit resolution (giving 2^{14} specifiable levels), but such DACs cannot settle rapidly enough to follow pixel-by-pixel changes along a single horizontal scan line. This is why dedicated waveform generators are typically restricted to generating horizontal gratings: in the latter case the outputs of the DACs need to change only between successive horizontal lines of the raster. In contrast, the DACs at the output of a frame store must be able to follow changes from pixel to pixel across a horizontal scan, and DACs that operate at this speed typically have a resolution of 8 or 10 bits. For the user who requires better resolution at the output of a frame store, a possible manoeuvre is to add the outputs of two parallel DACs, one providing a coarsely spaced set of voltage levels and the second providing a set of smaller, finely spaced, voltages (Watson *et al.*, 1986; Pelli and Zhang, 1991).[3]

[2] The vertically strung wires of the Trinitron display need one or two (depending on the screen size) horizontal wires for support, and these are visible as dark lines. The experimenter will normally be able to work round them, once their presence and function are recognized.

[3] An arrangement of this kind is incorporated in the Cambridge Research Systems VSG2/2 board (Cambridge Research Systems Ltd., 80 Riverside Estate, Sir Thomas Longley Rd., Rochester, Kent, ME2 4BH U.K.)

A frame store similar to those in common use is represented in Fig. 1. The maximum number of individually addressable pixels in the image will be determined by the x,y dimensions of this store. The number of 'planes' of memory (i.e. the number of bits per pixel) will determine the number of colours we can concurrently display. If we had only one plane of frame-store memory (i.e. a single bit per pixel), we could specify only two alternative outputs at each position on the screen. To sample colour space adequately we in fact need to be able to control each of the three guns with a resolution of 8 bits (giving us 256 different output levels); and if we wish to measure colour thresholds, we need at least 10 bits per gun (Cowan, 1983a).[4] It is possible to have a frame store 24 planes deep (i.e. 24 bits per pixel) and thus directly to specify three gun values at each pixel. But until recently the computers and graphics boards available to the average visual scientist have not been fast enough to redraw the contents of such a frame store between frames.

The more usual arrangement for recent colour vision research has been that shown in Fig. 1. What is stored for each pixel is not a set of gun values but a pointer to an entry in a look-up table (Rodieck, 1983; Watson *et al.,* 1986). If there are 8 planes of frame-store memory, then there will be only 256 possible entries in the look-up table, since this is the number of addresses that can be specified by 8 bits. But each entry in the look-up table can specify with high precision the voltages to be applied to the three guns of the monitor: eight bits per gun is the very minimum that the colour scientist will wish to have available. The number of planes of frame-store memory thus determines the length of the look-up table and, in turn, the number of combinations of luminance and chromaticity that can be simultaneously present on the screen. The total number of available colours – not just those concurrently present – depends on the resolution of the three digital-to-analogue converters at the output of the graphics board. From the user's point of view, this resolution is represented by the number of 'bits per gun' with which the programmer can specify each entry in the look-up table.

Most systems allow the programmer to alter, between frames, either the contents of the frame store or the part of the frame store that is displayed or the entries in the look-up table (Watson *et al.,* 1986). Provided the graphics board has sufficient memory, rapid substitution of images can be achieved by displaying different subsections of the framestore. Alternatively, if the same entry in the look-up table is assigned to all the pixels of a complex shape or to several shapes distributed across the screen, then all these sub-areas can be simultaneously changed by changing the one entry in the look-up table. If a target area and its background field are linked to different positions in the look-

[4] There are, however, poor man's ways of gaining the resolution needed to measure colour thresholds with an 8-bit graphics system. One is to reduce the contrast of the display by optically superposing on it a homogeneous background field (Cole *et al.,* 1993). Another is to use 'dithering' or 'half-toning', gaining an extra bit by setting alternate raster lines to different values (Mulligan, 1986; Mulligan and Stone, 1989; Pelli and Zhang, 1991). A third solution is discussed below under Setting up and calibration.

up table, then the target can be made to appear according to whether the two entries in the look-up table contain the same set of gun values (e.g. Dixon, 1991). When a technique of this kind is used to produce apparent movement, the method is called 'look-up table animation' or 'palette animation' (e.g. Baro and Hughes, 1991).

Pixel rates and video rates

An important factor to consider in choosing a graphics board is the 'pixel rate', the frequency at which new voltages can be sent to the monitor. The highest pixel rate available on a given board is limited by the response speed of the digital-to-analogue converters, which translate the numerical values of the look-up table into voltages on the RGB lines to the monitor. The pixel rate takes a typical value in the range 10-200 MHz.

There is a systematic relationship between pixel rate and the following variables:

- The frame rate (the frequency at which the monitor is set to redraw the whole screen)
- The number of pixels per line of the raster (the limiting horizontal resolution)
- The number of horizontal lines of the raster (the limiting vertical resolution).

The pixel rate sets the limit to the product of these three variables,[5] and the user will often have to compromise between frame rate and spatial resolution. To minimize visible flicker, the visual scientist may wish to use a frame rate of 75 Hz or more,[6] but the higher the chosen rate, the more limited is likely to be the spatial resolution. Thus the speed of the graphics board, and not merely the dimensions of its frame store, often limits the total number of pixels that can be plotted within one frame.

The bandwidth of the monitor should be matched to the pixel rate of the graphics board. The 'bandwidth' or 'video rate' of the monitor is the limiting frequency at which the electron flux of the monitor's guns can be modulated. If the pixel rate is too high, in comparison with the bandwidth of the monitor, then the scanning beam of the monitor will not faithfully translate into spatial modulations the temporal modulations of the output of the graphics board.

A monitor's scan rate is the product of frame rate and number of lines, i.e. the total number of horizontal sweeps per second. For graphics monitors, typical values are in the range of 30–80 kHz, and it is scan rate that manufacturers commonly use to characterize such displays. 'Multiscan' monitors have the capacity to operate at a range of scan rates, but the range of a given monitor will still be limited, and the buyer should match this range to the properties of the

[5] Allowance has additionally to be made for the duration of the fly-back, the return of the beam from the end of one line to the beginning of the next. The total of the fly-back times can amount to one third of the total time per frame.

[6] A significantly higher rate is needed for electrophysiological work, since individual cells in the mammalian visual system can follow frequencies well above the human critical-flicker-fusion frequency (Rodieck, 1983).

graphics board. At a given frequency, a monitor tuned to that fixed scan rate may offer better spatial resolution than the equivalent multiscan monitor from the same manufacturer, and so may be preferred if the experimenter is sure in advance of the scan rate to be used.

The pitch of a colour monitor is the width of each triad of phosphor dots or lines. A typical value for a 19-inch Sony GDM series monitor is 0.30 mm; for Hitachi 'Accuvue' monitors, typical values are 0.28 mm for a 16-inch display and 0.31 for a 19-inch display. There is no automatic relationship between the pixels of the framestore and the phosphor triads of the monitor: the user is left to determine what this relationship should be and to ensure that no aliasing occurs between the stimulus pattern and the pattern of phosphor triads (Bauer, 1992). In most colour work it would be inappropriate to have more pixels per line than phosphor triads; for it would then be impossible to control exactly the chromaticity of a given pixel.

It is not only the pitch of the phosphor triads and the pixel rate of the graphics board that determine the spatial resolution that the user can expect from a display. The spatial profile of a single displayed pixel will depend both on the focus of the electron beams and on the extent to which the electronics faithfully reproduce the temporal waveform specified by the digitized representation in the framestore. The DACs at the outputs of the graphics board, the cables, and especially the high-voltage amplifiers of the monitor may all act as low-pass filters, and this temporal filtering will be translated into spatial filtering. Naiman and Makous (1992), measuring a monochrome monitor, found that a nominally square pixel was not radially symmetric when displayed, but rather was elongated in the direction of the horizontal scan. Cables and amplifiers may also introduce distortions other than low-pass filtering (v. infra).

Gamma functions

The most celebrated non-linearity of the CRT is the 'gamma function'. The luminous output of a given phosphor is approximately a power function of the voltage applied to the corresponding gun. The exponent typically has a value of about 2.3 and is conventionally designated γ. This non-linearity arises between the gun voltage and the electron flux, whereas the relationship between electron flux and luminous output is essentially linear – provided there is no change in the EHT (Extra High Tension) accelerating voltage present at the screen.

Setting up and calibration

Brightness and contrast

Of the many external controls on a modern monitor, two of the most important are those called 'brightness' and 'contrast', yet these are often set with little

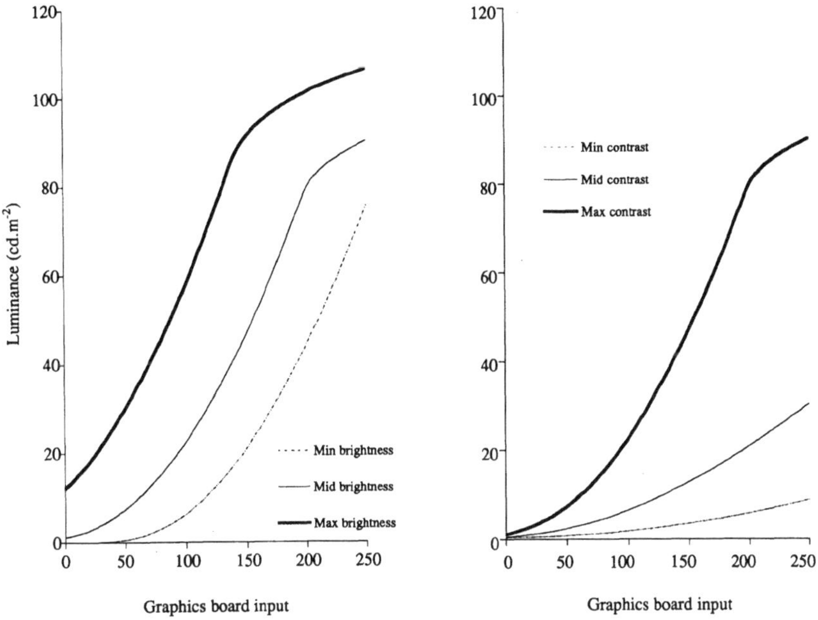

Fig. 2. Left: Gamma functions obtained from a Sony monitor with 'Contrast' set at maximum and 'Brightness' set at three different values. Right: Gamma functions obtained with 'Brightness' set at an intermediate value and 'Contrast' set at three different values.

thought by the newcomer. It is in fact crucial to set both of them carefully before investing the time and the effort demanded by a full calibration.

The very names of the brightness and contrast controls are misleading. The 'brightness' control nominally sets the offset, the outputs of the guns when the nominal input is zero, whereas the 'contrast' control sets the gain and thus the maximum luminance available. In practice, some interaction between the two controls may be observed (Fig. 2).

The visual scientist will usually wish to turn down the brightness until, with the graphics board connected, there is just no visible output when the nominal input is zero. For most purposes, the contrast setting should be the maximal setting that still gives additivity of the guns (v. infra).[7] But for some purposes, the experimenter may choose to set up the monitor differently. If thresholds are to be measured around a fixed background level, then a higher setting of the brightness and a lower setting of the contrast may offer the necessary resolution.

Once the brightness and contrast settings have been chosen, they should be sealed against interference by unthinking graduate students.

[7] An informal way to establish the appropriate level is to illuminate the full screen of the monitor and to increase the contrast setting until the illuminated area visibly expands. The expansion indicates a fall in the screen voltage, and thus in the velocity of the electron beam; the beam is then effectively less stiff and its deflection increases.

Calibration

In order to generate stimuli of known colour and luminance, the experimenter needs to know (a) the chromaticities of the three guns of the monitor and (b) the gamma function for each gun. Four main routes to calibration can be identified and are listed below from the costliest to the cheapest. The necessary apparatus varies in price over at least two orders of magnitude.

(i) Spectroradiometry. By means of a telespectroradiometer the experimenter can measure the radiant output of the monitor in a series of narrow wavelength bands across the spectrum and so calculate the XYZ tristimulus values of the CIE system. Two sources of error deserve mention. First, the light emitted from the display may be partially polarized, and a mirror or grating within the spectroradiometer may act as an analyzer. Check for this by making a second set of measurements with the spectroradiometer rotated through 90 degrees; or ensure that the light is depolarized by an integrating sphere before it enters the instrument. Second, remember that the output of a local region of the screen is a train of intense, brief pulses. The peaks of individual pulses may be 100 times greater than the mean luminance and may saturate the detector. And if these pulses are not synchronized with the sampling interval of the spectroradiometer, the total photon capture will vary from one sample to another, while the apparent chromaticity will also vary since the different phosphors typically have different time constants (v. infra). Some modern spectroradiometers can be automatically synchronized with the display being measured. Alternatively, the modulating detection system of the spectroradiometer can be by-passed and replaced by a simple integrator (Hanson and Verrill, 1991).

It is hardly worth owning a spectroradiometer if one does not also have access to a calibrated standard light source, and associated power supply, with which to calibrate the spectroradiometer itself.

(ii) Electronic colorimeters. Whereas a true spectroradiometer may cost several times as much as a complete colour graphics system, the prices of some electronic colorimeters are of the same order of magnitude as those of graphics systems. Such instruments directly estimate the CIE XYZ tristimulus levels, instead of first obtaining the spectral radiance distribution of the source. The colorimeter typically contains three detectors, which sample the light passed by three colour filters. The three detector/filter combinations have sensitivity spectra that simulate – with more or less precision – the CIE $\bar{x}$, $\bar{y}$, $\bar{z}$ colour-matching functions. Necessarily the colorimeter can be accurate only to the degree that these simulations are accurate; and different instruments vary in the extent to which they approach perfect simulations. For example, in the Minolta Chroma Meter CS100 – a relatively economical device adopted widely by visual scientists – a short-cut is taken in simulating the $\bar{x}$ function: the correspondng filter/detector combination is unimodal in its spectral sensitivity (whereas $\bar{x}$ is bimodal) and the short-wave mode of $\bar{x}$ is simulated by adding the $\bar{z}$ signal,

suitably scaled, to the long-wave lobe of $\bar{x}$. Terstiege and Gundlach (1991) have calculated – for gas-filled lamps rather than monitors – the theoretical average error of several colorimeters, calculated in terms of the CIELUV colour-difference formula. The most accurate was an LMT instrument[8] that incorporates the (expensive) device of partial filtering: small pieces of secondary filter are placed in series with the primary filters to improve the approximation to the $\bar{x}$, $\bar{y}$, $\bar{z}$ functions. In this sample of instruments, and for these sources, the Minolta was the least accurate colorimeter, while Topcon and Photo Research colorimeters were intermediate. The LMT C1200 colorimeter is also recommended in the survey of Berns *et al.*, (1993b).

Despite the inaccuracy of electronic colorimeters, they are stable instruments and they can serve well to check for drift of a research monitor in the field. The procedure used by the National Physical Laboratory (U.K.) is to calibrate a colorimeter, such as the Minolta Chroma Meter, in conjunction with a specific monitor (Hanson and Verrill, 1992). A gamut of stimuli are presented on the monitor and measurements are made concurrently with a reference telespectroradiometer and with the user's colorimeter, the same measuring geometry being adopted for the two instruments. Since each of the three phosphors of the monitor has a stable emission spectrum, the colorimeter can then be reliably used in the field to calibrate this particular monitor: the correction needed in a particular region of the chromaticity diagram will be constant. However, the corrections will not be valid for a different monitor with different emission spectra.

(iii) Visual colorimeters. There is much to be said for the use of older instruments, such as the Lovibond Tintometer,[9] in which a visual match is made between light that is drawn from the CRT screen and light that has been drawn from a reference source and has passed through calibrated, stable, glass colour filters (Chamberlin and Chamberlin, 1980). Although such visual measurements are relatively slow and the experimenter is unlikely to be the monozygotic twin of the CIE standard observer, it is difficult to make large errors with a visual instrument, and the calibrations are likely to be as valid as those made with a more expensive electronic instrument. One difficulty is that visual matches become very imprecise in the short-wave corner of the chromaticity diagram (Stiles, 1955), owing, probably, to the saturation of the post-receptoral channel that carries the signals of the short-wave cones (Polden and Mollon, 1980; Mollon and Estévez, 1988). So it is difficult to locate the blue phosphor directly. The way around this is to make measurements nearer the centre of the chromaticity diagram, where post-receptoral channels are more sensitive: if two different excitations of the blue phosphor are combined with two different values of the red/green ratio to give four points in the middle of the

[8] Lichtmesstechnik, Helmholtzstr. 9, D-10586 Berlin, Germany
[9] The Tintometer Limited, Waterloo Rd., Salisbury, SP1 2JY, U.K.

diagram, then the chromaticity of the blue phosphor can be found by triangulation.

Visual calibration of the three gamma functions would be clumsy and imprecise, and so a hybrid arrangement will often recommend itself – especially if the colour scientist needs only an approximate estimate of absolute luminance. Measure the chromaticities of the three guns with a visual colorimeter. With the same instrument measure the chromaticity of the white obtained when all guns are maximally excited. Given this information, reference to the CIE standard observer allows you to calculate the relative luminances of the three guns at maximum. The gamma function for each gun can then be measured with a relatively cheap physical detector, such as a silicon photodiode, and can be converted to relative luminance values by reference to the relative luminance at maximum. An approximate estimate of absolute luminance can be made with the visual instrument.

(iv) Use of manufacturers' values. Some of the larger manufacturers of monitors will supply chromaticities for their phosphors, and can provide 'factory-floor' values as opposed to nominal ones. The gamma functions can then be measured with an electronic photometer or a simple photodiode. If you have faith in the manufacturer's chromaticities and if your experiment needs only modest accuracy, then this approach to calibration may be an honourable one.

The calibrations should be made using the graphics board that is to be used in the experiments. Also, since the earth's magnetic field as well as those deriving from nearby apparatus may affect the electron beams, the calibration should be made with the monitor in the position and orientation in which it is to be used. Since the shadow mask and structural metalwork within the monitor may themselves have become magnetized, the monitor should be demagnetized – 'degaussed' – before any calibrations are begun. Modern monitors incorporate a degaussing circuit that operates when the monitor is turned on: an alternating current of the order of 1A, derived from the mains supply, is passed through coils mounted on the tube (Hutson *et al.,* 1990, pp. 187–189). The circuit incorporates resistors that have the property of increasing their resistance as they become hot: so the current through the degaussing coils falls to a low level within a few seconds. Some monitors additionally offer an external degaussing button. Other things being equal, the visual scientist will prefer to buy a monitor with the latter facility, since he or she may wish to leave the monitor turned on for a period to minimize warm-up effects (v. infra). It is also possible to purchase an external degaussing coil or wand, a portable electromagnet that is powered from the a.c. supply. Merely turning a warm monitor on and off may not achieve degaussing, since the 'positive temperature coefficient' resistors in the degaussing circuit must be given time to cool (Hutson *et al.,* 1990).

The 'gamma functions' of the monitor are seldom in fact found to be exactly of the form $L = V^{\gamma}$ (Cowan, 1983b), and in the older literature there is much discussion of how best to estimate the true function from a set of calibrated positions on the curve (for references, see Post (1992)). However, modern spectroradiometers, such as the Photo Research Spectrascan, and colorimeters, such as the Minolta CS100 Chroma Meter, can be interrogated via a serial interface; so it is feasible to set the host computer to measure the output of each gun at each of 256 levels. Our own recommendation is to establish a computer file containing the luminance values for each of the 256 values for each gun and to require the experimental program to seek the gun inputs closest to the values theoretically required.[10] The advantage of automated calibration is that the host computer's time is cheap and the measurements can usually be repeated or integrated until a smooth, monotonic function is obtained.

The non-additivities and non-uniformities of monitors

In principle, once one knows the chromaticities of the phosphors and the gamma function for each gun, one can calculate the gun values needed to generate any stimulus within the gamut of one's monitor (Cowan, 1983b). One's calculation will depend, however, on a number of assumptions, each of which may be mistaken (Cowan and Rowell, 1986; Brainard, 1989; Post, 1992).

(i) Phosphor constancy. This is the assumption that the luminous output generated by a given gun has the same spectral power distribution – and thus the same chromaticity – whatever the level of the signal sent to the gun. Cowan and Rowell (1986) and Brainard (1989) found that the assumption of phosphor constancy was generally valid, and that is our own experience. There are two simple ways in which an experimenter may be misled into believing that phosphor constancy does not hold. First, if the brightness knob is mis-set and a low background illumination is present, then the chromaticity of a given gun will appear to change at low gun inputs, owing to the proportionately greater contamination by the background. Second, a systematic but spurious change in chromaticity may be recorded if the Minolta Chroma Meter is used at low levels where the display is flashing and where the manufacturers explicitly do not guarantee the accuracy.

(ii) Independence of guns. This is the assumption that the luminous and chromatic output resulting from a given input to a given gun is independent of the signals sent to the other two guns. If we measure the X, Y, Z outputs when given voltages are sent to each gun separately, we can determine whether the sum of the three sets of X, Y, Z values is obtained when the specified voltages are sent to the three guns simultaneously. Cowan and Rowell (1986) found good

[10] This approach could be refined by not considering the guns independently but by seeking the triad of gun inputs that minimized the distance from the target combination of luminance and chromaticity, the error being computed in units appropriate to the experiment. Thus, if precision of luminance is most important, the best value for a given gun will not be independent of the values selected for the other guns.

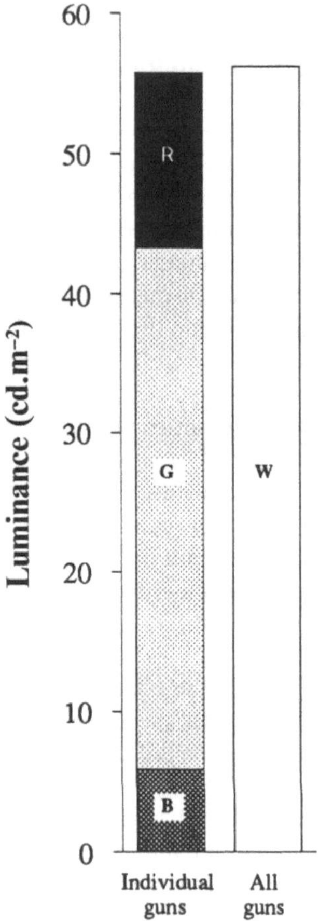

Fig. 3. Result of a luminance additivity test for a Mitsubishi HL20 monitor. The measurements were made with a Minolta Chroma Meter.

additivity for the (now obsolete) Tektronix monitor 690SR. The discrepancies were in the direction of sub-additivity and did not exceed 1%. Since the three tristimulus values were subject to similar sub-additivities, the error in chromaticity coordinates was much smaller, less than one part in 300. Brainard (1989) reported only small sub-additivities for a Barco 5351 monitor driven by a frame buffer manufactured by Number Nine Graphics Corporation: the subadditivities observed were of the order of 3%, were worst for the blue gun, and were worst at low levels. We ourselves have found good additivity for Barco CD351, Mitsubishi HL20 and Sony GVM1400 monitors driven by a Cambridge Research Systems VSG board (see Fig. 3). However, failures of additivity of guns may be observed in both cheap and expensive monitors. Berns *et al.*, (1993) found a serious failure of additivity for a Sony PVM1942Q monitor when it was driven to produce high-luminance images (186 cd.m^{-2}); at lower levels, the additivity became acceptable. In the case of cheap monitors, the non-additivity may arise from poor regulation of the accelerating voltage present at the screen (Rodieck, 1983). In the case of modern expensive monitors, the culprit may be circuits deliberately introduced to ensure compliance with regulations on X-ray emission: these circuits place limits on the total emission. In principle, non-additivities may also arise in the graphics driver board.

(iii) Spatial homogeneity. One assumption that definitely fails is that of spatial uniformity of gun outputs across the screen. Most monitors show a falling-off in the luminous output near the edges of the screen (Brainard, 1989; Post 1992; Cook *et al.,* 1993). The effect may be of the order of 20-25%, and is illustrated in Figure 4. It may be due in part to the reduced cross-section that the shadow-mask aperture presents to the electron beam. In addition to the variation in luminance level, there may also be a change in chromaticity, but any such effect is much smaller: Brainard (1989) found that most of the spatial variation could be accounted for by a single scale factor applied to all three gamma functions.

With a resolution of only 8 bits per gun and a look-up table of only 256 entries, it would be difficult to generate, by computational compensation, a truly uniform field: the quantization of the corrections would probably yield visible steps in parts of the field.

(iv) Spatial additivity. The luminous output from a given region of the screen may vary according to the illumination of other regions. Moreover, the chromaticity obtained from a given region may depend on the chromaticity of

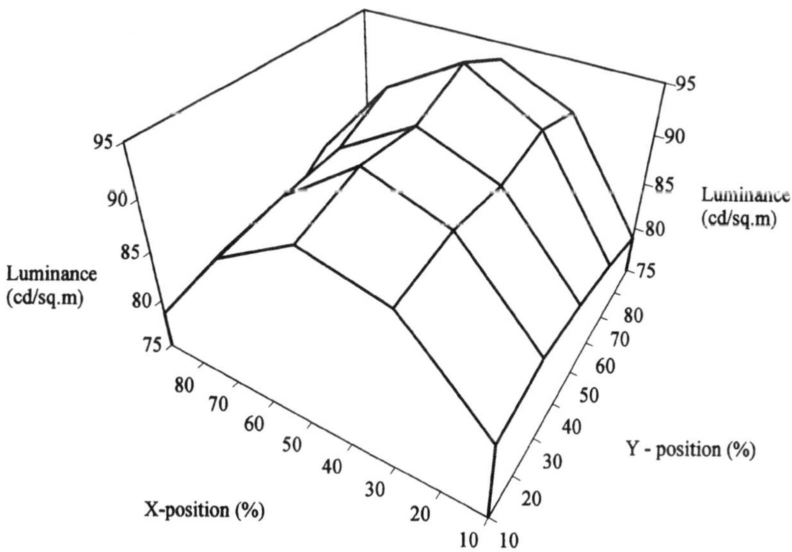

Fig. 4. Variation with screen position of the luminous output of a Sony monitor.

a surrounding region. Such effects are disastrous in experiments on colour contrast and constancy; yet they are only occasionally discussed in papers on those topics.

Brainard (1989) found only small spatial interactions in the case of a Barco 5351 monitor. He measured the output of a central 64 x 38 mm patch in the presence or absence of eight surrounding patches of the same size. The mean percentage change in fractional output was 2.5%. We have observed similarly good behaviour in the case of a Mitsubishi HL20 monitor (although the same monitor showed a spatial gradient in the absolute luminous output, increasing from the left to the right of the display). We have observed serious spatial sub-

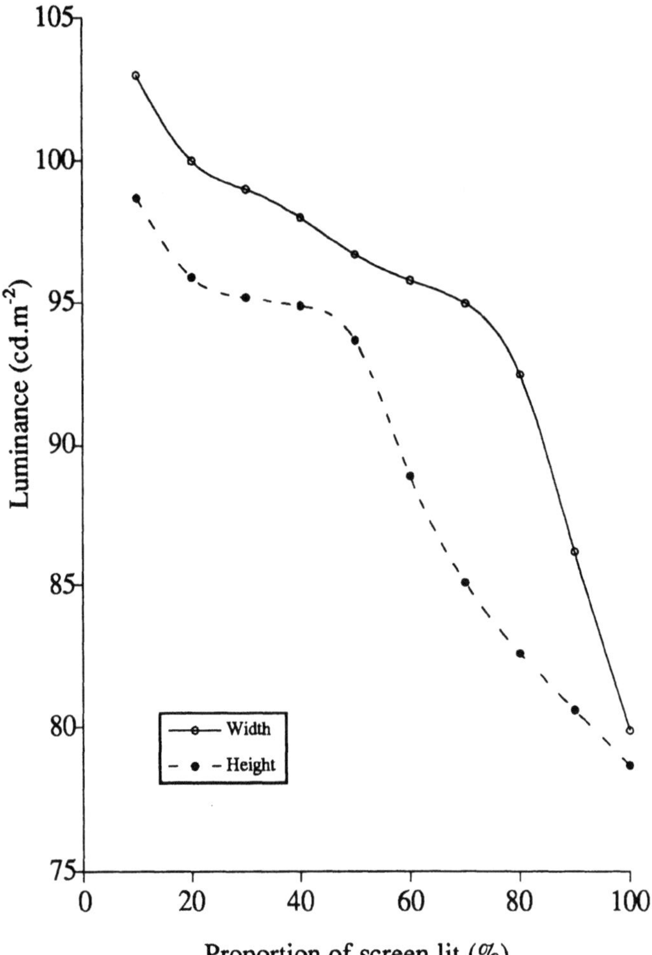

Fig. 5. Spatial sub-additivity for a Sony GVM-1400QM monitor. Luminance was measured at a fixed point in the bottom left-hand corner of the screen, and the width of an illuminated strip was increased horizontally or vertically to occupy increasing proportions of the total screen area.

additivities in the case of Sony and Eizo displays. Figure 5 illustrates the effect for a Sony GVM1400 monitor. Luminance was measured, with a Minolta Chroma Meter, at a fixed point in the bottom left corner of the display, while an increasingly large proportion of the screen was illuminated. For white light, at maximum output, there was a 20% variation in the measured luminance. Such sub-additivities may reflect limitations in the power supplies of monitors or may arise from circuits deliberately introduced to limit X-ray emission. Performing similar tests, Naiman and Makous (1992) report super-additivity in the case of a monochrome Apple display.

Spatial interactions may become more complex when the measured area and the surround area can vary independently in colour and luminance. Figure 6 shows measurements made for an Apple colour monitor by A.J. Shepherd. In this particular case, the central target was a red square and the surround was masked with matte black card. The nominal luminance and chromaticity of the target remained constant throughout the measurements. The graphs show the measured changes of the target as the surround signal is increased. Red, yellow, purple and white backgrounds (i.e. all backgrounds that require excitation of the red gun) lead to sub-additivity, whereas blue and green backgrounds give super-additivity. Analogous effects are found for other colours of the central square.

Very local interactions, those between neighbouring pixels, are likely to be super-additive in the horizontal direction. In the case of a monochrome display, Naiman (1991) and Naiman and Makous (1992) found that two vertically adjacent pixels were essentially independent, but there was a super-additive and asymmetric interaction between horizontally adjacent pixels: the luminance of a given pixel depended on that of the two preceding pixels in the raster. The super-additive component increased the total luminous flux by about 30%. Naiman and Makous suggest that the slew rate of the monitor's Z amplifier does not allow the voltage to attain full value during the time corresponding to the exposure of a single pixel in the horizontal scan. This horizontal interaction probably explains an otherwise bewildering phenomenon: if the experimenter programs two arrays of narrow bars of the same nominal luminance, one array vertical, the other horizontal (e.g. in an experiment on the McCollough effect), then the horizontal array will look much the brighter, even though the aspect ratio of the display has been adjusted to give equal numbers of pixels for a given horizontal or vertical distance. Once the artifact is recognized, it may be side-stepped in some experiments by using orthogonal arrays each at 45 degrees from the vertical.

Misconvergence and pseudo-misconvergence

Misconvergence is an error in the alignment of the landing of the 'red', 'green' and 'blue' electron beams and is apparent as coloured fringes at the edges of white regions. It is best examined by displaying a regular grid of white pixels.

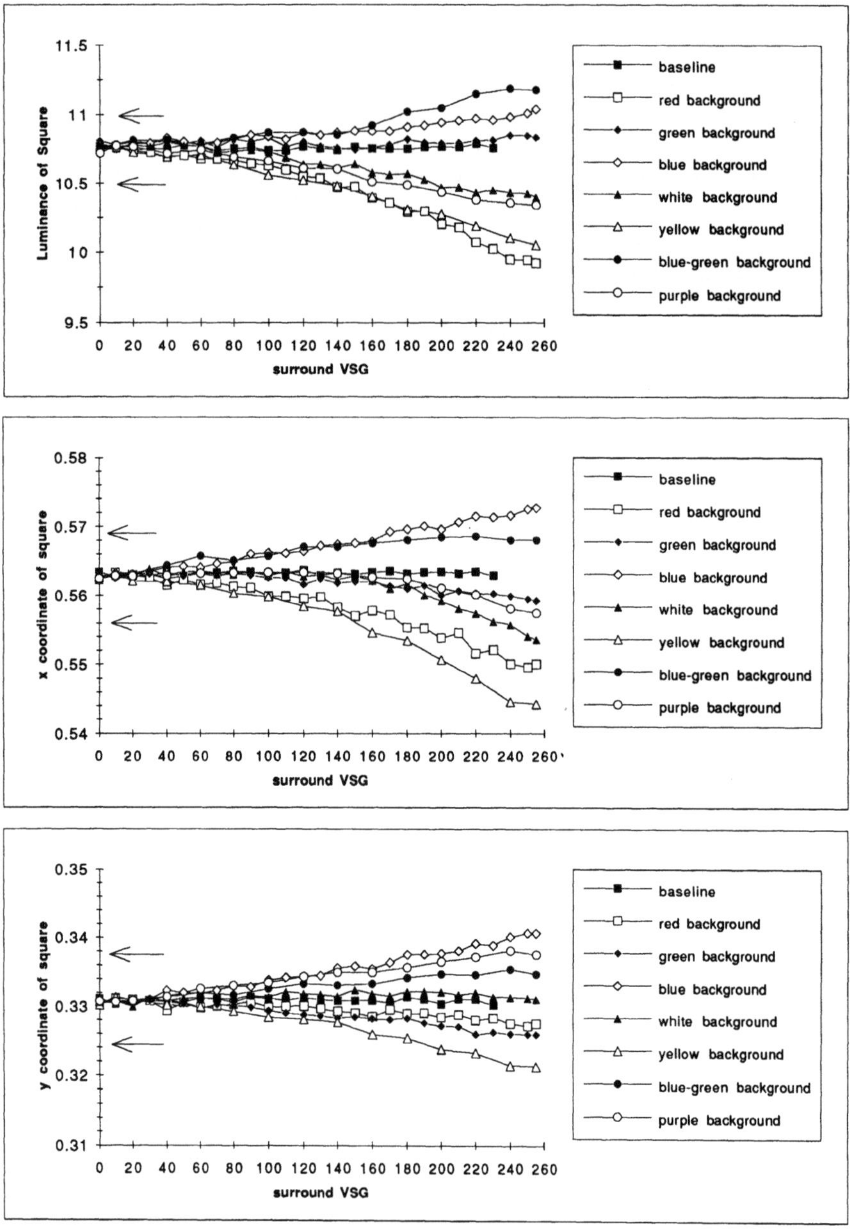

Fig. 6. Spatial interactions for an Apple colour monitor (MO401Z). A central square was illuminated (nominally) by only the red gun. The measurements show how the luminance (top panel) and chromaticity (middle and lower panels) of the central region varied as a function of the gun signals sent to a surrounding region. These measurements were made by A.J. Shepherd.

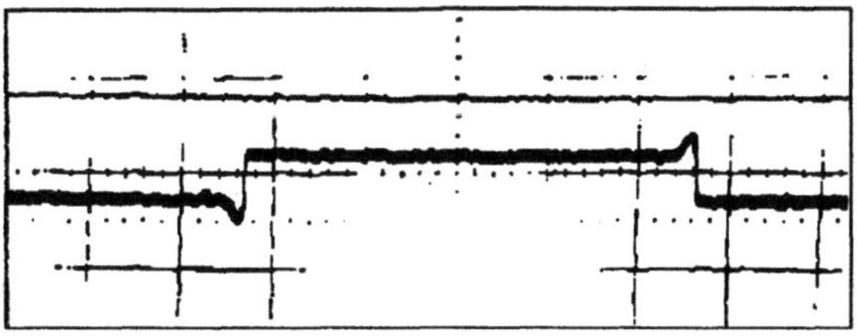

Fig. 7. Oscilloscope trace taken from the green gun input to a Sony graphics monitor from a VSG/2 graphics board under conditions of impedance mismatch. The undershoots and overshoots in the analogue signal led to edge artifacts in an 'equiluminant' random-dot stereogram.

Some monitors offer external adjustments of convergence, which adjust small direct currents that pass through an auxiliary set of electromagnets mounted within the convergence assembly. But in many cases, correction of misconvergence requires the adjustment of internal resistances or magnets. Since the adjustments notoriously interact, the user will usually wish to have such corrections carried out by the manufacturers or by an experienced technician. But is is important to distinguish true convergence errors from electronic artifacts that may simulate misconvergence. Figure 7 shows transient overshoots and undershoots recorded on the cable between a graphics board and a monitor when an impedance mismatch was present between the cable and the monitor. The image on the screen was intended to be an equiluminant red-green random-dot stereogram, and the recording was made from the cable to the green gun. The temporal transients on the line were translated into light and dark striations at the edges of individual picture elements. A pseudo-misconvergence of this kind should be suspected when (a) the primary error is in the horizontal direction and (b) it is constant across the face of the screen. Other sources of bright lines or colour shifts at borders are resonances intentionally introduced into amplifiers to sharpen their responses.

Temporal properties of monitors

In this section, working from short intervals to long, we consider several temporal aspects of the performance of monitors.

Time constants of phosphors

Long after the electron beam has hurried away along its scan, the phosphor dots or stripes continue to emit a decaying stream of photons. Some monitors have phosphors with time constants of tens or even hundreds of milliseconds: the experimenter should be alert to this problem if using simply the monitor supplied with a particular computer. When buying a specialist graphics monitor, the user will normally wish to select a screen with a short time constant, i.e. one of units of milliseconds. However, there are still problems to watch for, especially if very brief or moving coloured targets are to be used in the experiments.

First, the time constants may be markedly different for the red, green and blue phosphors. Vingrys and King-Smith (1986) have suggested how this may lead to a detectable luminance transient when a nominally equiluminant substitution is made between two stimuli.

Secondly, each of the phosphors may in fact be a mixture of different components with different time constants. The cautious experimenter may measure a trace with a photodiode and an oscilloscope and may conclude that the emission has fallen to 1% in one or two milliseconds, but he or she may miss an 'after-glow' from a second component that emits a similar total number of photons spread over hundreds of milliseconds.[11] A real example may lend substance to this warning. Jonides *et al.*, (1982) described experiments in which the subject appeared to integrate spatially, and not retinotopically, information drawn from successive fixations. They postulated a central store that maintained a single spatial representation of the visual world; but they were later obliged to admit that their results had arisen artifactually as a result of phosphor persistence (Jonides *et al.*, 1983).[12]

Beating with the host monitor

Unexplained flicker or drifting horizontal shadows on a good-quality graphics monitor may be due to the proximity of a second monitor – that of the host computer, for example. A sure way to detect such interaction is to alter the frame rate of the graphics display: the drift rate of the shadows will slow down as the frame rate of the host's monitor is approached.

Warm-up times

Monitors typically should be left on for an hour before being calibrated or used experimentally, and many experimenters leave them on continuously or arrange time switches to turn them on before the working day. Harris *et al.*, (1987) found that 12 minutes were sufficient to stabilize a Tektronix 608 monochrome monitor, but at least 3 hours were required in the case of a Digivision CD14 colour monitor. In Fig. 8 we illustrate the behaviour of a Sony monitor over time.

[11] In an earlier scandal, almost an entire generation of experimenters missed the long time constants of tachistoscope lamps, until it was shown (Mollon and Polden, 1978) that the fluorescent tubes in common use were coated with a fast short-wave phosphor and a much slower long-wave phosphor.

[12] A mixture of phosphor components (or alternatively a difference between manufacturers) may lie behind the very recent controversy about the persistence of the P31 phosphor. Making direct measurements with a photomultiplier, Westheimer (1993) found that the P31 phosphor decayed to 10% in 600 μsec and to 2% in 2 msec, but Groner *et al.*, (1993) were able to read the screen when a mechanical shutter was opened tens, or even hundreds, of milliseconds after the nominal display had ended. Sherr (1993, p. 91) gives 38 μsec as the time to decay to 10%.

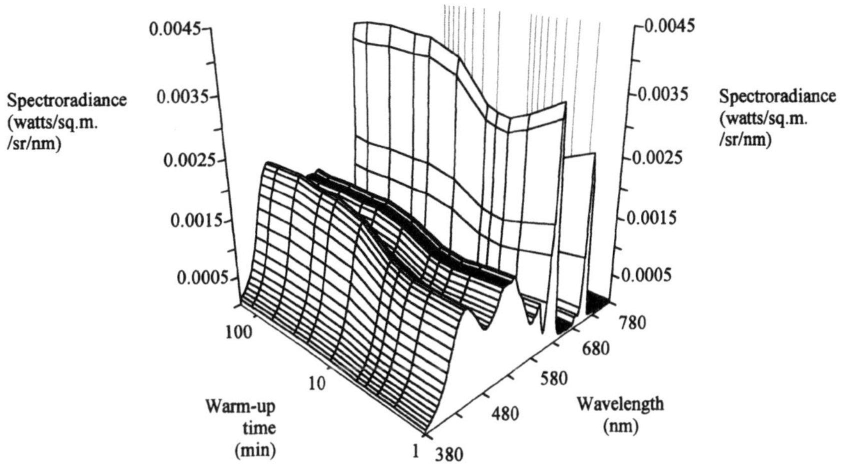

Fig. 8. Warm-up effects for a Sony GVM-1400QM monitor. The radiant output is shown as a function of time and wavelength. The measurements were made by Mr. A. Hansen, using a Photo Research Spectrascan spectroradiometer.

Long-term stability

Some colour graphics monitors show good temporal stability. Brainard (1989) made measurements on a Barco 5351 monitor at an interval of two months: there was only a small (5%) drop in luminous output and, since the three guns covaried, very little change in chromaticity. Lucassen and Walraven (1990), however, found a substantial shift in the gamma functions of a high-resolution 19-inch Hitachi monitor during the first six months of use, and the changes were not ones that could be corrected simply by translating the log gamma functions along the ordinate. They did find that the chromaticity coordinates of the three phosphors were stable.

Recommendations

Design the experiment so that the total illuminated area of the screen is the minimum necessary and remains as constant as possible. Wherever possible, confine stimuli to the central region. Use luminance and spatial noise to isolate chromatic pathways (Reffin *et al.*, 1991) rather than abutting 'equiluminant' regions. Check the initial calibrations by measuring the chromaticities of a sample of experimental colours generated from the stored gamma functions. Calibrate with the spatial array present that will be used in the experiment. Check calibrations often.

Acknowledgements

We are grateful for the advice of many colleagues, including Dr. D. Bauer, Professor C.R. Cavonius, Mr. A. Hanson, Mr. Y. Kim, Mr. B. Regan, Dr. J. Robson, Mr. T. Robson, A.J. Shepherd, Dr. S. Shevell, Dr. Y. Tadmor, Mr. J. Verrill, Dr. M. Webster, Dr. P. Whittle and Dr. S. Würger. Our examination of colour monitors has been supported in part by the SERC and in part by the National Measurement System Policy Unit, Department of Trade and Industry, U.K., through a contract from the National Physical Laboratory, U.K.

References

In addition to references cited in the text, we include here some further sources that will be of use to those who use colour monitors in visual research.

Baro, J.A. and Hughes, H.C. (1991). The display and animation of full-color images in real time on the Macintosh computer. Behav. Res. Meth. Instr. Comp. 23: 537–545.

Bauer, D. (1992). Evaluierung visueller Probleme bei Farbrasterschirmen. Wirtschaftsverlag, Bremerhaven.

Benzschawel, T. (1992). Colorimetry of self-luminous displays. In: Widdel, H. and Post, D.L. (eds.), Color in Electronic Displays: 39–92, Plenum, New York.

Berns, R.S., Motta, R.J. and Gorzynski, M.E. (1993a). CRT Colorimetry. Part I: Theory and Practice. Color Res. Appl. 18: 299–314.

Berns, R.S., Motta, R.J. and Gorzynski, M.E. (1993b). CRT Colorimetry. Part II: Metrology. Color Res. Appl. 18: 315–325.

Brainard, D.H. (1989). Calibration of a computer controlled color monitor. Color Res. Appl. 14: 23–34.

Braun, F. (1897). Ueber ein Verfahren zur Demonstration und zum Studium des zeitlichen Verlaufes variabler Ströme. Annalen der Physik und Chemie 60: 552–559.

Brettel, H. (1988). Quantitative luminance and colour representation with CRT displays. SPIE 1027 ("Image Processing II"): 107–112.

Chamberlin, G.J. and Chamberlin, D.G. (1980). Colour: Its Measurement, Computation and Application. Heyden, London.

Cole, G.R. and Hine, T. (1992). Computation of cone contrasts for color vision research. Behav. Res. Meth. Instr. Comp. 24: 22–27.

Cole, G.R., Hine, T. and McIlhagga, W. (1993). Detection mechanisms in L–, M-, and S-cone contrast space. J. Opt. Soc. Am. A 10: 38–51.

Cook, J.N., Sample, P.A. and Weinreb, R.N. (1993). Solution to spatial inhomogeneity on video monitors. Color Res. Appl. 18: 334–340.

Cowan, W.B. (1983a). Discreteness Artifacts in Raster Display Systems. In: Mollon, J.D. and Sharpe, L.T. (eds.), Colour Vision: Physiology and Psychophysics: 145–154, Academic Press, London.

Cowan, W.B. (1983b). An inexpensive scheme for calibration of a colour monitor in terms of CIE standard coordinates. Comp. Graph. 17: 315–321.

Cowan, W.B. and Rowell, N. (1986). On the gun independence and phosphor constancy of color video monitors. Color Res. Appl. 11 (Suppl.): 34–38.

Dixon, P. (1991). Tachistoscopic display using color-look-up-table display devices. Behav. Res. Meth. Instr. Comp. 23: 380–386.

Groner, R., Groner, M.T., Müller, P., Bischof, W.F. and Di Lollo, V. (1993). On the confounding effects of phosphor persistence in oscilloscopic displays. Vision Res. 33: 913–917.

Hanson, A.R. and Verrill, J.F. (1992). Standardisation in the measurement of the colour of self luminous displays. In: Proceedings of the 22nd Session of the CIE 1991, Melbourne, 87–88, CIE Central Bureau, Vienna.

Harris, J.P., Makepeace, A.P.W. and Troscianko, T.S. (1987). Cathode ray tube displays in psychophysiological research. J. Psychophysiol. 1: 413–429.

Hutson, G., Shepherd, P. and Brice, J. (1990). Colour Television. McGraw-Hill, Maidenhead.

Jackson, R.N., MacDonald, L.W. and Freeman, K.W. (1993). Computer Generated Colour – A Practical Guide to Presentation and Display. John Wiley, Chichester.

Jonides, J., Irwin, D.E. and Yantis, S. (1982). Integrating visual information from successive fixations. Science 215: 192–194.

Jonides, J., Irwin, D.E. and Yantis, S. (1983). Failure to integrate information from successive fixations. Science 222: 188.

Keller, P.A. (1991). The Cathode-Ray Tube: History and Applications. Palisades Press, New York.

Lucassen, M.P. and Walraven, J. (1990). Evaluation of a simple method for color monitor recalibration. Color Research and Application 15: 321–326. (reprinted in Lucassen, M.P. (1993). Quantitative Studies of Color Constancy, Ph.D. Thesis, Utrecht).

Mollon, J.D. and Estévez, O. (1988). Tyndall's paradox of hue discrimination. J. Opt. Soc. Am. A 5: 151–159.

Mollon, J.D. and Polden, P.G. (1978). On the time constants of tachistoscopes. Quar. J. Exp. Psychol. 30: 555–568.

Mulligan, J.B. (1986). Minimizing quantization errors in digitally-controlled CRT displays. Color Res. Appl. 11: S47–S51.

Mulligan, J.B. and Stone, L.S. (1989). Halftoning method for generation of motion stimuli. J. Opt. Soc. Am. A 6: 1217–1227.

Naiman, A.C. (1991). CRT spatial non-linearities and luminance linearization. In: Morris, R. (ed.), Raster Imaging and Digital Typography. Cambridge University Press, Cambridge.

Naiman, A.C. and Makous, W. (1992). Spatial non-linearities of grayscale CRT pixels. SPIE 1666 (Human Vision, Visual Processing, and Digital Display III): 41–56.

Pelli, D.G. and Zhang, L. (1991). Accurate control of contrast in microcomputer displays. Vision Res. 31: 1337–1350.

Polden, P.G. and Mollon, J.D. (1980). Reversed effect of adapting stimuli on visual sensitivity. Proc. Roy. Soc. Lond. B 210: 235–272.

Post, D.L. (1992). Colorimetric measurement, calibration, and characterization of self-luminous displays. In: Widdel, H. and Post, D.L. (eds.), Color in Electronic Displays: 299–312, Plenum, New York.

Reffin, J.P., Astell, S. and Mollon, J.D. (1991). Trials of a computer-controlled colour vision test that preserves the advantages of pseudoisochromatic plates. In: Drum, B., Moreland, J.D. and Serra, A. (eds.), Colour Vision Deficiencies X, Doc. Ophthalmol. Proc. Ser. 54: 69–76, Kluwer Acad. Publ., Dordrecht.

Rodieck, R.W. (1983). Raster-based colour stimulators. In: Mollon, J.D. and Sharpe, L.T. (eds.), Colour Vision: Physiology and Psychophysics: 131–144, Academic Press, London.

Sherr, S. (1993). Electronic Displays, 2nd ed. John Wiley, New York.

Sproson, W.N. (1983). Colour Science in Television and Display Systems. Hilger, Bristol.

Stiles, W.S. (1955). Interim Report to the Commission Internationale de l'Eclairage, Zurich, 1955, on the National Physical Laboratory's investigation of Colour-Matching. Optica Acta 2: 168–181.

Terstiege, H. and Gundlach, D. (1991). Characterizing the quality of colorimeters. SID 91 Digest: 641–644.

Travis, D. (1991). Effective Color Displays. Academic Press, London.

Vingrys, A.J. and King-Smith, E. (1986). Factors in using color video monitors for assessment of visual thresholds. Color Res. Appl. 11: S57–S62.

Walraven, J. (1992). Color basics for the display designer. In: Widdel, H. and Post, D.L. (eds.), Color in Electronic Displays: 3–38, Plenum, New York.

Watson, A.B., Nielsen, K.R.K., Poirson, A., Fitzhugh, A., Bilson, A., Nguyen, K. and Ahumada, A.J. (1986). Use of a raster framebuffer in vision research. Behav. Res. Meth. Instr. Comp. 18: 587–594.

Westheimer, G. (1993). Phosphor persistence in oscilloscopic displays. Vision Res. 33: 2337–2338.
Wheeler, T.R.H. and Clark, M.G. (1992). CRT Technology. In: Widdel, H. and Post, D.L. (eds.), Color in Electronic Displays: 221–251, Plenum, New York.

Department of Experimental Psychology
University of Cambridge
Downing St.
Cambridge CB2 3EB
U.K.

52. Discrimination ellipses in the MacLeod-Boynton diagram: Results for normal and colour-deficient subjects obtained with a CRT display

B.C. REGAN and J.D. MOLLON
(Cambridge, U.K.)

Abstract

A computer-controlled colour test has been extended to measure full discrimination ellipses for normal and colour-deficient observers. In this test, spatial and luminance noise are used to circumvent the problems that arise in tests that employ equiluminant targets and backgrounds.

Introduction

Computer-controlled tests of colour vision have the advantage that the difficulty of the chromatic discrimination can be adjusted according to the subject's responses, so as to give a quantitative measure of performance. In such tests, the task typically is to detect a spot or grating of one chromaticity on a field of different chromaticity but the same luminance (King-Smith *et al.*, 1983; Hart *et al.*, 1984; Heard *et al.*, 1987; Arden *et al.*, 1988; Cuvinot, 1992). However, a preliminary luminance equation must then be made for each subject. This is time consuming, is often difficult for the young or elderly to perform, and even when complete may not be accurate, for what is equiluminant in one part of the visual field may not be equiluminant in another. In addition, the boundary between equiluminant regions may be revealed by artefacts of the CRT monitor, such as misalignment of the guns, slew time of the amplifiers, the intrinsic pattern of phosphor dots, and changes over time in the gamma functions (Vingrys and King-Smith, 1986; Mollon and Baker, Chapter 51).

These difficulties are side-stepped in a computer-controlled test introduced by Mollon and Reffin (1989): spatial and luminance noise are added to the display, breaking the target and field into small patches, each with its own contour and varying randomly in luminance. These manoeuvres eliminate edge artefacts and mask any small average luminance differences between areas of different chromaticity, thus forcing observers to rely on chromatic information alone to detect the target. Although it is possible that luminance noise may raise the absolute level of chromatic thresholds, our conditions resemble those under

B. Drum (ed.), Colour Vision Deficiencies XII, pp. 445–451.
© 1995 *Kluwer Academic Publishers, Dordrecht.*

which colour serves to impose organization on real visual scenes (Mollon, 1989; Regan *et al.*, 1994).

Methods

In the present form of the test, the subject's task is to detect the orientation of a C-shaped target on a field of a different chromaticity (Fig. 1). The luminance of any given patch of the mosaic varies from trial to trial and is randomly assigned to one of six equally spaced and equally probable levels in the range 7.6 to 17.0 cd.m^{-2}. The outer diameter of the C subtends 4.3° of visual angle and the inner diameter subtends 2.2°; the gap subtends 1°. There are four possible orientations of the target and the subject indicates his or her choice by means of four buttons.

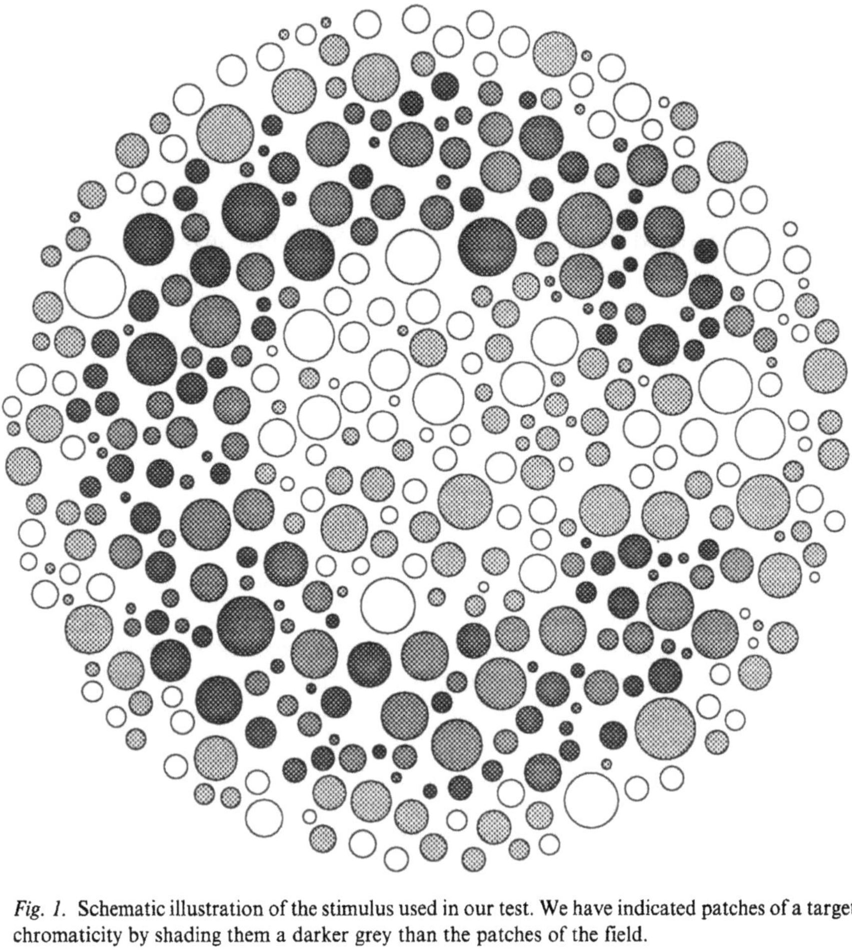

Fig. 1. Schematic illustration of the stimulus used in our test. We have indicated patches of a target chromaticity by shading them a darker grey than the patches of the field.

The chromaticity of the target is varied along a straight line in colour space that passes through the chromaticity of the field; and the threshold difference between test and field chromaticity is determined by a staircase procedure. In the standard form of the test, colour discrimination is probed along the protanopic, deuteranopic and tritanopic confusion axes (Reffin *et al.*, 1991; Regan *et al.*, 1994). In the present extension of the test to measure discrimination ellipses, we probe discrimination in twenty directions of colour space (chosen to be spaced at equal 18-degree intervals in the CIE (1976) *u', v'* diagram, which is an approximately uniform colour diagram for normals). We test discrimination on randomly chosen pairs of these directions, until all twenty axes have been probed. This method was chosen, rather than interleaving trials randomly on the full set of axes, as it ensures that subjects do not spend prolonged periods of time making judgments close to threshold levels. In all it takes about twenty minutes to obtain the data for one ellipse from an untrained subject.

Results and discussion

We fitted ellipses to the data using the method of least squares.[1] The centres of the ellipses were forced to coincide with the chromaticity of the background field used, and the remaining three parameters were free. The best-fitting ellipse was found by minimizing the sum of the squares of the log radial distances of the data points from the ellipse.

We have fitted ellipses in a number of different colour spaces, and here we present ellipses fitted in the MacLeod-Boynton colour diagram (MacLeod and Boynton, 1979). The two axes of the MacLeod-Boynton diagram are often taken to correspond to two independent subsystems of the early visual system: a phylogenetically ancient subsystem that compares the signal of the short-wave cones with some combination of the signals of the M and L cones, and a phylogenetically newer subsystem that compares the signals of the L and M cones (Krauskopf *et al.*, 1982; Mollon and Jordan, 1988). It is a feature of the MacLeod-Boynton diagram that the ordinate, representing short-wave cone excitation, can be scaled arbitrarily with respect to the abscissa. We began with the particular scaling suggested by MacLeod and Boynton, with 1 unit of S cone excitation being the excitation produced by a monochromatic 400 nm light. However, the ellipses then obtained for normal subjects were highly eccentric, with the long axis around five times the length of the short axis: any interaction between the putative subsystems is thus obscured by the horizontal elongation of the ellipses. Using an iterative procedure, we established the scaling of the ordinate that gave for five normals an average axis ratio for the central

[1] Ellipses are used as a convenient, atheoretical (and historically hallowed) way of summarizing discrimination data from a large variety of phenotypes. In the case of dichromats, of course, the loci of the limens are strictly straight lines.

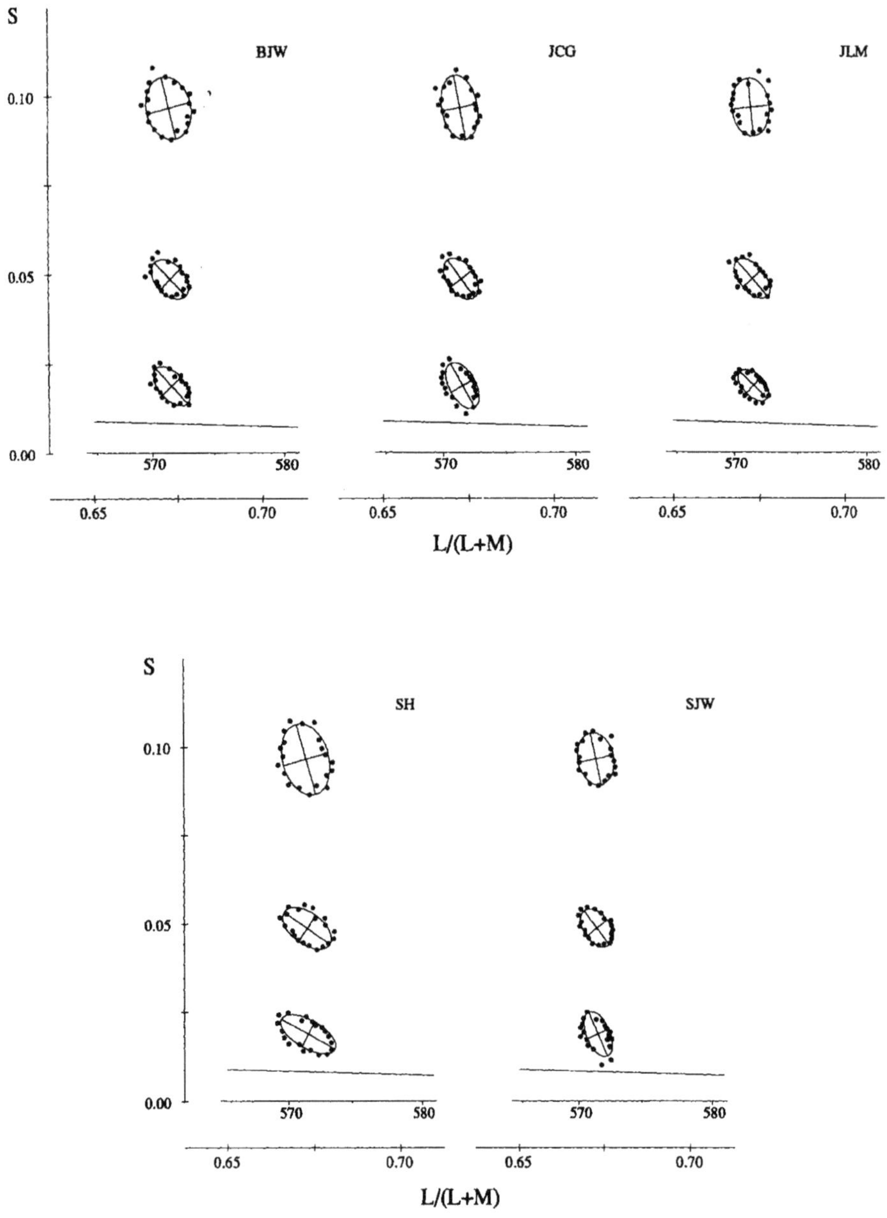

Fig. 2. Discrimination ellipses, plotted in the MacLeod-Boynton chromaticity diagram, for five young adults with normal colour vision. Each chart shows a short section of the spectrum locus, between 564 and 581 nm, and a part of the gamut of the colour monitor between the red and green phosphors. SH, JLM and JCG are female, and BJW and SJW are male. SH is thought to be a carrier of a protan form of deficiency and it is interesting that her ellipses are slightly larger than those of the other young adults in this sample; in addition, two of her ellipses appear to be rotated relative to those of other subjects.

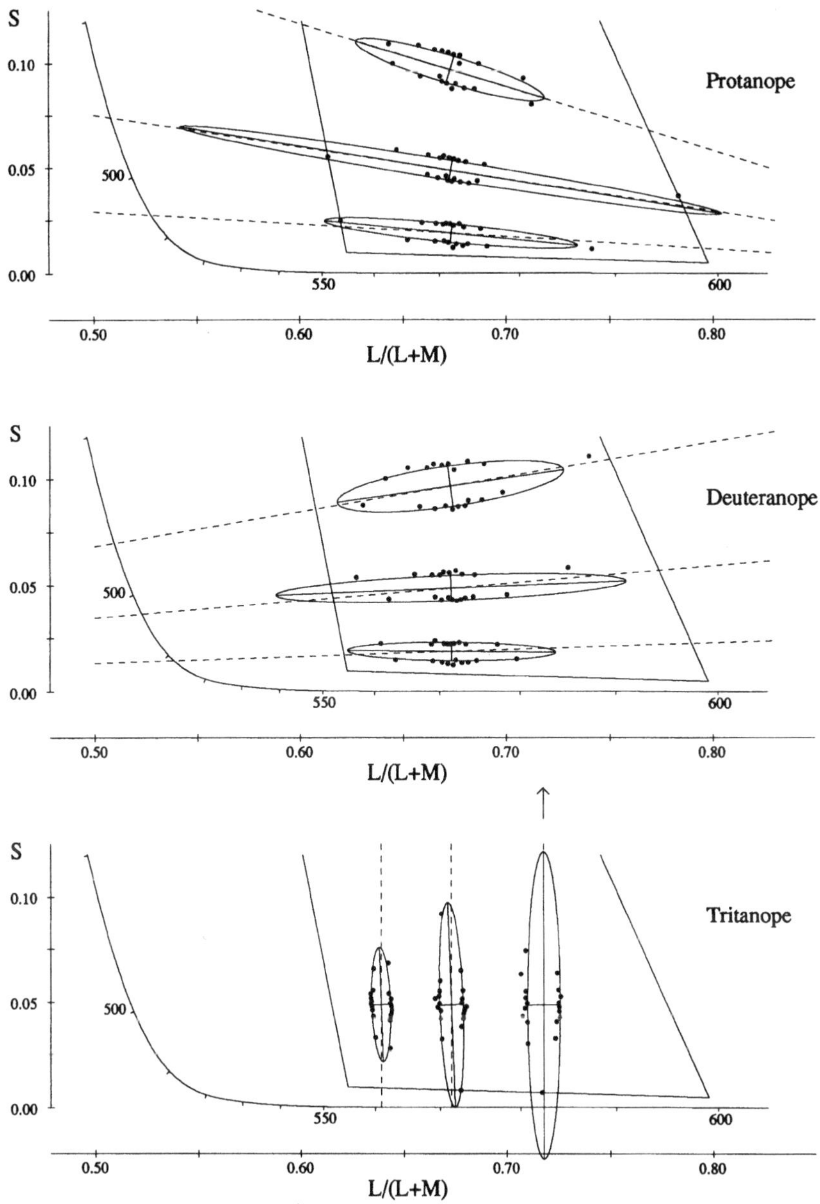

Fig. 3. Discrimination ellipses for three different types of dichromat, plotted in the MacLeod-Boynton chromaticity diagram. The dashed lines in each panel indicate the confusion lines theoretically predicted for this chromaticity diagram.

ellipse that was as close to unity as possible. The scaling thus obtained is 3.46 times that suggested by MacLeod and Boynton.[2]

Figure 2 shows the discrimination ellipses obtained for five observers with normal colour vision, as determined by the Nagel anomalscope (Model 1, Schmidt and Haensch). The results for these subjects show a pleasing uniformity, and the most remarkable feature of the figure is that the ellipses all point in a direction corresponding to a concomitant increase or decrease in M and S cone excitation. Our results resemble those of Boynton *et al.* (1983) in showing that the two subsystems are not independent at the site where chromatic thresholds are determined: thresholds are lower when the signals of S and L cones are varied in phase than when these signals are out of phase (Boynton *et al.*, 1983, 1986). Measuring thresholds for Gaussian-blurred flashes embedded in a bright white field, Cole *et al.* (1993) obtained ellipses suggesting independence of the two subsystems.

Using the same ellipse-fitting routine and using the same chromaticity diagram as in Fig. 2 but reducing the scale, we show in Fig. 3 the ellipses obtained for a protanope, a deuteranope and a tritanope. The ellipses for the congenital protanope point towards (1,0) and for the congenital deuteranope towards (0,0), as theoretically predicted for this chromaticity diagram (MacLeod and Boynton, 1979). The tritanope is the case described by Jordan *et al.* (1990). This patient lost short-wave cone function suddenly and bilaterally at the age of 40: her ellipses are parallel to the S axis, as expected for an observer who has no access to short-wave cone signals.

Acknowledgements

Supported by SERC grant F24689 and MRC grant G9311798N. We thank Drs. V. Bonnardel, G. Jordan, S. Shevell and M. Webster for comments.

References

Arden, G., Gündüz, K. and Perry, S. (1988). Color vision testing with a computer graphics system: preliminary results. Doc. Ophthalmol. 69: 167–174.

Boynton, R.M., Nagy, A.L. and Eskew, R.T. (1986). Similarity of normalized discrimination ellipses in the constant-luminance chromaticity plane. Perception 15: 755–763.

Boynton, R.M., Nagy, A.L. and Olson, C. (1983). A flaw in equations for predicting chromatic differences. Color Res. Appl. 8: 69–74.

Cole, G.R., Hine, T. and McIlhagga, W. (1993). Detection mechanisms in L-, M- and S-cone contrast space. J. Opt. Soc. Am. A, 10: 38–51.

[2] This rescaling is arbitrary. It will be expected to hold only for the spatiotemporal conditions of our measurements; and, as in the case of earlier data sets (Le Grand, 1949; Krauskopf & Gegenfurtner, 1992), our results suggest that a compressive transformation of the S axis would be needed to give ellipses of the same size at different background chromaticities.

Cuvinot, P. (1992). Mesures et développement de tests ophtalmologiques sur le GEPCO. Mémoire d'ingénieur, Conservatoire National des Arts et Métiers, Paris.

Hart, W.M., Hartz, R.K., Hagen, R.W. and Clark, K.W. (1984). Color contrast perimetry. Invest. Ophthalmol. Vis. Sci. 25: 400–413.

Heard, P.F., Stone, C.J., Gregory, R.L. and Marmion, V.J. (1987). A new computer graphics test for red/green colour anomaly. In: Verriest, G. (ed.). Colour Vision Deficiencies VIII, Doc. Ophthalmol. Proc. Ser. 46: 181–19, Dr. W. Junk, The Hague.

Jordan, G., Sarkies, N. and Mollon, J.D. (1990). Clinical evidence for the independence of two subsystems of colour vision. Perception 19: 360.

King-Smith, P.E., Chioran, G.M., Sellers, K.L. and Alvarez, S.L. (1983). Normal and deficient colour discrimination analysed by colour television. In: Mollon, J.D. and Sharpe, L.T. (eds.), Colour Vision: Physiology and Psychophysics: 167–172, Academic Press, London.

Krauskopf, J. and Gegenfurtner, K. (1992). Color discrimination and adaptation. Vision Res. 32: 2165–2175.

Krauskopf, J., Williams, D.R. and Heeley, D.W. (1982). Cardinal directions of color space. Vision Res. 22: 841–856.

Le Grand, Y. (1949). Les seuils différentiels de couleurs dans la théorie de Young. Revue d'Optique théor. exp. 28: 261–278.

MacLeod, D.I.A. and Boynton, R.M. (1979). Chromaticity diagram showing cone excitation by stimuli of equal luminance. J. Opt. Soc. Am. 69: 1183–1186.

Mollon, J.D. (1989). "Tho' she kneel'd in that Place where they grew...". J. Exp. Biol. 146: 21–38.

Mollon, J.D. and Jordan, G. (1988). Eine evolutionäre Interpretation des menschlichen Farbensehens. Die Farbe 35/36: 139–170.

Mollon, J.D. and Reffin, J.P. (1989). A computer-controlled colour vision test that combines the principles of Chibret and of Stilling. J. Physiol. (Lond.) 414: 5P.

Reffin, J.P., Astell, S. and Mollon, J.D. (1991). Trials of a computer-controlled colour vision test that preserves the advantages of pseudoisochromatic plates. In: Drum, B., Moreland, J.D. and Serra, A. (eds.), Colour Vision Deficiencies X, Doc. Ophthalmol. Proc. Ser. 54: 69–76, Kluwer Acad. Publ., Dordrecht.

Regan, B.C., Reffin, J.P. and Mollon, J.D. (1994). Luminance noise and the rapid determination of discrimination ellipses in colour deficiency. Vision Res. 34: 1279–1299.

Vingrys, A.J. and King-Smith, P.E. (1986). Factors in using color video monitors for assessment of visual thresholds. Color Res. Appl. 11: S57–S62.

Department of Experimental Psychology
Downing St.
Cambridge CB2 3EB
U.K.

53. Computer generated pattern on a video screen for the examination of colour vision deficiencies

HEINWIG LANG, WOLFGANG JAEGER and JANOS SCHANDA
(Darmstadt, Germany)

Abstract

In our test for the examination of colour vision deficiencies we determine colour contrast thresholds. On a computer monitor a coloured test pattern with random orientation can be presented on different coloured backgrounds. The colours of the pattern and background lie on dichromatic colour confusion lines in colour space. Colours, pattern size, presentation time and number of presentations are controlled by a specially designed computer program. This program also evalutes the test responses. Preliminary test results with different types of colour vision deficiencies are presented.

Introduction

A computer controlled Landolt ring test was originally developed for the investigation of visual contrast sensitivity for lightness, hue and chroma differences in small areas. We have modified this test for the examination of colour vision deficiencies. The colour of the test pattern can be moved on protan, deutan or tritan confusion lines through the colour of the background (Fig. 1). The background colour can be chosen within the colour space of the monitor. As the test uses the criterion of colour contrast threshold, it differs from a realization of the pseudo-isochromatic plates on video screen.

Measurement of colour contrast with isoluminant test patterns on a computer controlled display has been reported (Hart *et al.*, 1984; Arden *et al.*, 1988), and computer controlled tests of colour vision deficiencies were described by Mollon and Reffin (Mollon and Reffin 1989; Reffin *et al.*, 1991) and (Huber, 1989). The confusion lines of our test are defined in colour space. The protan confusion line for example is defined by D = C1, T = C2, saying that the fundamental tristimulus values D and T are constant, whereas P is variable.

B. Drum (ed.), Colour Vision Deficiencies XII, pp. 453–458.
© 1995 *Kluwer Academic Publishers, Dordrecht.*

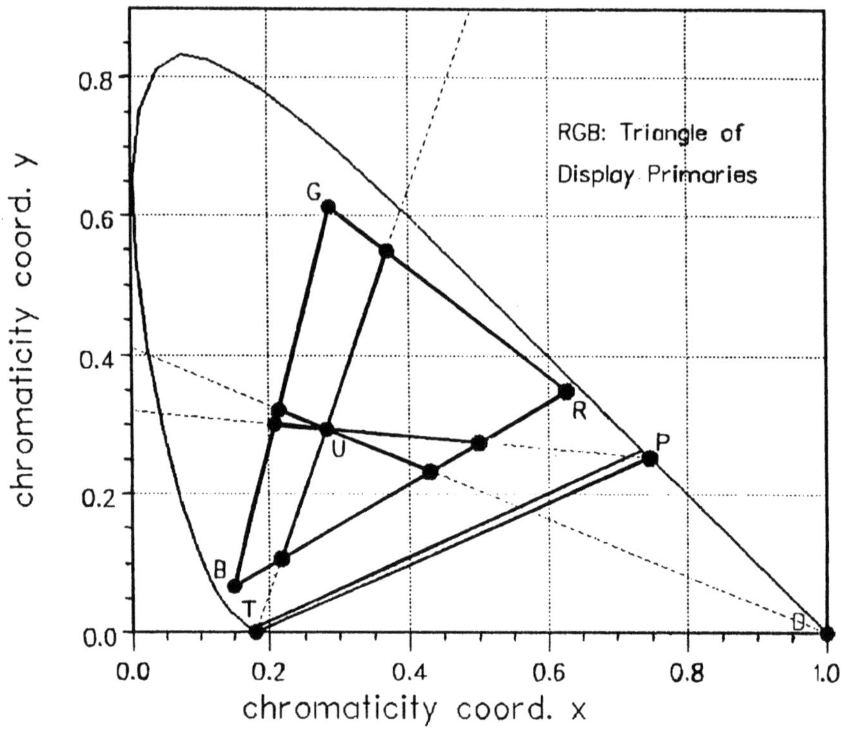

Fig. 1. CIE chromaticity diagram with spectral locus, triangle of display primaries (R, G, B) and chromaticities of fundamental primaries (P, D, T). Confusion lines through display neutral point (U) are dotted. Only the solid part inside the triangle can be realized on the display.

Test equipment and program description

The test has been developed for a personal computer with VGA graphics and a VGA monitor with 1024 x 768 pixels and is written in TURBO PASCAL. The background is a square filling the center and right part of the screen, at the left the menu for the test program is displayed. The test pattern (letters, numbers or a Landolt ring) appears in the middle of the background square. Three times the picture height is considered as the normal viewing distance for the tests. Under this condition ring diameters between 0.5 and 10° (gap widths between 0.2 and 3°) are available. Presentation time and the number of presentations for a test are variable. The appearance of the pattern is accompanied by a humming noise. The orientation of the ring gap is randomly chosen between 8 possible directions by the program. The subject has to indicate the perceived orientation by pressing an appropriate number key.

The VGA Graphics system allows a variation of the three monitor primaries red (R), green (G) and blue (B) in 64 steps (6 bit). The background colour can be chosen among the 64 * 64 * 64 colours each defined by three numbers between 0 and 63. For the generation of the colour of the test pattern itself the possible steps per colour channel have been extended to 256 (8 bit) by pixel dithering. This means that groups of four neighbouring pixels are set to slightly different colours giving the impression of a uniform area of intermediate colour. This method allows the realization of smaller colour differences near visual threshold between background and pattern.

Pixel dithering has two additional advantages. The edges between optotype and background are blurred, and as the adjacent pixels have slightly different luminances the influence of a possible luminance difference between optotype and background is reduced.

The monitor has been calibrated by measuring the tristimulus values of the phosphor primaries as well as the monitor transfer function (picture tube gamma). With these data the CIE tristimulus values X, Y, Z of every displayed colour can be calculated. The X, Y, Z values are transformed into the fundamental tristimulus values P, D, T of this colour using a triple of fundamental primaries, for example those derived by Pitt.

After starting of the program it is possible to enter phosphor and neutral point chromaticities of an alternative monitor or chromaticities of a triple of different fundamentals.

The program allows the pattern colour to be altered by varying only one of the fundamental values P, D or T. The colour moves in this case exactly along one of the dichromatic confusion lines in colour space. That means that optotype and background colours are not isolumiant for normal trichromats, but they are isoluminant for protanopes if both lie on the protan confusion line, and they are isoluminant for deuteranopic subjects if both colours lie on the deutan confusion line. In cases of anomalous trichromacy or other deficiencies caused by abnormal spectral cone sensitivities the luminances of optotype and background may appear different. Tests with protanomals and deuteranomals have however shown that the test can clearly differentiate between the two defects and that the results agree with those found with the Nagel anomaloscope.

Test procedure

Prior to a test run the person to be tested is informed about the keys she has to press for the different orientations and is trained with an easily perceivable luminance contrast. Then the test begins with small colour differences. The test is regarded as 'passed' if for four out of five presentations with constant colour contrast the answers are correct. If the test was not successful, the colour difference is set to a higher level. This procedure continues until the test is successful.

The whole procedure is performed for the 6 directions along the colour confusion lines with their intersection point lying in the background colour. Figure 1 shows these lines for the case of the monitor neutral point as

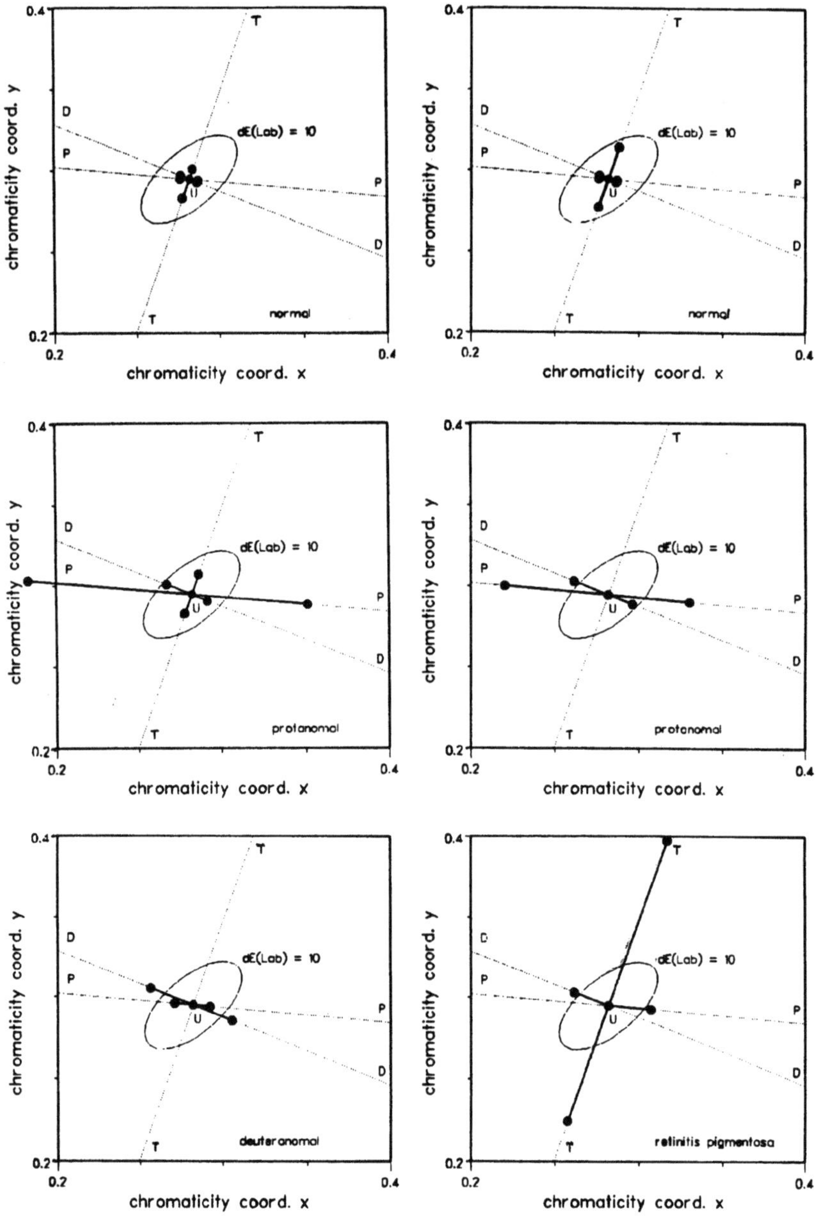

Fig. 2. Sections of the CIE chromaticity diagram near the display neutral point with test results. The top row shows results for a young colour normal and an aged colour normal person, the latter with reduced differentiation in tritan direction. The middle row and lower left diagrams are results from two protanomals and a deuteranomal. Lower right diagram is from a young man aged 15 suffering from retinitis pigmentosa.

background. The results of a complete test can be plotted as distances from the background colour in the 6 spatial directions.

Test results

Test results are shown in Fig. 2, representing a section of the CIE chromaticity diagram near the neutral point of the display, with the confusion lines intersecting in the neutral point. The measured thresholds in the different directions are indicated by black circles. The ellipse around the neutral point connects colours in an isoluminant plane with a colour difference of 10 dE-units in the approximately uniform CIE-LAB space.

The tests were performed in a dark surround with neutral background at a luminance of about 50 cd/m^2, Landolt ring as test pattern with a diameter of 3° (1° gap width), 1 second presentation time and 5 presentations per test.

Conclusion

The test can be performed without special equipment on a monitor driven by a personal computer with VGA graphic system. In a number of preliminary tests the method has proved its efficiency and ability to differentiate between different types of colour deficiencies. It is intended to develop a modified test for peripheral vision as proposed by Aulhorn (Aulhorn, 1960), especially for the analysis and early test of retinitis pigmentosa.

Acknowledgements

The authors would like to thank Prof. Krastel of the Augenklinik of the University of Heidelberg for his encouragement. He arranged the tests and gave many valuable stimuli for the improvement of the method.

References

Arden, G., Gündüz, K., Perry, S.(1988). Color vision with a computer graphics system: Preliminary results. Doc. Ophthalmol. 69: 167–174.

Aulhorn, E. (1960). Sehschärfeprüfung am Perimeter. Bericht über die 63. Zusammenkunft der Deutschen Ophthalmologischen Gesellschaft Berlin.

Hart, W.M., Hartz, R.K., Hagen, R.W. and Clark, K.W. Color contrast perimetry. Invest. Ophthalmol. Vis. Sci. 25: 400–413.

Huber, Chr. (1989). Colour Vision Testing on Video Screen Using the Macintosh II Computer. Eur. J. Implant. Ref. Surg. 1: 257–263.

Mollon, J.D. and Reffin, J.P. (1989). A computer-controlled colour vision test that combines the principles of Chibret and of Stilling. J. Physiol. (Lond.) 414: 20.

Reffin, J.P., Astell, S. and Mollon, J.D. (1991). Trials of a computer-controlled colour vision test that preserves the advantages of pseudoisochromatic plates. In: Drum, B., Moreland, J.D. and Serra, A. (eds.), Colour Vision Deficiencies X, Doc. Ophthalmol. Proc. Ser. 54: 69–76, Kluwer Acad. Publ., Dordrecht.

Heinwig Lang
BTS - Broadcast Television Systems GmbH
P.O. Box 11 02 61
D - 64278 Darmstadt

54. A new lantern test using light-emitting diodes

NORIKO TAKAHASHI, KAORU HAMANO, AKIKO TOYOGUCHI
and YASUO OHTA
(Tokyo, Japan)

Abstract

We have developed a new lantern test using light-emitting diodes to detect and classify color vision deficiencies.

The lantern color lights are 1.5 mm in diameter and subjects view them from a distance of 2.5 m. Dominant wavelengths of the lights were: red, 616 nm; yellow, 577 nm; and green, 555 nm. Paired, single and different-size color light tests were conducted. 142 subjects, with congenital color vision deficiencies, were tested. They were: 11 protanopes, 1 extreme protanomal, 33 protanomals, 25 deuteranopes, 17 extreme deuteranomals, 55 deuteranomals.

We found a correlation between the results of the lantern test and the classification obtained with the Neitz anomaloscope, and between the results of the lantern test and the Panel D-15 test. Paired light tests had more satisfactory results than single light tests.

Introduction

The lantern test was first implemented as a means of promoting the safe operation of shipping and railways. In recent years, however, the lantern test has become the basis for an aptitude test for the purpose of ensuring automobile and aircraft traffic safety.

In Japan, Ichikawa's lantern was the device used for the lantern test. However, this lantern is no longer manufactured, and therefore the lantern test is hardly ever performed.

In order to make practical assessment of the severity of color vision deficiencies, we developed and tested a new, small, easy-to-handle lantern.

B. Drum (ed.), Colour Vision Deficiencies XII, pp. 459–465.
© 1995 *Kluwer Academic Publishers, Dordrecht.*

Subjects and methods

Subjects

The present study included 142 subjects, aged 13 to 30 (mean = 20.03), with congenital color vision deficiencies who consulted the outpatient clinic at the Tokyo Medical College Hospital from 1992 to 1993. The type of deficiency was classified using a Neitz anomalscope. They were classified as 11 protanopes, 1 extreme protanomal, 33 protanomals, 25 deuteranopes, 17 extreme deuteranomals, 55 deuteranomals. Extreme anomals are defined as subjects having wide matching ranges that include one of the anomaloscope primaries.

Device

The lantern we developed is 16 cm wide, 34 cm long, and 20 cm high, and weighs 2.8 kg (Fig. 1).

On the front surface of the lantern, two apertures that transmit colored lights are vertically aligned. These two apertures are enclosed within a cylinder.

The colored-light apertures each have a diameter of 1.5 mm, and are separated by 7.5 mm.

The main switch, program-selecting dial, color monitoring window, and buttons for start, stop, and reset, are all on the left side of the lantern.

For the light source, we used several light-emitting diodes that produced red, yellow, and green light, respectively. Wavelengths were regulated by filtration of the yellow and green color components.

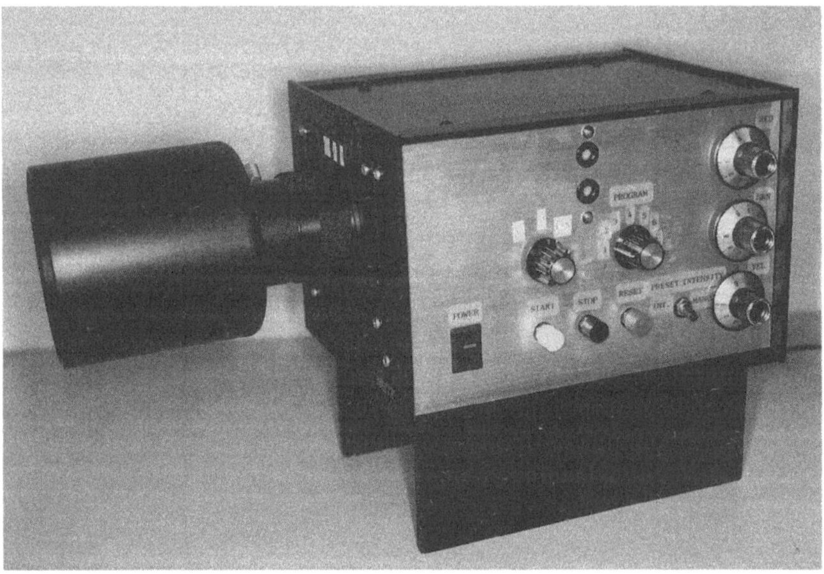

Fig. 1. The lantern device. Light-emitting diodes are used as light sources.

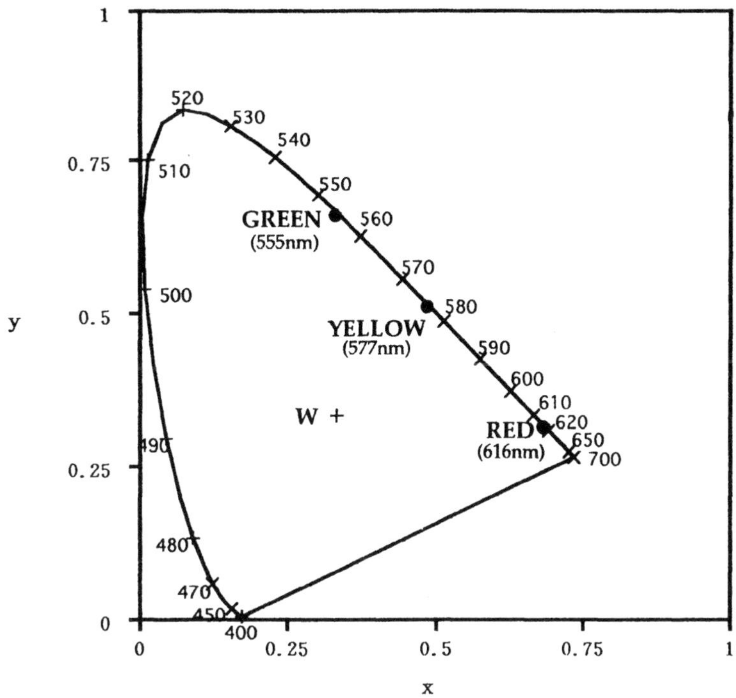

Fig. 2. The wavelengths and chromaticities of the lantern lights. The yellow and green diode lights are filtered.

Figure 2 shows the chromaticities of the color lights. Red is 616 nm, yellow is 577 nm and green is 555 nm. The luminance used was approximately 80 cd/m^2 for all three colors for all test conditions.

Tests were conducted for aperture diameters 1.5 mm, 3 mm, and 6 mm, with separations 7.5 mm, 15 mm, and 30 mm, respectively.

The lantern lights were exposed cyclically for 2 seconds with a 2-second inter-stimulus-interval.

Test method

Testing was conducted in a dark room with a floor illumination of approximately 3.0 lx. The subject sat 2.5 m from the lantern. At this position, the 1.5 mm diameter apertures subtended about 2 min. arc and were separated by 10.3 min arc.

Subjects were tested with correction of their refractive error. The subject was told that two of the three colored lights, red, yellow, or green would be shown. He or she was then instructed to name the colors, upper first and lower second. An error was counted if the subject gave a wrong color name for either or both of the colored lights presented.

The severity of the deficiency was based on the number of errors.

Testing was conducted using three different programs:

1 Program 1 presented paired-color lights through 1.5 mm apertures. Program 1 presented 15 sets of colored lights.
2. Program 2 presented colored lights in the same order as Program 1, but the angular size of the colored lights was changed. At first, tests were conducted with paired lights with 1.5 mm apertures. After a one minute rest, measurements were taken for lights with 3 mm apertures. After another one minute rest, testing was conducted using 6 mm apertures.
3. Program 3 involved further tests using single lights with 1.5 mm apertures; each color, red, yellow, and green was presented twice.

A total of 111 subjects were tested on Program 1. Another 31 subjects were tested on Program 2. Fifty-six out of the 111 subjects of Program 1 were tested on Program 3.

Results

Figures 3 and 4 show the test results of Program 1. Among 10 normal subjects tested, 1 subject made 1 error and the other 9 subjects made no errors.

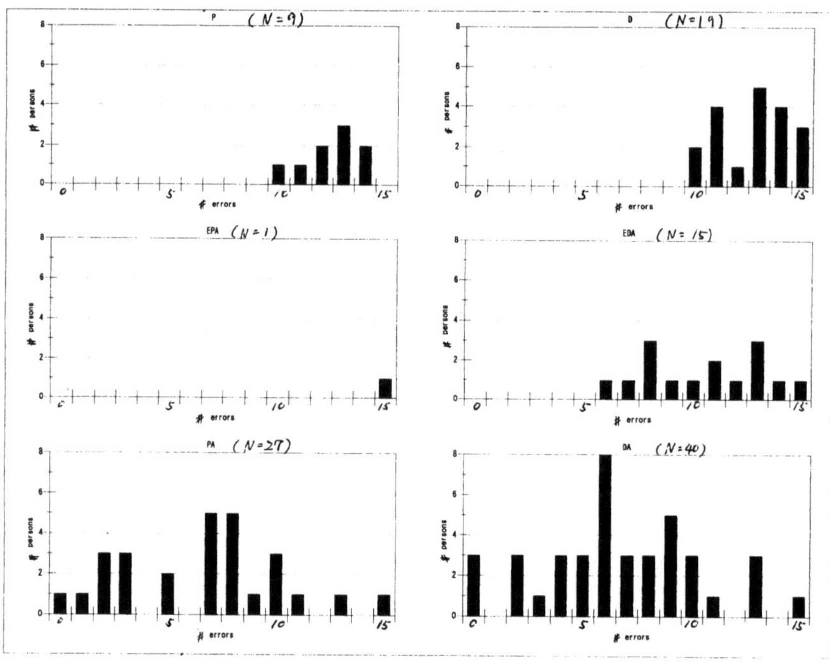

Fig. 3. Number of errors with the paired light test (Program 1) among 15 pairs. The vertical axis represents number of subjects. The horizontal axis represents number of lantern errors. P-protanopia, EPA-extreme protanomaly, PA-protanomaly, D-deuteranopia, EDA-extreme deuter-anomaly, DA-deuteranomaly.

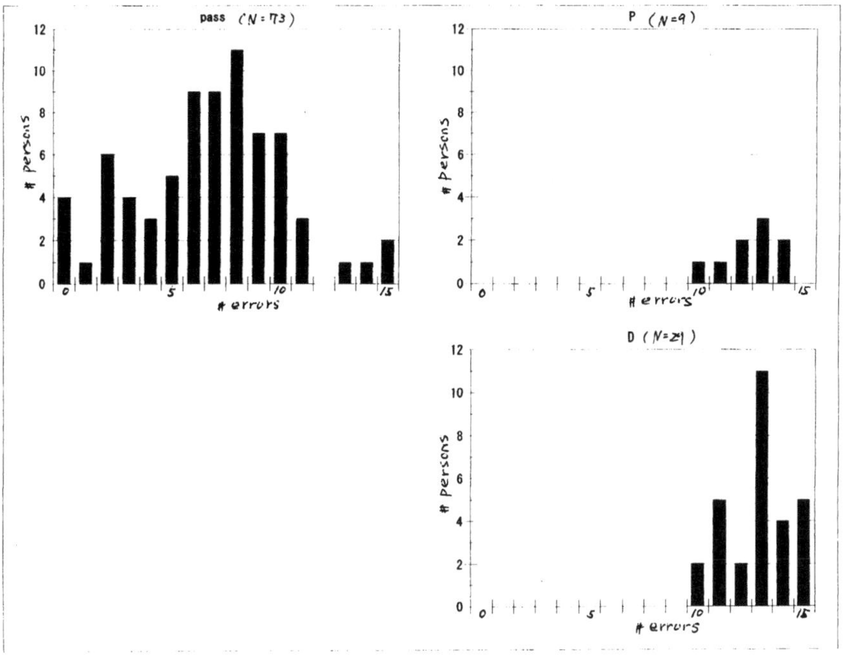

Fig. 4. Program 1 errors classified by Panel D-15 test results. The vertical axis represents number of subjects. The horizontal axis represents number of lantern errors. P-protan pattern on D15, D-deutan pattern on D15. Both P and D are groups failed on D15. The subjects were the same of program 1. 'Pass' includes 26 protanomals, 2 deuteranopes, 7 extreme deuteranomals and 38 deuteranomals on anomaloscope. 'P' includes 9 protanopes on anomalscope. 'D' includes 1 extreme protanomal, 1 protanomal, 17 deuteranopes, 8 extreme deuteranomals and 2 deuteranomals on anomaloscope.

Figure 3 shows the relation of lantern test errors to anomaloscopy results. In both protan and deutan color deficiencies, lantern errors increased with change of anomaloscope classification from anomaly to anopia. Lantern errors occurred with a similar frequency in both protan and deutan groups.

Figure 4 shows the relationship of Program 1 lantern errors to Panel D-15 test results. Subjects who failed the Panel D-15 test (P and D in Fig. 4) made greater number of errors on the lantern test.

Table 1 shows the test results of Program 2. The larger the light was, the fewer the number of errors. This demonstrated that a larger color light allowed individuals with color vision deficiency better discrimination between various color lights.

Figure 5 compares the test results of Program 3 with the anomalscopy results. All 10 normal subjects made no errors. In subjects with abnormal color vision, the results of this test were slightly less correlated to the anomaloscopy results than the results of the paired light test of Program 1.

Table 1. Results of program 2. Average number of errors using different sizes of color lights. Total number of subjects = 31; P(2), PA(6), D(6), EDA(2), DA(15). The larger the lights were, the fewer the number of errors ($\alpha = 0.01$).

Aperture size	Average number of errors (among 15 pairs)
1.5 mm	10.35 (SD = 2.21)
3 mm	9.03 (SD = 3.39)
6 mm	7.67 (SD = 2.76)

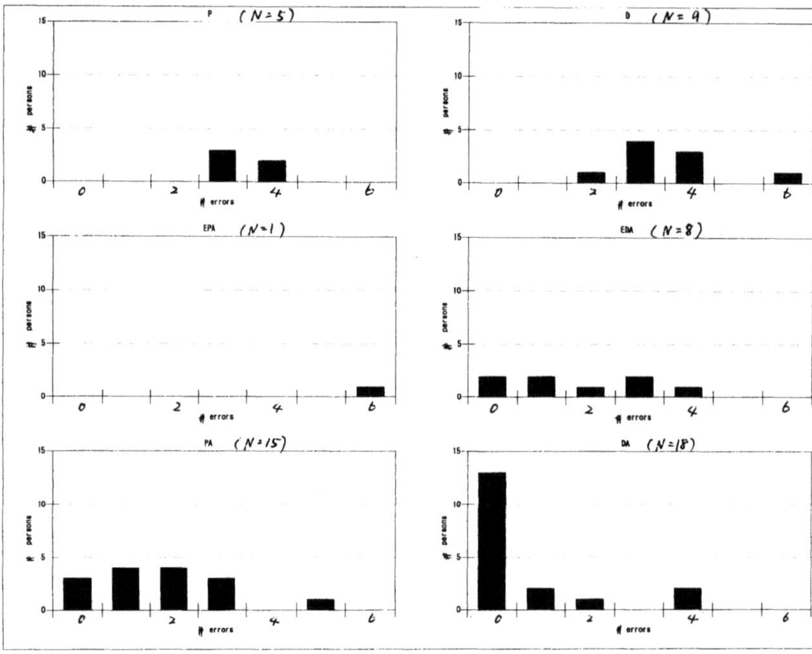

Fig. 5. Number of errors with single-color light test. Each color: red, yellow and green is shown twice. Other details as in Fig. 3.

Discussion

The Holmes-Wright lantern was based on the colors of traffic signal lights (Holmes and Wright, 1982).

The Farnsworth lantern uses colors on the dichromatic confusion lines on the CIE chromaticity diagram. Farnsworth recommended that the lantern test should be used with both pseudoisochromatic plates and the Panel D-15 test (Farnsworth, 1957).

In Japan, color deficiencies are detected with pseudoisochromatic plates and then they are classified according to type and severity of color vision defects with the Nagel anomaloscope or the Neitz anomaloscope. And the classification is adopted as the basis of occupational aptitude.

We produced a lantern test instrument using red, yellow and green lights which were on the straight line of the CIE chromaticity diagram. We tested the lantern device whether we can classify severity of color deficiencies by its number of errors.

Results of program 1 (shown in Figs. 3 and 4) indicate that the number of errors of the lantern test correlated with the groups classified using anomaloscope as well as with the results on the Panel D-15 test.

Results of program 2 suggest that it may be necessary to use larger colored lights because using larger colored lights resulted in fewer errors. However, in terms of the size of the testing device and the maintenance of luminance, 3 mm - 6 mm apertures are thought to be proper sizes.

Comparison of program 1 and 3 results suggest that testing with a paired light presentation produces more satisfactory results than single-color presentation.

Paired light tests are thought to be more appropriate to prepare standards for assessing color vision deficiency using this lantern test.

Acknowledgements

We thank Neitz Co. Ltd. for manufacturing the lantern device used in this study. We also thank all the subjects who participated in our testing.

References

Cole, B.L. and Vingrys, A.J. (1982). A survey and evaluation of lantern tests of color vision. Am. J. Optom. Physiol. Opt. 59: 346–374.

Farnsworth, C.D. (1957). Testing for color deficiency in industry. A.M.A. Arch. Indust. Health. 16: 100–103.

Holmes, J.G. and Wright, W.D. (1982). A new color-perception lantern. Col. Res. Appl. 7: 82–88.

Noriko Takahashi
Dept. of Ophthalmology
Tokyo Medical College
6-7-1, Nisi-Shinjuku
Shinjuku-ku
Tokyo, Japan

55. A new lantern test using light-emitting diode lamps

SHOKO TANABE[1], KAITIRO HUKAMI[2], SHINICHI YAMADE[3], KAZUO ICHIKAWA[1] and SHIGEO TAGAI[4]

([1] Nagoya, [2] Fukui, [3] Shiga and [4] Tokyo, Japan)

Abstract

A lantern test using LED lamps has been devised. In this model, two lights each 4 mm in diameter placed 15 mm apart in tandem are presented for 2 seconds at a test distance of 5 m. The color of the light is either red (630 nm), green (555 nm), or yellow (580 nm). All possible combinations of three colors are included in one trial, which consists of nine pairs of lights. The color pairs are presented in random sequence and the number of errors in the second trial is scored. Test results are printed out.

Anomalous trichromats usually make from 3 to 7 errors out of a maximum of 9 errors in one trial, although some mild cases make fewer or no errors. Dichromats usually make 7 or more errors and they never recognize the colors correctly even if the viewing distance is shortened to 30 cm.

The lantern test is persuasive as well as important, especially for occupational purposes. This model is portable and easy to operate. In addition, the three colors are located on the confusion lines for both protanopia and deuteranopia, that provides a theoretical basis for its use as a color vision test.

Introduction

A lantern test is a unique method among clinical color vision tests as its form of testing is practical. Besides, it uses colored lights of high stability while many other tests use pigment colors. Therefore, it is regarded as an important clinical test especially for occupational purposes (Farnsworth, 1957; Cole, 1993), though from the diagnostic point of view, it is not essential.

We have devised a new lantern test using light-emitting diode lamps because Ichikawa's lantern (Ichikawa, 1968) which was used in Japan for many years is out of production. This is tentatively named JFC lantern.

B. Drum (ed.), Colour Vision Deficiencies XII, pp. 467–472.
© 1995 *Kluwer Academic Publishers, Dordrecht.*

Description

Testing unit (Fig. 1, left)

Two lights, each 4mm in diameter placed 15 mm apart, are presented for 2 seconds. Test lights are red, green, and yellow provided by light-emitting diode lamps. All of the test lights are adjusted to equal brightness; approximately 20 millicandela. the peak wavelengths are 630 nm for red, 580 nm for yellow, and 555 nm for green respectively. The half height bandwidth is 30 nm each.

The nine possible combinations of the three colors comprise one trial. The combinations appear in random sequence.

Control unit (Fig. 1, right)

Test procedure is regulated remotely using a radio control box. There are keys for presentation, representation, input, print and clear.

Each pair of lights is presented by the 'START' key, so that interval of presentation is arbitrary. Answers are recorded with input keys; 'R' for red 'G' for green, and 'Y' for yellow, each of which is placed in one of two rows corresponding to the top and bottom lights.

The presentation time can be altered inside the unit if needed.
Test results are printed out by the 'PRINT' key.

Subjects and method

Eighty subjects, 9 protanopes, 11 protanomalous trichromats, 26 deuteranopes, and 34 deuteranomalous trichromats were examined.

Each subject performed two trials. The number of errors of the second trial of 9 color pairs was scored.

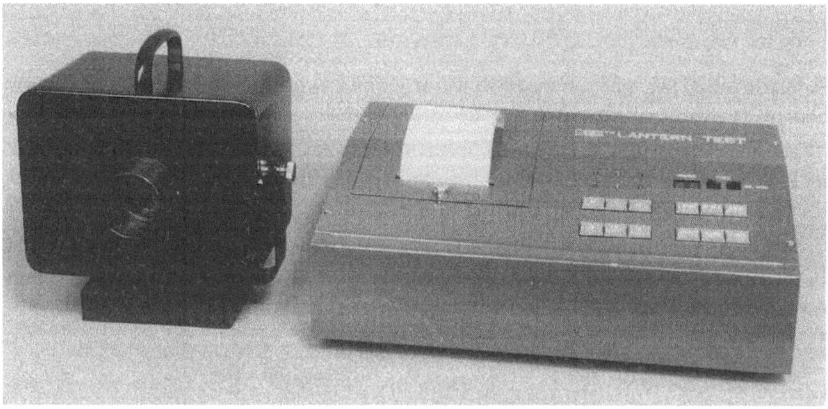

Fig. 1. JFC lantern, testing unit (left) and control unit (right).

Viewing distance was 5 m, when each light subtended about 2.8 min arc. Additionally, 49 subjects (4 P, 7 PA, 18 D, and 20 DA) were tested with a smaller viewing distance of 30 cm (visual angle 45.8 min arc).

Results

Results of the standard procedure are given in Table 1. Dichromats made from 4 to 9 errors and mean score was 7.0. Anomalous trichromats made from 0 to 8 errors and mean score was 3.5.

Table 1. Number of errors recorded for 80 color defective subjects (35 dichromats and 45 anomalous trichromats) at a test distance of 5 m.

Number of errors	Dichromats	Anomalous trichromats
0	–	4
1	–	8
2	–	5
3	–	8
4	2	4
5	4	6
6	8	4
7	5	1
8	11	5
9	5	–
mean	7.0	3.5

In trials with the smaller distance (Table 2), dichromats made from 3 to 9 errors and mean score was 6.4 and anomalous trichromats made from 0 to 7 errors and mean score was 1.9.

Table 2. Number of errors recorded for 49 color defective subjects (22 dichromats and 27 anomalous trichromats) at a test distance of 30 cm.

Number of errors	Dichromats	Anomalous trichromats
0	–	13
1	–	4
2	–	1
3	1	1
4	4	2
5	3	2
6	2	3
7	4	1
8	6	–
9	2	–
mean	6.4	1.9

Subjects who made many errors, i.e. dichromats and anomalous trichromats of probably severe degree, could not recognize the colors correctly even when viewed with the larger visual angle. Subjects with relatively mild defects showed improvement of recognition when test distance was shortened and the visual angle increased (Table 3).

Table 3. Comparison of the number of errors made by 48 color defective subjects at 5 m and 30 cm. 27 anomalous trichromats (top) and 21 dichromats (bottom in parentheses).

5 m \ 30 cm	Number of errors									
Number of errors	0	1	2	3	4	5	6	7	8	9
0	2									
1	4	1								
2	2	2								
3	2	2								
4	1		1		1					
					(1)	(1)				
5	2					1	1			
6					1	1				
				(1)	(2)				(1)	(1)
7				1			1	1		
							(1)	(1)	(1)	(1)
8							1			
					(1)	(2)		(1)	(4)	
9							(1)	(1)		

Discussion

Lantern tests were primarily designed to assess recognition of signal lights of small visual angle in maritime, aviation, and railway services. It is also valuable as an item of color vision test batteries for quantitative classification (Farnsworth, 1957; Cole, 1993; Ichikawa, 1968).

It is unavoidable that the lantern test is considered a model of signal lights. In this case, only 'no errors' would be generally accepted as passing, because signal lights must be recognized correctly by all means. Color naming of small test lights is a difficult task for color defective persons, especially for protans (Vingrys and Cole, 1993). Mere presentation of colored lights is enough for detecting most anomalous subjects. Further devices to induce errors, for example, to reduce brightness or saturation, are not needed.

Since few subjects make no errors, scores other than passing and failing may

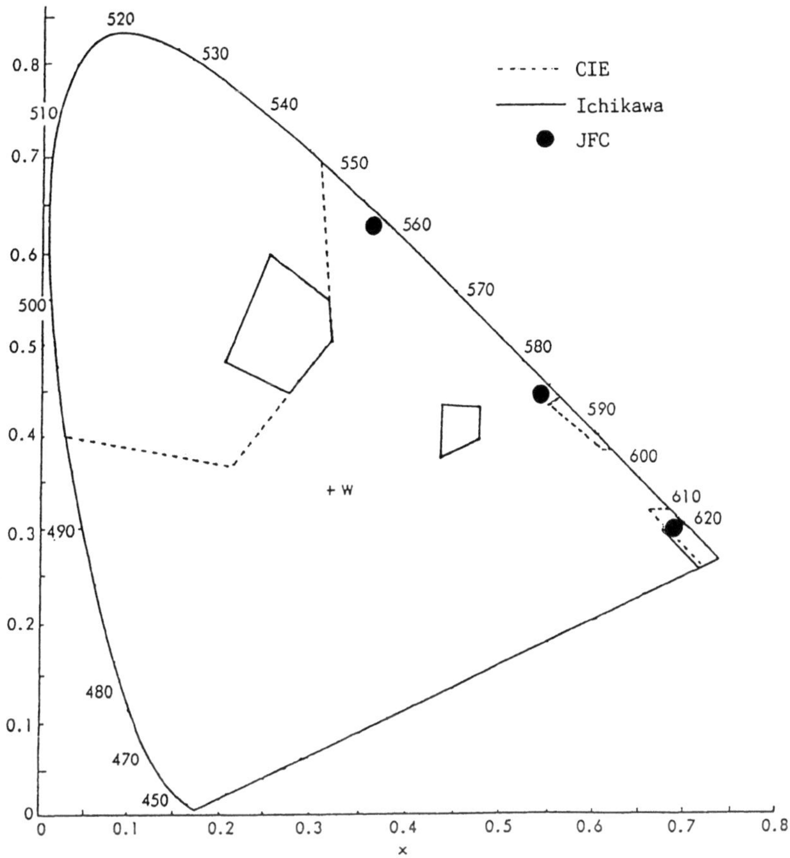

Fig. 2. Location of the colors of JFC lantern. • JFC lantern, —— Ichikawa's lantern, ---- Ranges recommended by CIE.

be appropriately adopted as the boundary between mild and moderate degree. Some people insist on ad hoc tests with practical things such as colored wires, signal lights, actually used color samples and so on. However, if such ad hoc tests are employed, various testing conditions simulating every possible work situation should be prepared, that can never be obtained. Consequently, tests with firm theoretical bases should be preferred, even though they are somewhat circuitous. Such is the reason for which the colors of the Farnsworth lantern were chosen (NAS/NRC Committee on Vision, 1981).

The fact that the three colors of the JFC lantern are on the confusion line of protanopia and deuteranopia and are similar to the colors of the anomaloscope (Fig. 2) would commend the JFC lantern as a color vision test.

Conclusion

A JFC lantern test using LED lamps was introduced. The test procedure is remotely controlled and the test result is printed out.
The test lights, red, green and yellow are on the confusion line of protanopia and deuteranopia that provides a theoretical basis as a color vision test.
In order to determine the standard of judgement with this model, a larger number of subjects should be examined.

References

Committee on Vision Assembly of Behavioral and Social Science, National Research Council (1981): Procedures for Testing Color Vision, p. 77.

Ichikawa, H. (1968). A New Standardization of Color Perception Lantern, J. Clinical Ophthalmol. 22: 503–508.

Vingrys, A.J. and Cole, B.L. (1993). The ability of colour defective observers to recognize an optimized set of red, green and white signal lights, In: Drum, B (ed.), Colour Vision Deficiencies XI, Doc. Ophthalmol. Proc. Ser. 56: 87–95, Kluwer Acad. Publ., Dordrecht.

Farnsworth, D. (1957). Testing for Color Deficiency in Industry, AMA Arch. Industrial Health 16: 10–103.

Cole, B.L. (1993). Does defective colour vision really matter? Drum, B. (ed.), Colour Vision Deficiencies XI, Doc. Ophthalmol. Proc. Ser. 56: 67–86, Kluwer Acad. Publ., Dordrecht.

Shoko Tanabe
Japanese Red Cross Nagoya First Hospital
3-35 Michishitacho Nakamuraku Nagoya
Japan

56. Heterochromatic luminance matches in automated Rayleigh and Moreland equations

J. SOMMERHALDER, M. PELIZZONE, B. ROSSILLION and A. ROTH
(Geneva, Switzerland)

Abstract

We have developed an automated color vision examination consisting of Rayleigh and Moreland matches on a computer-controlled anomaloscope. With this system, it is possible to vary automatically the radiance (and saturation) of the test field when a given ratio of primaries is presented in the mixture field. The aim of these experiments was to test different methods of heterochromatic luminance matches so that a correct radiance of the test field could be used during the automated color vision examination itself.

Luminance matches were obtained on six normal subjects with two different methods: (1) by direct comparison of the two fields and (2) by the flicker method. Our computer-controlled anomaloscope was modified to allow both methods to be used on the same instrument. In the Moreland equation, luminance matches were easy to achieve. Both methods yielded very consistent results over the whole blue-green range. In the Rayleigh equation, luminance matches were more difficult to achieve. As expected, the flicker method was more precise than direct comparison, especially at the ends of the red-green range where differences in hue between the two fields are pronounced.

While both methods yielded consistent results, they have their specific strengths and weaknesses when considering them as preliminaries to an automated color vision examination.

Introduction

Color-normal subjects as well as patients with inherited or acquired color vision defects can be characterised by their performance on a simple test, called 'the two-equation method', combining the red-green Rayleigh equation with the blue-green Moreland equation (Roth, 1990; Roth *et al.*, 1989, 1991; Hermès *et al.*, 1989). Both equations use precisely defined primaries and yield accurate results. Pokorny and Smith (1984, 1990) demonstrated elegantly the power of graphical analysis of color matching equations and concluded 'Many of the

B. Drum (ed.), Colour Vision Deficiencies XII, pp. 473–480.
© 1995 *Kluwer Academic Publishers, Dordrecht.*

color matching abnormalities observed accompanying ophthalmic and systemic disease may be understood within the framework of plausible physiological mechanisms. Two equations allow the clinician to classify abnormalities of color matching according to the von Kries mechanisms of Absorption, Alteration and Reduction.' Thus, the 'two-equation method' is certainly one of the most powerful procedures to test color vision, but unfortunately its use has remained limited to a few specialised centres disposing of the required technology and know-how.

We have developed an automated color vision examination consisting of Rayleigh and Moreland equations on a computer-controlled anomaloscope (Pelizzone *et al.,* 1989, 1991, 1993a, b), now commercially available from Interzeag (Schlieren, Switzerland). Our present research work is focused on designing an optimal automated color vision examination that could be systematically applied to normals as well as to color-defective subjects as a routine procedure.

In clinical color matching equations, such as those mentioned above, only two parameters are varied: the ratio of primaries in the mixture field and the luminance of the test field. In an automated procedure, the control program will of course explore the whole range of ratios of primaries. But what is the correct matching luminance of test field for each ratio of primaries? A patient's colour vision and V_λ status is not known BEFORE testing. The aim of the experiments reported in this paper was to test different methods to achieve heterochromatic luminance matches between mixture and test fields, so that a correct radiance of the test field could be used during the automated color vision examination itself. Such a procedure will economise test-time by freeing the patient to consider 'luminosity' during subsequent 'hue-judgements' and reduce the incidence of biased colour-naming.

Methods

Heterochromatic luminance matches were obtained with two different methods: (1) by direct comparison of mixture and test fields and (2) by flickering mixture and test fields. Experiments were conducted on our laboratory computer-controlled anomaloscope (for details see Pelizzone *et al.,* 1991), which was slightly modified to allow both methods to be used on the same instrument. In its usual configuration this instrument uses a mirror to split the 2° observation area into two halves side-by-side. On one side the mixture of primary lights is projected, on the other the test light. For the present experiments, this (fixed) mirror was replaced by a 'rotating mirror' which alternated the mixture of primaries with the test light on the whole observation area. The 'rotating mirror' assembly consisted of a cruciform chopper blade with a highly reflecting surface, driven by a DC motor (ESCAP 23D21-216E) with a Hewlett Packard (HEDS5310 and HCTL-1000) optical encoding of position.

For direct heterochromatic luminance matches, the rotating mirror was placed precisely at a predetermined position to get the usual bipartite observation field and the subject was asked to adjust the radiance of the test field by using the mouse buttons until it was apparently 'equally luminous' to the mixture field. For flicker heterochromatic luminance matches, the rotating mirror was set for 15 Hz (Moreland equation), or 25 Hz (Rayleigh equation), and the subject was asked to adjust the radiance of the test field by using the mouse buttons until flicker was minimal.[1]

Six normal trichromats (4 females, 2 males) participated as subjects. They made heterochromatic luminance matches with both methods on the Rayleigh equation (670 nm + 546 nm = 589 nm) and on the Moreland equation (490 nm + 436 nm = 480 nm + 580 nm). Each run consisted of 9 luminance matches at 9 different ratios of primaries (10-90, 20-80, 30-70,...,90-10). At least, three runs were made on each subject and in each condition.

Results and discussion

Rayleigh equation – direct luminance matches

In the Rayleigh equation, normal subjects found at first direct heterochromatic luminance matches easy to perform, but they soon discovered that they had trouble achieving reproducible values. We observed very different individual results. One part of the subjects – the more experienced ones (e.g. subject JS in Fig. 1a) – could achieve quite reliable luminance matches over the whole red-green range. Inexperienced subjects (e.g. subject DD in Fig. 1a) were inconsistent and had problems in repeating the same luminance match, especially at the ends of the red-green range where they tended to compensate strong hue differences by increasing the radiance of the test field. This is reflected by the 'banana' shape of the luminance vs mixture curve as well as by substantial standard deviations in the measurements. They matched for example an almost pure red with a very bright yellow. All subjects made their best luminance matches in the neighbourhood of the color match, i.e. when hue differences were minimal.

Figure 2a presents group results on 6 normal subjects. The large variability in individual measurements is reflected in the large standard deviations of the group mean values (Fig. 2a). As noted above, the best agreement is reached in the neighbourhood of the color match. The calculated values for a luminance match, predicted from photo-current measurements on the observation screen and V_λ, are systematically lower than the mean subject values.

In the Rayleigh equation, normal subjects can use direct heterochromatic luminance matches to estimate the correct radiance of the test field in the neighbourhood of the color match, where it is important for the color vision test. We found it helpful to instruct the subjects first to turn down completely

[1] The two flicker frequencies were used for both equations. We have retained the one which yielded the most precise matches.

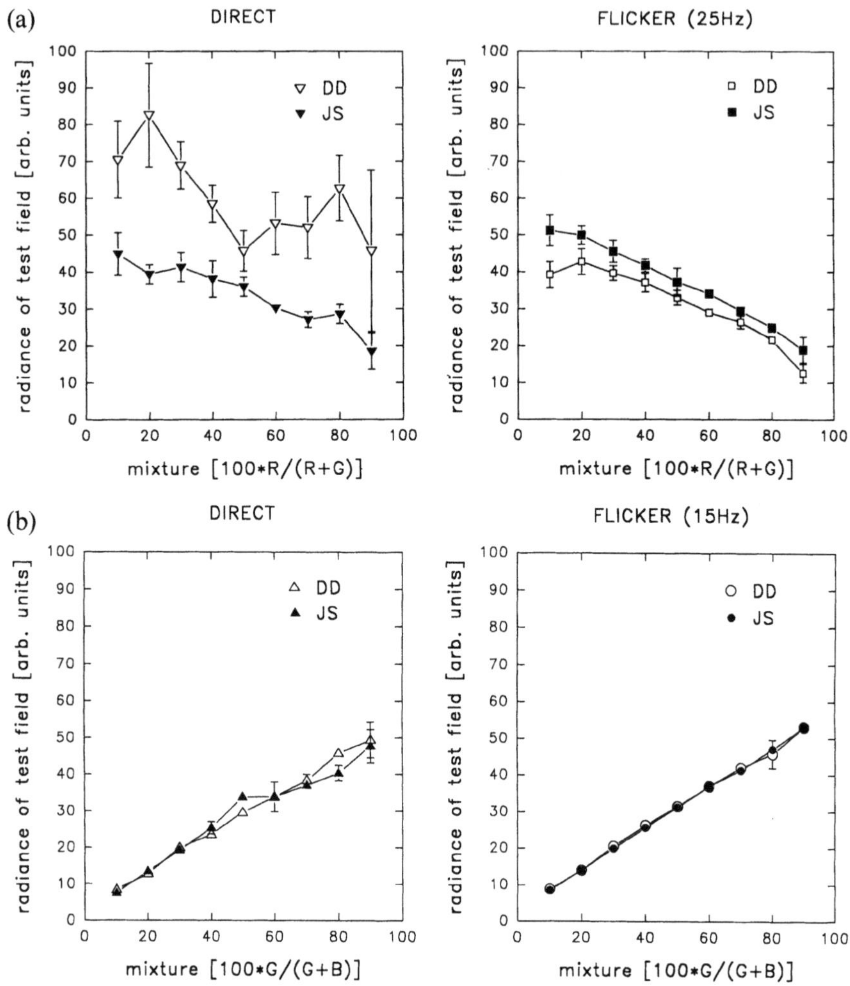

Fig. 1. Direct and flicker heterochromatic luminance matches between test and mixture fields for individual subjects. (a) Rayleigh equation. (b) Moreland equation. Symbols represent mean values. Error bars ± 1SD.

the radiance of the test field and then progressively to increase it, and to stop as soon as both fields are apparently 'equally luminous'. With this procedure, one gets a minimal value of the luminance of the test field for direct luminance matches, which is usually more reproducible and in better agreement with the calculated expectations, particularly with strong hue differences between the mixture and test fields. Patients like red-green dichromats have no problem with direct heterochromatic luminance matches in the Rayleigh equation (because they are not disturbed by hue differences) and so their luminance values can be safely used during the color vision test.

Rayleigh equation – flicker luminance matches

In the Rayleigh equation, normal subjects found that flicker luminance matches were more difficult and less intuitive than direct luminance matches. It took them a while to experience the perception of minimum flicker. Once they were used to the task, normal subjects achieved very reliable luminance matches over the whole red-green range. Two subjects who showed large individual differences with direct luminance matches performed quite similarly with flicker luminance matches (Fig. 1a). Note also the smaller standard deviations when compared to direct luminance matches.

Figure 2a presents group results on 6 normal subjects. All normal subjects made very similar flicker luminance matches in the whole red-green range. This is reflected by the small standard deviations of the group data. The calculated values for a luminance match, predicted from photo-current measurements on the observation screen and V_λ, are in excellent agreement with the group mean values. This is not surprising since V_λ values were determined using flicker luminance matches.

In the Rayleigh equation, flicker luminance matches would be the better way to determine the correct luminance of the test field, but the task of flicker luminance matches proved to be inappropriate for routine clinical use since training is required for reproducible results.

Moreland equation, direct luminance matches

In the Moreland equation, normal subjects found direct heterochromatic luminance matches very easy to perform. They had no problem in achieving reproducible values. Individual results were very similar across subjects (Fig. 1b). Only minor inaccuracies were found towards the green end for some subjects.

Figure 2b presents group results on 6 normal subjects. The calculated values for a luminance match, predicted from photo current measurement on the observation screen and V_λ, are in excellent agreement with the group mean values.

In the Moreland equation, direct heterochromatic luminance matches are an excellent and easy way to determine the correct radiance of the test field. Hue differences are much less disturbing than in the Rayleigh equation.

Moreland equation, flicker luminance matches

In the Moreland equation, normal subjects confirmed that flicker luminance matches were more difficult to perform. They preferred the easier static luminance matches. Individual results were almost identical across subjects (Fig. 1b).

Figure 2b presents group results on 6 normal subjects. The calculated values for a luminance match, predicted from photo current measurement on the

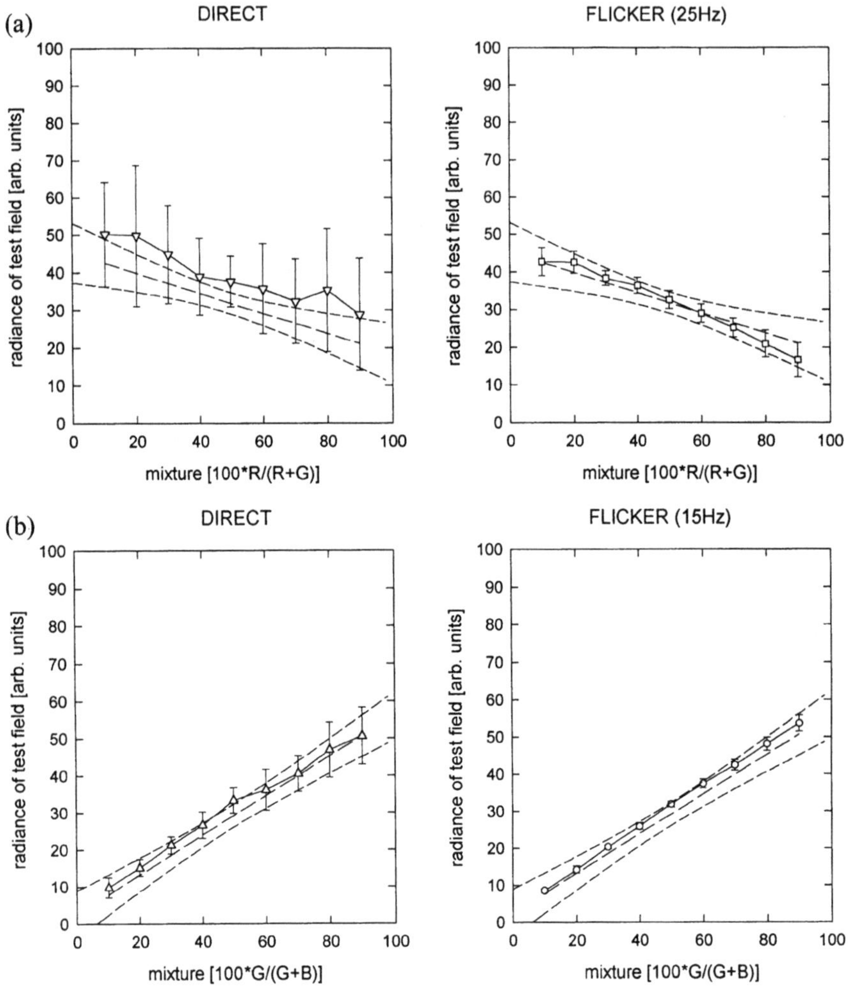

Fig. 2. Mean values on six normal observers for direct and flicker heterochromatic luminance matches between test and mixture fields. (a) Rayleigh equation. (b) Moreland equation. Symbols represent mean values. Error bears ± 1Sd. Dashed lines indicate calculated values ± estimated error predicted from photo current measurement on the observation screen and V_λ.

observation screen and V_λ, are in excellent agreement with the group mean values. For the whole blue-green range, the group mean data are within the estimated error limits of the calculated values.

In the Moreland equation, flicker luminance matches yield results that are even more reproducible than those obtained with direct luminance matches. However, the task of flicker luminance matches proved again to be inappropriate for routine clinical use.

Conclusion

Dircct and flicker heterochromatic luminance matches can both be used to determine the correct radiance of the test field as the ratio of primaries is varied in the mixture field. All subjects found direct luminance matches more easy to perform than flicker luminance matches. However, direct luminance matches led to incorrect luminance estimates at both ends of the red-green range of Rayleigh equation for some subjects. Flicker luminance matches were more reproducible than direct luminance matches, but the training required is not convenient for routine clinical use.

In conclusion, the optimal automated clinical color vision examination begins, for each individual and for each test, by a few direct luminance matches for a series of predetermined ratios of primaries. These luminance matches are used by the computer to set appropriate test field radiances during the automated color vision examination. The whole procedure takes a little more time, but guarantees an optimal result in measuring the true colour vision performance of the subject.

Acknowledgements

We thank Prof. Jack Moreland for many helpful comments on this work. This work was financially supported by the Swiss National Science Foundation (grant no. 32-33931.92).

References

Hermès, D., Roth, A. and Borot, N. (1989). The two equation method. II. Results in retinal and optic nerve disorders. In: Drum, B. and Verriest, G. (eds.), Colour Vision Deficiencies IX, Doc. Ophthalmol. Proc. Ser. 52: 325–337, Kluwer Acad. Publ., Dordrecht.

Pokorny, J. and Smith, V.C. (1984). Metameric matches relevant for assessment of color vision. I. Theoretical considerations. In: Verriest, G. (ed.), Colour Vision Deficiencies VII, Doc. Ophthalmol. Proc. Ser. 39: 83–94, Dr. W. Junk, The Hague.

Pokorny, J. and Smith, V.C. (1990). Color matching as a clinical tool. Theory of modification by disease. In: Ohta, Y. (ed.), Color Vision Deficiencies: 255–267, Kugler and Ghedini, Amsterdam.

Pelizzone, M., Roth, A. and Hermès, D. (1989). First trials on an automated examination of color vision. In: Drum, B. and Verriest, G. (eds.), Colour Vision Deficiencies IX, Doc. Ophthalmol. Proc. Ser. 52: 523–528, Kluwer Acad. Publ., Dordrecht.

Pelizzone, M., Sommerhalder, J. Roth, A. and Hermès, D. (1991). Automated Rayleigh and Moreland matches on a computer-controlled anomaloscope. In: Drum, B., Moreland, J. and Serra, A. (eds.), Colour Vision Deficiencies X, Doc. Ophthalmol. Proc. Ser. 54: 151–159, Kluwer Acad. Publ., Dordrecht.

Pelizzone, M., Sommerhalder, J., Roth, A. and Hermès, D. (1993a). Automated Rayleigh and Moreland matches. Optimization of stimulation parameters for normal observers. In: Drum, B. (ed.), Colour Vision Deficiencies XI, Doc. Ophthalmol. Proc. Ser. 56: 345–355, Kluwer Acad. Publ., Dordrecht.

Pelizzone, M., Sommerhalder, J. and Roth, A. (1993b). Détermination des seuils de discrimination dans les épreuves d'égalisation colorée. Ophtalmologie (in press).

Roth, A. (1990). The power of metameric color equations in testing color vision. In: Ohta, Y. (ed.), Color Vision Deficiencies: 181–190, Kugler and Ghedini, Amsterdam.

Roth, A., Pelizzone, M., Hermès, D. and Sommerhalder, J. (1989). Neuere Ueberlegungen und Entwicklungen zur klinischen Untersuchung des Farbensehens - die Zwei-Gleichungsmethode. Fortschr. Ophthalmol. 86: 374–379.

Roth, A., Pelizzone, M., Hermès, D. and Sommerhalder, J. (1990). L'examen de la vision des couleurs par la méthode des deux équations métamériques. Ophtalmologie 4: 197–205.

Roth, A., Pelizzone, M., Sommerhalder, J., Hermès, D. and Simonda, F. (1991). The two equation method: III. Results in normal subjects above 50 years of age. Correlation with lens opacity. In: Drum, B., Moreland, J. and Serra, A. (eds.), Colour Vision Deficiencies X, Doc. Ophthalmol. Proc. Ser. 54: 353–359, Kluwer Acad. Publ., Dordrecht.

Marco Pelizzone
Clinique d'Ophtalmologie
Hôpital Cantonal Universitaire
CH - 1211 Genève 14, Switzerland

57. Automated Moreland equations on 7° and 2° fields

B. ROSSILLION, M. PELIZZONE, J. SOMMERHALDER and A. ROTH
(Geneva, Switzerland)

Abstract

We tested automated Moreland equations with 7° and 2° fields on a population of normal subjects. Before each test, we determined by direct comparison the radiance of the test field for luminance matches with different ratios of primaries in the mixture field. This relationship was used in the subsequent color vision examination to set automatically the correct radiance of the test field. The task appeared to be notably easier with the large field. Accordingly, the matching ranges of the Moreland equation were significantly smaller with the 7° than with the 2° field. A learning effect can be ruled out because the large field was systematically used first. The population distribution of the match mid-points was also narrower with the large field.

Introduction

We have developed an automated color vision examination consisting of Rayleigh and Moreland equations on a computer-controlled anomaloscope (Pelizzone *et al.*, 1989, 1991, 1993b), now commercially available from Interzeag (Schlieren, Switzerland). Our present research work is focused on designing an optimal color vision examination, based on the 'two-equation method' (Roth, 1990; Roth *et al.*, 1989, 1990, 1991; Hermès *et al.*, 1989), that could be used as a routine clinical procedure. To achieve this goal, one has to know the influence of several stimulation parameters (Pelizzone *et al.*, 1993a) and one has to determine a set of optimal parameters that can be recommended as a 'good first choice' to begin with.

Field size is of course a very important parameter in testing color vision. In the Rayleigh equation, the preferred field size is 2° since the pioneering work of Francheschetti (1928) on the clinical classification of Daltonism.[1] The same

[1] Variations of field size in the Rayleigh equation can however be an important help in clinical populations (Pokorny and Smith, 1984).

B. Drum (ed.), Colour Vision Deficiencies XII, pp. 481–488.
© 1995 *Kluwer Academic Publishers, Dordrecht.*

field size might not be optimal for the Moreland equation. According to our experience most subjects report spontaneously that the Moreland equation with a 2° field is much more difficult than the Rayleigh equation. The first time they are tested, they give hesitant blue-green responses around the match region, resulting in artificially large matching ranges. The matching ranges of these subjects usually converge towards smaller and stable values after a few repetitions of the test, thus demonstrating a learning effect. Interestingly, the same subjects perform Rayleigh equation with a 2° field without problem, showing no learning effect. Blue cones are scarce compared to red and green cones in the central retina and the Moreland equation with a 2° field might be a 'too difficult' psychophysical task for the average color-normal observer. Ease of use being essential for a routine clinical test, we investigated the merits of a large field in the Moreland equation.

Moreland (1978) made a population study on his custom anomaloscope with the Moreland equation with field sizes of 2 and 11° which demonstrated that the matching ranges were systematically smaller with the larger field. The aim of the experiments reported in this paper was to achieve a new population study that would establish the influence of the field size on the results of the automated Moreland equation on our computer-controlled anomaloscope. All the subjects of the population study were also tested on Rayleigh equation so that we could determine statistically the limits of normality for the automated 'two equation method' on our computer-controlled anomaloscope.

Methods

In all the experiments reported here, the field size was varied by changing circular stops. The viewing distance was kept constant.

In a preliminary study, we tested 17 colour-normal observers (age range 22–63 years). Each observer performed first three automated Moreland equations with the 7° field followed by three automated Moreland equations with the 2° field. Before each test, the subject made direct heterochromatic luminance matches between the mixture and test fields for different ratios of primaries in the mixture field (Sommerhalder *et al.*, Chapter 56). These data were used in the subsequent color vision test to set automatically the radiance of the test field.

For the population study, we tested 100 young normal subjects (49 females, 51 males; age range 19–42). Each subject made first an automated Rayleigh equation with the 2° field, then an automated Moreland equation with the 7° field and finally an automated Moreland equation with the 2° field. Before each test, direct heterochromatic luminance matches between the mixture and test field were performed. Only one eye per subject was tested.

Results and discussion

Preliminary study

All 17 subjects reported that the test with the larger field was much easier and accordingly, the matching ranges were systematically smaller with the larger field (Fig. 1). For 15 out of 17 subjects, the mean Matching Range (MR) with the 7° field was significantly smaller ($t = 6.0$, $p < 0.001$) than the mean MR with the 2° field, in spite of the fact that any learning effect would favour the smaller field. Interestingly, MRs with both field sizes were significantly correlated ($r = 0.82$, $p < 0.001$ in Fig. 1). This implies that subjects having small MRs retain small MRs with both fields while subjects having large MRs retain large MRs with both fields. Therefore the individual specificity of the test is not lost although the test was judged to be much easier with the large field. These results convinced us it was worth undertaking a population study using both field sizes on the Moreland equation.

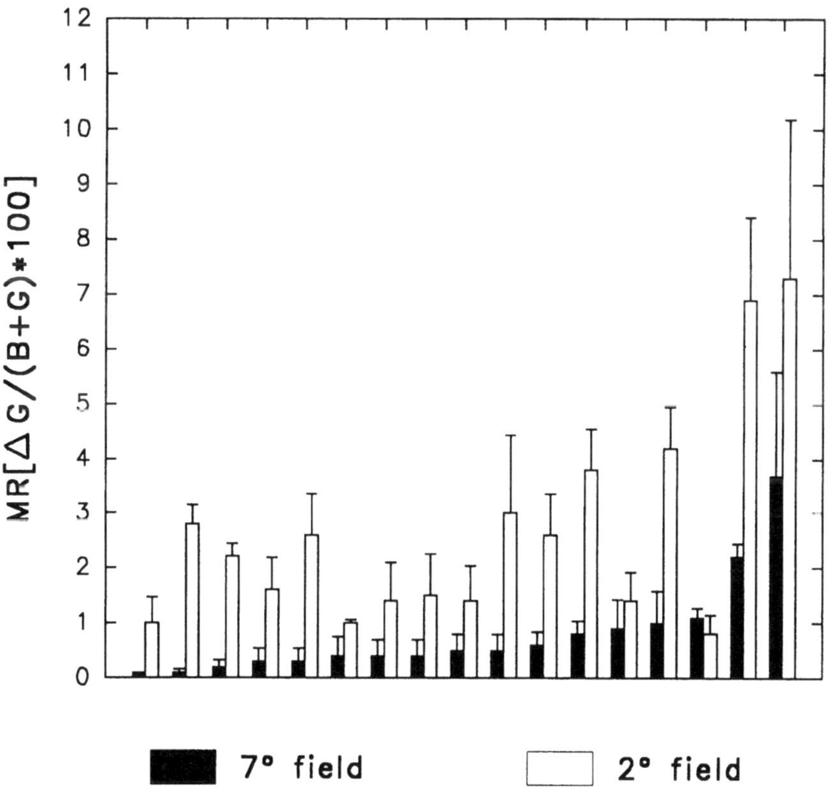

Fig. 1. Preliminary study. Comparison of the Matching Ranges (MR) of the automated Moreland equation with 7° (black bars) and 2° (white bars) observation fields. Bars represent the mean values of 3 measurements on each subject. Error bars represent standard deviations of the 3 measurements.

Population study

Histograms[2] of MRs measured on 100 normal subjects for automated Moreland equations with fields of 7° and 2° are presented in Fig. 2. Log-normal distributions, i.e. distributions that are not symmetrical around their most frequent value, were chosen as best fits to the MR data. The two distributions are significantly different ($t = 5.6$, $p<0.0001$), thus confirming that MRs with the larger field are significantly smaller. Defining the limits of normality as the

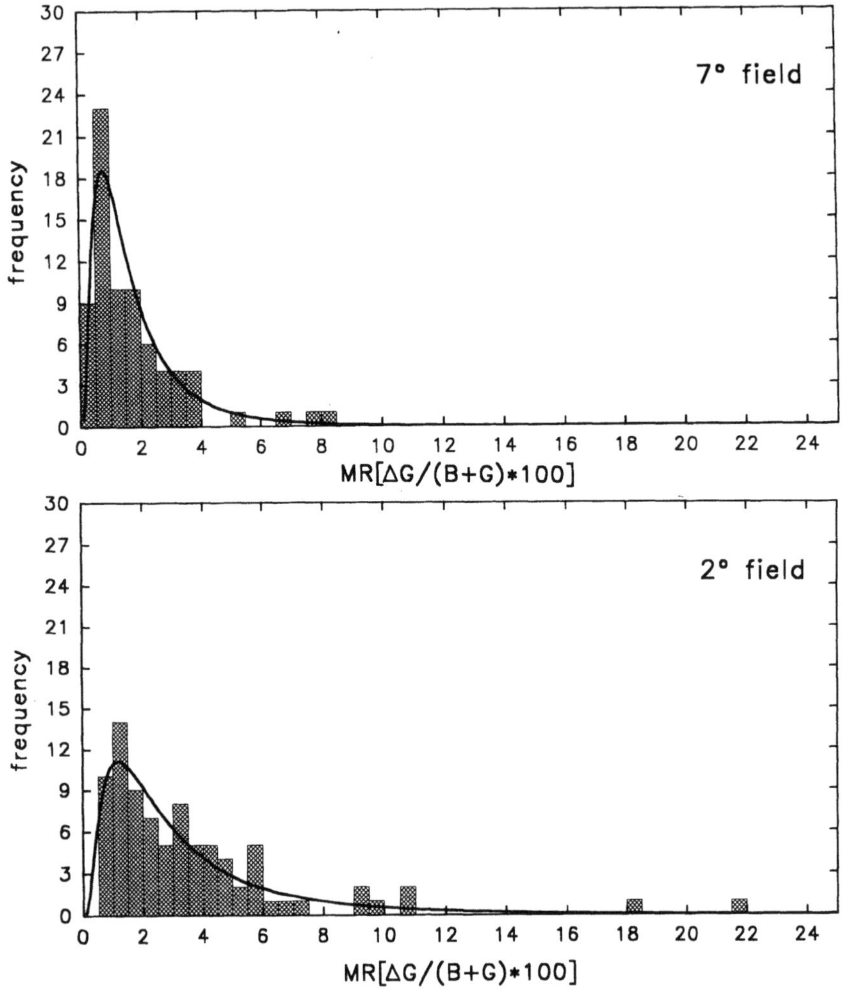

Fig. 2. Population study. Distribution of Matching Ranges (MR) on 100 young color-normal observers for the automated Moreland equation with 7° and 2° observation fields.

[2] The finite size of the smallest steps in the iterative algorithm determining the matching range resulted in artifactual clustering of data for extremely small matching ranges. We discarded therefore these data when fitting a log-normal distribution to the measurements.

95% of the corresponding distribution, we get MR $\leq$ 5.1 units on the G/(B+G)*100 mixture scale for the 7° field and, respectively, MR $\leq$ 11 units for the 2° field. This means that color discrimination around the 480 nm desaturated cyan is better with the larger field. We interpret this observation as a consequence of increasing the number of receptors (especially blue receptors) involved when increasing by more than one order of magnitude the retinal surface used for the test. Finally, for each individual subject, MRs with both fields are also significantly correlated ($r = 0.47$, $p = 0.001$), thus confirming that the individual specificity of the test is not lost by increasing the field size.

Histograms of Mid-Match Points (MMP) measured on 100 normal subjects for the automated Moreland equation with 7° and 2° fields are presented in Fig. 3. The distribution being symmetrical around their most frequent values,

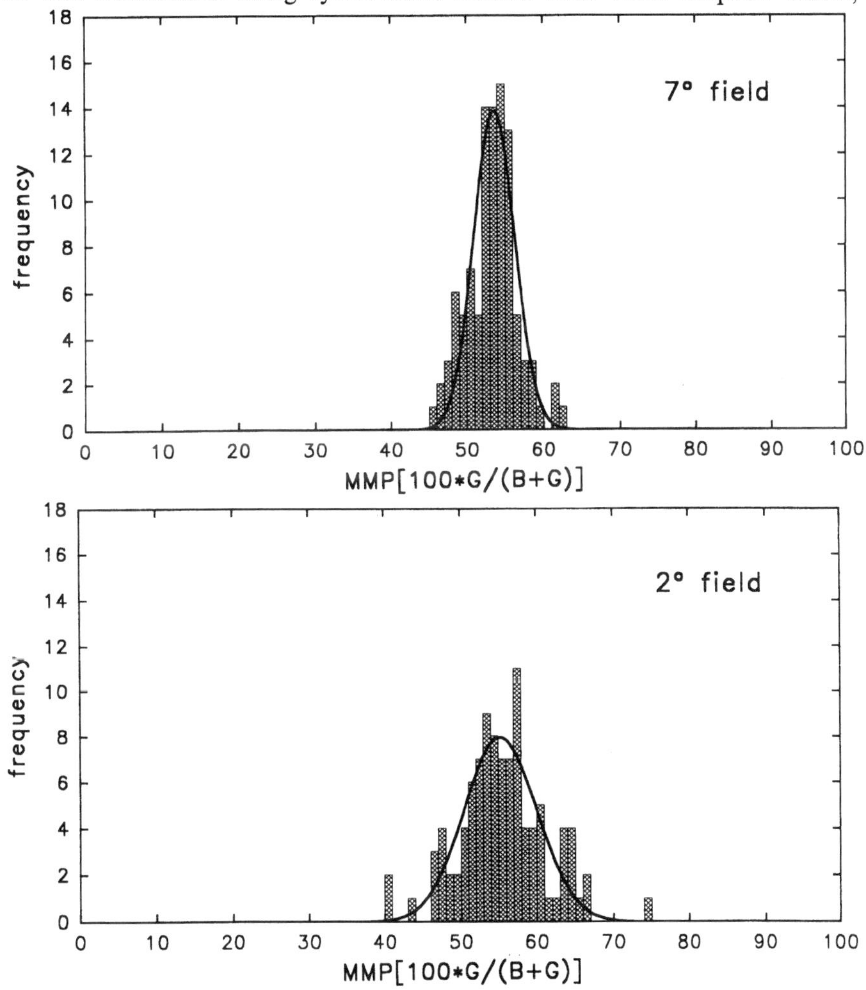

Fig. 3. Population study. Distribution of Mid-Match Points (MMP) on 100 young color-normal observers for the automated Moreland equation with 7° and 2° observation fields.

normal distributions were chosen as best fits to the MMP data. The mean MMP for the Moreland equation is 53.6 units on the G/(G+B)*mixture scale for a 7° field and 55.2 for the 2° field. These values are significantly different ($t = 5.2$, $p < 0.0001$), thus indicating that a color match with the smaller field requires significantly more green primary than with the larger field. The interpretation of this observation can be attributed to the distribution of the macular pigment on the retina. With the choice of our primaries for the Moreland equation, tabulations of the absorbance spectrum of the macular pigment reveal that our 490 nm-green primary is slightly more absorbed than our 436 nm-blue primary (relative transmittance 0.3890 at 490 nm and 0.4131 at 436 nm; Table X (2.4.6), p. 721, in Wyszecki and Stiles, 1982). The optical density of the macular pigment being maximal at the centre of the retina and decreasing progressively towards the periphery, the average influence of the macular pigment is stronger with the smaller field and, as a consequence, a match with the smaller field requires more green primary. It is interesting to compare these data with those reported by Zaidi *et al.*, (1989), measuring Moreland matches on 9 color-normal observers with 2° and 8° fields. These investigators used a 500 nm-green primary and blue primaries ranging from 410 nm to 470 nm. As expected from macular pigment absorbances for the primaries they used, they observed that the smaller field required more of the blue primary. Both sets of data confirm that the choice of the exact wavelength of the primaries is critical in the Moreland equation and that the macular pigment really affects color-matching data substantially.[3]

The distribution of MMPs is significantly narrower for the 7° field than for the 2° field ($F = 3.16$, $p < 0.01$). This can be explained as a consequence of two factors: (1) the match with the 7° field is easier and therefore less variable and (2) the influence of the individual variability of the macular pigment is reduced with the larger field. Finally, for each individual subject, MMPs with both fields are also significantly correlated ($r = 0.66$, $p < 0.001$), thus confirming that the individual specificity of the test is not lost by increasing the field size.

On the basis of this population study, we defined the limits of normality for automated Rayleigh and Moreland equations on our computer-controlled anomaloscope as the 95% limits of the distributions. These values are summarised in table 1.

Conclusion

For the average-color normal observer, the Moreland equation is a 'more easy' psychophysical task with a 7° field than with a 2° field. In our population study on 100 young color-normal observers, this resulted in narrower population

[3] It should be noted that the choice of our primaries (436 nm-blue and 490 nm-green) for the Moreland equation results from careful studies designed to minimise combined population variance contributions of macular *and* lens pigment *and* of matching range (see Moreland, 1990 and references therein). The observed shift is thus a consequence of this choice.

Table 1. Limits of normality for automated Rayleigh and Moreland equations.

Moreland equation:	
2°	7°
$45.7 \leq MMP \leq 64.8$	$48.2 \leq MMP \leq 59$
$MR \leq 11$	$MR \leq 5.1$

Rayleigh equation:
2°
$40.0 \leq MMP \leq 55.2$
$MR \leq 3.3$

MMP: Mid-Match Point; MR: Matching Range

distributions of the MRs and MMPs with the larger field. The Moreland equation with a 7° field yields results in which the population variability is comparable to that of the Rayleigh equation with a 2° field.

In the future we will evaluate the 'clinical sensitivity' of the Moreland equation with 7°. One possible measure of clinical sensitivity is to express the results of a given patient relative to that of a population of color-normal observers and to choose the correct statistical metrics.

Acknowledgements

We thank Prof. Jack Moreland for numerous helpful comments on this work. This work was financially supported by the Swiss Science Foundation (grant no. 32-33931.92).

References

Franceschetti, A. (1928). Die Bedeutung der Einstellungsbreite am Anomaloskop für die Diagnose der einzelnen Typen der Farbensinnstörungen: nebst Bemerkungen über ihren Vererbungsmodus. Schw. Med. Wschr. 52: 1273–1278.

Hermès, D., Roth, A. and Borot, N. (1989). The two equation method. II. Results in retinal and optic nerve disorders. In: Drum, B. and Verriest, G. (eds.), Colour Vision Deficiencies IX, Doc. Ophthalmol. Proc. Ser. 52: 325–337, Kluwer Acad. Publ., Dordrecht.

Moreland, J.D. (1978). Optimisation of stimuli for tritanomaloscopy. Mod. Probl. Ophthalmol. 19: 162–166.

Moreland, J.D. (1990). The clinical utility of anomaloscopy. In: Ohta, Y. (ed.), Color Vision Deficiencies: 127–143, Kugler and Ghedini, Amsterdam.

Pelizzone, M., Roth, A. and Hermès, D. (1989). First trials on an automated examination of color vision. In: Drum, B. and Verriest, G. (eds.), Colour Vision Deficiencies IX, Doc. Ophthalmol. Proc. Ser. 52: 523–528, Kluwer Acad. Publ., Dordrecht.

Pelizzone, M., Sommerhalder, J., Roth, A. and Hermès, D. (1991). Automated Rayleigh and Moreland matches on a computer-controlled anomaloscope. In: Drum, B., Moreland, J. and

Serra, A. (eds.), Colour Vision Deficiencies X, Doc. Ophthalmol. Proc. Ser. 54: 151–159, Kluwer Acad. Publ., Dordrecht.
Pelizzone, M., Sommerhalder, J., Roth, A. and Hermès, D. (1993a). Automated Rayleigh and Moreland matches. Optimization of stimulation parameters for normal observers. In: Drum, B. (ed.), Colour Vision Deficiencies XI, Doc. Ophthalmol. Proc. Ser. 56: 345–355, Kluwer Acad. Publ., Dordrecht.
Pelizzone, M., Sommerhalder, J. and Roth, A. (1993b). Détermination des seuils de discrimination dans les épreuves d'égalisation colorée. Ophtalmologie (in press).
Pokorny, J. and Smith, V.C. (1984). Metameric matches relevant for assessment of color vision. I. Theoretical considerations. In: Verriest, G. (ed.), Colour Vision Deficiencies VII, Doc. Ophthalmol. Proc. Ser. 39: 83–94, Dr. W. Junk, The Hague.
Roth, A. (1990). The power of metameric color equations in testing color vision. In: Ohta, Y. (ed.), Color Vision Deficiencies: 181–190, Kugler and Ghedini, Amsterdam.
Roth, A., Pelizzone, M., Hermès, D. and Sommerhalder, J. (1989). Neuere Ueberlegungen und Entwicklungen zur klinischen Untersuchung des Farbensehens – die Zwei-Gleichungsmethode. Fortschr. Ophthalmol. 86: 374–379.
Roth, A., Pelizzone, M. Hermès, D. and Sommerhalder, J. (1990). L'examen de la vision des couleurs par la méthode des deux équations métamériques. Ophtalmologie 4: 197–205.
Roth, A., Pelizzone, M., Sommerhalder, J., Hermès, D. and Simona, F. (1991). The two equation method: III. Results in normal subjects above 50 years of age. Correlation with lens opacity. In: Drum, B., Moreland, J. and Serra, A. (eds.), Colour Vision Deficiencies X, Doc. Ophthalmol. Proc. Ser. 54: 353–359, Kluwer Acad. Publ., Dordrecht.
Sommerhalder, J., Pelizzone, M., Rossillion, B. and Roth, A. (1994). Heterochromatic luminance matches in automated Rayleigh and Moreland equations. In: Drum, B. (ed.), Colour Vision Deficiencies XII, Doc. Ophthalmol. Proc. Ser. 57: 473–480, Kluwer Acad. Publ., Dordrecht (this volume).
Wyszecki, G. and Stiles, W.S. (1982. Color Science. John Wiley and Sons, New York.
Zaidi, Q., Pokorny, J. and Smith, V.C. (1989). Sources of individual differences in anomaloscope equations for tritan defects. Clin. Vis. Sci. 4: 89–94.

Marco Pelizzone
Clinique d'Ophtalmologie
Hôpital Cantonal Universitaire
CH - 1211 Genève 14, Switzerland.

58. Effect of lamp voltage on Nagel anomaloscope settings

C.R. CAVONIUS
(Dortmund, Germany)

Abstract

While it has long been known that Nagel anomaloscope matches are influenced by the mains voltage (e.g. Schmidt, 1955) there seem to be no reports of the magnitude of this effect. Because of a recent increase in the nominal mains voltage in Germany from 220 to 230 volts, with a permitted range of 207 to 253 volts in the future, it seemed prudent to see whether this could influence routine tests of color vision.

Between 190 and 250 volts a normal trichromat's R-G settings increase by approximately 0.05 Nagel units/volt, in a highly linear manner ($r^2 = 0.98$). Thus, a 10-volt increase will raise the R-G setting by $\approx$ 0.5; and a change from 207 to 253 volts corresponds to just over two R-G units. The accompanying change in the brightness setting (–0.0056 units/volt) is trivial. While it is unlikely that these variations would lead to a clinical misdiagnosis, they could result in an artifactual bimodality in Nagel settings when comparing populations, especially if different populations are tested at different sites.

Introduction

The nominal mains voltage in Germany was recently raised from 220 V to 230 V, with a permitted range of 207 V – 244 V until the year 2003, and 207 V – 253 V thereafter. Since the color temperature of the light source influences matches on the Nagel anomaloscope (indeed, this was among the considerations that prompted Trendelenberg to advocate the Anomalquotient in 1929), the question arises as to whether a change in voltage of this magnitude would significantly influence Nagel matches.

The Nagel model I presents a bipartite field, one half of which (in modern instruments) contains a mixture 546 nm and 671 nm light. The subject adjusts the relative proportions of these lights so as to match the hue of a 589 nm light in the other half-field. If other factors are held constant, the wavelengths of these stimuli are established by the geometry of the instrument and cannot be

B. Drum (ed.), Colour Vision Deficiencies XII, pp. 489–494.
© 1995 *Kluwer Academic Publishers, Dordrecht.*

influenced by the color temperature of the source. However, their relative radiances can be changed: increasing color temperature increases the radiance of the green stimulus to that of the red. (Of course, the yellow also changes, but it will be seen that this has virtually no effect upon the luminance equation of the two half-fields.) Since the empirical relationship between lamp voltage and color temperature is known (Wensel *et al.,* 1934), it is possible in principle to calculate the change in the relative radiant emittances of the primaries as the voltage changes. Despite this, it seemed prudent to make direct observations of the effect of changing voltage, because the empirical voltage-temperature relationship was established for a certain type of lamp: for any other lamp one should know in addition the actual color temperature at two voltages. In retrospect, any errors that this and other approximations might introduce into the calculation are probably trivial, but this was not obvious *a priori.*

Method

Measurements were made with a Schmidt & Haensch Nagel Model I, constructed in 1958. Since the optics of the Model I (Nagel, 1907) have remained essentially unchanged for the better part of a century, the results should apply to other examples of this anomaloscope. The lamp housing was different from the current model, but the lamp, a 220 V, 100 W cine-projection lamp with 4 nearly-vertical spiral filaments (AVI, pp-001), is the same as those that Schmidt & Haensch currently supply. The lamp was connected to the mains through a variable transformer so that its voltage could be adjusted between 160 V and 250 V, and the voltage was monitored on a precision analog meter. Because of the known sensitivity of Nagel R-G settings to temperature (Jordan and Mollon, 1993) the air temperature next to the prism housing was maintained within the range of 22.6° C – 22.8° C by adjusting a window in the laboratory. The approximate chromaticity of the light from the ground-glass screen between the Nagel lamp and the entrance slits was measured at various voltages with a Minolta CS-100 Chroma Meter. With the R-G and brightness slits at their 'normal' settings of 40 and 15, respectively, the green, yellow, and red lines peaked at 546, 587.5, and 666.5 nm, estimated by viewing through a double monochromator, which in turn had been checked at the 546 and 577 Hg and the 644 Cd lines.

In order to see whether a very different type of tungsten source would give different results, anomaloscope matches were also made at several voltages with light from a domestic 60 W frosted bulb, reflected into the entrance slits of the Nagel.

Results

The means and standard deviations of R-G settings of a normal trichromat are shown by the filled circles and associated error bars in Figure 1. Values from 160 V to 250 V are based on 20 settings each. The solid line is the regression (r^2 = 0.98), calculated for the data from 190 V to 250 V. Although the mean at 260 V appears to be consistent with those at lower voltages, it was not used in the calculations because it is based on only 5 settings, made late on the afternoon of December 31, when the laboratory building was deserted; on subsequent days the maximum output of the variable transformer never reached 260 V. Below 190 V the matches were never completely satisfactory, although it is difficult to describe the subjective difference: it was as if the upper (i.e. the mixture) field were slightly desaturated. Between 190 V and 250 V (which is a larger variation than one is apt to encounter in practice) the data have a slope of 0.047. Thus, a 50-volt change in line voltage changes the R-G setting by 2.3 Nagel scale units. Since this is less than the gap that is usually reported between normal and

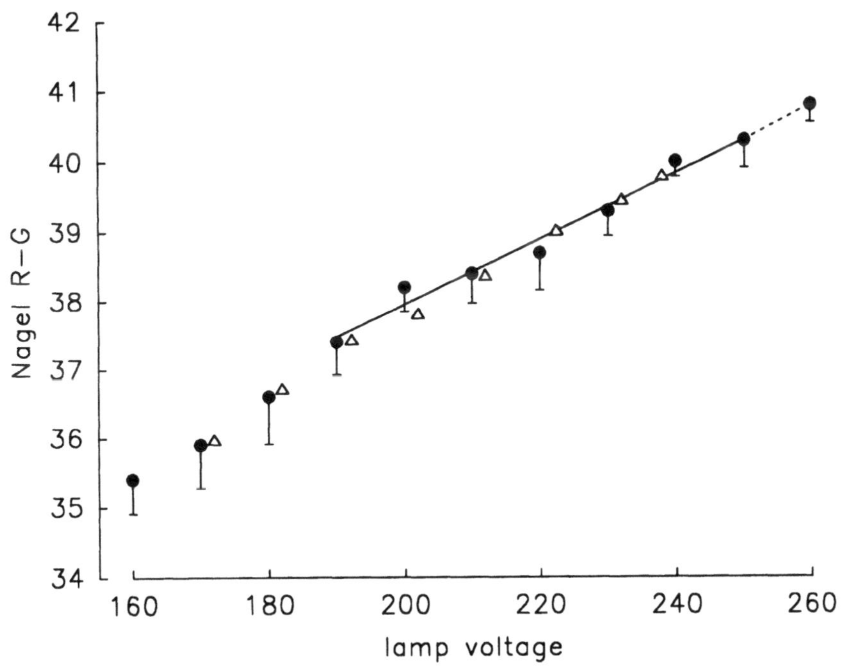

Fig. 1. Nagel R-G settings of two normal trichromats, showing the effect of changing lamp voltage. Circles and line: results of the present study; error bars show 1 SD. The regression line is calculated for the data from 190 V to 250 V. Triangles: data from Young (1974), adjusted as described in the text, and plotted 2 V to the right of their actual position for the sake of clarity.

anomalous settings, it is unlikely that even so extreme a change in voltage would lead to an error in clinical diagnosis.[1]

Brightness settings of the 589-nm field are scarcely affected by voltage: the slope of the function relating these settings to voltage is –0.0056. This means that a 50-volt increase will cause a 0.28-unit decrease in brightness settings, which would be completely lost in the variance of individual settings.

After making these measurements, I learned from Professor Moreland that the very first Canadian M.Sc. thesis in physiological optics (Young, 1974) contained similar data. Young's results are shown as triangles in Fig. 1. In order to compare the two sets of data, Young's voltages were doubled to compensate for the North American 110 V standard. After doing this, Young's results were on average 0.9 Nagel units above the present ones, and were moved down by this amount. This adjustment is of course completely arbitrary, but it is the slope that is of interest here, and not the absolute values. With this adjustment, Young's data resemble closely those of the present study, including the change of slope below 190 V. (A difference of 0.9 units between the two sets of data could simply be due to the fact that they were obtained on different instruments – Heinsius (1973) describes two Nagel anomaloscopes that differed by 4 units – or to the even larger range that may separate normal trichromats.)

The results (not shown) of using a conventional 60 W bulb as the light source also have a slope similar to that shown in Fig. 1; thus, even antique instruments, for which the user supplies the source, should behave in a similar manner.

The change in chromaticity as lamp voltage is varied between 250 V and 150 V is shown in Fig. 2. The relatively small change, at an oblique angle to the spectrum locus, agrees with the small change in R-G settings. The inset shows that a linear regression describes satisfactorily the data between 190 and 250 V; the fact that below 190 V the data drop away from this line agrees with the more rapid drop in R-G settings in this region (see Fig. 1), as the loci of the chromaticities become more nearly parallel to the spectrum locus. The apparent discrepancy between the data and the locus of a blackbody radiator is in large measure due to the greatly expanded y axis: the largest separation between the two functions is less than 0.002 y units.

Discussion

The practical message in these results is that Nagel settings are remarkably robust in the presence of large voltage swings, so that in most applications the latter can be ignored. If a research application demands high reproducibility –

[1] In the worst case – that of the gap that separates normal and protanomalous observers – the Fachnormenausschuß Farbe specifies limits of 0.07 and 0.6, respectively, for the limits of their Anomalquotienten (D.I.N., 1957). Taking 40 as representative of normal R-G settings, this corresponds to a range of 2.5 scale units, which is uncomfortably close to the 50-volt variability. However, a prudent examiner would in any case do further tests on an individual whose Nagel setting fell in this region.

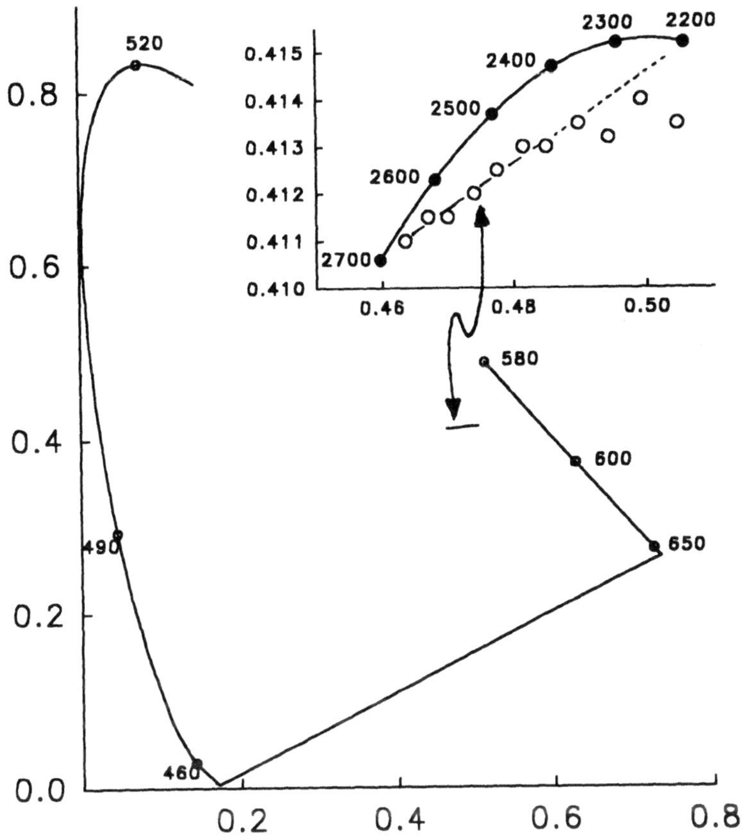

Fig. 2. Loci of the chromaticity of the Nagel source (lamp and diffusing glass) as voltage is changed from 250 V (left) to 150 V (right). Inset shows the same loci, greatly expanded, at 10-volt intervals, as well as those of a blackbody radiator at the temperatures shown. Note that the y axis is expanded 6 times as much as the x axis.

if, for example, a comparison is to be made between two populations measured at different locations – a voltage stabilizer or a variable transformer and meter should be considered. If in field studies these solutions prove to be inconvenient, voltage readings should be taken so that in case of a voltage deviation the measured R-G values can be adjusted back to the laboratory's standard voltage by adding one-half a scale unit to the measured values for each 10-volt drop in the mains voltage. Jordan and Mollon (1993) have shown that the Nagel can be used as a thermometer. The present results extend its versatility to measurements of voltage, although the economics of this practice does not recommend itself.

Acknowledgement

Supported by the Deutsche Forschungsgemeinschaft/British Council Project 313-ARC-VI-92.
I wish to thank J.D. Moreland and J.M. Pokorny for helpful suggestions.

References

D.I.N. (1957). Fachnormenausschuß Farbe. Anweisung für die Untersuchung des Farbensehens mit dem Anomaloscope nach Nagel. Die Farbe 6: 115–119.
Heinsius, E. (1973). Farbsinnstörungen und ihre Prüfung in der Praxis. Enke, Stuttgart.
Jordan, G. and Mollon, J.D. (1993. The Nagel anomaloscope and a seasonal variation of colour vision. Nature 363: 546–549.
Nagel, W. (1907). Zwei Apparate für die augenärtzliche Funktionsprüfung. Z. Augenheilk. 17: 201–222.
Schmidt, I. (1955). Some problems related to testing color vision on the Nagel anomaloscope. J. Opt. Soc. Am. 45: 514–521.
Trendelenburg, W. (1929). Zur Diagnostik des abnormen Farbensinnes. Klin. Mbl. Augenheilk. 83: 721–732.
Wensel, H.T., Judd, D.B. and Roeser, W.F. (1934). Establishment of a scale of color temperature. Bur. Std. J. Res. 12: 527–536.
Wyszecki, G. and Stiles, W.S. (1967). Color Science. John Wiley, New York.
Young, W.B. (1974). Some temporal aspects of colour vision and a pilot study with a new interference filter anomaloscope. Unpublished M. Sc. thesis. University of Waterloo, Canada.

C.R. Cavonius
Institut für Arbeitsphysiologie
Abteilung Sinnes- und Neurophysiologie
Ardeystraße 67
D-44139 Dortmund
Germany

59. Design and use of the Holmgren Wool test

JENNIFER BIRCH and NITA PATEL
(London, U.K.)

Abstract

The Holmgren Wool test was commissioned by the Swedish railway authority in 1875 and became a standard test for recruits in many national railway systems. Several different versions of the test have been produced and the test is still commercially available. A new test, purchased in 1990, has 49 wool skeins (3 test skeins and 46 comparison skeins) compared with an older test available in our laboratory having 55 skeins (3 test skeins and 52 comparison skeins). The new test has different coloured wools and a different method of examination is introduced which reduces the number of skeins compared with each test sample. We have evaluated the design of both tests by matching the wools with Munsell papers and obtaining the C.I.E. chromaticity co-ordinates, for source C, from published tables. These data have been compared with established isochromatic zones for protan and deutan colour deficiency. The results show that neither test is correctly designed and that neither of these versions of the Holmgren Wool test has potential for identifying red-green colour deficiency. This conclusion is illustrated by results obtained for six colour deficient observers and three normal trichromats.

Introduction

The Holmgren Wool test was commissioned by the Swedish railway authority in 1875 following a railway accident near the town of Lagerlunda in November of that year. Although colour deficiency could not be proved to be the cause of the accident, the disaster emphasised the need for colour vision screening of railway personnel. The Holmgren Wool test was subsequently accepted as a standard test by a number of national railway systems and was adopted by the British Board of Trade in 1895. Several different versions of the test have been produced and the test is still commercially available. The original test contained 125 wool skeins but modern versions of the test have about 50 wools. Each test contains three test skeins which are yellow-green, pink (rose) and dark red. Each

B. Drum (ed.), Colour Vision Deficiencies XII, pp. 495–500.
© 1995 *Kluwer Academic Publishers, Dordrecht.*

test skein is presented in turn and the task is to select skeins which resemble it in colour, but are either lighter or darker in shade, from the remaining wools. Each wool skein is attached to a disk having a numerical code which indicates to the examiner that the selection is either correct (normal colour vision) or incorrect (defective colour vision).

The accuracy of the original Holmgren test was brought into question almost immediately when it was found that only 1% of male recruits failed the test compared with other surveys showing a prevalence of more than 6% red-green colour deficiency (Topley, 1959). The shortcomings of the test were also publicly exposed in the celebrated case of the seaman John Trattles (Boltz, 1952; Cole and Vingrys, 1982). In the course of his application for a First Mate certificate, Trattles took the Holmgren test six times between 1904 and 1907 passing on three attempts and failing on the other three. During the debate on the Trattles' case in the British House of Lords, it was revealed that approximately a third of railway recruits who failed the Holmgren test passed on re-examination.

In spite of these adverse reports, the Holmgren Wool test is still favoured by some industries in the belief that revised versions of the test have been modified to produce an accurate vocational test. We have evaluated the design of two modern versions of the Holmgren test to ascertain whether the test has potential for screening or grading the severity of red-green colour deficiency.

Method

A Holmgren Wool test purchased in 1990 has different coloured wools and a different examination procedure than an older test available in our Department. The older test has 55 wool skeins (3 test skeins and 52 comparison skeins). The examination is made by presenting each test skein in turn and asking the subject to select skeins which are the same colour, but either lighter or darker, from the remaining 52 skeins. A numerical code is attached to each skein which identifies the wool as either a corrected (matching) or incorrect (confusion) choice for each of the three test skeins. Incorrect choices indicate red-green colour deficiency. There are 12 extra skeins which are neither matching nor confusion for any of the test skeins and should not be selected by either normal or colour deficient observers. The new test has 49 wool skeins (3 test skeins and 46 comparison skeins). A separate small number of the comparison skeins are presented with each test skein. All these wools are either matching or confusion for a particular test skein. There are no extra skeins. A new concept is introduced in that the third test skein is not used for colour vision screening but aims to classify protan (dark confusion) and deutan (light confusion) colour deficiency.

In order to evaluate the design of both tests, each of the wools were matched with Munsell samples, using standard Source C (400 lux) as provided by the MacBeth Easel lamp, and the chromaticity co-ordinates obtained from published tables (Kelly *et al.,* 1943). The co-ordinates of matching and confusion wools for each test skein were then plotted on the C.I.E. chromaticity diagram (1931) and compared with isochromatic zones for protanopes and deuteranopes (Pitt, 1935). Colour matching data obtained by Pitt (1935) divides the C.I.E. diagram into 17 isochromatic zones for protanopes and 23 zones for deuteranopes.

Skeins designated as matching were considered appropriately designed if they were within two Munsell hue steps (0.5) of test skein while differing in Munsell value and chroma. Skeins designated as confusion were considered as appropriately designed if the C.I.E. co-ordinates were positioned within two protan or deutan isochromatic zones with reference to the test skein.

Finally, both Holmgren tests were completed by six colour deficient observers and three observers with normal colour vision.

Results

The results of the colour matching procedure are summarised for each of the three test skeins in Tables, 1, 2 and 3. Neither the three test skeins nor the comparison skeins were identical in the two tests. An indication of the Munsell notations is included for comparison in Tables 1, 2 and 3. These results show that the design of both tests is very inconsistent. Only approximately half of the matching and confusion skeins for test skein 1 in both the Old and the New test fulfil the design criteria adopted for this study. In addition, four of the twelve extra skeins included in the Old test are within two isochromatic zones and could be chosen as confusion skeins. The results are similar for test skein 2 except that only two of the ten confusion skeins in the New test are within two isochromatic zones. None of the five confusion skeins for test skein 3 in the Old test satisfies the design criterion adopted for this study. Although nine of the available eleven confusion skeins for test skein 3 of the New test are within two isochromatic

Table 1. Munsell and Isochromatic Data for the Holmgren Wool test. Test skein (1) –light green.

	Old test, 2.5 GY 5/6	New test, 2.5 G 6/6
Matching skeins	10	10
Within 2 Munsell hue steps	5	3
Munsell notations	Y, GY, G, BG	GY, G, BG
Confusion skeins	10	10
Within 2 protan/deutan isochromatic zones	6	4
Munsell notations	Y, YR, N6	Y, YR, R, N6
Extra skeins	12	–
Munsell notations	YR, R, B, G	–
Within 2 Munsell hue steps (matching)	0	–
Within 2 protan/deutan isochromatic zones (confusion)	4	–

Table 2. Munsell and isochromatic data for the Holmgren Wool test. Test skein (2) – rose.

	Old test, 5 RP 7/8	New test, 5 RP 5/10
Matching skeins	5	5
Within 2 Munsell hue steps	2	3
Munsell notations	R, RP	R, RP
Confusion skeins	5	10
Within 2 protan/deutan isochromatic zones	4	2
Munsell notations	YR, R, G, B, BG	GY, Y, YR, RP, P, PB
Extra skeins	12	–
Munsell notations	YR, R, B, G	–
Within 2 Munsell hue steps	0	–
Within 2 protan/deutan isochromatic zones (confusion)	2	–

Table 3. Munsell and isochromatic data for the Holmgren Wool test. Test skein (3) – red.

	Old test, 7.5 RP 7/8	New test, 5 RP 3/14
Matching skeins	5	–
Within 2 Munsell hue steps	2	–
Munsell notations	R	–
Confusion skeins	5	11
Protan (dark)	–	6
Munsell notations	YR, R	G, YR, R, RP
Deutan (light)	–	5
Munsell notations	YR, R	GY, YR, R
Within 2 protan/deutan isochromatic zones	0	9
Protan		6 dark/–
Deutan		3 dark/3 light
Extra skeins	12	–
Munsell notations	YR, R, B, G	–
Within 2 Munsell hue steps (matching)	2	–
Within 2 protan/deutan isochromatic zones (confusion)	2	–

zones, the classification of protans and deutans according to dark and light confusions is not correctly realised.

The results for 6 representative colour deficient subjects (four anomalous trichromats and two dichromats) and three normal trichromats are shown in Table 4. Colour deficient subjects are expected to select both matching and confusion skeins; normal trichromats should select matching skeins only.

Colour deficient observers rarely selected confusion colours in either the Old or the New Holmgren Wool tests.

Table 4. The Holmgren Wool test. Results for 3 protans, 3 deutans and 3 normal trichromats.

		Old test, (3 tests skeins, 52 others)									New test, (3 test skeins, 46 others)					
		T1 (Light Green)			T2 (Rose)			T3 (Red)			T1 (Light Green)		T2 (Rose)		T3 [Red] Confusion only	
Matching/Confusion and Extra skeins		M	C	Ex	M	C	Ex	M	C	Ex	M	C	M	C	Dark	Light
	Number	10	10	12	5	5	12	5	5	12	10	10	5	10	6[a]	5[b]
PROTAN	PA1 (slight)	2	1	–	2	–	–	–	–	2	2	–	2	–	1	1
	PA2 (moderate)	6	1	–	2	–	–	–	–	–	3	–	–	–	1	–
	P	–	1	–	2	–	–	–	–	–	–	–	–	1	–	1
DEUTAN	DA1 (slight)	4	3	–	2	3	3	3	1	4	2	1	3	–	1	3
	DA2 (moderate)	5	–	–	4	–	–	5	–	5	4	–	–	4	2	2
	D	8	–	–	4	2	–	4	–	7	7	–	3	2	3	2
NORMALS	1)	5	–	–	4	–	–	3	–	–	4	–	2	–	–	–
	2)	5	1	–	3	–	–	3	–	2	4	–	2	–	–	1
	3)	7	–	–	5	–	–	2	–	2	3	–	–	–	–	–

[a] = Protan
[b] = Deutan

In the Old test identification of protan colour deficiency relied on selection of a single confusion colour for test skein 1 and one of the three deutan observers passed the test. One of the three normal trichromats and one protanomalous observer obtained identical results. Both normals and colour deficient observers tended to select some of the extra colours for test skein 3.

Two of the three protan observers were not identified as colour deficient with the New test since the third test skein is normally used for classification only.

Conclusions

The design criteria (within two Munsell hue steps for matching skeins and within two isochromatic zones for confusion skeins) selected for this study are fairly broad. The fact that a large number of the comparison skeins in both the Old and New Holmgren Wool tests do not fulfil this criteria shows that neither test has been adequately designed. The results obtained for a small group of representative observers confirm this conclusion. Neither of the two Holmgren Wool tests is accurate for colour vision screening and identification of some colour deficient observers may rest on the selection of one or two confusion skeins out of a total of twenty designated confusion skeins in each test. Examiners will therefore find that the written instructions provided are irrelevant and will have difficulty interpreting the test results. The design of the New test does not represent any improvement in screening efficiency compared with the Old test and the aim of classifying protan and deutan colour deficiency is not correctly realised. Our analysis of the test skeins and confusion colours does not suggest that these tests have a role either for grading the severity of colour deficiency or for use in vocational guidance.

The Holmgren Wool test is of historic interest but is not recommended for clinical use.

References

Boltz., C.L. (1952). A Statue to Mr. Trattles and other Scientific Topics. Butterworths, London.

Cole, B.L. and Vingrys, A.J. (1982). A survey and evaluation of lantern tests of colour vision. Am. J. Optom. Physiol. Opt. 59 (4): 346–374.

Kelly, K.L., Gibson, K.S. and Nickerson, D. (1943). Tristimulus specification of the Munsell book of color from spectrophotometric measurements. J. Opt. Soc. Am. 33 (7): 355–376.

Pitt, F.H.G. (1935). Characteristics of dichromatic vision. Report 14, Medical Research Council, London.

Topley, H. (1959). Sight testing for the merchant navy. Brit. J. Physiol. Opt. 16: 36–47.

Mrs J. Birch
Senior Lecturer in Clinical Optometry
Department of Optometry & Visual Science
City University
Dame Alice Owen Building
311–321 Goswell Road
London EC1V 7DD, U.K.

60. Protans and PAPI: Recognition of a two colour code by persons with defective colour vision

BARRY L. COLE and JENNIFER D. MADDOCKS
(Melbourne, Australia)

Abstract

In some countries there is controversy as to whether normal or near normal colour vision is necessary for pilots of aircraft. It has been argued that information displayed by aviation colour codes is always displayed redundantly and that in modern aviation there is less dependence on information acquired visually from outside the aircraft. One aviation signal system that displays information solely by colour coding is the Precision Approach Path Indicator (PAPI) which signals that an approach to land is too high or too low by means of the number of red and white signals. A Farnsworth lantern was modified to simulate PAPI signals viewed from one nautical mile under conditions close to 1.5 km visibility. Normal observers were able to identify the red and white signals and distinguish them from blank signals without difficulty except that some subjects made a few errors confusing a yellowish white (simulating the colour of a white PAPI signal attenuated by the atmosphere) with red. Most colour vision defective observers made errors confusing red and white, and protan observers frequently failed to see the red signal, calling it a blank. Although a few protans could identify the signal colours without error it is concluded that the majority of colour vision defective observers are not able to identify a red/white signal system reliably despite the code being a simple two category code and the fact that brightness differences between the red and white signals were an available cue. The results are surprising since past reports suggest that colour vision defectives are able to recognise red with only a few errors so that a two colour red/white code might have been expected to be within their capacity. The number of red = white and white = red errors made by deutans was also surprising.

Introduction

Occupational colour vision standards have been in place for well over a century (Vingrys and Cole, 1986) and have long been subject to controversy and debate

B. Drum (ed.), Colour Vision Deficiencies XII, pp. 501–510.
© 1995 *Kluwer Academic Publishers, Dordrecht.*

(Vingrys and Cole, 1988). In recent years, they have had to withstand the critical scrutiny of equal employment opportunity legislation which has to some extent exposed weaknesses and inconsistencies in existing standards (Cole, 1993). If colour vision standards are to survive as effective means of reducing risk or avoiding work inefficiencies due to an inability to recognise colour codes, they must be well supported by credible evidence and argument.

If they cannot be properly supported they should indeed be swept aside on the grounds that they unreasonably discriminate against those who have defective colour vision. On the other hand, it is not in the public interest to do so if the result is an increased risk of accident or a significant loss of occupational efficiency.

In an earlier paper (Cole, 1993), a framework for a consistent set of colour vision standards for a wide range of occupations that involve colour discrimination or colour recognition was proposed.

One occupation for which it is generally agreed that there should be a colour vision standard in the interests of public safety is aviation. The International Civil Aviation Organisation requires member nations to maintain a colour standard to ensure that pilots can recognise the colours of signal lights used in aviation. This is reasonable since the studies of Dille and Booze (1976) and Harper (1964) provide evidence of an increased frequency of accidents for colour vision defective pilots and there is laboratory evidence to show that persons with defective colour vision make errors and are slower in recognising aviation signals and colour coded instrument displays (Vingrys and Cole, 1986).

The member nations meet the requirement to have a colour vision standard for pilots in different ways. In some countries, pilots with defective colour vision are essentially excluded from commercial flying. In other countries, the standard is much more liberal and some persons with defective colour vision are permitted to hold senior commercial pilot's licences. In Australia, and the USA, the Farnsworth lantern is used to determine whether an applicant for a pilots licence with defective colour vision can fly. The Farnsworth lantern is a liberal test and passes one third of people with defective colour vision (Cole and Vingrys, 1983a).

In the USA a practical test can be given on appeal after failing the Farnsworth lantern and this permits a greater proportion of colour vision defectives to hold a commercial pilot's licence (Steen *et al.*, 1974).

In Australia, the colour vision standard has been subject to legal appeal (Cole, 1993) on the grounds that defective colour vision is not a risk factor because colour signals are always redundant in modern aviation and that the exclusion of colour vision defectives is discriminatory. The appeals have been partially successful, and the Australian Civil Aviation Authority has agreed to introduce a practical test as a test of last resort for those who fail the lantern test.

We were asked to develop a suitable practical test. The Director of Medical Aviation, Dr. Robert Liddell, suggested that one critical aviation signal system that involves no redundancy is the Precision Approach Path Indicator (PAPI).

PAPI is an array of four signal lights each projecting a red beam in the lower field and a white beam in the upper field. The signals are angled systematically so that two appear red and two white when the aircraft is on the correct approach path for landing. The number of white signals increases if the approach is too high and the number of red signals increases if the approach is too low.

This signal system offers no visual redundancy: there is no alternative to recognising the red and white signals in order to determine visually whether a landing approach is on the correct path.

There is an alternative system, T-VASIS, which does not have the same international acceptance but does offer redundancy. A row of lights extends above or below a horizontal bar of lights forming a ⊥ or ⊤ to indicate a low or high approach. While this sytem does not require colour recognition, when the approach is too low all the signal lights in the T turn red to signal a dangerously low approach.

The risk for colour defective pilots using T-VASIS is that they may fail to appreciate the need for an immediate response to a dangerous situation if they are unable to recognise the red colour, or in the case of protan observers, fail to see them at all.

The geometry of PAPI and T-VASIS signal systems is shown in Fig. 1.

Of course it is possible that colour vision defective observers may have little difficulty with recognising the red and white colour of the PAPI or T-VASIS signals. It is a simple two colour code and Heath and Schmidt (1959) suggest that dichromats may be able to correctly interpret a two colour code. Moreover, colour vision defective observers can usually recognise red signals with few errors (Vingrys and Cole, 1993). Lastly, there is a brightness cue, since in PAPI the white signals are required to be no less than twice and no more than 6.5 times as bright as the red signal. This brightness difference is further exaggerated for protans.

On the other hand, the recognition of the red and white lights in PAPI is not always an easy task since atmospheric scatter, and the use of low voltage at night to dim the lights, can shift the colour of the white signal toward yellow. This often poses a problem for colour normal observers and may cause even more difficulty for colour defective observers.

Method

We modified a Farnsworth lantern so that it presented only red and white signals with chromaticities representative of the red and white found in an actual PAPI.

We introduced some uncertainty by having two whites, one with a high colour temperature to simulate a white signal operated at full voltage observed under good visibility conditions and a yellowish-white representing a white signal attenuated by poor visibility conditions. The chromaticities of the colours are given in Table 1.

Since there were only two stimuli, the correct response can occur 50% of the

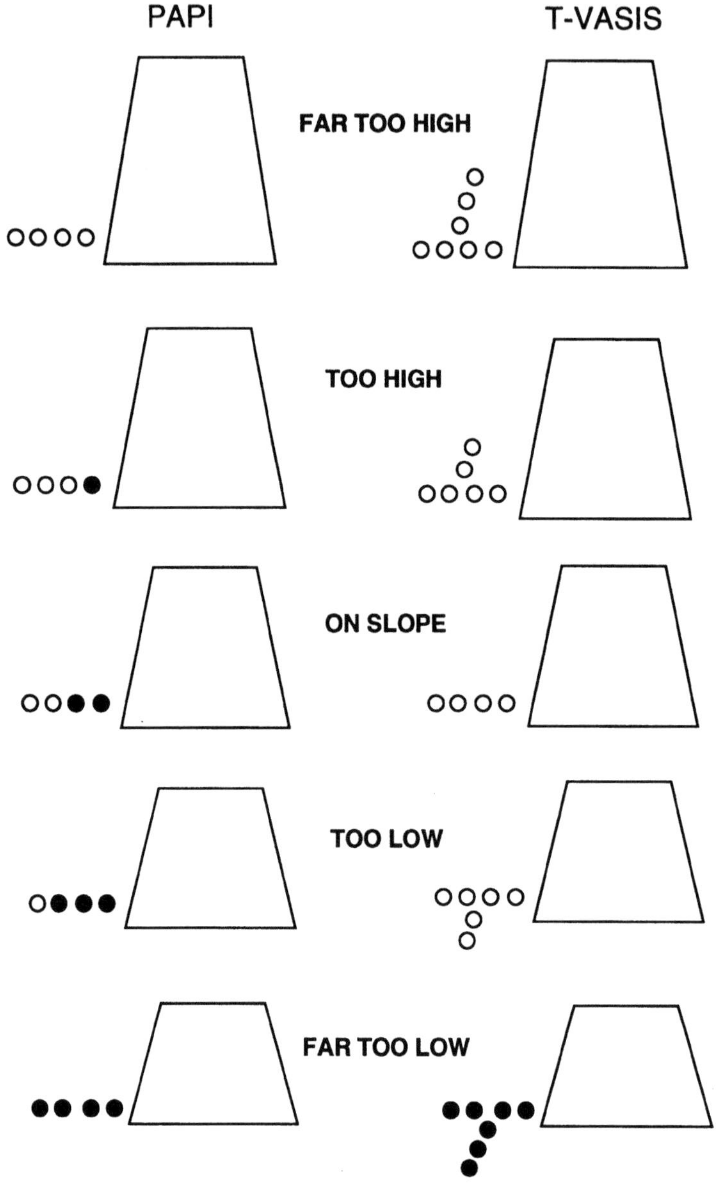

Fig. 1. The geometry of the signal arrays in PAPI and T-VASIS to indicate departures from the correct flight path in landing an aircraft. The filled circles represent a red signal, the open circles a white.

time by guessing. In order to increase uncertainty, a blank was introduced. The blank was intended to confound the protans who might identify stimuli they cannot see as 'red' if the only choices were red and white.

Table 1. The CIE x, y chromaticity coordinates of the colours used.

	High intensity		Low intensity	
	x	y	x	y
Red	0.685	0.315	0.684	0.315
White 1	0.428	0.401	0.441	0.411
White 2	0.529	0.419	0.535	0.420

The use of ND filters to reduce intensity to the low intensity shifted the colours slightly

Table 2. Combinations of pairs of colours and blanks in the nine positions of the Farnsworth lantern.

	1	2	3	4	5	6	7	8	9
Top	W1	R	R	Blank	W2	R	W1	Blank	W2
Bottom	Blank	R	W2	R	Blank	W1	R	W2	R

The pairing of colours and the blanks and the order of the pairs is given in Table 2.

The angular size of the stimuli was set at 1.8 min arc replicating PAPI signals seen from 1 nautical mile. This required the Farnsworth lantern to be positioned at 3.8 m.

The intensities of the stimuli were intended to simulate the illuminances of PAPI signals when viewed from 1 nautical mile under conditions of 1.5 km visibility. However, a pilot trial with normal subjects showed that this set too difficult a task. As a result, it was decided to use 2 levels of illuminances, one in which the red stimuli illuminances were 10 times the chromatic threshold and one in which they were six times the chromatic threshold. In the PAPI, the white signal is required to be at least twice and no more than 6.5 times that of the red and this intensity difference was preserved in the simulation.

Table 3 gives the illuminances of the stimuli.

Thirteen protanomals, 12 protanopes, 14 deuteranomals and 13 deuteranopes participated in the experiment. The control group comprised 38 subjects matched for age.

Table 3. Illuminances in the plane of the eye of the test stimuli used in the validation trial.

Colour	Illuminance μlux	
	High intensity	Low intensity
Red	5.5	3.3
White 1	35.3	17.6
White 2	17.6	8.8

The subjects were dark adapted for 6.5 minutes and had the colours demonstrated to them. They were read standard instructions. Five runs of nine pairs of colours were shown at each intensity level, each run starting at a predetermined random starting point. The high and low intensity levels were shown first to alternate subjects.

Our expectations were that the protans would often fail to see the red stimulus because of their reduced luminosity to red light and they would therefore make errors calling the red 'blank' and perhaps calling the blank 'red'. This was expected to occur only at the low intensity level, since protans sensitivity to red light is only reduced to half (Cole and Vingrys, 1983b) and the higher intensity level was ten times the chromatic threshold. It was not expected that protans would often confuse red and white because previous work had shown that colour vision defectives recognise red with few errors and the protans would also be assisted by the exaggerated difference in the apparent brightness of the red and white.

We expected the deutans to make few errors because red is reasonably reliably recognised by colour vision defective observers and if it is recognised the white will be identified as the other colour since there were only two alternatives.

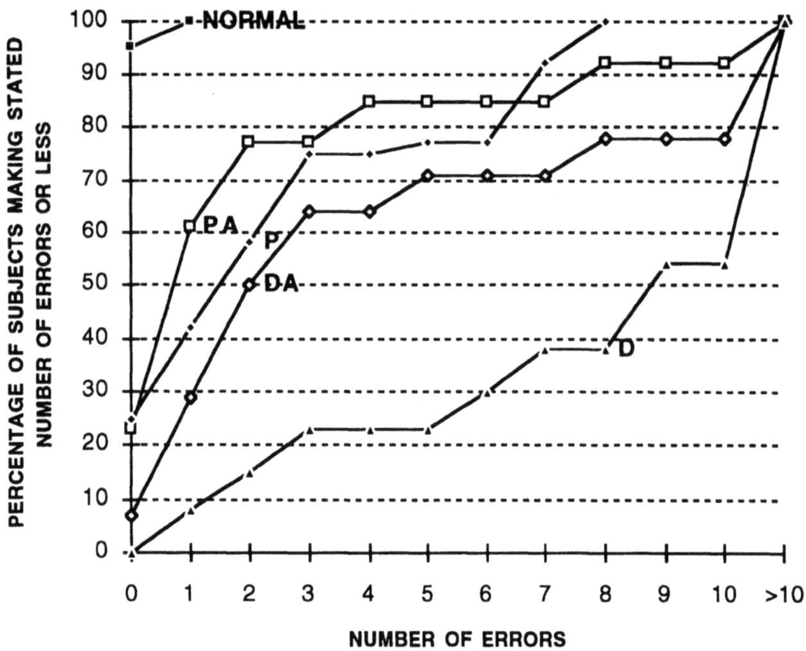

Fig. 2. Cumulative percentage of subjects making errors recognising red and white signal colours simulating the colours of aviation visual approach path signal lights. High intensity.

Results

Figures 2 and 3 shows the cumulative percentage of subjects making errors from zero errors to more than 10. Figures 4 and 5 shows the kinds of errors by the various classes of defective colour vision.

In evaluating the error rates, account should be taken of the fact there were only three choices (red, white, blank) so that 33% correct responses would be obtained by guessing. If the stimulus is seen, the 'blank' response is excluded as an alternative so that a 50% correct response could occur by guessing.

Three protanomals and three protanopes made no errors at the high intensity while three protanomals and one protanope made no errors at the low intensity. This error free performance was sustained at both the high and low intensities by only one protanomal and one protanope.

Deutans found the task more difficult as is evident in Figs. 2 and 3: only one deuteranomal made no errors at the high intensity and one deuteranope made no errors at the low intensity.

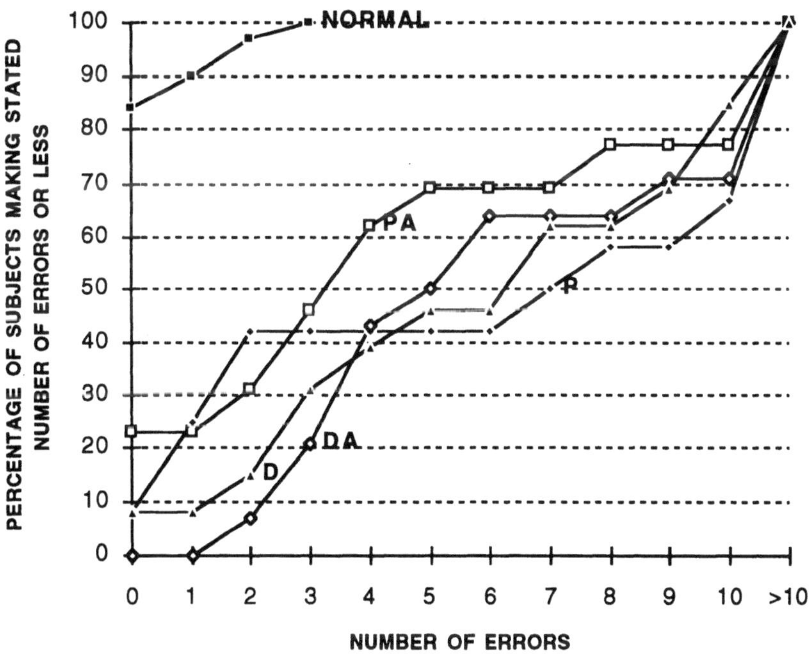

Fig. 3. Cumulative percentage of subjects making errors recognising red and white signal colours simulating the colours of aviation visual approach path signal lights. Low intensity.

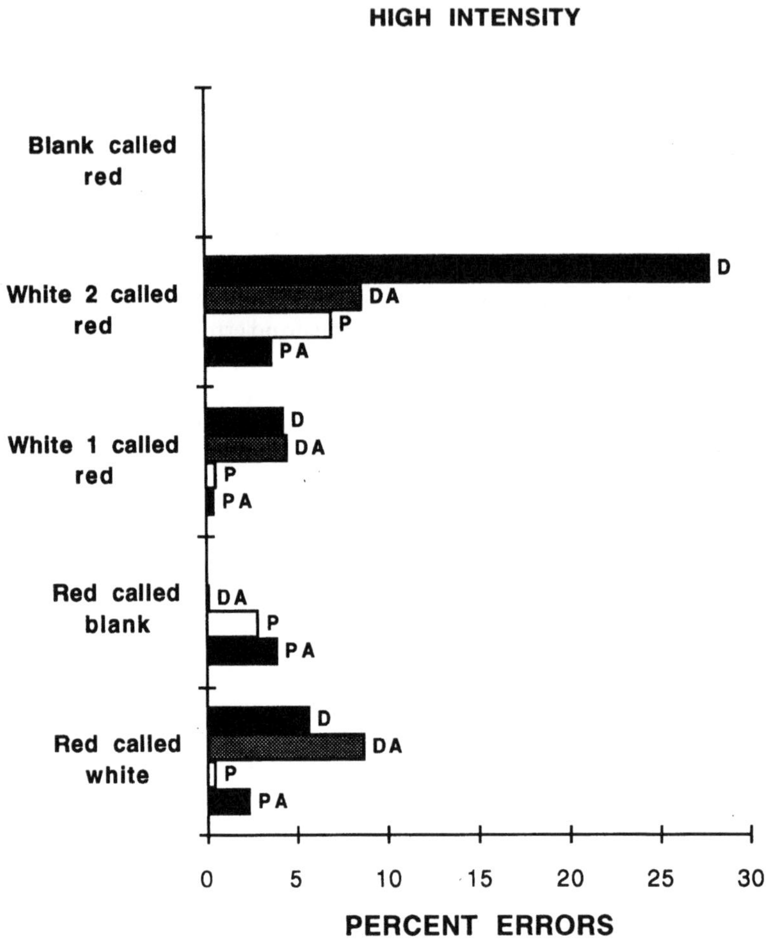

Fig. 4. Percentage errors made by colour vision defective observers at the high intensity. Colour normal subjects made few errors: two colour normal observers each made one error confusing white 2 with red.

Discussion

The results are surprising: firstly the deutans made more errors than expected and performed less well than the protans.

The second unexpected result was that red-white confusions did occur (Figs. 4 and 5) despite there being only two choices and the expectation based on earlier published reports that colour vision defectives can recognise red reasonably reliably.

Deutans make more red = white and white = red errors than protans, and deutans have particular difficulty with the yellowish white (W2).

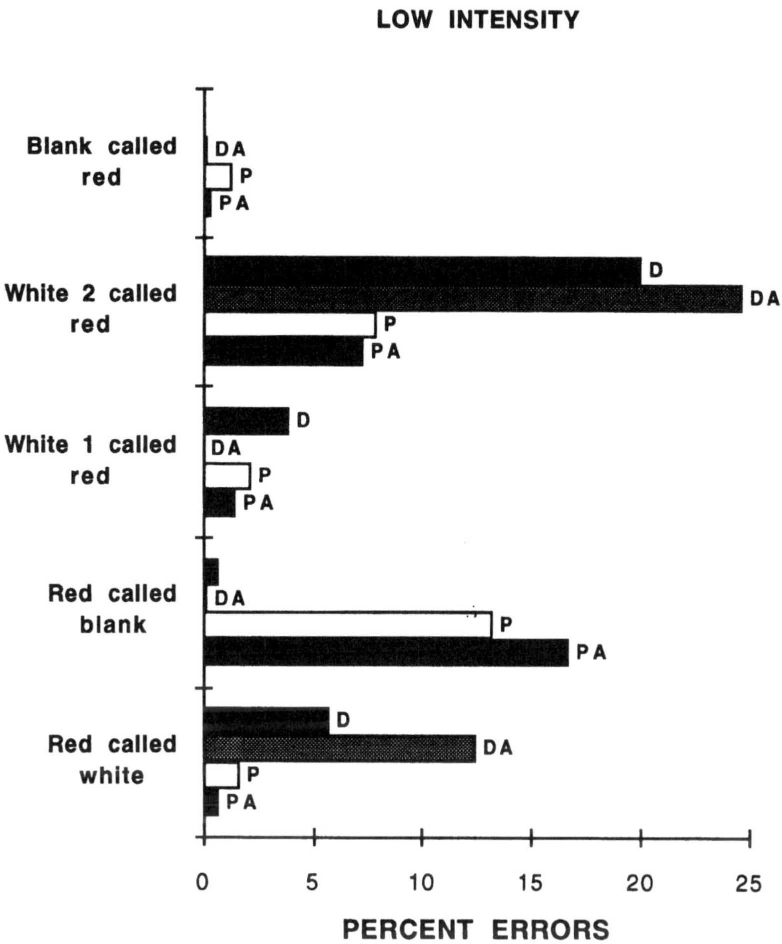

Fig. 5. Percentage errors made by colour vision defective observers at the low intensity. Five colour normal observers made errors: two made one error (in 90 presentations), two made two errors and one subject made three errors. In each case except one, the error was to confuse white 2 with red. In the one exception red was confused with white.

The protans sometimes failed to see the red, calling it 'blank' on 3 to 4% of occasions at the high intensity level despite it being 10 times above the normal chromatic threshold. At the lower level of intensity (twice the normal chromatic threshold), protans fail to see the red stimulus on 13% to 17% of occasions.

If the simulation was to be used as a 'practical' test to determine whether a person with defective colour vision had sufficient capacity to discrimate the red and white signals of PAPI, the test would pass some protanomals and protanopes but most deuteranomals or deuteranopes would fail. If the test involved presentation of the red and white stimuli at both the high and low intensities and up to three white 2 = red errors were allowed because some normals make this kind of error, about 10% of protanomals and 20% of protanopes would 'pass' and all deutans would 'fail'.

Conclusions

Almost all colour vision defective observers have difficulty with a simple two colour red/white signal code even though past reports suggest that most can recognise red quite well and in this experiment they were assisted by a brightness difference between the red and white.

Colour vision defective observers cannot be expected to recognise the colours of PAPI and T-VASIS approach path signal systems which is further evidence that defective colour vision is a risk factor in aviation.

Acknowledgements

This project was funded by the Australian Civil Aviation Authority. We gratefully acknowledge the advice and guidance given by the Director of Medical Aviation, Dr. Liddell in the planning and development of the experiment.

References

Cole, B.L. (1993). Does defective colour vision really matter? In: Drum, B. (ed.), Colour Vision Deficiencies XI, Doc. Ophthalmol. Proc. Ser. 56: 67–86, Kluwer Acad. Publ., Dordrecht.

Cole, B.L. and Vingrys, A.J. (1983a). Who fails lantern tests? Doc. Ophthalmol. 55: 157–173.

Cole, B.L. and Vingrys, A.J. (1983b). Do protanomals have difficulty seeing red lights? Proc. CIE 20th Session, Amsterdam, 1983. CIE Publication 56, E04, Paris.

Heath, G.G. and Schmidt, I. (1959). Signal color recognition by color defective observers. Am. J. Optom. Arch. Am. Acad. Optom. 36: 421–437.

Steen, J.A., Collins, W.E. and Lewis, M.F. (1974). Utility of several clinical tests of color defective vision in predicting daytime and night-time performance with aviation signal gun. Aerospace Med. 45: 467–472.

Vingrys, A.J. and Cole, B.L. (1986). Origins of colour vision standards in the transport industry. Ophthal. Physiol. Opt. 6: 369–375.

Vingrys, A.J. and Cole, B.L. (1988). Are standards of colour vision in the transport industries justified? Ophthal. Physiol. Opt. 8: 1257–1274.

Vingrys, A.J. and Cole, B.L. (1993). The ability of colour defective observers to recognise an optimised set of red, green and white signal lights. In: Drum, B. (ed.), Colour Vision Deficiencies XI, Doc. Ophthalmol. Proc. Ser. 56: 87–95, Kluwer Acad. Publ., Dordrecht.

61. A new windows-based computer program for analysing and plotting results of the Farnsworth-Munsell 100 hue test

W. DAVID THOMSON and JENNIFER BIRCH

(London, U.K.)

Abstract

The Farnsworth-Munsell 100 hue test is a well established method for detecting and differentiating colour deficiencies. However, scoring and analysing test results can be time-consuming and laborious. A number of computer-based systems have been produced to speed up and simplify the scoring and analysis stages. A new computer program which greatly simplifies data entry and analysis of test results will be described. The program runs on PCs under *Microsoft Windows* and exploits the graphics user interface provided by the *Windows* environment. A graphical representation of the Munsell chips is displayed on the screen and chip order can be rapidly changed using a mouse. The program calculates the error scores and displays the results in a variety of polar and linear formats. Data may be filtered using various algorithms and a number of statistical indices are calculated. The program also establishes a provisional diagnosis. Data can be filed on disc, linked to other programs (e.g. spreadsheets, wordprocessors) or output to a wide variety of printers and plotters.

Introduction

The Farnsworth-Munsell 100 hue test is a well established method for detecting and differentiating congenital and acquired colour vision deficiencies. However, scoring and analysing test results can be time-consuming and laborious.

A number of computer-based systems have been produced in an attempt to speed up the scoring and analysis stages. These range from fully-automated dedicated systems to simple computer programs requiring chip order to be entered from the keyboard. While the dedicated systems allow very rapid acquisition of chip order, few clinics or laboratories undertake sufficient 100 hue testing to make this option cost-effective. Conversely, simple computer programs requiring keyboard input of chip order do not produce a significant time saving over manual analysis.

B. Drum (ed.), Colour Vision Deficiencies XII, pp. 511–515.
© 1995 *Kluwer Academic Publishers, Dordrecht.*

This paper describes a novel computer program which has been developed to speed up and simplify 100 hue data input (chip order) and to provide a powerful set of analytical and administrative tools.

The program runs on any PC capable of running Microsoft Windows (3.1) and exploits the graphical user interface provided by the Windows environment to produce a simple and attractive user interface. Results can be saved on disc, displayed in polar or linear format and filtered/analysed according to a variety of algorithms. The program is also capable of producing high quality graphical output on a wide range of displays, printers and plotters.

Data input

A graphical representation of one of the four boxes of Munsell chips is displayed on the screen (see Fig. 1). Chip order is entered by *clicking* or *dragging* the mouse on the appropriate *scroll bar*. This method has proved to be significantly faster than methods involving keyboard entry.

Having changed a chip (A) from (N1) to a new number (N2), the program automatically searches the box to find the chip (B) that has the same number (N2) and changes the number of chip (B) to N1, i.e. the chips are swapped. This method reduces the number of chips that need to be changed and ensures that no chip numbers are duplicated.

Error scores are continuously updated and displayed below the scrollbar (see Fig. 1). Having entered the chip order for one box, the next or previous box can be displayed by selecting the *NEXT* or *PREV* options.

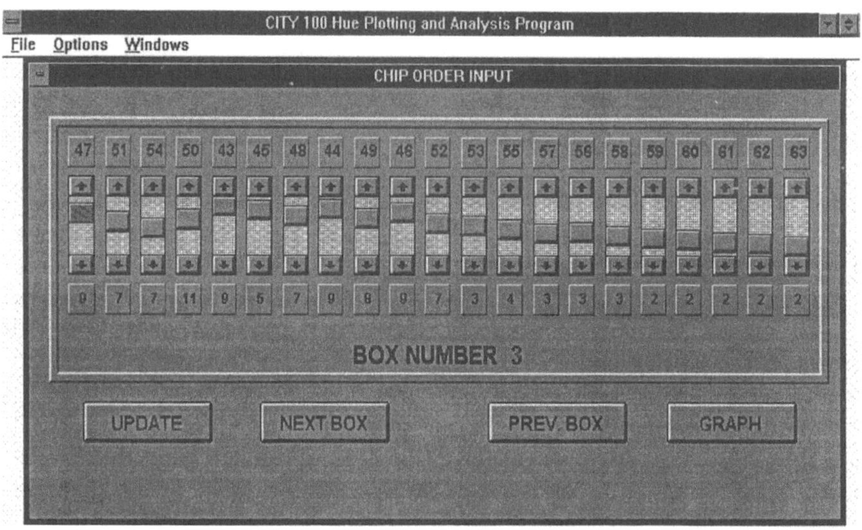

Fig. 1. Screen dump showing the novel method for entering chip order. The number of each chip is entered by *clicking* or *dragging* the mouse on the appropriate scroll bar

Data display and analysis

By selecting the *GRAPH* option, the data is displayed graphically on the screen in either polar or linear format (see Figs. 2, 3 and 4). Data can be plotted according to either the 'standard' or the Kinnear method and filtered according to the algorithms described by Dain and Birch (1987) or Kitahara (1984). Unfiltered and filtered data for a tritanopic subject are shown in Figs. 2 and 3 respectively.

The contents of the analysis box is determined from a set-up screen. It may contain the total error score, the square root of the total error score and a C and S-index calculated by the vector analysis method described by Vingrys and King Smith (1988). Other forms of analysis e.g. Kitahara (1985) will be included in future versions of the program.

Various options and defaults can be selected or set from a *Set-up screen*. For example, to specify the maximum scale value to be used on the graphs or the type of analysis to be displayed or the type of filter to be applied to the data and whether data is plotted according to the standard or Kinnear method.

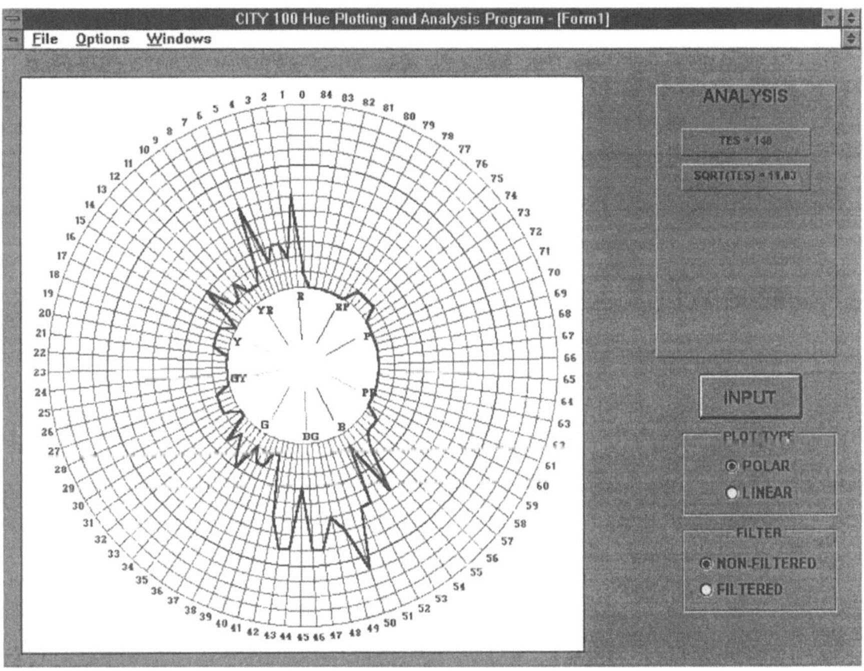

Fig. 2. Screen-dump showing how 100 hue results can be displayed on the screen in the conventional polar format. Results for a tritanopic subject are shown

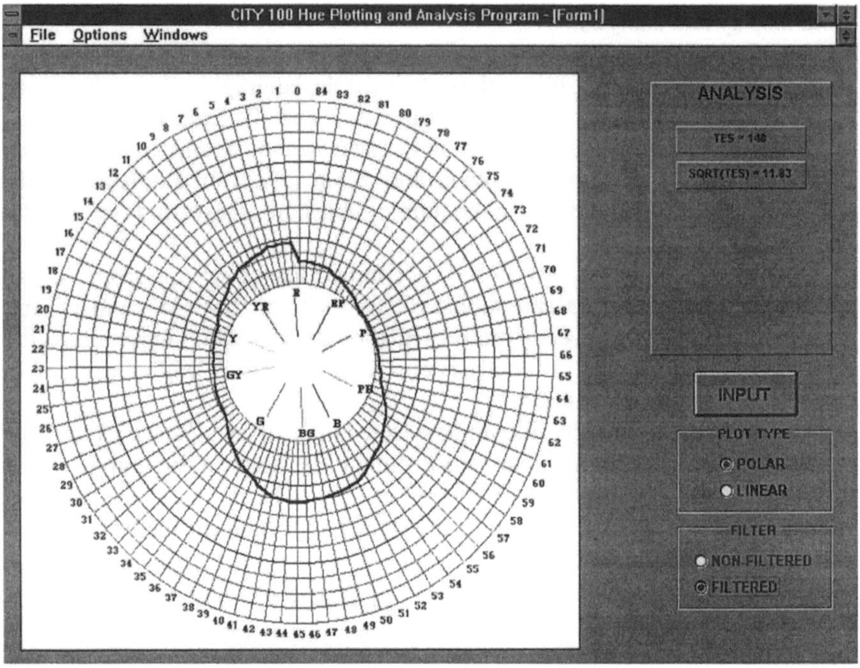

Fig. 3. By selecting the *FILTERED* option, the filters described by Dain and Birch (1987) or Kitahara (1984) can be applied to the data. The plot shows the result of applying the Dain and Birch filter to the data shown in fig. 2.

Other features

Multiple windows

The program allows multiple Windows to be displayed on the screen simultaneously. This permits different plots to be juxtaposed on the screen for comparison.

Printing/plotting

A hard copy of the chip order, graphs and analysis can be readily obtained using any printer or plotter supported by Windows 3.1.

Exporting graphics

Graphs can be saved on disc in bitmap or metafile format and exported to other Windows packages such as word processors or graphics programs. This allows results to be readily incorporated into reports or publications.

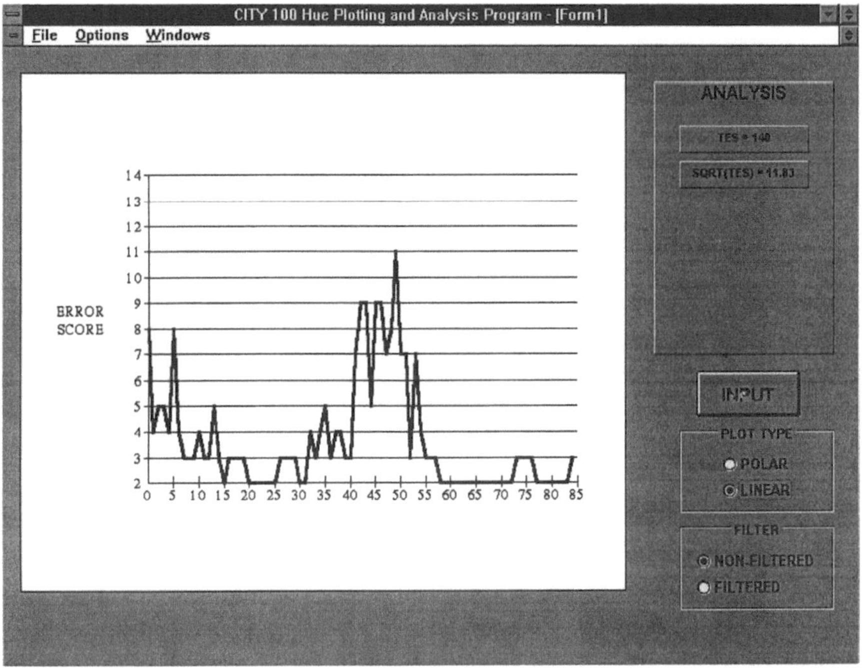

Fig. 4. By selecting the *LINEAR* option, the data is displayed in a linear format. The plot presents the same data shown in Fig. 2.

Filing

Results can be saved to disc for future reference or analysis.

References

Dain, S.J. and Birch, J. (1987). An averaging method for the interpretation of the Farnsworth-Munsell 100-Hue Test. Ophthal. Physiol. Opt. 7: 267–280.

Kinnear, P.R. (1970). Proposals for scoring and assessing the 100-hue test. Vision Res. 10: 423–433.

Kitahara, K. (1984). An analysis of the Farnsworth-Munsell 100 hue test. In: Verriest, G. (ed.), Colour Vision Deficiencies VII, Doc. Ophthalmol. Proc. Ser. 39: 233–238, Kluwer Acad. Publ., Dordrecht.

Kitahara, K. (1985). A new analysis of the Farnsworth-Munsell 100 hue test. Acta Soc. Ophthalmol. Jpn. 89: 544–547.

Vingrys, A.J. and King-Smith, P.E. (1988). A quantitative scoring technique for panel tests of color vision. Invest. Ophthalmol. Vis. Sci. 29: 50.

62. Macular pigment contributes to variance in 100 hue tests

J.D. MORELAND[1] and S.L. DAIN[2]
([1] *Keele, U.K.* and [2] *Kensington, Australia*)

Abstract

The F-M 100 hue test was performed in 'daylight' (430 Lux, 5500 K) by 10 normal trichromats viewing binocularly through artificial 'macular' filters with peak (458 nm) absorbances of 1 and 2 and through a neutral filter of approximately equivalent luminous transmittance. The peak absorbance change 0-1 is the approximate population range for macular pigment. Increases in 'macular' absorbance were accompanied by systematic increases in total error score and in bipolarity of the cap error plot and by rotation of the bipolar axis towards the tritan direction. Selective changes in cap chromaticity were characterised by dramatic increases in ellipticity of the 100 hue locus. Chromatic crowding near the vertices of the 'ellipse' correlated with increases in cap errors, accounting for the associated changes in bipolarity. Chromatic precession around the 'ellipse' accounted for the change in axis.

Introduction

Macular pigment forms a yellow filter having an approximately exponential retinal distribution with concentration peaking centrally by an amount which varies between individuals by 1 log unit or more (Moreland and Bhatt, 1984). This large variability affects colour matching significantly as well as other aspects of colour vision testing (Wright, 1946; Moreland and Kerr, 1979; Moreland and Krastel, 1984; Moreland and Viénot, 1988).

The 100 hue test, with its finely graded set of 85 coloured Munsell papers partitioned between four trays, poses a hue-sorting task. Any variation in prereceptoral spectral transmission would change chromatic spacing in the set and could affect task performance. However, there seems to be no publication on the effect of macular pigment variation on this test even though that due to age-related lens yellowing is well documented (Verriest, 1963). Additionally, the bipolar error axes of congenital dichromats show diagnostically inconvenient large variability (Farnsworth, 1957; Dain and Birch, 1987) which invites

B. Drum (ed.), Colour Vision Deficiencies XII, pp. 517–522.
© 1995 *Kluwer Academic Publishers, Dordrecht.*

speculation on a contributory role for macular pigment.

In this paper the effect of macular pigment variability on the 100 hue test is assessed by performing the test with simulation filters.

Materials and Method

Uniform 'daylight' illumination (430 Lux, 5500 K) of the test was provided by ceiling fluorescent tubes (Philips Colour 55) and an incandescent tungsten lamp (Macbeth) all run from a stabilised AC supply. Artificial macular pigment was prepared by extracting fresh carrot in benzene and adding acetone to shift the absorbance peak to 458 nm. This yielded a solution whose spectrum agreed closely in shape with the macular pigment average (Wyszecki and Stiles, 1982;

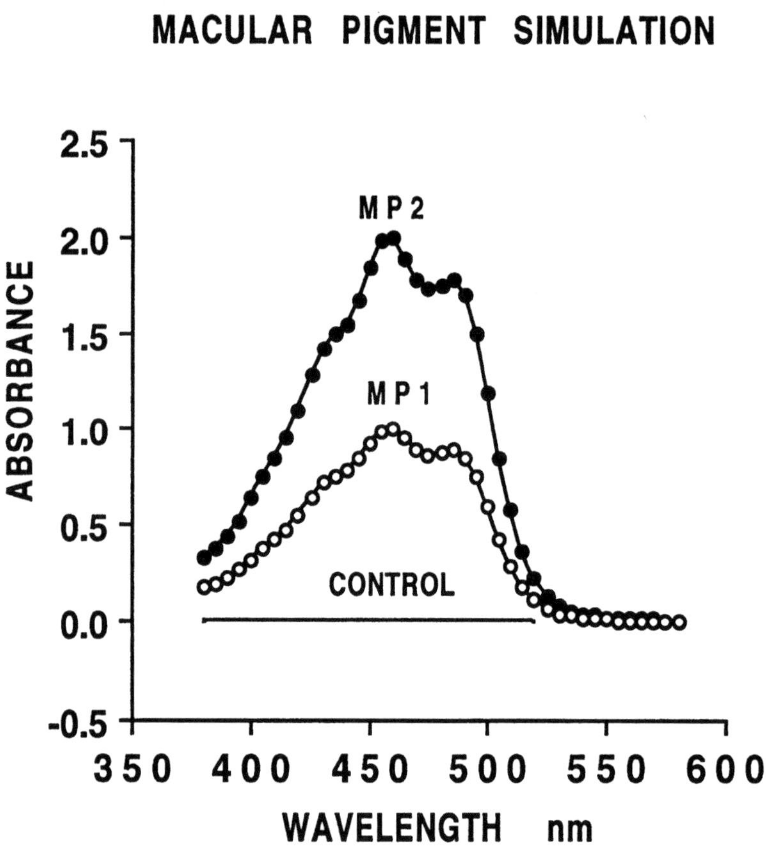

Fig. 1. Relative absorbance spectra of a neutral control filter and two filters, MP1 and MP2, closely simulating macular pigment. The CONTROL and MP1 are closely matched for luminous transmittance (84% and 83%) and their difference spectrum represents the population range in macular pigment concentration. The CONTROL-MP2 step was added for 'interest'.

Moreland *et al.*, 1987). A pair of large optical cells (dia. 50 mm), mounted in an optometric trial frame and affording a wide field of view, were filled with solution and sealed. This simulation filter, MP1, had a peak absorbance of about 1 (Figure 1) and a luminous transmittance of 83%. Neutral control filters with a luminous transmittance of about 84% were provided by discs of PMMA plastic, mounted in pairs in the trial frame. An additional test 'macular' filter, MP2, with peak absorbance of 2 (luminous transmittance 80%) was also used.

Reflectances of all Munsell papers and transmittances of the control and 'macular' filters were measured with a Hitachi U3410 spectrophotometer, and the power distribution of the illumination was measured with an Optronics 746IR-D spectroradiometer using a 5 nm bandwidth at 5 nm intervals. These data were used to calculate, for the 1931 CIE Standard Observer, the effective chromaticities of the 85 Munsell papers, as seen through the control, MP1 and MP2 filters.

Ten normal trichromats (screened using clinic records and the Ishihara test) performed the 100 hue test binocularly through the control, MP1 and MP2 filters. Each of the three conditions, in order of increasing concentration, was tested on 3 separate days for each subject. The experiment was conducted just before the XIth IRGCVD Symposium began in Sydney in June 1991.

Results

Increases in peak 'macular' absorbance (from 0 to 1 to 2) were accompanied by:

a) increases in mean total error score (from 33.9 to 75.7 to 157.4),
b) in bipolarity of the mean cap error plot (ratio of the 2nd and 0th Fourier harmonic amplitudes: from 0.27 to 0.37 to 0.83),
c) rotation of the bipolar axis towards the tritan direction (caps: from 40 to 43 to 46),
d) movement of the 100 hue locus towards the yellow part of the spectrum locus,
e) increases in ellipticity of the 100 hue locus with chromatic crowding near the vertices of the 'ellipse' and chromatic precession clockwise around the locus. See Figs. 2 and 3.

Discussion

Differences in the effective illumination level for filters and control were too small (1% and 4% for filters MP1 and MP2) to affect the total error score and the lowest effective level was high enough (> 340 Lux) to avoid the tritan-like errors associated with low illumination (< 5 Lux). Bipolar chromatic crowding correlated with high local errors and accounted for the associated changes in bipolarity. Chromatic precession clockwise around the 100 hue locus accounted for the changes in axis (compare Figs. 2 and 3). These remarks relate to changes

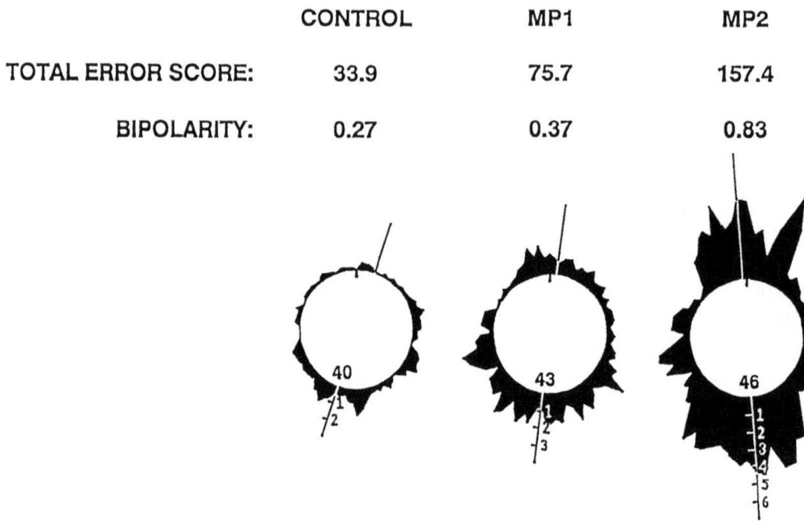

Fig. 2. Mean polar error plots for 10 normal trichromats under the CONTROL, MP1 and MP2 conditions. Line: bipolar (2nd Fourier harmonic) axis with error scale and nearest cap number designating direction. Both total error score and bipolarity increase with macular pigment and the axis rotates 3 caps towards the tritan direction for each pigment change.

associated with the global means for the 10 subjects. When the 100 hue responses are analysed individually we confirm high correlations with macular pigment absorbance of the square root of the total error score ($r = 0.996$, $p < 0.001$) and of bipolar axis ($r = 0.967$, $p < 0.01$; slope 3.7 cap $\pm$ 1.0 SE) but not of the relative bipolar amplitude ($r = 0.397$, n.s.; slope 0.12 ± 0.19 SE). The latter discrepancy reflects especially the difference between Fourier analyses of 'noisy' individual low-score 100 hue error plots and of the essentially 'smoother' global mean for the 10 subjects.

The population limits for change in pigment concentration for normal trichromats is represented approximately by the filter change from Control to MP1. If we assume that the same population limits apply to congenital dichromats then the 3-cap shift found here would account for about half the range in axis reported (Farnsworth, 1957; Dain and Birch, 1987) for dichromats. Of course, the limits for those dichromatic populations are not known, nor can we estimate the effect of any life-long habituation or adaptational responses of dichromats to their individual macular pigment complement. However, it seems unlikely that the residual axis variability could be attributed to age since, although total error score and bipolarity vary with age in normals, the direction of the bipolar axis itself remains remarkably constant (Verriest, 1963; Moreland, 1989). Speculation is constrained by small size of the population in the present study. The residual may well reflect experimental error but a contribution from genetic variation cannot be excluded.

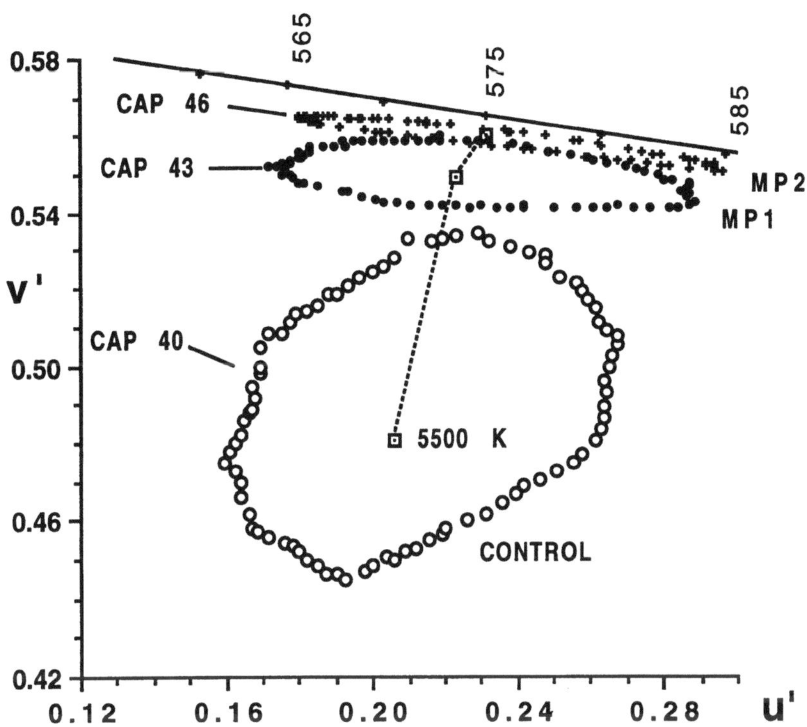

Fig. 3. Change of 100 hue chromaticities with macular pigment concentration. As concentration increases, the effective source (squares) and the 100 hue locus are displaced towards the yellow region (565–585 nm) of the spectrum locus. The displacement is accompanied by compression of the 100 hue locus with bipolar chromatic crowding and chromatic precession around the locus. Caps 40, 43 and 46 denote the bipolar error axes in Fig. 2.

Conclusion

A significant proportion of the directional uncertainty in the 100 hue bipolar axes of dichromats may be attributable to population variance in macular pigment concentration.

Acknowledgments

This research and the travel associated with it was supported by grants from the Royal Society, the Visual Research Trust and the Wellcome Trust.

References

Dain, S.J. and Birch, J. (1987). An averaging method for the interpretation of the Farnsworth-Munsell 100 Hue test. Ophthalmic Physiol. Opt. 7: 267–280.

Farnsworth, D. (1957). The Farnsworth-Munsell 100 Hue test. Manual, Munsell Color Company, Baltimore.

Moreland, J.D. (1989). Unipolar, bipolar and other errors in the Farnsworth-Munsell 100 hue test. In: Drum, B. & Verriest, G. (eds.), Colour Vision Deficiencies IX, Doc. Ophthalmol. Proc. Ser. 52: 397–405, Kluwer Acad. Publ., Dordrecht.

Moreland, J.D. and Bhatt, P. (1984). Retinal distribution of macular pigment. In: Verriest, G. (ed.), Colour Vision Deficiencies VIII, Doc. Ophthalmol. Proc. Ser. 39: 127–132, Nijhoff/Junk, Dordrecht.

Moreland, J.D. and Kerr, J. (1979). Optimization of a Rayleigh-type equation for the detection of tritanomaly. Vision Res. 19: 1369–1375.

Moreland, J.D. and Krastel, H. (1984). Effects of colour filters on vision. Proc. Int. Soc. Eye Res. 3: 53.

Moreland, J.D. and Viénot, F. (1988). Macular pigment multiplicity. Perception 17: A42b, 380.

Moreland, J.D., Krastel, H., Jaeger, W., Pelizzone, M. and Roth, A. (1987). Experimental trials on a modified Engelking equation. In: Verriest, G. (ed.), Colour Vision Deficiencies VIII, Doc. Ophthalmol. Proc. Ser. 46: 195–202, Nijhoff/Junk, Dordrecht.

Verriest, G. (1963). Further studies on acquired deficiency of color discrimination. J. Opt. Soc. Am. 53: 185–195.

Wright, W.D. (1946). Researches in Normal and Defective Colour Vision. Kimpton, London.

Wyszecki, G. and Stiles, W.S. (1982). Color Science, 2nd ed. Wiley, New York.

Prof J.D. Moreland
Department of Communication & Neuroscience
Keele University
Staffordshire, ST5 5BG U.K.

63. Evaluation of light sources for the D-15 color vision test

JEFFERY K. HOVIS and PAUL NEUMANN
(Waterloo, Ontario, Canada)

Abstract

Colorimetric analysis was carried out on 9 light sources to determine whether these sources were suitable for illuminating the D-15 color vision test when evaluating congenital red-green color deficiencies. Cap location under the various illuminants was determined for both normal trichromatic and dichromatic color spaces. Predicted ordering for red-green dichromats was analyzed using the Color Difference Vectors scoring method. Of the 9 sources, 3 were similar to Illuminant C in terms of cap location in both trichromatic and dichromatic color space. These sources had a CIE General Color Rendering Index of at least 90 and correlated color temperatures between 5000 K and 6700 K.

Introduction

Regardless of which color vision tests are included in a test battery, proper illumination is a prerequisite. However, obtaining the proper light source has been a continuing nuisance for clinicians. The problem arises either from the apparent lack of availability or cost considerations. With the emergence of several 'natural daylight' fluorescent lamps into the general marketplace, availability and cost may be less of a problem. Nevertheless, the issue remains as to whether these daylight lamps are suitable for color vision testing.

Although the effect of illuminant quality on color vision tests has been investigated for various pseudoisochromatic plates (Dain et al., 1993; Chioran and Sheedy, 1983) and the F-M 100 Hue test (Dain and Honson, 1987), few studies have investigated the effects of illuminant quality on the D-15 test (Lange *et al.*, 1980; Higgins *et al.*, 1978). Because the D-15 can be an important component of a color vision test battery (Cole, 1993), we determined the suitability of 9 common sources for evaluation of red-green color vision defects with the D-15 test using a colorimetric analysis.

B. Drum (ed.), Colour Vision Deficiencies XII, pp. 523–529.
© 1995 *Kluwer Academic Publishers, Dordrecht.*

Method

Lamps

Light sources used in this study were either 80 cm fluorescent tubes or tungsten filament bulbs. The fluorescent tubes were installed in overhead fixtures that held two lamps, and the tungsten lamps were installed in typical desk lamp fixtures. Combinations of the fluorescent tubes were created by placing one lamp of each type in the overhead fixture. No special arrangements were made to control the voltage.

Colorimetric calculations

Spectral irradiance (2 nm intervals) of the light sources were measured with a LI-1800 portable spectroradiometer (Li-Cor, Lincoln, NE). Irradiance measurements were accurate to 7.6% at any interval and wavelength settings were accurate to ± 0.5 nm. Reflectances of the individual D-15 caps were measured with a Zeiss DMR-21 dual beam spectroradiometer (Carl Zeiss, FDR) in the reflectance mode. Reflectance values at any single wavelength were accurate to ± 0.3% and wavelength settings were accurate to ± 0.1 nm. The wavelength range measured by both instruments was 380 to 830 nm. Judd's 1951 color matching functions (Judd, 1951) were used to calculate the tristimulus values of each cap. The precision of the tristimulus calculations was ±0.6%. Correlated color temperature, color rendering indices, and color differences between adjacent caps were also determined (CIE 1974,1986). Dichromatic bistimulus values and spectral sensitivity functions were calculated using the protanopic and deuteranopic copunctal points given by Vos and Walraven (1971).

Predicted cap order

Cap arrangement for dichromats was based on the magnitude of the bistimulus ratios of each cap. Although the ratios provide a theoretical basis for cap ordering, it is possible that two or more adjacent caps with different bistimulus ratios are less than 1 JND apart. In this case, there is the potential for 'atypical' dichromatic arrangements. These possible variations from the theoretical dichromatic ordering will be referred to as transpositions.

To determine the number of transpositions between adjacent caps, a basic line element theory was applied (Wyszecki and Stiles, 1982) using distances calculated from Pitt's (1944) wavelength discrimination data as the basis for a 1 JND step. Caps that were within 0.5 of the JND distance were considered as transpositions.

Results

Table 1 lists the light sources, correlated color temperatures, and General Color Rendering Indices (GCRI's) after approximately 2500 hrs of operation. Most GCRI's were greater than 90 but correlated color temperatures were generally lower than specified values.

Table 1. General color rendering index and correlated temperatures of the light sources.

Light source	Stated color temperature (°K)	Measured color temperature after 2,500 hrs of operation (°K)	General color rendering index after 2,500 hrs of operation
Vita-Lite plus; 34 watt fluorescent lamp: (Duro-Test Corp.)	5500	4700	91
Color Classer 75; 40 watt fluorescent lamp: (Duro-Test Corp.)	7500	6570	90
Combination of 1 Vita-Lite lamp & 1 Color Classer: (Duro-Test Corp.)	Not Measured	5390	92
Chroma 50; 40 watt fluorescent lamp: (General Electric Corp.)	5000	4460	89
Chroma 75: 40 watt fluorescent lamp: (General Electric Corp.)	7500	6700	94
Combination of 1 Chroma 50 & 1 Chroma 75: (General Electric Corp.)	Not Measured	5510	91
Macbeth Easel lamp with 60 watt tungsten bulb & daylight filter: (Macbeth Lighting Corp.)	5500	5190	94
Tungsten bulb; 60 watts: (Phillips Corp.)	2750	2530	99
40 Watt cool white fluorescent lamp: (Phillips Corp.)	4200	3790	61

Luminous reflectance

The dichromatic and trichromatic relative luminous reflectances of the caps were within ± 2.5% of each other under all light sources. This small difference is considered to be negligible for the testing conditions (Judd, 1948).

Color differences

Table 2 lists the mean ΔE's and standard deviation between adjacent caps in L*U*V* space. Note that the mean ΔE values for the test light sources are less than mean for Illuminant C. In order to determine whether these differences in the means were important, we determined the precision of our ΔE values. Given that our tristimulus units are accurate to ±0.6%, the mean ΔE values can vary up to 2.5 units. Thus, within the accuracy of our calculations, all but the tungsten and cool white fluorescent lamps produce a cap spacing that is similar to Illuminant C in trichromatic space.

Table 2. Mean ΔE's and standard deviation of adjacent under the various sources.

Light source	Mean ΔE	Standard deviation
Illuminant C	11.2	3.82
Chroma 75	10.6	3.63
Color Classer 75	10.3	3.56
Combination of Vita-Lite & Color Classer lamps	10.1	4.05
Combination of Chroma 50 & Chroma 75 lamps	9.94	3.72
Vita-Lite plus	9.87	4.51
Macbeth Easel lamp	9.83	3.78
Chroma 50	9.60	4.10
Tungsten bulb	7.79	5.72
Cool white fluorescent lamp	7.70	3.09

Cap arrangement

The theoretical cap ordering was evaluated using the Color Difference Vector method proposed by King-Smith and Vingrys (1988). Tables 3 and 4 list the confusion angle, C-index, S-index, and Total Error Score along with the number of transpositions for deuteranopes and protanopes. Note that the confusion angles vary slightly from data reported in the literature because we used the U* and V* values calculated for each cap under the different light sources in the Color Difference Vector Analysis, and we used Judd's color matching functions instead of the CIE 1931 color matching functions.

Discussion

Light sources that are suitable for illuminating the D-15 should result in a cap spacing that is similar to Illuminant C in both trichromatic and dichromatic color space. Because the spacing of the caps is generally closer together in trichromatic space for the tungsten and cool white fluorescent lamps, these lamps would be unsuitable.

Table 3. Color Vector Difference Analysis and number of transpositions for the different light sources for the deuteranope.

Light Source	Confusion angle (range ± 6°)	Total error score	C-index	S-index	Number of transpositions
Illuminant C	−13.83 (−19.83 – −7.83)	36.89	3.98	3.96	0
Chroma 75	−11.91 (−17.91 – −5.91)	36.01	3.87	3.83	0
Color Classer 75	−12.45 (−18.45 – −6.45)	35.49	3.81	3.78	0
Vita & Classer 75 Combination	−10.02 (−16.02 – −4.02)	40.01	4.31	4.28	0
Chroma 50 & Chroma 75 Combination	−9.92 (−15.92 – −3.92)	37.25	4.01	3.98	0
Vita-Lite	−11.80 (−17.80 – −5.80)	46.18	4.99	4.96	1
MacBeth Lamp	−12.35 (−18.35 – −6.35)	40.55	4.37	4.35	0
Chroma 50	−12.82 (−18.82 – −6.82)	41.61	4.51	4.49	1
Tungsten Bulb	−8.39 (−14.39 – −2.39)	40.12	4.34	4.34	5
Cool White Fluorescent	−10.79 (−16.79 – −4.79)	43.36	4.68	4.66	0

Cap spacing in the dichromatic spaces can be evaluated by comparing the confusion angle, C-index, S-index, Total Error Score, and the number of transpositions. Tables 3 and 4 indicate that for the MacBeth lamp and remaining fluorescent lamps, the Total Error Scores, C-indices, and S-indices are similar. Thus, these parameters are not affected appreciably by the different sources. However, the confusion angles and number of transpositions show more variation across the sources.

In evaluating the confusion angle, the important aspect is whether the sources can provide a clear distinction between a protanope and deuteranope. To determine whether a clear separation is possible, we assumed that the standard deviation about the confusion angle for both dichromats was 2° (Atchison *et al.*, 1991). If there was no overlap of the estimated third standard deviation values (±6°), then the source was considered to be suitable in terms of the confusion angle. Using this criterion, only the MacBeth, Chroma 50, Duratest 75, and Chroma 75 lamps are acceptable.

The number of transpositions indicates how many caps fall within the same zone of confusion and reflects how close adjacent caps are in dichromatic space.

Table 4. Color Vector Difference Analysis and number of transpositions for the different light sources for the protanope.

Light Source	Confusion angle (range ± 6°)	Total error score	C-index	S-index	Number of transpositions
Illuminant C	2.43 (8.43 – –3.57)	40.83	4.39	4.36	5
Chroma 75	2.04 (8.04 – –3.96)	40.73	4.39	4.37	4
Color Classer 75	1.14 (7.14 – –4.86)	40.08	4.32	4.30	4
Vita-Classer 75 Combination	0.74 (6.74 – –5.26)	45.13	4.87	4.86	3
Chroma 50 & Chroma 75 Combination	0.47 (6.47 – –5.53)	41.61	4.48	4.46	2
Vita-Lite	–1.98 (4.02 – –7.98)	41.23	4.46	4.45	2
MacBeth Lamp	0.56 (6.56 – –5.44)	36.72	3.96	3.94	4
Chroma 50	–0.02 (5.98 – –6.02)	42.83	4.62	4.60	2
Tungsten Bulb	–6.09 (–12.09 – –0.09)	46.96	5.08	5.08	8
Cool White Fluorescent	–1.95 (–7.95 – –4.05)	44.61	4.48	4.46	4

For the Chroma 50 lamp, the number of transpositions is reduced from 5 to 2 for a protanope. This reduction suggests that this lamp would be unsuitable because cap spacing under this light is greater than Illuminant C for protan observers. For the MacBeth, Dura-test 75 and Chroma 75, the number of transpositions are within 1 of the number predicted for Illuminant C for both protanopes and deuteranopes. Thus, these three lamps would be the most suitable for illuminating the D-15 test.

Our analysis indicates that because of normal lamp aging, fluorescent lamps with a stated color temperature of 7500 K and GCRI of at least 90 are best suited for use with the D-15. Filtered tungsten lamps are also acceptable, provided the color temperature remains above 5000 K.

References

Atchison, D.A., Bowman, K.J. and Vingrys, A.J. (1991). Quantitative scoring methods for D15 panel tests in the diagnosis of congenital color vision deficiencies. Optom. Vis. Sci. 68: 41–48.
C.I.E. (1974). Method of measuring and specifying colour rendering properties of light sources (2nd ed.), Publ. C.I.E. No. 13.2 (TC-3.2), Bureau Central de la Commission Internationale de l'Eclairage, Paris.
C.I.E. (1986). Colorimetry (2nd ed.), Publ. C.I.E. No. 15.2, Bureau Central de la Commission Internationale de l'Eclairage, Paris.
Cole, B.L. (1993). Does defective colour vision really matter? In: Drum, B. (ed.), Colour Vision Deficiencies XI, Doc. Ophthalmol. Proc. Ser. 56: 67–86, Kluwer Acad. Publ., Dordrecht.
Chioran, G.M. and Sheedy, J.E. (1983). Pseudoisochromatic plate design – Macbeth or tungsten illumination. Am. J. Optom. Physiol. Opt. 60: 204–216.
Dain, S.J. and Honson, V.J. (1989). Selection of an optimal light source for the FM 100 hue test. In: Drum, B. and Verriest, G. (eds.), Colour Vision Deficiencies IX, Doc. Ophthalmol. Proc. Ser. 52: 425–432, Kluwer Acad. Publ., Dordrecht.
Dain, S.J., Honson, V.J. and Curtis, C.T. (1993). Suitability of fluorescent tube light sources for the Ishihara test as determined by colorimetric methods. In: Drum, B. (ed.), Colour Vision Deficiencies XI, Doc. Ophthalmol. Proc. Ser. 56: 327–333, Kluwer Acad. Publ., Dordrecht.
Higgins, K.E., Moskowitz-Cook A. and Knoblauch, K. (1978). Color vision testing: An alternative 'source of Illuminant C. In: Verriest, G. (ed.), Colour Vision Deficiencies IV, Mod. Probl. Ophthalmol. 19: 113–121, Karger, Basel.
Judd, D.B. (1948). Color perceptions of deuteranopic and protanopic observers. J. Res. Natl. Bur. 41: 247–271.
Judd, D.B. (1951). Report of U.S. Secretariat Committee on colorimetry and artificial daylight. C.I.E. Proc., Vol. 1, Part 7. (Stockholm, 1951). Bureau Central de la Commission Internationale de l'Eclairage, Paris.
Lange, J.E., Morey, J.A. and Richards, O.W. (1980). Cool white, Ultralume, and Vita-Lite fluorescent lamps for use in color vision testing. Am. J. Optom. Physiol. Opt. 57: 400–403.
Pitt, F.H.G. (1944). The nature of normal trichromatic and dichromatic vision. Proc. R. Soc. Lond. (Biol.) 132: 101–117.
Vingrys, A.J. and King-Smith, P.E. (1988). A quantitative scoring technique for panel tests of colour vision. Invest. Ophthalmol. Vis. Sci. 29: 50–63.
Vos, J.J. and Walraven, P.L. (1971). On the derivation of the foveal receptor primaries. Vision. Res. 11: 799–818.
Wyszecki, G. and Stiles, W.S. (1982). Color Science: Concepts and Methods, Quantitative Data and Formulae, (2nd ed.), J. Wiley and Sons, New York.

Jeffery K. Hovis
School of Optometry
University of Waterloo
Waterloo, Ontario N2L 3G1
Canada

64. Panel D-15 test in color vision deficiencies at reduced illumination levels

HITOSHI KUDO, FUKUKO OBARA and YASUO OHTA
(Tokyo, Japan)

Abstract

We evaluated the performance of the Panel D-15 test in color vision deficiencies at reduced illumination. We used seven light levels ranging from 0.9 to 900 lux. Forty observers with congenital X-linked color vision defects served in the experiment: three protanopes, ten protanomals, ten deuteranopes, seven extreme deuteranomals, ten deuteranomals. And ten color normals also served in the experiment. All color normals passed the test at 3.5 lux and 30% failed at 0.9 lux. fifty-six percent of color vision deficients failed the test at 3.5 lux and 92% failed at 0.9 lux. Some protanomals showed a deutan pattern in the test at a certain illumination level or levels. Under reduced illumination, the performance of the Panel D-15 test in color vision deficiencies changed from protan or deutan to scotopic, anarchic pattern as illumination was decreased.

In summary, it is possible to classify mild color vision deficiencies on the basis of the lowest illumination level at which a subject passes the Panel D-15 test.

Introduction

The Panel D-15 test is an excellent color vision test that can easily be used to distinguish between severe and mild anomalies. For a definitive diagnosis, however, an anomaloscope is needed. On the other hand, in everyday life, color discrimination of dye colors is far more important than colored lights. Therefore, test methods using dye colors instead of colored lights may be more relevant to the determination of the severity of color vision defects. We performed the Panel D-15 test on color vision defectives to determine how their color vision would change at reduced illumination levels, and we also did experiments to determine whether it would be possible to use the Panel D-15 test for finer classification of color vision defects.

B. Drum (ed.), Colour Vision Deficiencies XII, pp. 531–538.
© 1995 *Kluwer Academic Publishers, Dordrecht.*

Methods

Forty congenital X-linked color vision deficient subjects (38 males and 2 females, aged 15–24 years) who visited the Tokyo Medical College Hospital for consultation and who were also succesfully classified by anomaloscope, and ten normal control subjects were included in this study. All observers gave informed consent.

The observers with color vision defects as classified by anomaloscopy consisted of three with protanopia, ten with protanomaly, ten with deuteranopia, seven with extreme deuteranomaly and ten with deuteranomaly. The color vision defectives were divided into two groups based on the results of the Panel D-15 test at 900 lux: one group was for observers who passed the test and the other for those who failed. The control group consisted of two males and eight females, aged 24–35 years.

The Ishihara plates, the Okuma plates, the Tokyo Medical College plates, the Standard Pseudoisochromatic Plates part 1 and the Panel D-15 test were performed on observers under 900 lux illumination and anomaloscopy was also performed. For the observers who passed the Panel D-15 test at 900 lux, the Panel D-15 test was repeated at lower illumination levels, i.e. at 225, 56, 25, 14, 3.5 and 0.9 lux. The light source used for the test was EDL D_{65} fluorescent lamp. The illumination level was lowered stepwise using a special test booth equipped with a dimmer. Observers who failed the Panel D-15 test at 900 lux were re-tested at 1500 lux.

The test results were classified as pass, protan, deutan, tritan, unclassified, scotopic or anarchic pattern (Fig. 1).

Results

Changes in color vision at reduced illumination levels were observed in both color vision defectives and normal observers. Fifteen color vision defectives failed the Panel D-15 test at 900 lux: they consisted of all protanopes and deuteranopes and two of seven of extreme deuteranomals. All of them showed typical protan or deutan patterns. The classification of all fifteen subjects remained unchanged even at 1500 lux.

On the other hand, 25 observers passed the Panel D-15 test at 900 lux: they consisted of all ten observers with protanomaly, all ten observers with deuteranomaly and five of seven observers with extreme deuteranomaly. The number of observers who failed the test increased as illumination was reduced. All normal observers passed the test at 3.5 lux or more. At 0.9 lux, three (30%) failed the test. Color defective observers began to fail the test at 225 lux and the number who failed increased as the illumination level decreased. Fourteen observers out of 25 (56%) failed the test at 3.5 lux and 23 (92%) at 0.9 lux (Fig. 2).

Normal observers showed only tritan pattern errors. Color defectives, however, showed various patterns of errors: at 0.9 lux, 64% showed anarchic or scotopic patterns (Fig. 3).

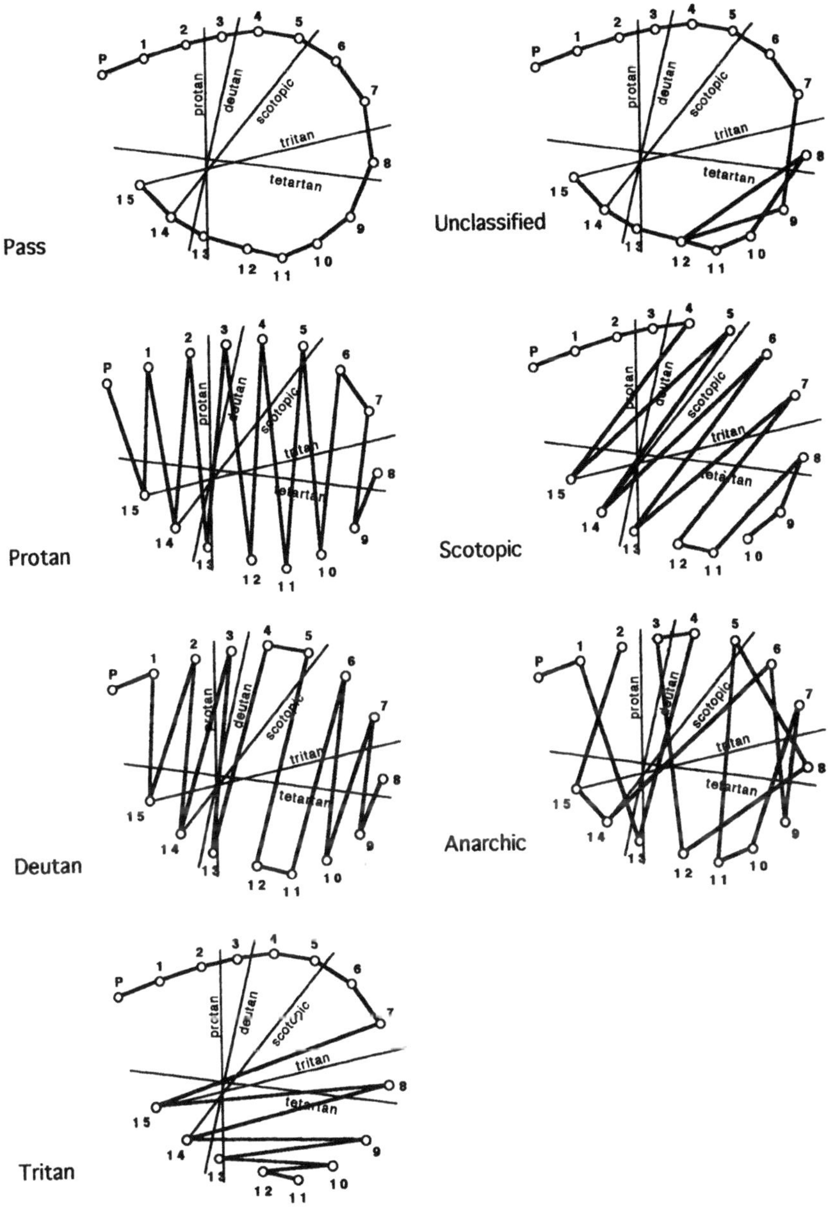

Fig. 1. Classification of typical results of Panel D-15 test

Among those failing the Panel D-15 test, protanomalous observers showed either protan or deutan patterns, and extreme deuteranomalous and deuteranomalous observers showed deutan and scotopic patterns at 3.5 lux. At 0.9 lux most of the protanomalous and extreme deuteranomalous observers

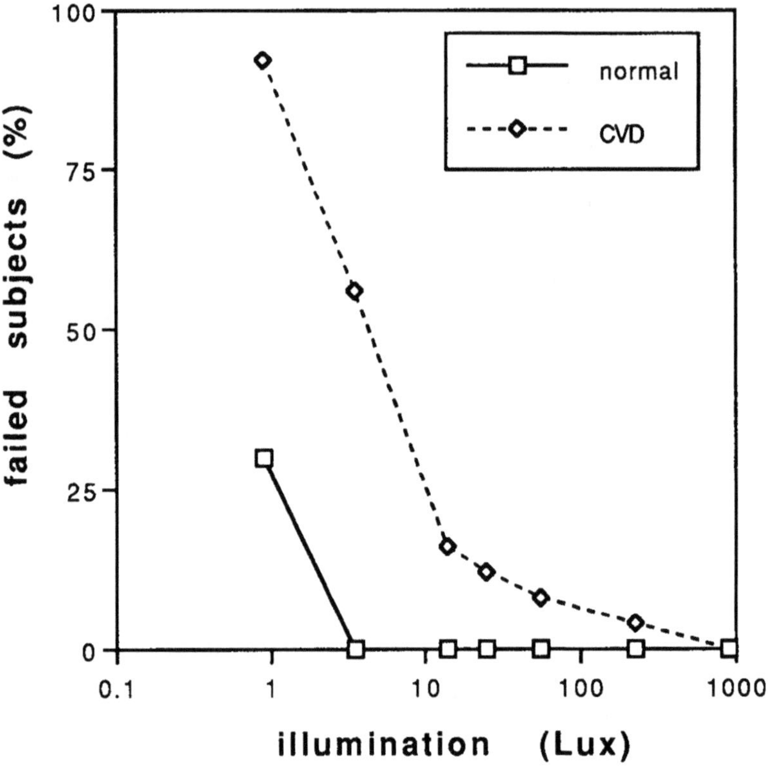

Fig. 2. Percentages of observers who failed Panel D-15 test as a function of illumination. ○ = normal control subjects; ▲ = color vision deficiencies

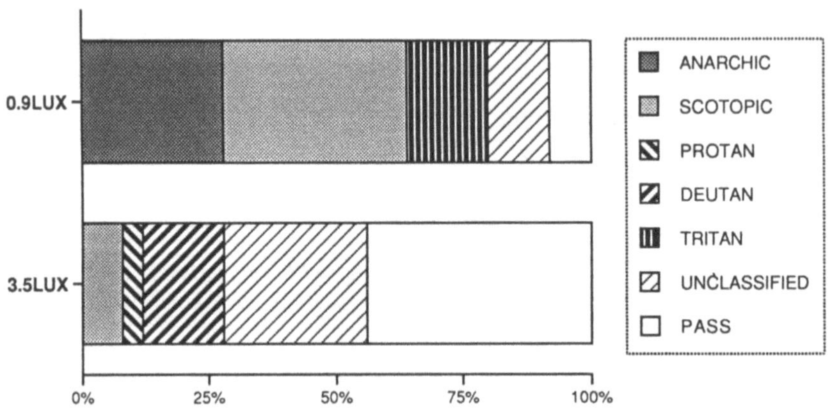

Fig. 3. The results of the Panel D-15 test in the color vision deficiencies at 0.9 lux and 3.5 lux

showed anarchic or scotopic patterns, 40% of deuteranomalous observers showed scotopic patterns and another 40% showed tritan patterns while 30% of normal observers showed tritan patterns (Fig. 4).

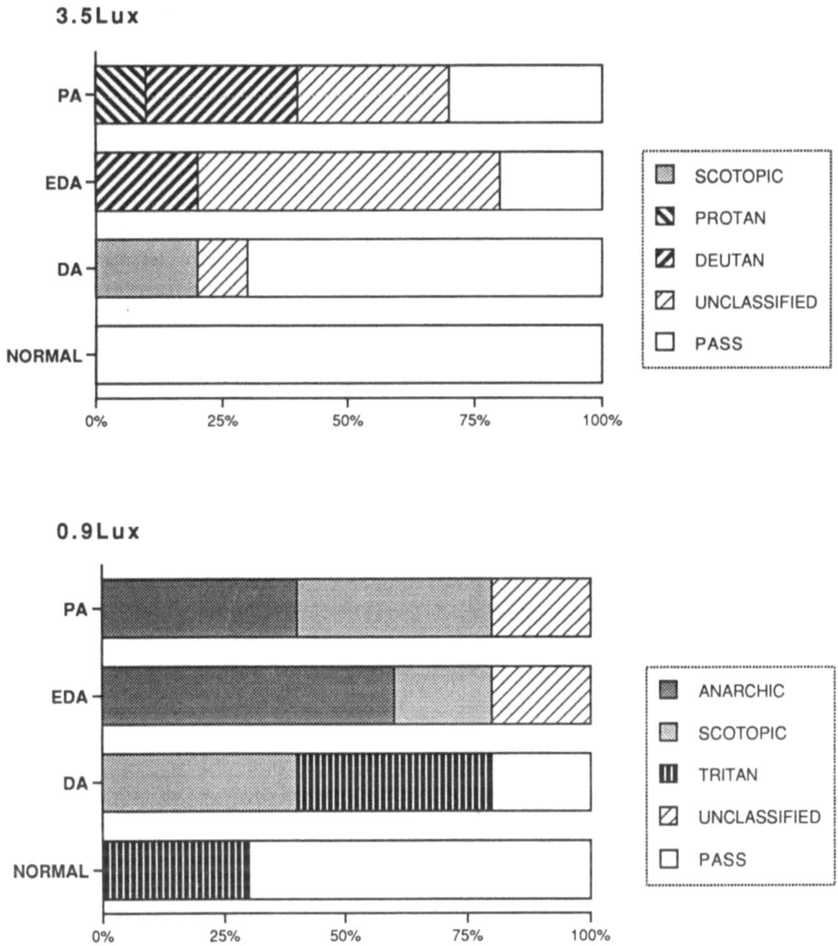

Fig. 4. The classification of the Panel D-15 test on the observers who failed Panel D-15 test at 0.9 lux and 3.5 lux

Discussion

Deterioration in color discrimination was observed in color vision defectives and in normal observers at reduced illumination. No normal observers failed the Panel D-15 test at 3.5 lux, whereas 56% of the color vision deficiencies failed the test at that level. Although normal observers are known to make color confusions similar to those seen in Type III acquired violet-yellow defects (Everdingen *et al.*, 1991, Smith *et al.*, 1991), only a few color vision defectives made such color confusions in our study. This could be explained by the fact that only one confusion axis can be detected in the Panel D-15 test. In order to detect color confusion in the Panel D-15 test, the minor axis should be shorter than 16% of the length of the major axis of the ellipse formed by the perceived color space coordinates of the color caps of the Panel D-15 test in the uniform color space coordinates (Farnsworth, 1943). Conversely, no color confusion can be recognized in this test if the minor axis of the ellipse is longer than 16% of the length of the major axis.

In a previous study using the Minimalist test (Mollon *et al.*, 1991), we reported that normal subjects showed not only blue-yellow defects but also red-green defects at low illumination levels (Kudo, *et al.*, 1993). According to the study, normal observers showed deterioration for color discrimination along the blue-yellow and red-green axes at reduced illumination levels. Moreover, the blue-yellow defect seemed to begin to occur at a higher illumination level than the red-green defect. The defects similar to blue-yellow defect and red-green defect must also occur in observers with color vision deficiencies.

In observers with protanomaly or deuteranomaly, the axis of color confusion rotated clockwise from the protan axis to the deutan axis and then to the scotopic axis at reduced illumination levels. Although reduction in color space along the blue-yellow axis is most sensitive to illumination level decrease, mild reduction along other axes may occur. Such mild reduction of the color space associated with reduced illumination levels causes color confusion which can be detected by the Panel D-15 test, since the color space is inherently reduced along red-green axis in red-green color defectives. When the illumination level is further lowered, strong reduction in color space along the blue-yellow axis occurs and this reduction may cause the direction of the minor axis of the color space in color vision deficiencies to rotate clockwise, toward the blue-yellow axis. The same may happen to control observers, too. But since their color spaces are larger than those of color vision defectives, they are not so seriously affected by reduced illumination. Moreover, color confusion in them is first detected along the blue-yellow axis because the color space reduced most in this direction (Fig. 5).

In summary, mild color vision defectives may show similar color confusions to strong color vision deficiencies at reduced illumination levels. The method of the reduced illumination Panel D-15 test may not be adequate to classify the type of color defect, for at certain illumination levels, protanomalous observers may be wrongly judged as deuteranomalous. For the evaluation of color discriminating ability, however, the method is useful. If a person is able to discriminate colors, or his or her color space is big enough, the person may not be influenced by the effects of the reduced illumination and he or she may be able to discriminate colors at lower illumination levels.

Conclusion

Mild color vision deficient subjects have various sizes of color spaces that may represent different severities of color vision defect and it may be possible to classify them on the basis of the lowest illumination level at the Panel D-15 test is passed.

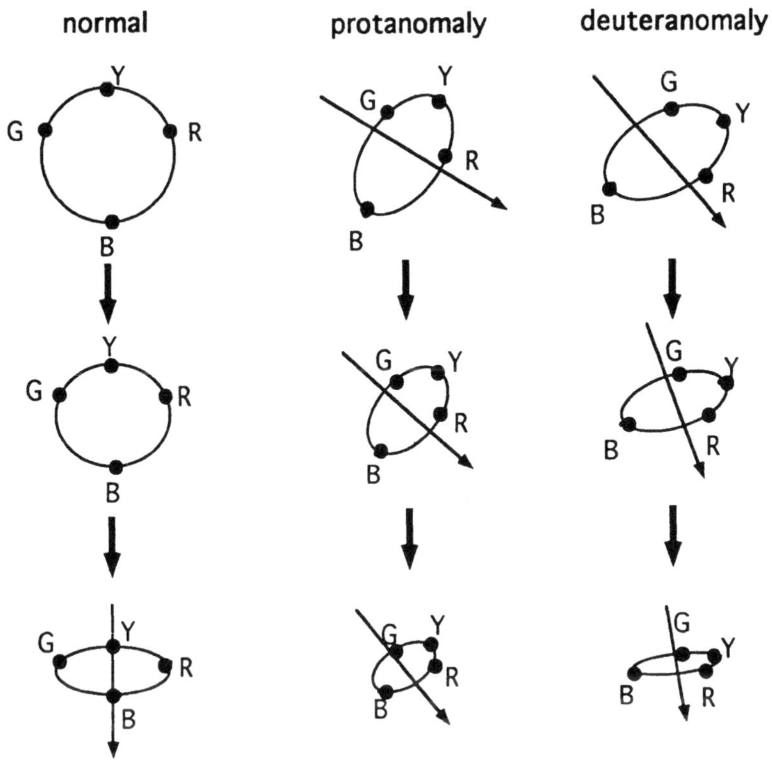

Fig. 5. A pictorial representation of a diagram of color spaces of color normal, protanomalous and deuteranomalous observers. R = red, G = green, B = blue, Arrow = color confusion axis. Illumination level lowers toward the bottom of the diagram. The confusion axes of both protanomalous and deuteranomalous observers rotate clockwise as illumination level decreases

Although the Panel D-15 test is an excellent test, it cannot evaluate severity and types of color vision defects correctly unless it is performed at proper illumination levels.

References

Everdingen, J.A.M., Smith, V.C. and Pokorny, J. (1991). Sensitivity of tritan screening tests as evaluated in normals at reduced levels of illumination. In: Drum, B., Moreland, J.D. and Serra, A. (eds.), Colour Vision Deficiencies X, Doc. Ophthalmol. Proc. Ser. 54: 167–175, Kluwer Acad. Publ., Dordrecht.

Farnsworth, D. (1943). The Farnsworth-Munsell 100 hue and dichotomous test for color vision. JOSA 33: 568–578.

Kudo, H., Smith, V.C. and Pokorny, J. (1993). Sensitivity of screening tests for S-cone discrimination. In: Drum, B. (ed.), Colour Vision Deficiencies XI, Doc. Ophthalmol. Proc. Ser. 56: 119-127, Kluwer Acad. Publ., Dordrecht.

Mollon, J.D., Astell, S. and Reffing, J.A. (1991). A minimalist test of colour vision. In: Drum, B.,

Moreland, J.D. and Serra. A. (eds.), Colour Vision Deficiencies X, Doc. Ophthalmol. Proc. Ser. 54: 59–67, Kluwer Acad. Publ., Dordrecht.
Smith, V.C., v. Everdingen, J.A.M. and Pokorny, J. (1991). Sensitivity of arrangements tests as evaluated in normals at reduced levels of illumination. In: Drum, B., Moreland, J.D. and Serra, A. (eds.), Colour Vision Deficiencies X, Doc. Ophthalmol. Proc. Ser. 54: 177–185, Kluwer Acad. Publ., Dordrecht.

Hitoshi Kudo
Department of Ophthalmology
Tokyo Medical College
6-7-1 Nishi-Shinjuku
Tokyo 160 Japan

Author index

B. Drum (ed.), Colour Vision Deficiencies XII, pp. 539–540.
© 1995 *Kluwer Academic Publishers, Dordrecht.*

Cumulative author index for all IRGCVD proceedings volumes

B. Drum (ed.), Colour Vision Deficiencies XII, pp. 541–558.
© 1995 *Kluwer Academic Publishers, Dordrecht.*

Author	Year	Volume	Pages
Harwerth, R.S.	1989	IX	353–364
Haughey, A.	1976	III	158–160
Haughey, A.E.	1976	III	158–160
Havener, W.H.	1989	IX	301–308
He, J.C.	1994	XII	185–191
Heard, P.F.	1987	VIII	181–194
Heckmann, B.	1991	X	323–329
Hedin, A.	1974	II	64–66
Hedin, A.	1974	II	185–189
Hedin, A.	1976	III	231–236
Hedin, C.-E.	9191	X	95–106
Heel, L. v.	1980	V	256–260
Heinsius, E.	1972	I	106–110
Helve, J.	1972	I	122–129
Hemminger, H.J.	1976	III	147–150
Henry, C.J.	1984	VII	227–231
Hermes, D.	1984	VII	141–146
Hermès, D.	1984	VII	259–274
Hermès, D.	1993	XI	345–355
Hermès, D.	1989	IX	523–528
Hermès, D.	1987	VIII	203–215
Hermès, D.	1989	IX	317–323
Hermès, D.	1989	IX	325–337
Hermès, D.	1991	X	151–159
Hermès, D.	1991	X	353–359
Hermès, D.	1991	X	421–427
Heron, J.R.	1991	X	441–450
Hewlett, S.E.	1989	IX	345–352
Hibino, H.	1987	VIII	21–29
Hibiya, Y.	1993	XI	13–17
Higgins, K.E.	1978	IV	113–121
Higgins, K.E.	1989	IX	243–250
Higgins, K.E.	1989	IX	413–423
Hill, A.R.	1976	III	264–272
Hill, A.R.	1978	IV	64–66
Hill, A.R.	1978	IV	136–141
Hill, A.R.	1978	IV	187–189
Hill, A.R.	1980	V	164–171
Hill, A.R.	1980	V	261–267
Hill, A.R.	1980	V	320–326
Hill, A.R.	1980	V	381–384
Hill, A.R.	1984	VII	181–187
Hill, A.R.	1989	IX	447–453
Hine, T.	1993	XI	51–57
Hine, T.	1993	XI	397–405
Hisato, G.	1991	X	561–568
Hita, E.	1984	VII	211–219
Hogg, C.H.	1994	XII	3–12
Hogg, C.J.	1994	XII	3–12
Holst, G.C.	1974	II	284–290
Holst, G.C.	1976	III	257–263
Honan, W.P.	1991	X	441–450
Honson, V.J.	1989	IX	425–432
Honson, V.J.	1991	X	29–36
Honson, V.J.	1993	XI	327–333
Hood, D.C.	1989	IX	233–241
Hoshino, M.	1984	VII	329–334
Hovis, J.K.	1991	X	85–93
Hovis, J.	1994	XII	523–529
Huie, K.E.	1987	VIII	237–242
Huie, K.	1993	XI	357–364
Hukami, K.	1980	V	200–203
Hukami, K.	1984	VII	199–209
Hukami, K.	1994	XII	467–472
Humanski, R.A.	1991	X	273–282
Huppert, W.E.	1987	VIII	413–422
Huppert, W.E.	1989	IX	261–272
Hurvich, L.M.	1974	II	56–63
Hurvich, L.M.	1974	II	200–209
Hynes, R.	1989	IX	413–423
Ibañez Vallejo, M.C.	1994	XII	221–226
Ichikawa, H.	1974	II	265–271
Ichikawa, H.	1980	V	200–203
Ichikawa, H.	1980	V	273–277
Ichikawa, H.	1984	VII	199–209
Ichikawa, H.	1984	VII	329–334
Ichikawa, H.	1987	VIII	133–140
Ichikawa, K.	1984	VII	329–324
Ichikawa, K.	1987	VIII	133–140
Ichikawa, K.	1994	XII	467–472
Iinuma, I.	1976	III	151–157
Iivanainen, A.	1991	X	21–28
Imamura, P.M.	1976	III	315–319
Iordanis, P.	1976	III	161–164
Ishihara, R.	1993	XI	251–257
Israel. A.	1972	I	76–81
Izutsu, Y.	1978	IV	126–130
Izutsu, Y.	1980	V	184–188
Jacobs, G.H.	1984	VII	49–54
Jacobs, G.H.	1989	IX	27–34
Jacobs, G.H.	1991	X	341–346
Jacobs, G.H.	1991	X	397–403
Jacobs, G.H.	1991	X	405–411
Jacobs, G.H.	1993	XI	25–31
Jacobs, G.H.	1993	XI	107–112
Jacoby, R.	1994	XII	277–283
Jaeger, W.	1972	I	145–147
Jaeger, W.	1976	III	121–130
Jaeger, W.	1976	III	135–142
Jaeger, W.	1976	III	143–146

Documenta Ophthalmologica Proceedings Series

1. J. François (eds.): *Symposium on Light-Coagulation.* Argon Laser and Xenon Arc (Ghent, Belgium, 1972). 1973 ISBN 90-6193-141-X
2. J.T. Pearlman (ed.): *10th ISCERG Symposium* (Los Angeles, Calif., USA, 1972). 1973 ISBN 90-6193-142-8
3. H.E. Henkes (ed.): *Photography, Electro-Ophthalmology and Echo-Ophthalmology in Ophthalmic Practice.* 1973 ISBN 90-6193-143-6
4. E. Dodt & J.T. Pearlman (eds.): *11th ISCERG Symposium* (Bad Nauheim, Germany, 1973). 1974 ISBN 90-6193-144-4
5. W.J. Holmes (ed.): *Public Health Ophthalmology.* Papers Presented at the Conference on the Prevention of Impaired Vision and Blindness (Paris, France, 1974). 1975 ISBN 90-6193-145-2
6. A. Th. M. van Balen (ed.): *First International Symposium on Artificial Lensimplantation* (Utrecht, The Netherlands, 1974). 1975 ISBN 90-6193-146-0
7. A.F. Deutman (ed.): [Symposium on] *New Developments in Ophthalmology.* (Nijmegen, The Netherlands, 1975) 1976 ISBN 90-6193-147-9
8. O. Hockwin (ed.): *Progress of Lens Biochemistry Research.* In Honour of Prof. Dr. med. J. Nordmann. 1976 ISBN 90-6193-148-7
9. J.J. de Laey (ed.): *International Symposium on Fluorescein Angiography* (Ghent, Belgium, 1976). 1976 ISBN 90-6193-149-5
10. R. Alfieri & P. Solé (eds.): *12th ISCERG Symposium* (Clermont-Ferrand, France, 1974). 1976 ISBN 90-6193-150-9
11. E. Auerbach (ed.): *Experimental and Clinical Amblyopia. 13th ISCERG Symposium* (Kibbutz Ginossar, Israel, 1975). 1977 ISBN 90-6193-151-7
12. E.L. Greve (ed.): *Symposium on Medical Therapy in Glaucoma* (Amsterdam, The Netherlands, 1976) 1977 ISBN 90-6193-152-5
13. T. Lawwill (ed.): *ERG, VER and Psychophysics. 14th ISCERG Symposium* (Louisville, USA, 1976). 1977 ISBN 90-6193-153-3
14. E.L. Greve (ed.): *2nd International Visual Field Symposium* (Tübingen, Germany, 1976). 1977 ISBN 90-6391-154-1
15. J. François & A. De Rouck (eds.), J.T. Pearlman and J. Kelsey (co-eds.): *Electrodiagnosis, Toxic Agents and Vision. 15th ISCEV Symposium* (Ghent, Belgium, 1977). 1978 ISBN 90-6193-155-X
16. H.-J Merté (ed.): *Genesis of Glaucoma.* Contributions of the Wessely Symposium in Munich (October 1974) with final Considerations by H. Goldmann. 1978 ISBN 90-6193-156-8
17. A.F. Deutmann & J.R.M. Cruysberg (eds.): *5th International Congress on Neurogenetics and Neuro- Ophthalmology* (Nijmegen, The Netherlands, 1977). 1978 ISBN 90-6193-159-2
18. O. Hockwin & W.B. Rathbun (eds.): *Progress in Anterior Eye Segment. Research and Practice.* In Honour of Prof. J.E. Harris. 1979 ISBN 90-6193-158-4
19. E.L. Greve (ed.): *3rd International Visual Field Symposium* (Tokyo, Japan, 1978). 1979 ISBN 90-6193-160-6
20. J. François, S.I. Brown & M. Itoi (eds.): *Proceedings of the Symposium of the International Society for Corneal Research* (Kyoto, Japan, 1978). 1979 ISBN 90-6193-157-6

Documenta Ophthalmologica Proceedings Series

21. J. François, E. Maumenee & I. Esente (eds.): *First International Congress on Cataract Surgery* (Florence, Italy, 1978). 1979 ISBN 90-6193-162-2
22. E.L. Greve (ed.): *Glaucoma Symposium.* Diagnosis and Therapy (Amsterdam, The Netherlands, 1979). 1980 ISBN 90-6193-164-9
23. E. Schmöger & J.H. Kelsey (eds.): *Visual Electrodiagnosis in Systemic Diseases.* Proceedings of the 17th ISCEV Symposium (Erfurt, GDR, 1979) 1980 ISBN 90-6193-163-0
24. A. Hamburg (ed.): *Symposium on Uveal Melanomas.* On the Occasion of the Snellen Medal Presentation to Dr W.A. Manschot (Utrecht, The Netherlands, 1979). 1980 ISBN 90-6193-722-1
25. H. Zauberman (ed.): *Proceedings of the Conference on Subretinal Space* (Jerusalem, Israel, 1979). 1981 ISBN 90-6193-721-3
26. E.L. Greve & G. Verriest (eds.): *4th International Visual Field Symposium* (Bristol, UK, 1980). 1981 ISBN 90-6193-165-7
27. H. Spekreijse & P.A. Apkarian: *Visual Pathways: Electrophysiology and Pathology.* 18th ISCEV Symposium (Amsterdam, The Netherlands, 1980). 1981 ISBN 90-6193-723-X
28. H.C. Fledelius, P.H. Alsbirk & E. Goldschmidt (eds.): *3rd International Conference on Myopia* (Copenhagen, Denmark, 1980). 1981 ISBN 90-6193-725-6
29. J.M. Thijssen & A.M. Verbeek (eds.): *Ultrasonography in Ophthalmology.* Proceedings of the 8th SIDUO Congress (Nijmegen, The Netherlands). 1981 ISBN 90-6193-724-8
30. L. Maffei (ed.): *Pathophysiology of the Visual System.* Proceedings of a Workshop (Pisa, Italy 1980). 1981 ISBN 90-6193-726-4
31. G. Niemeyer & Ch. Huber (eds.): *Techniques in Clinical Electrophysiology of Vision.* 19th ISCEV Symposium (Horgen-Zürich, Switzerland, 1981). 1982 ISBN 90-6193-727-7
32. A.Th.M. van Balen & W.A. Houtman (eds.): *Strabismus Symposium* (Amsterdam, The Netherlands, 1981). 1982 ISBN 90-6193-728-0
33. G. Verriest (ed.): *Colour Vision Deficiencies VI.* Proceedings of the 6th Symposium of the International Research Group on Colour Vision Deficiencies (Berlin- Steglitz, Germany, 1981). 1982 ISBN 90-6193-729-9
34. A. Roucoux & M. Crommelinck (eds.): *Physiological and Pathological Aspects of Eye Movements.* Proceedings of a Workshop (Pont d'Oye Castle, Habay-la-Neuve, Belgium, 1982). 1982 ISBN 90-6193-730-2
35. E.L. Greve & A. Heijl (eds.): *5th International Visual Field Symposium* (Sacramento, Calif., USA, 1982). 1983 ISBN 90-6193-731-0
36. R. Birngruber & V.-P. Gabel (eds.): *Laser Treatment and Photocoagulation of the Eye.* Proceedings of an International Symposium (Munich, Germany, 1982). 1984 ISBN 90-6193-732-9
37. H.E.J.W. Kolder (ed.): *Slow Potentials and Microprocessor Applications.* 20th ISCEV Symposium (Iowa City, USA, 1982). 1983 ISBN 90-6193-733-7

Documenta Ophthalmologica Proceedings Series

38. J.S. Hillman & M.M. Le May (eds.): *Ophthalmic Ultrasonography.* Proceedings of the 9th SIDUO Congress (Leeds, UK, 1982). 1984 ISBN 90-6193-734-5
39. G. Verriest (ed.): *Colour Vision Deficiencies VII.* Proceedings of the 7th Symposium of the International Research Group on Colour Vision Deficiencies (Geneva, Switzerland, 1983). 1984 ISBN 90-6193-735-3
40. J.R. Heckenlively (ed.), G.H.M. van Lith & T.Lawwill (ass. eds): *Pattern Electroretinogram, Circulatory Disturbances of the Visual System and Pattern-Evoked Responses.* 21st ISCEV Symposium (Budapest, Hungary, 1983). 1984 ISBN 90-6193-503-2
41. E.C. Campos (ed.): *Sensory Evaluation of Strabismus and Amblyopia in a Natural Environment.* In Honour of Prof. Bruno Bagolini. 1984 ISBN 90-6193-508-3
42. A. Heijl & E.L. Greve (eds.): *6th International Visual Field Symposium* (Santa Margherita Ligure, Italy, 1984). 1985 ISBN 90-6193-524-5
43. E.L. Greve, W. Leydhecker & C. Raitta (eds.): *2nd European Glaucoma Symposium* (Hyvinkää, Finland, 1984). 1985 ISBN 90-6193-526-1
44. P.C. Maudgal & L. Missotten (eds.): *Herpetic Eye Diseases.* Proceedings of an International Symposium (Leuven, Belgium, 1984). 1985 ISBN 90-6193-527-X
45. B. Jay (ed.): *Detection and Measurement of Visual Impairment in Pre-Verbal Children.* Proceedings of a Workshop (London, UK, 1985). 1986 ISBN 0-89838-789-2
46. G. Verriest (ed.): *Colour Vision Deficiencies VIII.* Proceedings of the 8th Symposium of the International Research Group on Colour Vision Deficiencies (Avignon, France, 1985). 1987 ISBN 0-89838-801-5
47. P.L. Emiliani (ed.): *Development of Electronic Aids for the Visually Impaired.* Proceedings of a Workshop on the Rehabilitation of the Visually Impaired (Florence, Italy, 1984). 1986 ISBN 0-89838-805-8
48. K.C. Ossoinig (ed.): *Ophthalmic Echography.* Proceedings of the 10th SIDUO Congress (St. Petersburg Beach, Florida, USA, 1984). 1987 ISBN 0-89838-873-2
49. E.L. Greve & A. Heijl (eds.): *7th International Visual Field Symposium* (Amsterdam, The Netherlands, 1986). 1987 ISBN 0-89838-882-1
50. D. BenEzra, S.J. Ryan, B.M. Glaser & R.P. Murphy (eds.): *Ocular Circulation and Neovascularization.* Proceedings of the First International Symposium (Jeruzalem, Israel, 1986). 1987 ISBN 0-89838-892-9
51. J.M. Thijssen, J.S. Hillman, P.E. Gallenga & G. Cennamo (eds.): *Ultrasonography in Ophthalmology 11.* Proceedings of the 11th SIDUO Congress (Capri, Italy, 1986). 1988 ISBN 0-89838-378-1
52. B. Drum & G. Verriest (eds.): *Colour Vision Deficiencies IX.* Proceedings of the 9th Symposium of the International Research Group on Colour Vision Deficiencies (Annapolis, Md., USA, 1987). 1989 ISBN 0-89838-403-6

Documenta Ophthalmologica Proceedings Series

53. R. Sampaolesi (ed.): *Ultrasonography in Ophthalmology 12.* Proceedings of the 12th SIDUO Congress (Iguazú Falls, Argentina, 1988). 1990 ISBN 0-7923-0765-8
54. B. Drum, J.D. Moreland & A. Serra (eds.): *Colour Vision Deficiencies X.* Proceedings of the 10th Symposium of the International Research Group on Colour Vision Deficiencies (Cagliari, Italy, 1989). 1991 ISBN 0-7923-0948-0
55. P. Till (ed.): *Ophthalmic Echography 13.* Proceedings of the 13th SIDUO Congress (Vienna, Austria, 1990). 1993 ISBN 0-7923-1808-0
56. B. Drum (ed.): *Colour Vision Deficiencies XI.* Proceedings of the 11th Symposium of the International Research Group on Colour Vision Deficiencies (Sydney, Australia, 1991). 1993 ISBN 0-7923-1864-1
57. B. Drum (ed.): *Colour Vision Deficiencies XII.* Proceedings of the 12th Symposium of the International Research Group on Colour Vision Deficiencies (Tübingen, Germany, 1993). 1994 ISBN 0-7923-2889-2

KLUWER ACADEMIC PUBLISHERS – DORDRECHT / BOSTON / LONDON

MIX
Papier aus verantwortungsvollen Quellen
Paper from responsible sources
FSC® C105338

If you have any concerns about our products,
you can contact us on
ProductSafety@springernature.com

In case Publisher is established outside the EU,
the EU authorized representative is:
Springer Nature Customer Service Center GmbH
Europaplatz 3, 69115 Heidelberg, Germany

Printed by Libri Plureos GmbH
in Hamburg, Germany